Denver Fire Department Deployment Policy Analysis

In 1972 the mayor and fire chief of Denver agreed that a study of the Fire Department was in order in the face of mounting costs. The challenge was to minimize cost while keeping the same level of fire protection. An exponential model was used to predict future trends via simulation methods.

Consumers' Surplus and Louisiana Fish Farming

What happens to a stable industry when a new, innovative production technique is introduced? One way to gauge the effect is to measure the gain to consumers in price reduction, called consumers' surplus. It is used to measure the effect on the Louisiana crawfish industry when fish farms using private ponds compete with standard fisheries using lakes and rivers.

Demography and the Social Security System

Many people mistakenly believe that the taxes they pay to the social security system are put into a savings account to be held until their retirement. Not so! The demographics show that either the system will prove to be a great burden on the work force or the benefits paid to retirees will be greatly reduced.

Minimum Cost Production in the Chemical Industry

Jojoba oil is used for producing amides which, in turn, are an essential ingredient in many chemical processes. Researchers have tried various methods to maximize the amide yield while conserving cost. The chemical industry uses the cost function, which is a function of the two variables temperature and time, to minimize the cost of producing amides.

Calculus with Applications

Raymond F. Coughlin
David E. Zitarelli

Temple University

SAUNDERS COLLEGE PUBLISHING

Philadelphia   Fort Worth   Chicago
San Francisco   Montreal   Toronto
London   Sydney   Tokyo

Text Typeface: Times Roman
Compositor: General Graphic Services, Inc.
Director of EDP: Tim Frelick
Acquisitions Editor: Robert B. Stern
Project Editor: Sally Kusch
Managing Editor: Carol Field
Copy Editor: Elaine Honig
Manager, Art & Design: Carol Bleistine
Art Assistant: Doris Bruey
Text Designer: Caliber Design Planning, Inc.
Cover Designer: Lawrence R. Didona
Text Artwork: Vantage Art, Inc.
Layout Artist: York Production Services
Production Manager: Merry Post

Cover Credit: A computer-generated image of the inside of a "Wente-Torus" of constant mean curvature. It was produced by James T. Hoffman of the Geometry, Analysis, Numerics and Graphics group at the University of Massachusetts, Amherst © 1987, 1988.

Printed in the United States of America

CALCULUS WITH APPLICATIONS

0-03-011283-4

Library of Congress Catalog Card Number:

901   015   98765432

Preface

Students in engineering and the physical sciences rarely ask the question "Why do we have to study this subject?" But other students are asking it in increasing numbers as more and more programs require more and more mathematics. The goal of this book is to answer the question for students who are majoring in business, a social science, or a biological science. Although the table of contents is fairly standard, the means of achieving the goal differ from other texts in two notable ways.

First, every section contains several referenced applications of mathematics in the form of examples or exercises. Each one of these applications is accompanied by a footnote so that the student can pursue it in greater depth or the instructor can assign it as part of a special project. Because such applications deal with real-life situations, calculators are frequently necessary to carry out the computations. Here is a representative list of the referenced exercises, along with the section where they can be found.

Application	Source	Section
stocks and bonds	*Forbes Magazine*	1–3
income tax	*Wall Street Journal*	2–3
telecommunications	*Forecasting Public Utilities*	3–2
regeneration of trees	*Ecology*	4–5
learning curves	*Decision Sciences*	5–2
pollution control	*American Journal of Agricultural Economics*	6–4
gifted students	*Journal of Mathematical Psychology*	7–1
cancer treatment	*Physics in Medicine and Biology*	8–2
new technologies	*Management Science*	9–2
sound transmission	(text) *Musical Acoustics: Piano and Wind Instruments*	10–4
swimming records	*The American Scientist*	11–3
insurance	*Journal of Risk and Insurance*	12–2

In addition, almost every chapter has a Case Study that treats a particular application in greater detail than do the referenced examples. The topics have been chosen not only to illustrate instances in which calculus has played a crucial role in solving an important problem, but to highlight areas of particular human interest.

Even if the instructor does not have time to cover the referenced examples or Case Studies, it is hoped that their relevance and wealth (in both diversity and number) will make a lasting impression.

Organization   The text can be divided into three parts: differential calculus (Chapters 1 through 5), integral calculus (Chapters 6 and 7), and special topics (Chapters 8 through 12).

Differential Calculus

Chapter 1 covers the algebraic techniques needed for the subsequent material. Functions are defined and studied, and various properties of polynomial functions and rational functions are illustrated. Chapter 2 explains the limit of a function by discussing velocity, rate of change of a moving object, and the tangent to a graph. The definition of the derivative is used to calculate the derivatives of several functions.

Chapter 3 presents various techniques for computing the derivatives of functions. Chapter 4 applies these techniques in a wide variety of problems.

Chapter 5 introduces exponential and logarithmic functions. Their derivatives are obtained by calculator experiments aimed at suggesting general rules. Those rules are then proved rigorously.

Integral Calculus

Chapter 6 introduces integration via antidifferentiation. After indefinite integrals have been defined, the Fundamental Theorem of Calculus relates the derivative to the integral. Then the integral is used to compute areas bounded by curves.

Chapter 7 presents two techniques of integration, the use of tables of integrals, and numerical integration.

Special Topics

Chapter 8 extends the definition of the derivative and the definition of the integral to functions of more than one variable. Partial derivatives are defined and applied to the sketching of surfaces, while double integrals are evaluated for functions of two variables.

Differential Equations

An equation involving properties about an unknown function and its derivatives is called a differential equation. Chapter 9 discusses some of the basic concepts of differential equations.

Trigonometry

The trigonometric functions, studied in Chapter 10, allow us to create models in such diverse disciplines as astronomy, ecology, navigation, and music, when we must consider "periodic" functions, which are functions that repeat their functional values periodically.

Taylor Series

The two most widely used methods of finding a functional value of functions such as ex, $\ln x$ and $\sin x$, is to look the value up in a table or to use a calculator. In Chapter 11 we see what steps the calculator uses to compute this number. It approximates the number $e^{1.1}$ by substituting 1.1 into a particular polynomial.

Probability and Calculus

The theory of probability, studied in Chapter 12, is the branch of mathematics that helps find patterns in seemingly haphazard occurrences. Some problems in probability study experiments that have a finite number of possible outcomes. When the outcomes in an experiment can be any number in an interval, calculus is used to discover the properties of the experiment.

Format   We have tried to keep the length of each section to material that can be covered in a typical 50-minute class. Sometimes, however, the topic has dictated more extensive coverage. Each numbered section has been partitioned into subsections.

There are two kinds of examples. One explains a new skill that is being encountered for the first time and is labeled simply EXAMPLE. The other illustrates a skill which has been explained previously. It uses the following format:

EXAMPLE | **Problem**

SOLUTION

The student should be able to make a good attempt at solving the problem independently before reading its solution.

Each chapter ends with a review of the terms, notation, and formulas that were introduced in the chapter. It also contains review problems that test a student's ability to solve problems in a cumulative manner. These problems can also be used profitably as a review for tests.

Exercises   The exercises are divided into three sets by color bullets. There is a slight distinction between the problems in the first two sets. The first set of problems provides practice on the skills introduced in the examples; they appear in the same order as the examples. The second set requires either new thinking or additional skills. Their order need not match the order of the presentation in the text.

The remaining problems serve several different purposes. Most are drawn from areas outside mathematics and are accompanied by complete references to the literature. Other problems present theoretical material for those instructors who want to provide a deeper insight into the subject, while some are aimed at challenging the better students.

The answers to all odd-numbered problems are provided in the back of the book.

Supplements

A number of supplements are available for use with this text. They are as follows:

The **Instructor's Manual** contains answers to all odd- and even-numbered problems along with lecture suggestions for every chapter.

A **Student Solutions Manual and Study Guide** contains answers and solutions to every other odd-numbered problem as well as helpful suggestions, examples, and additional problems for the student to work.

The **computer supplement** available for this text covers many of the important topics in each chapter with graphical and computational support. The software requires an IBM or IBM-compatible computer with at least 512K. It will operate with either a color or monochrome graphics adapter.

Also available for this book is a **computerized test bank** which permits the user to choose, edit, add, or delete items. The test bank consists of approximately 1,250 test items, and is available for the IBM-PC and IBM-compatible computers with 256K, dual disk drives, graphics card, graphics monitor, and graphics printer. The printer can be either parallel or serial if the serial printer has a parallel driver.

In addition to the computerized testbook a **complete booklet of prepared tests** has been written with five forms per chapter of 25 questions each. Both open response as well as multiple choice questions are included..

Acknowledgments

It is a pleasure to acknowledge the help we received from many people. Our biggest support came from our wives (Judy and Anita) and our children (Virginia, Christina, and Sara; Paul and Nicole), who have tolerated our total immersion in this project for several years.

Our reviewers played an important part in transforming a manuscript used at Temple University into a book that is appropriate for many types of institutions of higher learning. These reviewers deserve special thanks for their unstinting, constructive criticism of the entire manuscript:

Gail Broome, Providence College
Garret Etgen, University of Houston
Lou Hoelzle, Bucks County Community College
Ron Jeppson, Moorhead State University
Lester Levi, Middle Tennessee State
Lawrence Maher, North Texas State University
Giles Maloof, Boise State University
Jack Porter, University of Kansas
Holly Puterbaugh, University of Vermont
Gordon Schilling, University of Texas at Arlington
Daniel Symancyk, Anne Arundel Junior College

A very important part of the reviewing process is the checking of errors in problems and examples. For their aid in this process, we would like to thank the following people:

Louis Hoelzle, Bucks County Community College
Daniel Symancyk, Anne Arundel Junior College
Gary Van Velsir, Anne Arundel Community College

We extend special gratitude to our friend Vincent Damiano, who is a Senior Project Engineer for Chevron USA, Inc. He read the entire manuscript and suggested several realistic problems.

The staff at Saunders College Publishing has been a pleasure to work with. We especially thank our editor, Robert Stern, for the interest, encouragement, guidance, and persistence he has bestowed upon this project. We also thank the project editor, Sally Kusch, and the art director, Carol Bleistine, not only for their professional expertise but for their patience and humor as well. Our appreciation also goes to Elaine Honig for her excellent copyediting and for compiling the Index of Applications.

Raymond F. Coughlin
David E. Zitarelli

Philadelphia, PA
January 1989

Contents

8 Functions of Several Variables / 337

9 Differential Equations / 387

10 Trigonometric Functions / 419

11 Infinite Series / 463

Appendix Tables

Index of Applications

Business and Economics

Biological Sciences

1 Functions

CHAPTER OVERVIEW

There are two parts of calculus, differentiation and integration, both of which are based on the concepts of functions and limits. This chapter presents the relevant material on functions, and the next chapter introduces limits.

Section 1–1 discusses graphs of equations, and Section 1–2 studies straight lines and their graphs. In Sections 1–3 and 1–4 the equations are regarded as functions. Since not all equations are functions, a criterion for distinguishing between them is elaborated. The material is applied to an area called cost analysis.

Section 1–4 introduces power functions, polynomial functions, and rational functions. Two additional kinds of functions are described: those that are defined by different formulas over different intervals and those that are defined implicitly by a graph.

Section 1–5 discusses five operations on functions: the four rational operations and composition. It also introduces quality control. Section 1–6 presents some methods for solving polynomial equations, including a brief review of factoring and the quadratic formula. This section describes how to find the points of intersection of the graphs of functions.

CASE STUDY PREVIEW

''Which one is stronger, an ant or an elephant?'' The Case Study will not answer this proverbial question, but it will provide some food for thought by examining an analogous question about human beings.

1

Powerlifters can lift vast amounts of weight. Their meets are divided into weight classes, and each weight class has a champion. But powerlifting meets also feature a "Champion of Champions" award, and the person who lifts the largest weight may not necessarily win it. The reason is the so-called "Schwartz Formula," a handicapping scheme that distributes points according to a person's bodyweight. The Case Study will describe how it is defined in terms of a polynomial function.

1–1 THE CARTESIAN COORDINATE SYSTEM

About 1630 two Frenchmen, René Descartes and Pierre Fermat, independently revolutionized the way of expressing scientific laws. The idea they hit upon has come to be called "analytic geometry" or "algebraic geometry," because it is a blend of algebra and geometry. Descartes is usually given credit for being the founder of this field, and the term "Cartesian coordinate system" reflects human-kind's debt to him, but Fermat deserves at least as much credit for his pioneering work.* Sections 1–1 and 1–2 are devoted to their discovery.

The Graph of an Equation

Consider the formula $y = 3x + 1$. It is an algebraic expression that describes the relationship between the variables x and y, but it can also be viewed as a geometric object, namely, a line in the plane. To accomplish this both Descartes and Fermat introduced a *coordinate system*, which we will describe briefly.

We start with the **real line.** It is a straight line with one point marked as the **origin,** corresponding to the number 0 (see Figure 1a). A point to the right of 0 is marked and it corresponds to the number 1. The distance from 0 to 1 is called a **unit distance.** Points that are located at one-unit intervals to the right of 1 are

*For the historical details, see Chapter XVII, Section 11, of Carl B. Boyer, *A History of Mathematics*, Princeton, N.J., Princeton University Press, 1985.

FIGURE 1

(a)

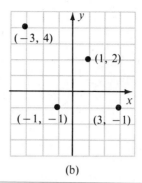
(b)

FIGURE 2

marked 2, 3, and so on. Negative numbers are located to the left of 0. In this way, each real number a corresponds to the point that is $|a|$ units to the right of 0 if a is positive or to the left if a is negative. Additional points, corresponding to some rational and irrational numbers, are marked off in Figure 1b.

Next construct a vertical line perpendicular to the horizontal real line at the origin, and mark off a unit length on it. The positive direction is upward. Usually the horizontal line is called the **x-axis** and the vertical line the **y-axis.** Every point in the plane is represented by an ordered pair of numbers. The ordered pair (1, 2) corresponds to the point that is one unit to the right of the y-axis and two units above the x-axis. (See Figure 2a.) In the ordered pair (1, 2), 1 is called the **x-coordinate** and 2 the **y-coordinate.**

In Figure 2b additional points are plotted in the coordinate system. The numbers in the ordered pairs are also called the first and second **coordinates.** This coordinate system is called the **Cartesian coordinate system** after Descartes. The choice of the letters x and y is arbitrary; we will change them when a particular problem dictates. The Cartesian coordinate system divides the plane into four distinct **quadrants** denoted by the Roman numerals shown in Figure 2a.

Consider the equation $y = 3x + 1$. If $x = 1$, then $y = 4$. These values are treated as the point (1, 4) in the Cartesian coordinate system. The point (1, 4) is said to "satisfy" the equation $y = 3x + 1$. The set of all points that satisfy an equation is called the **graph** of the equation. The graph is obtained from a table of values of x and the corresponding values of y. Example 1 illustrates the procedure for drawing the graph of an equation.

EXAMPLE 1

Consider the equation $y = 3x + 1$. To find its graph, make a table of some values of x and the corresponding values of y. If $x = -2$, then $y = 3x + 1 = 3(-2) + 1 = -5$. This means that the point $(-2, -5)$ lies on the graph. The table lists four additional points corresponding to $x = -1, 0, 1,$ and 2.

x	-2	-1	0	1	2
y	-5	-2	1	4	7
point	$(-2, -5)$	$(-1, -2)$	$(0, 1)$	$(1, 4)$	$(2, 7)$

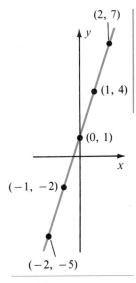

(2, 7)

(1, 4)

(0, 1)

(−1, −2)

(−2, −5)

FIGURE 3

Next, plot the points $(-2, -5)$, $(-1, -2)$, $(0, 1)$, $(1, 4)$, and $(2, 7)$ on a Cartesian coordinate system. By letting x be *any* real number in the equation $y = 3x + 1$, there will be an infinite number of points $(x, 3x + 1)$ on the graph. They constitute the graph of $y = 3x + 1$, which is the line drawn in Figure 3.

The values for x in the table were chosen arbitrarily. The same line would result if other choices were made.

Example 1 illustrates how algebra and geometry were molded into one mathematical discipline called analytic geometry. The essence is that equations in two variables correspond to curves in two dimensions. The procedure for sketching the graph of an equation is called the **plotting method.** It consists of three steps:

1. Make a table of values of x and y.
2. Treat the values as ordered pairs on a coordinate system and plot them.
3. Connect the points with a smooth curve.

The graph of the equation is the curve that results from step 3. In Section 1–2 we will see how to proceed in the opposite direction, from the graph to the equation.

This simple idea allows algebraic equations to be pictured geometrically, reflecting the old adage that "a picture is worth a thousand words." The next example illustrates it.

EXAMPLE 2

Problem

Sketch the graph of the equation $4x + 2y = 1$ by the plotting method.

SOLUTION Construct a table for the equation $4x + 2y = 1$ by choosing a value for x and substituting it into the equation to get the corresponding value of y. If $x = 1$, then $4(1) + 2y = 1$, so $y = -\frac{3}{2}$. Therefore $(1, -\frac{3}{2})$ is a point on the graph of $4x + 2y = 1$. The table provides four additional points.

x	-2	-1	0	1	2
y	$\frac{9}{2}$	$\frac{5}{2}$	$\frac{1}{2}$	$-\frac{3}{2}$	$-\frac{7}{2}$
point	$(-2, \frac{9}{2})$	$(-1, \frac{5}{2})$	$(0, \frac{1}{2})$	$(1, -\frac{3}{2})$	$(2, -\frac{7}{2})$

Plot these points and draw the curve through them. The graph is shown in Figure 4.

An equation in two variables whose graph is a line is called a **linear equation.** Examples 1 and 2 show that $y = 3x + 1$ and $4x + 2y = 1$ are linear equations.

Applications

For many businesses the total cost C (in dollars) of producing a commodity and the number x of units produced can be written as a linear equation. For instance,

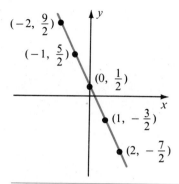

$(-2, \frac{9}{2})$

$(-1, \frac{5}{2})$

$(0, \frac{1}{2})$

$(1, -\frac{3}{2})$

$(2, -\frac{7}{2})$

FIGURE 4

let

$$C = 500 + 4x$$

If no units are produced, then $x = 0$, so $C = \$500$. This amount is called the fixed cost. If $x = 25$ units are produced, then the total cost is $C = \$600$.

Often the total revenue R (in dollars) derived from selling the commodity can be written as a linear equation. For instance, let

$$R = 8x$$

If 25 units are produced, then $R = \$200$, in which case the revenue is $\$400$ less than the cost. The business will have to produce more units in order to turn a profit.

EXAMPLE 3 | **Problem**

Sketch the graphs of the cost equation and the revenue equation on the same coordinate axis.

SOLUTION Use the plotting method to sketch the graph of the cost $C = 500 + 4x$.

x	0	50	100	150
C	500	700	900	1100
point	(0, 500)	(50, 700)	(100, 900)	(150, 1100)

Sketch the graph of $R = 8x$ for the same values of x.

x	0	50	100	150
R	0	400	800	1200
point	(0, 0)	(50, 400)	(100, 800)	(150, 1200)

The graphs of the two equations are shown in Figure 5.

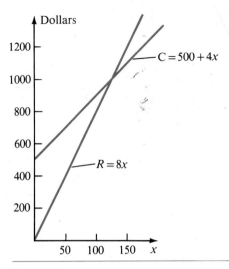

FIGURE 5

Most large firms have a planning group which constructs sales forecasts. These forecasts are critical for production scheduling, plant design, and financial planning. Graphs are often used for viewing the relationship between sales growth and external financial requirements. Example 4 examines the Arinson Products Company, whose owner, John Arinson, sought funds to maintain exclusive control of the company.*

EXAMPLE 4 Let x represent the sales growth rate of the Arinson Products Company and let y represent the external funds required (in thousands of dollars), where

$$y = 494x - 16$$

Problem

(a) Draw the graph of the equation. (b) If the sales growth rate is 5%, what external funding is required? (c) If $107,500 is available for external funding, what is the sales growth rate?

SOLUTION (a) If $x = 0.1$ (corresponding to a 10% sales growth rate), then $y = 494(0.1) - 16 = 33.4$. The table lists several other values of x and the corresponding values of y.

x	-0.2	-0.1	0	0.1	0.2
y	-114.8	-65.4	-16	33.4	82.8

The graph is sketched in Figure 6.

*B. Campsey and E. Brigham, *Introduction to Financial Management*, Hinsdale, Ill., The Dryden Press, 1985, pp. 190–204.

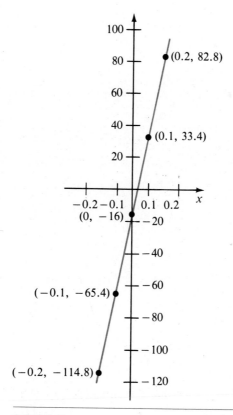

FIGURE 6

(b) For a sales growth rate of 5% set $x = 0.05$. Then $y = 494(0.05) - 16 = 8.7$. Since the units for y represent thousands of dollars, the amount of external funds required is $8700.

(c) If $107,500 is available for external funding, set $y = 107.5$. The sales growth rate is obtained by substituting $y = 107.5$ into the equation and solving for x.

$$494x - 16 = y$$
$$494x - 16 = 107.5$$
$$x = \frac{107.5 + 16}{494} = 0.25$$

The sales growth rate is $x = 25\%$.

What's in a Name?

All of the graphs in this section have been lines. Of course, these are not the only graphs of algebraic equations according to the method of Descartes and Fermat;

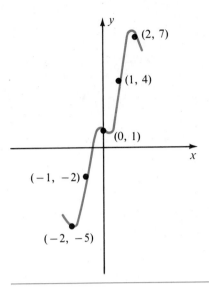

FIGURE 7

there are also circles, parabolas, and many others. It is useful to be able to picture a graph when its equation is given.

So far, graphs have been drawn by the plotting method, but this method has its shortcomings. Given the table of five points for the equation $y = 3x + 1$ in Example 1, it is conceivable that its graph could be the curve in Figure 7. In Section 1–2 we will see why it must be a straight line.

EXERCISE SET 1–1

• In Problems 1 to 4 fill in the table for the given values. Supply your own values of x in Problems 3 and 4.

1. $y = 2x - 3$

x	−2	−1	0	1	2	3
y						
point						

2. $y = x - 2$

x	−2	−1	0	1	2	3
y						
point						

3. $y = 4 - x$

x	
y	
point	

4. $3x - y = 2$

x	
y	
point	

In Problems 5 to 10 use the plotting method to sketch the graph of the equation.

5. $y = 2x - 3$ 6. $y = x - 2$

7. $y = 4 - x$

8. $3x - y = 2$

9. $2x + y = 6$

10. $4x + y = 7$

In Problems 11 to 14 the graphs of the given equations are straight lines. Plot two points and then sketch the graph.

11. $x - y = 3$

12. $2x + y = 3$

13. $3x - y = -1$

14. $3x + y = 1$

In Problems 15 to 18 sketch the graphs of the total cost C and the total revenue R on the same coordinate axis.

15. $C = 25 + 3x$
    $R = 8x$

16. $C = 40 + x$
    $R = 5x$

17. $C = 100 + 3x$
    $R = 13x$

18. $C = 300 + x$
    $R = 31x$

Problems 19 and 20 refer to the equation $y = 494x - 16$ given in Example 4.

19. (a) If the sales growth rate is 15%, what external funding is required? (b) If $33,400 is available for external funding, what is the sales growth rate?

20. (a) If the sales growth rate is -5%, what external funding is required? (b) If $82,800 is available for external funding, what is the sales growth rate?

•• In Problems 21 to 24 use the plotting method to sketch the graph of the equation.

21. $x + 4y = 7$

22. $x + 3y = 3$

23. $3x + 4y = -12$

24. $3x - 4y = 6$

In Problems 25 to 34 the equations are nonlinear. Sketch the graph using the plotting method.

25. $y = x^2 - 3$

26. $y = x^2 + 1$

27. $y = -x^2$

28. $y = -2x^2$

29. $y = 1 - 2x^2$

30. $y = 1 - x^2$

31. $y = x^3$

32. $y = x^3 - 8$

33. $y = \sqrt{x}$

34. $y = \sqrt{x} - 1$

In Problems 35 and 36 sketch the graphs of the two equations on the same coordinate axis.

35. $\begin{aligned} x + y &= 120 \\ 100x + 40y &= 6000 \end{aligned}$

36. $\begin{aligned} 45x + 9y &= 45 \\ 10x + 6y &= 20 \end{aligned}$

Problems 37 to 40 refer to the equation $5F - 9C = 160$, where F is degrees Fahrenheit and C is degrees Celsius.

37. Sketch the graph of the equation.

38. Find the degrees Celsius when $F = 68$ degrees.

39. Find the degrees Celsius when $F = 98.6$ degrees.

40. Find the degrees Fahrenheit when $C = 20$ degrees.

Problems 41 and 42 refer to a company whose total cost is $C = 0.5x + 20{,}000$, where x is the number of sales and the fixed cost is $20,000.

41. If the company sells $60,000 worth of goods, what is the total cost?

42. If the company sells $30,000 worth of goods, what is the total cost?

43. A person's height can be estimated by the length of the tibia bone, which runs from the knee to the ankle. The formula for the height y [in centimeters (cm)] in terms of the length x (in cm) of the tibia is

$$3y - 8x = 211$$

(a) Draw the graph of the equation.
(b) How tall is a person whose tibia is 37 cm long?
(c) How long is the tibia if the person is 185 cm tall?

44. A person's height can be estimated by the length of the radius bone, which runs from the wrist to the elbow. The formula for the height y (in cm) in terms of the length x (in cm) of the radius is

$$2y - 7x = 166$$

(a) Draw the graph of the equation.
(b) How tall is a person whose radius bone is 24 cm long?

(c) How long is the radius bone if the person is 174 cm tall?

45. What is the meaning of the value of x at which the cost line and the revenue line meet in Figure 5?

••• 46. The amplifier in a stereo system must be strong enough to produce loud peak volumes level (106 decibels) in the speakers. The strength needed depends on the character of the room. For instance, a "live" room is one in which a hand clap will cause a fluttering echo. For Acoustic Research loudspeakers,* the linear relationship between the power y (in watts) and the volume of a "live" room (in cubic feet) is given by

$$y = \frac{x}{180} + \frac{80}{9}$$

(a) What power is needed for a room measuring 2000 cubic feet?

(b) What power is needed for a room measuring 6500 cubic feet?

(c) If a system produces 40 watts. what size room will allow it to produce loud peak levels?

47. The percentage share of the national radio audience for FM radio stations (contrasted with AM radio stations) in any year since 1972 can be approximated by the formula

$$S = 3.514286Y - 226.2953$$

where S is a percent and Y is the last two digits in the year.† (For instance, for the year 1989, Y = 89.)

(a) What was the approximate percentage share in 1985?

(b) In what year was the percentage share about 50%?

---

*Source: Teledyne Acoustic Research of Canton, Massachusetts. This company manufactures the Acoustic Research loudspeakers.

†This equation was derived from data supplied by Statistical Research, Incorporated.

1–2 LINES

There are two basic aspects to the Descartes–Fermat invention of analytic geometry. One is a fixed frame of reference, called the Cartesian coordinate system. The other is the relationship between equations and curves. Section 1–1 showed how to begin with a linear equation in two variables and draw its graph. This section illustrates the opposite direction: given the graph of a line in some form, construct its equation.

General Form

Recall from Section 1–1 that a linear equation in two variables is an equation whose graph is a line. We examine three forms of linear equations.

It can be proved that an equation in two variables x and y is linear if and only if it can be written in the form $Ax + By = C$, where A, B, C are numbers such that A and B are not both zero. This form of a linear equation is called the *general form*. The equation $4x + 3y = 12$ is linear, but $x^2 - y = 0$ and $\sqrt{x} - y = 0$ are nonlinear.

Definition

The **general form** of a linear equation is

$$Ax + By = C$$

where A, B, and C are real numbers, with A and B not both equal to 0.

Example 1 illustrates a method of graphing linear equations in general form that is more effective than the plotting method shown in Example 2 of Section 1–1. It makes use of the intercepts of a line. The **x-intercept** of a line is the point where the line crosses the x-axis, and the **y-intercept** is the point where it crosses the y-axis.

EXAMPLE 1 | **Problem**

Sketch the graph of the linear equation

$$3x + 2y = 4$$

SOLUTION Find the y-intercept by setting $x = 0$. Then $y = 2$, so the y-intercept is $(0, 2)$. Find the x-intercept by setting $y = 0$. Then $x = \frac{4}{3}$, so the x-intercept is $(\frac{4}{3}, 0)$. Draw the line through the two intercepts. This is shown in Figure 1.

Slope-Intercept Form

Every equation of the form $y = mx + b$, where m and b are numbers, is linear since it can be written in the general form $mx - y = -b$. Therefore the graph of the equation $y = 3x + 1$ is a line with $m = 3$ and $b = -1$; its graph cannot be the curve shown in Figure 7 of Section 1–1. The expression "the line $y = mx + b$" refers to the graph of the linear equation $y = mx + b$.

The numbers m and b in the linear equation $y = mx + b$ play an important role. The property measured by m is intuitively described as "steepness." It is called the *slope* of the line. To measure the slope, start at some point (x_1, y_1) on

FIGURE 1   Graph of $3x + 2y = 4$.

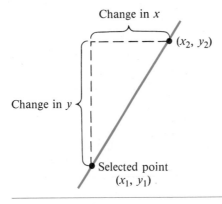

FIGURE 2   Slope.

the line, proceed in the vertical direction some distance, then proceed in the horizontal direction until the line is intersected at a point (x_2, y_2). (See Figure 2.) The vertical distance is called the ''change in y'' (or the ''rise'') and the horizontal distance is called the ''change in x'' (or the ''run''). The slope is then defined to be the ratio of the change in y to the change in x (or the ratio of the rise to the run).

_____ **Definition** _____

Let (x_1, y_1) and (x_2, y_2) be distinct points on a line with $x_1 \neq x_2$. The **slope** of the line is the number

$$m = \frac{\text{change in } y}{\text{change in } x} = \frac{y_2 - y_1}{x_2 - x_1}$$

The points $(1, -1)$ and $(4, 5)$ lie on the line $y = 2x - 3$. (See Figure 3.) The slope of the line is

$$m = \frac{\text{change in } y}{\text{change in } x} = \frac{5 - (-1)}{4 - 1} = \frac{6}{3} = 2$$

Figure 4 shows the graphs of several lines. As the values of x increase from left to right, those with positive slope tend upward (or increase) while those with negative slope tend downward (or decrease).

Example 2 investigates the slope of two pairs of points on the same line.

EXAMPLE 2 | **Problem**

Find the slope of the line $y = 3x + 1$ through these pairs of points.
(a) $(0, 1)$ and $(3, 10)$
(b) $(-1, -2)$ and $(4, 13)$

SOLUTION (a) The slope of the line through the points $(0, 1)$ and $(3, 10)$ is

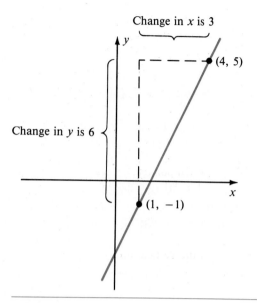

FIGURE 3   Slope of line $y = 2x - 3$.

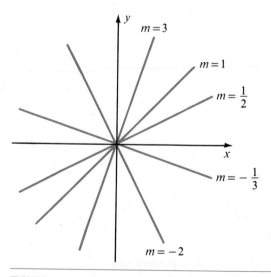

FIGURE 4   Lines with various slopes.

$$m = \frac{10 - 1}{3 - 0} = \frac{9}{3} = 3$$

(b) The slope of the line through the points $(-1, -2)$ and $(4, 13)$ is

$$m = \frac{13 - (-2)}{4 - (-1)} = \frac{13 + 2}{4 + 1} = \frac{15}{5} = 3$$

In general the slope of a line is independent of the two points chosen on the line. Example 2 verifies this fact. Another way of viewing the slope can be seen from a table of points.

x	-2	-1	0	1	2
y	-5	-2	1	4	7

Each time the value of x increases by 1, the value of y increases by the slope 3.

The number b in the linear equation $y = mx + b$ can be obtained by setting $x = 0$. Then, $y = m \cdot 0 + b = b$, so the line passes through the point $(0, b)$. The number b is referred to as the y-intercept of the line $y = mx + b$ even though formally the y-intercept is the point $(0, b)$.

The linear equation $y = mx + b$ is called the *slope-intercept form* of a line because m is the slope and b is the y-intercept.

_____ **Definition** _____

The **slope-intercept form** of a line is

$$y = mx + b$$

where m is the slope and b is the y-intercept.

_____

EXAMPLE 3 | **Problem**

Sketch the graph of the equation $y = -2x + 3$.

SOLUTION The y-intercept is $b = 3$, so the line passes through $(0, 3)$. The slope is $m = -2$. Since m is negative the line slopes downward (from left to right). The graph is sketched in Figure 5.

FIGURE 5   Graph of $y = -2x + 3$.

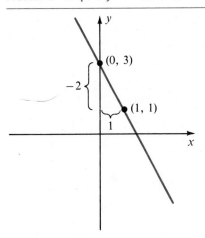

Parallel Lines

Two lines are **parallel** if they do not meet in either direction when extended indefinitely. Since parallel lines have the same steepness, it follows that two lines are parallel if and only if they have the same slope.

EXAMPLE 4 | **Problem**

Write the equation of the line that is parallel to the line $y = -2x + 2$ and passes through the origin.

SOLUTION The graph of the linear equation $y = -2x + 2$ is a line with slope $m = -2$. The new line will have the same slope $m = -2$. Since the new line passes through $(0, 0)$, its y-intercept is $b = 0$. Substituting the values $m = -2$ and $b = 0$ into the slope-intercept form $y = mx + b$ yields $y = -2x$. This is the equation of the line parallel to $y = -2x + 2$ and passing through the origin. The lines are sketched in Figure 6.

There are two special kinds of lines. **Horizontal lines** are parallel to the x-axis and **vertical lines** are parallel to the y-axis. The slope of a horizontal line can be determined by choosing any two points $(x_1, 0)$ and $(x_2, 0)$ on the x-axis. By definition the slope of the line through these points is

$$m = \frac{0 - 0}{x_2 - x_1} = 0$$

Thus the slope of the x-axis is 0. Since parallel lines have the same slope, it follows that *the slope of any horizontal line is zero*.

FIGURE 6 Parallel lines.

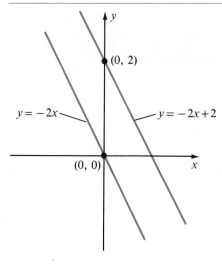

Next, consider vertical lines. If $(0, y_1)$ and $(0, y_2)$ are two points on the y-axis, the slope is

$$m = \frac{y_2 - y_1}{0 - 0}$$

Since division by 0 is not permitted, the slope of the y-axis is not defined. It follows that *the slope of any vertical line is not defined.*

EXAMPLE 5 | **Problem**

Write the equations of the horizontal and vertical lines that pass through the point $(2, 3)$.

SOLUTION The slope of the horizontal line is $m = 0$. Since the horizontal line passing through $(2, 3)$ is parallel to the x-axis, its y-intercept is $b = 3$. Substituting $m = 0$ and $b = 3$ into the equation $y = mx + b$ yields $y = 3$.

The equations of vertical lines cannot be obtained from the linear equation $y = mx + b$ because m is not defined. The vertical line through $(2, 3)$ is located two units to the right of the y-axis. Some additional points on it are $(2, 2)$, $(2, 1)$, $(2, 0)$, and $(2, -1)$. The value of x is always 2. Hence the equation of the line is $x = 2$.

Both lines are sketched in Figure 7.

Example 5 suggests the following standard forms for all horizontal and vertical lines:

horizontal lines: $y = c$,   a number
vertical lines: $x = k$,   a number

Table 1 summarizes the results on the slopes of lines.

FIGURE 7   Horizontal and vertical lines.

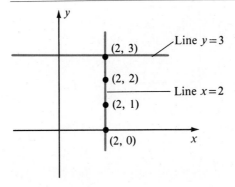

Table 1

Line:	Rising	Falling	Horizontal	Vertical
Slope:	Positive	Negative	Zero	Not defined
Graph				

Point-Slope Form

The equation of a line that is neither vertical nor horizontal can be expressed in several forms. We have already indicated that the slope-intercept form gives the equation of a line when its slope and y-intercept are known. But what if the slope m and another point (x_1, y_1) on the line are known? Then an arbitrary point (x, y) on the line can be used with (x_1, y_1) to compute the slope $m = (y - y_1)/(x - x_1)$. So $y - y_1 = m(x - x_1)$. This form is aptly named the *point-slope form*.

_____ **Definition** _____

The **point-slope form** of a line is

$$y - y_1 = m(x - x_1)$$

where m is the slope and (x_1, y_1) is the given point on the line.

_____

Example 6 shows how to convert from the point-slope form to the slope-intercept form.

EXAMPLE 6 | **Problem**

Find the point-slope form and the slope-intercept form of the line that passes through the point (3, 2) and has slope 4.

SOLUTION We have $x_1 = 3$, $y_1 = 2$, and $m = 4$. The equation of the line in point-slope form is

$$y - 2 = 4(x - 3)$$

Use the properties of algebra indicated on the right to write the equation of the line in slope-intercept form.

$y - 2 = 4(x - 3)$	(point-slope form)
$y = 4(x - 3) + 2$	(add 2 to each side)
$y = 4x - 12 + 2$	(distributive law)
$y = 4x - 10$	(add -12 and 2)

Therefore $m = 4$ and $b = -10$. The graph is sketched in Figure 8.

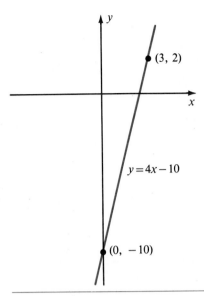

FIGURE 8   Line through (3, 2) with *m* = 4.

Example 7 shows how to determine the equation of a line when two points on the line are known.

EXAMPLE 7 | **Problem**

Find an equation of the line that passes through the points (1, 2) and (−1, 5).

SOLUTION The slope of the line is

$$m = \frac{5 - 2}{-1 - 1} = \frac{3}{-2} = -\frac{3}{2}$$

Using the point-slope form with (1, 2) (we could have chosen (−1, 5) instead) the equation is

$$y - 2 = -\frac{3}{2}(x - 1)$$

The equation in slope-intercept form is found by solving for *y*.

$$y = -\frac{3}{2}x + \frac{7}{2}$$

Here is a summary of the main items in this section.

1. The slope is $m = \dfrac{y_2 - y_1}{x_2 - x_1}$ (unless $x_1 = x_2$).

2. The slope measures steepness. As x increases from left to right, y increases if the slope is positive and y decreases if the slope is negative.

3. Horizontal lines have form $y = c$; their slope is 0. Vertical lines have form $x = k$; their slope is undefined.

4. Three principal forms for the equation of a line follow.

Name of Form	Equation	Variables
general	$Ax + By = C$	A, B, C real numbers A, B not both 0
slope-intercept	$y = mx + b$	$m = $ slope $b = y$-intercept
point-slope	$y - y_1 = m(x - x_1)$	$m = $ slope $(x_1, y_1) = $ point

EXERCISE SET 1–2

• In Problems 1 to 4 sketch the graph of the line.

1. $5x - 2y = 10$
2. $5x + 3y = 15$
3. $3x + 4y = 12$
4. $3x - y = 6$

In Problems 5 to 10 find the slope of the line passing through the two points.

5. $(1, 2), (2, 4)$
6. $(3, 1), (4, 3)$
7. $(-1, 2), (3, -4)$
8. $(-3, 8), (3, -8)$
9. $(1.6, 4.1), (2.5, 5.9)$
10. $(\frac{3}{2}, 5), (-1, -\frac{3}{2})$

In Problems 11 to 14 sketch the graph of the line.

11. $y = 4x - 5$
12. $y = 2x + 1$
13. $y = -2x + 3$
14. $y = -3x - 2$

In Problems 15 and 16 determine if the lines are parallel.

15. Line 1 passes through $(0, 1)$ and $(1, 2)$; line 2 passes through $(-1, 6)$ and $(6, -1)$.

16. Line 1 passes through $(1, 3)$ and $(5, -1)$; line 2 passes through $(1, 1)$ and $(2, 0)$.

In Problems 17 and 18 determine the slope of the line.

17. (a) $x = 2$ (b) $y = -7$
18. (a) $x = 15$ (b) $y = 2$

In Problems 19 and 20 find the slope-intercept form of the line passing through the given points.

19. $(2, 1), (3, -6)$
20. $(2, -1), (-3, 6)$

•• In Problems 21 to 34 find an equation of the line from the information given. Sketch the graph of the line.

21. $m = 2, b = 7$
22. $m = -2, b = 7$
23. The horizontal line that passes through $(-5, -3)$.
24. The vertical line that passes through $(-5, -3)$.
25. The slope is undefined and the x-intercept is -2.
26. The slope is 0 and $b = -4$.
27. $m = -2$; the line passes through $(6, -1)$.
28. $m = 1$; the line passes through $(6, -1)$.
29. $b = -1$ and the line is parallel to $y = 2x + 1$.
30. The line passes through $(1, 1)$ and is parallel to the line $x - 3y = 4$.
31. The line passes through $(-2, 0)$ and $(0, 2)$.
32. The line passes through $(2, 0)$ and $(0, -2)$.

33. The line passes through $(-2, 0)$ and $(-2, -2)$.

34. The line passes through $(0, 4)$ and $(2, 4)$.

In Problems 35 and 36 determine if the slope of the line is positive, negative, zero, or undefined.

35. (a)   (b)

   (c)

36. (a)   (b)

   (c)

37. (a) Put the equation $3x + 2y = 12$ in slope-intercept form. (b) What is the slope? (c) What is the y-intercept?

38. (a) Put the equation $5x + 3y = 15$ in slope-intercept form. (b) What is the slope? (c) What is the y-intercept?

In Problems 39 and 40 determine the equation of the line passing through the two intercepts.

39. (a)   (b)

40. (a)   (b)

    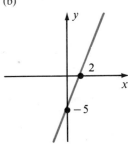

In Problems 41 and 42 write the equation (in point-slope form) of the line that has the given table.

41.

x	-1	0	1
y	5	1	-3

42.

x	-2	0	2
y	5	1	-3

43. For tax purposes, some firms assume that the current value y of an asset is related to the asset's age x by a linear equation. Suppose a firm buys a work station with ten microcomputers and assumes that after x years the work station's worth is given by

$$y = 60,000 - 15,000x$$

(a) Draw the graph of the equation.
(b) What is the work station's worth after three years?
(c) What economic interpretation can be given to the y-intercept?

44. Prove that the slope of the line $y = mx + b$ is m. (*Hint:* Let (x_1, y_1) and (x_2, y_2) be points on the line with $x_1 \neq x_2$. Show that

$$\frac{(y_2 - y_1)}{(x_2 - x_1)} = m$$

45. Is it true that every equation in the variables x and y in which the highest power of x is 1 and the highest power of y is 1 must be linear?

46. Figure 9 shows the power necessary to achieve loud peak volume levels (106 decibels) for a pair of Acoustic Research loudspeakers.* The three different types of rooms ("dead," "average," and "live") require amplifiers of different sizes to produce loud peak volume levels. Let y be the power (in watts) and x be the volume of the room (in cubic feet). For each type of room, express y in terms of x in slope-intercept form.

47. The petroleum industry uses "decline curve analysis" to estimate reserves of oil and gas wells, even though more sophisticated techniques are available. The reason for the continued emphasis on this method is that the required data are almost always available. Recently two petroleum engineers developed a simple graphing technique for evaluating production data from a linear equation.† Let y represent the loss ratio (a technical term measured here in months) and x represent the time (also in months). The line which the engineers discovered for a particular

Power per channel required for a pair of acoustic, research loudspeakers playing at 106 dB S.P.L.

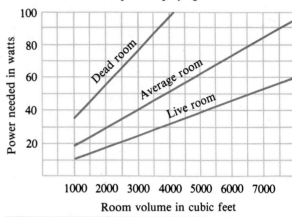

FIGURE 9   Problem 46.

set of data has slope 0.5 and y-intercept 7.5. (a) What is the equation of the line? (b) Sketch the line on the interval of 0 months to 72 months. (c) What is the loss ratio after 6 years?

48. This problem deals with four "rules of thumb" relating the height and weight of adults to those of children.‡ Write each relationship as an equation. (This requires the introduction of appropriate variables.) (a) The height of a male adult is double the height at age 2. (b) The height of a female adult is double the height at age $1\frac{1}{2}$. (c) The weight of a male adult is five times the weight at age 2. (d) The weight of a female adult is five times the weight at age $1\frac{1}{2}$.

---

*Source: Teledyne Acoustic Research of Canton, Massachusetts.

†D. A. Rowland and C. Lin, "New Linear Method Gives Constants of Hyperbolic Decline," *Oil and Gas Journal*, January 14, 1985, pp. 86–90.

‡Wilton M. Krogman, *Child Growth*, University of Michigan Press, 1972, p. 42.

1–3 LINEAR AND QUADRATIC FUNCTIONS

In Section 1–1 we sketched the graphs of linear equations. Here we introduce the fundamental concept of a function and regard a linear equation as a function. We also introduce quadratic functions and give a criterion for distinguishing between equations and functions.

Linear Functions

The equation $y = 1 - 4x$ has the property that for each value of x there corresponds precisely one value of y. In such a case y is called a *function* of x, written $y = f(x)$ or $f(x) = 1 - 4x$. [$f(x)$ is read "f of x."]

_____ Definition _____

A **function** f is a rule that assigns to each element x in a set exactly one element $f(x)$ in a (possibly different) set. The element $f(x)$ is called the **image** of x, and x is called the **preimage** of $f(x)$.

Consider the function $f(x) = 1 - 4x$. The **plotting method** for drawing its graph was described in Section 1–1. First, make a table of points satisfying the equation.

x	-1	$-\frac{1}{2}$	0	$\frac{1}{2}$	1
$f(x)$	5	3	1	-1	-3
point	$(-1, 5)$	$(-\frac{1}{2}, 3)$	$(0, 1)$	$(\frac{1}{2}, -1)$	$(1, -3)$

Then plot the points on a Cartesian coordinate system and connect them with the line drawn in Figure 1.

A **linear function** is a function that can be written in the form $f(x) = mx + b$, where m and b are given numbers. We saw in Section 1–1 that the graph of a linear function is a line with slope m and y-intercept b.

The table for the linear function $f(x) = 1 - 4x$ lists the images of five values of x. (Actually, only two images are necessary for drawing the graph of a line.) Since $y = 9$ when $x = -2$, the image of -2 is 9. We write $f(-2) = 9$. This image can be obtained directly by substituting the value of -2 for x wherever x occurs in the equation $f(x) = 1 - 4x$.

$$f(-2) = 1 - 4(-2) = 1 + 8 = 9$$

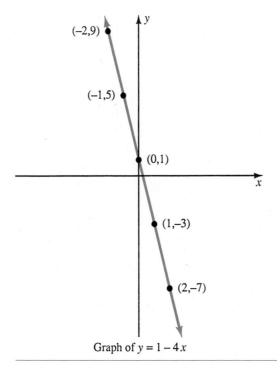

Graph of $y = 1 - 4x$

FIGURE 1

Example 1 computes some images of another function. Parts (a) and (b) show that it is possible for $f(a) = f(b)$ when $a \neq b$.

EXAMPLE 1 | Let f be the function defined by the equation

$$f(x) = 2x^2 - 4x + 5$$

Problem

Find (a) $f(2)$, (b) $f(0)$, and (c) $f(-\frac{1}{2})$.

SOLUTION (a) Since $f(2)$ is the image of 2, it is calculated by substituting 2 for x wherever x occurs in the formula for $f(x)$.

$$f(2) = 2(2)^2 - 4(2) + 5 = 5$$

Thus

$$f(2) = 5$$

(b) $f(0) = 2(0)^2 - 4(0) + 5 = 5$, so

$$f(0) = 5$$

(c) $f(-\frac{1}{2}) = 2(-\frac{1}{2})^2 - 4(-\frac{1}{2}) + 5 = 2(\frac{1}{4}) + 2 + 5$, so

$$f(-\frac{1}{2}) = \frac{15}{2}$$

Example 2 illustrates the procedure for finding the image of an algebraic expression.

EXAMPLE 2 | Let

$$f(x) = 2x^2 - 4x + 5$$

Problem

Find and simplify each of these expressions.

(a) $f(a)$, (b) $f(x + 3)$, (c) $f(x + h)$, and (d) $\dfrac{f(x + h) - f(x)}{h}$, for $h \neq 0$.

SOLUTION (a) The value of $f(a)$ is found by substituting a for x wherever x occurs in the formula for $f(x)$.

$$f(x) = 2x^2 - 4x + 5$$
$$f(a) = 2a^2 - 4a + 5$$

(b) To obtain $f(x + 3)$, substitute $x + 3$ for x wherever x occurs in the formula for $f(x)$.

$$\begin{aligned} f(x + 3) &= 2(x + 3)^2 - 4(x + 3) + 5 \\ &= 2x^2 + 12x + 18 - 4x - 12 + 5 \\ &= 2x^2 + 8x + 11 \end{aligned}$$

(c) To obtain $f(x + h)$, substitute $x + h$ for x wherever x occurs in the formula for $f(x)$.

$$\begin{aligned} f(x + h) &= 2(x + h)^2 - 4(x + h) + 5 \\ &= 2x^2 + 4xh + 2h^2 - 4x - 4h + 5 \end{aligned}$$

(d) Using part (c) and the definition of $f(x)$ gives $f(x + h) - f(x)$

$$\begin{aligned} &= 2x^2 + 4xh + 2h^2 - 4x - 4h + 5 - (2x^2 - 4x + 5) \\ &= 4xh + 2h^2 - 4h = h(4x + 2h - 4) \end{aligned}$$

Then

$$\frac{f(x + h) - f(x)}{h} = \frac{h(4x + 2h - 4)}{h}$$

Since $h \neq 0$ the h-factors can be canceled, so

$$\frac{f(x + h) - f(x)}{h} = 4x + 2h - 4$$

In many applications it is advantageous to use letters other than f to represent the function. *Fortune Magazine,* which is known for its many rankings of businesses, provides an example. Figure 2 displays "The World's 50 Biggest Industrial

Rank		Company	Headquarters	Industry	Sales	Net Income
1	1	General Motors	Detroit	Motor vehicles and parts	102,813,700	2,944,700
2	2	Exxon	New York	Petroleum refining	69,888,000	5,360,000
3	3	Royal Dutch/Shell Group	The Hague/London	Petroleum refining	64,843,217	3,725,779
4	6	Ford Motor	Dearborn, Mich.	Motor vehicles and parts	62,715,800	3,285,100
5	7	International Business Machines	Armonk, N.Y.	Computers	51,250,000	4,789,000
6	4	Mobil	New York	Petroleum refining	44,866,000	1,407,000
7	5	British Petroleum	London	Petroleum refining	39,855,564	731,954
8	12	General Electric	Fairfield, Conn.	Electronics	35,211,000	2,492,000
9	10	American Tel. & Tel.	New York	Electronics	34,087,000	139,000
10	8	Texaco	White Plains, N.Y.	Petroleum refining	31,613,000	725,000
11	16	IRI	Rome	Metals	31,561,709	197,118
12	14	Toyota Motor	Toyota City (Japan)	Motor vehicles and parts	31,553,827	1,717,733
13	31	Daimler-Benz	Stuttgart	Motor vehicles and parts	30,168,550	831,600
14	11	E.I. du Pont de Nemours	Wilmington, Del.	Chemicals	27,148,000	1,538,000
15	20	Matsushita Electric Industrial	Osaka	Electronics	26,459,539	946,571
16	18	Unilever	Rotterdam/London	Food	25,141,672	973,983
17	9	Chevron	San Francisco	Petroleum refining	24,351,000	715,000
18	30	Volkswagen	Wolfsburg (W. Germany)	Motor vehicles and parts	24,317,154	286,133
19	21	Hitachi	Tokyo	Electronics	22,668,085	679,609
20	15	ENI	Rome	Petroleum refining	22,549,921	342,275
21	19	Chrysler	Highland Park, Mich.	Motor vehicles and parts	22,513,500	1,403,600
22	28	Philips' Gloeilampenfabrieken	Eindhoven (Netherlands)	Electronics	22,471,263	414,418
23	32	Nestlé	Vevey (Switzerland)	Food	21,153,285	994,566
24	●	Phillip Morris	New York	Tobacco	20,681,000	1,478,000
25	29	Siemens	Munich	Electronics	20,307,037	629,353
26	27	Nissan Motor	Yokohama	Motor vehicles and parts	20,141,237	158,477
27	41	Fiat	Turin	Motor vehicles and parts	19,669,581	1,449,556
28	36	Bayer	Leverkusen (W. Germany)	Chemicals	18,768,914	391,745
29	38	BASF	Ludwigshafen (W. Germany)	Chemicals	18,640,985	416,418
30	13	Amoco	Chicago	Petroleum refining	18,281,000	747,000
31	●	Renault	Paris	Motor vehicles and parts	17,661,021	(844,233)
32	40	Hoechst	Frankfurt	Chemicals	17,509,344	568,152
33	24	Elf Aquitaine	Paris	Petroleum refining	17,287,058	617,833
34	●	RJR Nabisco	Atlanta	Tobacco	16,998,000	1,064,000
35	42	Samsung	Seoul	Electronics	16,522,664	182,175
36	49	Boeing	Seattle	Aerospace	16,341,000	665,000
37	43	Mitsubishi Heavy Industries	Tokyo	Motor vehicles and parts	15,932,973	299,241
38	34	United Technologies	Hartford, Conn.	Aerospace	15,669,157	72,727
39	●	Procter & Gamble	Cincinnati, Ohio	Soaps, cosmetics	15,439,000	709,000
40	39	Occidental Petroleum	Los Angeles	Food	15,344,100	181,100
41	●	Peugeot	Paris	Motor vehicles and parts	15,152,869	518,350
42	●	Toshiba	Tokyo	Electronics	15,036,390	268,916
43	46	Imperial Chemical Industries	London	Chemicals	14,867,911	817,031
44	33	Petrobrás (Petróleo Brasileiro)	Rio de Janeiro	Petroleum refining	14,701,534	2,073,456
45	17	Atlantic Richfield	Los Angeles	Petroleum refining	14,585,802	615,116
46	37	Tenneco	Houston	Petroleum refining	14,558,000	(39,000)
47	26	USX	Pittsburgh	Petroleum refining	14,000,000	(1,833,000)
48	48	Kuwait Petroleum	Safat (Kuwait)	Petroleum refining	13,911,716	184,562
49	25	Total Cie Française des Pétroles	Paris	Petroleum refining	13,821,041	(68,006)
50	●	Thyssen	Duisberg (W. Germany)	Metals	13,818,174	150,463

FIGURE 2 The World's 50 Biggest Industrial Corporations. (Adapted from "The World's 50 Biggest Corporations," *Fortune Magazine*, August 3, 1987, pp. 23–25.)

Corporations'' in 1986 (with the 1985 ranking on the right side of the left-hand column). The list describes a function F (for *Fortune*), whose domain is the set $\{1, 2, \ldots, 50\}$, defined by the rule that $F(n)$ is the sales of the nth biggest corporation in 1986. It follows from Figure 2 that $F(1) = 102,813,700$ and $F(21) = 22,513,500$.

Domain and Range

A function is a correspondence between two sets in which each element x in one set, called the **domain,** has associated with it precisely one element in another set, called the **range.** Most of the functions in this book will be defined by an equation, with the domain and range being sets of real numbers. Unless otherwise specified the domain of a function that is defined by an equation consists of all numbers for which the equation is defined. For example, the domain of a linear function $f(x) = mx + b$ is the set of all real numbers. However, the domain of the function $f(x) = \dfrac{3}{2 - x}$ consists of all numbers different from 2 because division by 0 is undefined. A **function** f is **defined at a number** a if $f(a)$ exists. Notice that the function $f(x) = \dfrac{3}{2 - x}$ is defined at 0 because $f(0) = \dfrac{3}{2}$, but it is not defined at 2.

Example 3 computes the domain of another standard type of function.

EXAMPLE 3 | **Problem**

Find the domain of the function f defined by the equation $f(x) = \sqrt{1 - x}$.

SOLUTION The square root is only defined when the radicand $1 - x$ is nonnegative, so $1 - x \geq 0$. Therefore the domain of $f(x)$ is the set of all numbers x such that $x \leq 1$. As an interval, the domain is $(-\infty, 1]$.

Quadratic Functions

A function defined by the equation

$$f(x) = ax^2 + bx + c$$

where a, b, and c are given real numbers with $a \neq 0$, is called a **quadratic function.** Quadratic functions are not linear functions; their graphs are parabolas, not lines. Chapter 4 will show how to sketch the graphs of quadratic functions by using the techniques of calculus, but for now we rely on the plotting method.

EXAMPLE 4 | **Problem**

Sketch the graph of the quadratic function $f(x) = 2x^2 - 4x + 1$ by finding the images of -1, 0, 1, 2, and 3.

SOLUTION Set $x = -1$. Then $f(-1) = 2(-1)^2 - 4(-1) + 1 = 7$, so the point $(-1, 7)$ lies on the parabola. Similarly, $f(0) = 1$, $f(1) = -1$, $f(2) = 1$, and $f(3) = 7$. Therefore the points $(0, 1)$, $(1, -1)$, $(2, 1)$, and $(3, 7)$ also lie on the parabola. The graph is sketched in Figure 3.

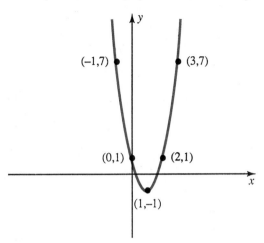

FIGURE 3 Graph of $f(x) = 2x^2 - 4x + 1$.

FIGURE 4 (a) The equation $y = x^2$ represents a function. (b) The equation $y^2 = x$ does not represent a function.

(a)

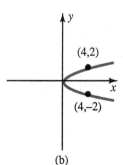

(b)

All parabolas are similar in shape to the one shown in Figure 3. Moreover, the graph of any quadratic function $f(x) = ax^2 + bx + c$ is a parabola that opens upward or downward, has a lowest or highest point (the **vertex**), and has a vertical line of symmetry passing through the vertex. Example 4 shows that the graph of $f(x) = 2x^2 - 4x + 1$ is a parabola that opens upward and has a vertical line of symmetry passing through the vertex (the lowest point on the graph) at $(1, -1)$.

Functions Versus Equations

There is a crucial distinction between functions and equations because not every equation represents a function. Consider the equations $y = x^2$ and $y^2 = x$. Their graphs, which are both parabolas, are drawn in Figure 4. The points $(4, 2)$ and $(4, -2)$ lie on the graph of the equation $y^2 = x$, so if a function $g(x)$ represented this equation, then we would have $g(4) = 2$ and $g(4) = -2$. By definition, a function of x cannot have a value x in the domain with two images. Therefore $y^2 = x$ is an equation that does not represent a function of x. However, for the equation $y = x^2$ each value of x has precisely one image so it represents the function $f(x) = x^2$.

The **vertical line test** provides a visual means of determining whether a curve represents the graph of a function.

_____ **Vertical Line Test** _____

A curve represents the graph of a function if and only if every vertical line intersects the curve at exactly one point or at no points.

Apply the vertical line test to the curves in Figure 4. In part (a) no vertical lines cross the curve in two or more points, so the curve represents the graph of the function $f(x) = x^2$. Many vertical lines intersect the curve in part (b) in two points, so the curve does not represent a function. Example 5 applies the vertical line test to other curves.

EXAMPLE 5   **Problem**

Which of these curves represents the graph of a function?

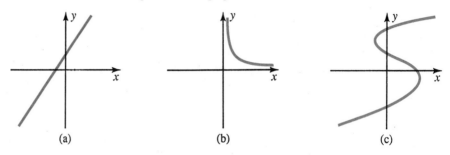

(a)   (b)   (c)

SOLUTION The curve in (a) is a straight line. Each vertical line intersects the curve at precisely one point, so the curve represents the graph of a linear function.

The curve in (b) is drawn in the first quadrant only. The vertical lines through positive values of x meet the curve at one point. The vertical lines through the other values of x do not meet the curve. Thus the curve represents the graph of a function.

Look at the y-axis in (c). It is a vertical line that intersects the curve at more than one point, in fact, at three points. Therefore the curve does not represent the graph of a function.

Cost Analysis

Linear equations can be applied to some manufacturing problems in the following way. Let the cost of producing an item be denoted by C. If C depends on the number x of items produced, then C is called the **cost function,** denoted by $C(x)$. The cost function is composed of two parts, the **fixed cost,** which does not depend on x (such as the manufacturer's insurance), and the **variable cost,** which changes as x changes.

$$C(x) = \text{variable cost} + \text{fixed cost}$$
$$C(x) = (\text{variable cost per item})x + \text{fixed cost}$$

The **marginal cost** is the change in the cost when one more item is produced. If the variable cost per item is constant, then $C(x)$ is a linear function whose slope is the marginal cost.

As an illustration of these terms consider the cost of receiving cable reception where there is a one-time installation fee of $20 and a charge of $0.30 per day (including fractions of a day). To write the cost function, let x represent the time in days. Since the variable cost is $0.30x$ and the fixed cost is the flat rate of $20, the cost function is $C(x) = 0.3x + 20$, where $x \geq 0$. The cost for one regular year is $C(365) = 0.3(365) + 20 = 129.5$, or $129.50.

EXAMPLE 6

A printer quotes a price of $5 for producing a 200-page book and $8 for producing a 600-page book.

Problem

Assuming a linear relationship that depends on the number of pages, (a) What is the cost function? and (b) How much will it cost to produce a 300-page book?

SOLUTION (a) Since the cost depends on the number of pages, let x denote the number of pages. There is a linear relationship between $C(x)$ and x, so $C(x) = mx + b$, where m is the variable cost per book and b is the fixed cost. We have $C(200) = 5$ and $C(600) = 8$, so two points on the line are $(200, 5)$ and $(600, 8)$. Therefore the variable cost is $m = (8 - 5)/(600 - 200) = 3/400$. By the point-slope form, using the point $(200, 5)$, we get

$$C(x) - 5 = \left(\frac{3}{400}\right)(x - 200)$$

$$C(x) = \left(\frac{3}{400}\right)x + \left(\frac{7}{2}\right)$$

From this form we see that the fixed cost is $3.50 since $C(0) = 7/2$. (b) Once the cost function is known, the task of finding the cost of producing a book of any number of pages is straightforward. If $x = 300$, then the cost is $5.75 since $C(x) = (3/400)300 + (7/2) = 23/4$.

EXERCISE SET 1–3

• In Problems 1 to 4 compute $f(2)$, $f(0)$, and $f(-\frac{1}{2})$ for the given function.

1. $f(x) = 2x - 3$
2. $f(x) = -3x + 4$
3. $f(x) = -x^2 + x + 3$
4. $f(x) = x^2 + 4x - 5$

In Problems 5 to 8 find and simplify each of these expressions for the given function. (a) $f(a)$
(b) $f(x + 3)$ (c) $f(x + h)$

(d) $\dfrac{f(x + h) - f(x)}{h}$, for $h \neq 0$.

5. $f(x) = 2x - 3$
6. $f(x) = -3x + 4$
7. $f(x) = -x^2 + x + 3$
8. $f(x) = x^2 + 4x - 5$

Problems 9 and 10 refer to Figure 2 in the text, which lists the 50 biggest corporations in the world in 1986. Let F be the function defined by the rule that $F(n)$ was the sales of the nth biggest corporation that year.

9. Compute $F(2)$
10. Compute $F(3)$

In Problems 11 to 14 find the domain of the stated function.

11. $f(x) = \sqrt{x + 1}$ 12. $f(x) = \sqrt{2x + 1}$

13. $f(x) = \sqrt{1 - 2x}$ 14. $f(x) = \sqrt{2 - x}$

In Problems 15 to 18 sketch the graph of the quadratic function by finding the images of $-2, -1, 0, 1,$ and 2.

15. $f(x) = x^2 + 2x - 5$

16. $f(x) = x^2 - 2x + 3$

17. $f(x) = 3x^2 - 5x - 4$

18. $f(x) = 2x^2 - 3x - 5$

In Problems 19 and 20 determine which curves represent the graph of a function.

19.

(a)

(b)

(c)

20.

(a)

(b)

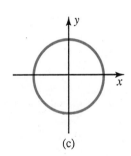

(c)

•• In Problems 21 to 24 evaluate $f(x + h) - f(x)$ for the given function.

21. $f(x) = 2x - 3$ 22. $f(x) = -3x + 4$

23. $f(x) = -x^2 + x + 3$

24. $f(x) = x^2 + 4x - 5$

In Problems 25 to 28 evaluate $\dfrac{f(x + h) - f(x)}{h}$ for the given function, with $h \neq 0$.

25. $f(x) = 4 - x$ 26. $f(x) = -1 - 2x$

27. $f(x) = -2x^2 - 4x$

28. $f(x) = 6x^2 + 3x$

In Problems 29 to 38 find the domain of the stated function.

29. $f(x) = \dfrac{3}{2 - x}$

30. $f(x) = \sqrt{x^2 + 1}$

31. $f(x) = \sqrt{1 - x^2}$

32. $f(x) = \dfrac{1}{x}$

33. $f(x) = \sqrt{x^2 - 1}$

34. $f(x) = \sqrt{x^3 - 1}$

35. $f(x) = \dfrac{1}{x(x - 2)}$

36. $f(x) = \dfrac{x}{x^2 - 9}$

37. $f(x) = \dfrac{\sqrt{x - 3}}{x - 4}$

38. $f(x) = \dfrac{x + 3}{\sqrt{x + 4}}$

In Problems 39 and 40 find all preimages of the given value for the function $f(x) = x^2 + 1$.

39. (a) 5   (b) 0   (c) 1.25

40. (a) 2   (b) −1   (c) 3.25

In Problems 41 and 42 sketch the graph of the equation and determine if it represents a function.

41. $y = \sqrt{x}$

42. $y^2 = 2x$

••• 43. Meteorologists use the equation $d^3 = 216t^2$ as a model to describe the size and intensity of four types of violent storms: tornadoes, thunderstorms, hurricanes, and cyclones.* Here d is the diameter of the storm (in miles) and t is the time (in hours) the storm travels before dissipating.
   (a) Is d a function of t? If so, solve for $d(t)$.
   (b) The world's worst recorded monsoon lasted for 24 hours on November 13–14, 1970, in the Ganges Delta Islands in Bangladesh. More than one million people died. What was the storm's diameter?
   (c) In the United States about 150 tornadoes occur each year, mostly in the central plains states, with the highest frequency in Iowa, Kansas, Arkansas, Oklahoma, and Mississippi. If a tornado's diameter is 2 miles, how long would it be expected to last?

44. Figure 5 lists the top 50 corporations in 1986 located outside the United States.† Let I be the function defined by the rule that $I(n)$ is the sales of the nth biggest corporation that year.
   (a) What is $I(1)$?
   (b) What is the image of 4?
   (c) What is the preimage of 22,668,085?

45. Does the Fortune 500 list of the biggest corporations in 1986 define a function if there is a tie for ninth place?

46. This problem refers to the number of chirps that a cricket makes. Biologists have shown that although different species of crickets produce different numbers of chirps, the number of chirps produced by each species is a linear function of the temperature. For instance, the number of chirps per minute that the cricket *Gryllus pennsylvanicus* makes is given by $C(t) = 4t - 160$, where t is in degrees Fahrenheit.‡
   (a) What is the domain of this function? (Make sure to take into account physical considerations.)
   (b) How many chirps per minute does this cricket make at 68°?
   (c) If this cricket makes 160 chirps per minute, what is the temperature?

---

*E. E. David and J. G. Truxal, *The Man-Made World,* New York, McGraw-Hill Book Company, 1971, pp. 176–177. The authors thank Glenn Allinger of Montana State University for supplying this excellent source of applications.

†Terence Paré and Wilton Woods, "The International 500: The Fortune Directory of the Biggest Industrial Corporations Outside the U.S.," *Fortune Magazine,* August 3, 1987, pp. 214–233.

‡Philip S. Callahan, *Insect Behavior,* New York, Four Winds Press, 1977.

Rank 1986	1985	Company	Country	Sales ($ thousands)
1	1	Royal Dutch/Shell Group	Neth./Britain	64,843,217
2	2	British Petroleum	Britain	39,855,564
3	5	IRI	Italy	31,561,709
4	3	Toyota Motor	Japan	31,553,827
5	16	Daimler-Benz	W. Germany	30,168,550
6	7	Matsushita Electric Industrial	Japan	26,459,539
7	6	Unilever	Neth./Britain	25,141,672
8	15	Volkswagen	W. Germany	24,317,154
9	8	Hitachi	Japan	22,668,085
10	4	ENI	Italy	22,549,921
11	13	Philips' Gloeilampenfabrieken	Netherlands	22,471,263
12	17	Nestlé	Switzerland	21,153,285
13	14	Siemens	W. Germany	20,307,037
14	12	Nissan Motor	Japan	20,141,237
15	22	Fiat	Italy	19,669,581
16	19	Bayer	W. Germany	18,768,914
17	20	BASF	W. Germany	18,640,985
18	30	Renault	France	17,661,021
19	21	Hoechst	W. Germany	17,509,344
20	10	Elf Aquitaine	France	17,287,058
21	23	Samsung	South Korea	16,522,664
22	24	Mitsubishi Heavy Industries	Japan	15,932,973
23	38	Peugeot	France	15,152,869
24	31	Toshiba	Japan	15,036,390
25	27	Imperial Chemical Industries	Britain	14,867,911
26	18	Petrobrás (Petróleo Brasileiro)	Brazil	14,701,534
27	28	Kuwait Petroleum	Kuwait	13,911,716
28	11	Total Cie Française de Pétroles	France	13,821,041
29	37	Thyssen	W. Germany	13,818,174
30	32	Nippon Oil	Japan	13,690,884
31	26	General Motors of Canada	Canada	13,335,100
32	36	BAT Industries	Britain	13,210,379
33	33	Nippon Steel	Japan	13,034,715
34	41	Honda Motor	Japan	12,481,447
35	42	Volvo	Sweden	11,795,263
36	48	CGE (Cie Générale d'Électricité)	France	11,681,337
37	43	Lucky-Goldstar	South Korea	11,433,748
38	54	Saint-Gobain	France	11,222,417
39	49	Daewoo	South Korea	11,204,191
40	9	Pemex (Petróleos Mexicanos)	Mexico	11,032,902
41	40	Canadian Pacific	Canada	10,807,709
42	47	NEC	Japan	10,562,396
43	45	Ford Motor of Canada	Canada	10,308,776
44	46	Idemitsu Kosan	Japan	10,255,663
45	58	Robert Bosch	W. Germany	10,003,662
46	62	Mazda Motor	Japan	9,678,888
47	53	Ruhrkohie	W. Germany	9,637,210
48	52	Mitsubishi Electric	Japan	9,401,584
49	34	Petróleos de Venezuela	Venezuela	9,249,726
50	61	Thomson	France	9,045,864

FIGURE 5 The Top 50 Corporations Located Outside the United States, 1986. (Adapted from "The International 500 Ranked by Sales," *Fortune Magazine*, August 3, 1987, 214–233.)

47. In 1820 the median age in the United States was 16.7 years. This means that half of the population was younger than that age, and half was older. Table 1 lists the median age in the United States for each year from 1980 to 1987.§

   (a) Define a function $y = f(x)$ in which x is a year and y is the corresponding median age.
   (b) Evaluate $f(1990)$.
   (c) For what value of x is $y = 29.1$?

TABLE 1

Year	Median Age
1987	32.1
1986	31.8
1985	31.5
1984	31.2
1983	30.9
1982	30.6
1981	30.3
1980	30.0

§Randolf E. Schmid, "Median Age in U.S. Reaches 32.1 Years, Highest on Record," Associated Press, April 6, 1988.

1–4 POLYNOMIAL AND OTHER FUNCTIONS

In this section we define polynomial functions and sketch their graphs by the plotting method. Then we introduce rational functions and sketch their graphs. The concept of a vertical asymptote plays an important role in graphing rational functions. We also introduce functions that are defined by different formulas over different intervals and functions that are defined by a curve instead of a formula.

Polynomial Functions

The graph of the quadratic function $f(x) = x^2$ is called a **parabola.** The graph of the function $f(x) = x^4$ has a very similar shape, although it is slightly different. The two curves are contrasted in Figure 1. The graphs of the functions $f(x) = x^n$ for $n = 6, 8, 10, \ldots$ are similar.

FIGURE 1 Contrast of $f(x) = x^2$ vs. $f(x) = x^4$.

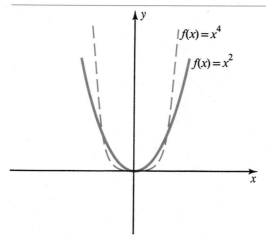

However, the graph of $f(x) = x^n$ is quite different when n is an odd number. For $n = 1$ the graph of $f(x) = x$ is a line. The graph of $f(x) = x^3$ is sketched in Figure 2. The graphs of the functions $f(x) = x^n$ for $n = 5, 7, 9, \ldots$ have similar shapes.

Collectively, functions of the form $f(x) = x^n$, where n can be any real number, are called **power functions.** Since n does not have to be a natural number, some additional examples of power functions are

$$n = 0 \qquad n = -1 \qquad n = \frac{1}{2}$$

$$f(x) = 1 \qquad f(x) = \frac{1}{x} \qquad f(x) = \sqrt{x}$$

A **polynomial function** of the variable x is defined by the formula

$$f(x) = a_n x^n + a_{n-1} x^{n-1} + \cdots + a_2 x^2 + a_1 x + a_0$$

where n is a nonnegative integer, the coefficients a_i are real numbers, and $a_n \neq 0$. The integer n is called the **degree of the polynomial.** The domain of a polynomial function consists of all real numbers. A polynomial function of degree 2 is a quadratic function, a polynomial function of degree 1 is a linear function, and a polynomial function of degree 0 is a constant function. The function

$$f(x) = x^4 - 8x^3 + 22x^2 + 9$$

FIGURE 2   Graph of $f(x) = x^3$.

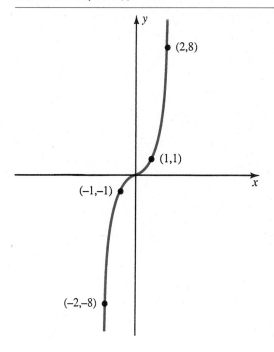

is a polynomial function of degree 4 with $a_4 = 1$, $a_3 = -8$, $a_2 = 22$, $a_1 = 0$, and $a_0 = 9$. Some other examples of polynomial functions are

$$g(x) = 12 \qquad F(x) = 4x - 3.5 \qquad G(x) = 3x^2 - 1.5x + 2$$

Some functions which are not polynomial functions are

$$h(x) = \frac{2}{x} \qquad P(x) = \sqrt{x}$$

Example 1 shows how to sketch the graph of a polynomial function by the plotting method.

EXAMPLE 1 | **Problem**

Sketch the graph of the polynomial function

$$f(x) = x^4 - 8x^3 + 22x^2 - 24x + 9$$

SOLUTION Evaluate $f(x)$ at the values $x = -2, -1, 0, 1,$ and 2. The results are displayed in the table.

x	-2	-1	0	1	2
$f(x)$	225	64	9	0	1

These values suggest that the graph is the curve shown in Figure 3a. However, plotting additional points from an expanded table (rounded to one decimal place) leads to the correct graph of the polynomial function shown in Figure 3b.

x	0.5	1.5	2.5	3	3.5	4	5
$f(x)$	1.6	0.6	0.6	0	1.6	9	64

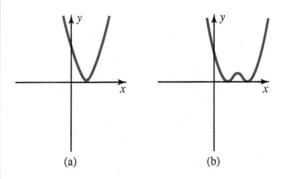

(a)   (b)

Example 1 shows that sometimes it is necessary to plot many points to obtain an accurate graph of a function. Yet the question persists: How do we know that the curve in Figure 3b is the correct graph of the given polynomial? The techniques of calculus will answer this question.

Rational Functions

Consider the power function $f(x)$ defined by the formula

$$f(x) = \frac{1}{x}$$

Its graph, which is called a **hyperbola,** consists of two disconnected branches. (See Figure 4.) The function is not defined at $x = 0$. Notice that as the values of x get closer and closer to 0, the values of $f(x)$ get larger and larger in absolute value. When a number v has the property that the values of a function $f(x)$ get larger and larger in absolute value as the values of x approach v, the line $x = v$ is called a **vertical asymptote** of the graph of $f(x)$. Figure 4 shows that the line $x = 0$ is a vertical asymptote of the function $f(x) = 1/x$.

A **rational function** is the quotient of two polynomial functions. Therefore $f(x)$ is a rational function if it can be written in the form

$$f(x) = \frac{g(x)}{h(x)}$$

for some polynomials $g(x)$ and $h(x)$. The function $f(x) = 1/x$ is a rational function with $g(x) = 1$ and $h(x) = x$.

The domain of a rational function $f(x) = g(x)/h(x)$ consists of all real numbers for which $h(x) \neq 0$. Each number v for which $h(v) = 0$ will produce a vertical asymptote of $f(x)$ at $x = v$ if $g(v) \neq 0$. Example 2 shows how to find vertical asymptotes.

FIGURE 4   Graph of $f(x) = \dfrac{1}{x}$.

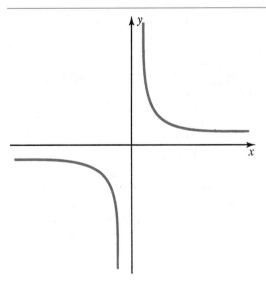

EXAMPLE 2

Problem

Find the vertical asymptotes of the function

$$f(x) = \frac{x - 2}{x^2 - 1}$$

SOLUTION Set $g(x) = x - 2$ and $h(x) = x^2 - 1$. If the rational function $f(x)$ is going to have a vertical asymptote, it will occur where $h(x) = 0$. Set $h(x) = x^2 - 1 = 0$. Then $x = 1$ or $x = -1$. Also, $g(1) \neq 0$ and $g(-1) \neq 0$. Therefore the lines $x = 1$ and $x = -1$ are the vertical asymptotes of the graph of $f(x)$.

Consider the rational function $f(x) = \dfrac{x^2 - 4}{x + 2}$. Set $g(x) = x^2 - 4$ and $h(x) = x + 2$. The only possible vertical asymptote will be the line $x = -2$ since -2 is the only number with $h(-2) = 0$. However, $g(-2) = 0$, so the line $x = -2$ is *not* a vertical asymptote of $f(x)$.

A knowledge of the vertical asymptotes of a rational function eases the task of sketching its graph. Example 3 illustrates the procedure.

EXAMPLE 3

Problem

Sketch the graph of the rational function

$$f(x) = \frac{x}{x - 1}$$

SOLUTION Set $g(x) = x$ and $h(x) = x - 1$. The only vertical asymptote of $f(x)$ is the line $x = 1$ since $h(1) = 0$ but $g(1) \neq 0$. Draw it as a dashed line. Construct a table of ordered pairs for $x < 1$.

x	-2	-1	0	0.5	0.9	0.99
$f(x)$	$\frac{2}{3}$	$\frac{1}{2}$	0	-1	-9	-99

Notice that as the values of x approach 1, the images of $f(x)$ get larger and larger in absolute value. This shows the behavior of $f(x)$ to the left of the vertical asymptote. A table for $x > 1$ shows the behavior to the right of the vertical asymptote.

x	4	3	2	1.5	1.1	1.01
$f(x)$	$\frac{4}{3}$	$\frac{3}{2}$	2	3	11	101

Taken together, the graph consists of the two branches that result from plotting the points in these tables. (See Figure 5.)

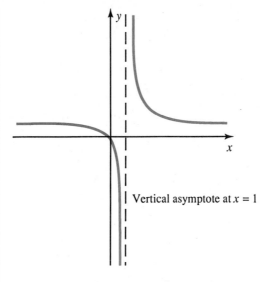

FIGURE 5 Graph of $f(x) = \dfrac{x}{x - 1}$.

Other Functions

We introduce now two additional kinds of functions. The first concerns functions that are defined by different formulas over different intervals.

EXAMPLE 4 | **Problem**

Sketch the graph of the function defined by

$$f(x) = \begin{cases} 2x + 1 & \text{if } x < 2 \\ x^2 - 1 & \text{if } x \geq 2 \end{cases}$$

SOLUTION The notation means that $f(x) = 2x + 1$ for all values of x such that $x < 2$. This is the linear function with slope 2 and y-intercept 1. The graph of $f(x)$ in the interval $(-\infty, 2)$ is part of this line. An open circle at the right end point of the line indicates that this point is not part of the function, but rather, serves as a boundary. The graph of $f(x)$ in the interval $[2, \infty)$ is part of the quadratic function $f(x) = x^2 - 1$. The entire graph consists of the two branches shown in Figure 6.

A function can be defined by a graph as well as by a formula. This sometimes occurs when the equation is either unknown or too complicated to express easily. For each value of x for which there is one point (x, y) on the curve, the image $f(x)$ is defined to be equal to y. This defines the function f implicitly. Example 5 examines one such function.

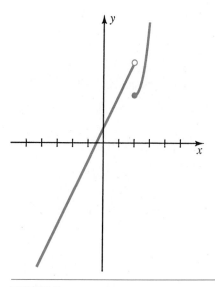

FIGURE 6

EXAMPLE 5

A firm's financial leverage is defined as its debt/assets ratio, or D/A ratio. The D/A ratio is usually given as a percent. There is a relation between the D/A ratio and the firm's expected earnings-per-share (EPS). Figure 7 gives the graph for Santa Clara Industries on January 1, 1985.

Problem

(a) Approximate $f(30)$. (b) What D/A ratio has the maximal expected EPS?

SOLUTION (a) The notation $f(30)$ stands for the image of $x = 30\%$. It corresponds to the value of y from the point $(30, y)$ on the graph. From Figure 7

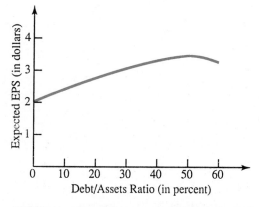

FIGURE 7 Financial Leverage vs. D/A Ratio for Santa Clara Industries. (Adapted from "Financial Leverage," in B. Campsey and E. Brigham, *Introduction to Financial Management,* Hinsdale, IL, The Dryden Press, 1985, pp. 534–541.)

it appears that $y = 3$, so $f(30) = 3$ and the expected EPS is \$3. (According to Santa Clara Industries' tables, $f(30) = \$2.97$. This degree of accuracy cannot be inferred from the graph, however.)

(b) The maximal expected EPS of \$3.36 occurs at the point $(50, 3.36)$. Thus $f(50) = 3.36$. The preimage of 3.36 is 50, so the D/A ratio is 50%.

Application

In the previous section we examined linear cost functions. Here we discuss a kind of nonlinear cost function that arises in manufacturing.

EXAMPLE 6

A manufacturer has determined that the cost of producing x items is $C(x) = \dfrac{x^2 + 25}{x}$, where fractional parts of an item can be produced.

Problem

(a) What is the domain of $C(x)$? (b) Sketch the graph of $C(x)$ to determine how many items should be produced in order to minimize cost.

SOLUTION (a) The domain of $C(x)$ does not include $x = 0$ because division by 0 is undefined. In addition, since it makes no sense to talk about a negative number of items, it is meaningless to let x assume negative values. Therefore the domain of $C(x)$ consists of all positive real numbers, or the interval $(0, \infty)$.

(b) Set $g(x) = x^2 + 25$ and $h(x) = x$. Then $h(x) = 0$ and $g(x) \neq 0$, so the line $x = 0$ is a vertical asymptote of $f(x)$. The relevant graph of $C(x)$ is obtained by the plotting method from the table.

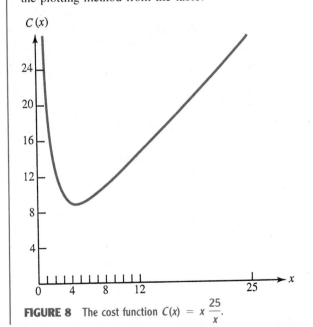

FIGURE 8 The cost function $C(x) = x \dfrac{25}{x}$.

x	1	2	3	4	5	6	7	25
$C(x)$	26	$\frac{29}{2}$	$\frac{34}{3}$	$\frac{41}{4}$	10	$\frac{61}{6}$	$\frac{74}{7}$	26

The graph, which is sketched in Figure 8, indicates that the minimum cost occurs at $x = 5$ items.

In Chapter 4 we will describe how to use calculus to find the minimum cost in Example 6.

EXERCISE SET 1–4

• In Problems 1 to 4 sketch the graph of the stated function by the plotting method.

1. $f(x) = 2x^3 - 9x^2 + 12x$

2. $f(x) = x^3 - 9x^2 + 24x - 20$

3. $f(x) = x^4 - 4x^3 + 4x^2 + 1$

4. $f(x) = x^4 - 6x^2$

5. What is the degree of the polynomial in Problem 1?

6. What is the degree of the polynomial in Problem 4?

In Problems 7 to 10 find the vertical asymptotes of each function.

7. $f(x) = \dfrac{x}{x - 5}$

8. $f(x) = \dfrac{3x}{2x - 5}$

9. $f(x) = \dfrac{x - 2}{x^2 - 9}$

10. $f(x) = \dfrac{3 - x}{x^2 - 4}$

In Problems 11 to 16 sketch the graph of the stated function.

11. The function defined in Problem 7.

12. The function defined in Problem 8.

13. $f(x) = \begin{cases} 3x + 2 & \text{if } x \le 0 \\ x^2 + 2 & \text{if } x > 0 \end{cases}$

14. $f(x) = \begin{cases} 2x + 1 & \text{if } x \ge 3 \\ x^2 - 2 & \text{if } x < 3 \end{cases}$

15. $f(x) = \begin{cases} x^2 - 1 & \text{if } x < 0 \\ 2x - 1 & \text{if } x \ge 0 \end{cases}$

16. $f(x) = \begin{cases} 1 - x^2 & \text{if } x > 2 \\ 1 - 2x & \text{if } x \le 2 \end{cases}$

Problems 17 to 20 refer to Figure 7, which shows the relation between the D/A ratio and the expected EPS for Santa Clara Industries.

17. Approximate $f(0)$.

18. Approximate $f(60)$.

19. What D/A ratio produces an expected EPS of approximately $3?

20. What D/A ratio produces an expected EPS of approximately $2.75?

In Problems 21 and 22 the cost of producing x items is given. Fractional parts of an item can be produced.
(a) What is the domain of $C(x)$?
(b) Sketch $C(x)$ to determine how many items should be produced in order to minimize cost.

21. $C(x) = \dfrac{x^2 + 16}{x}$

22. $C(x) = \dfrac{x^2 + 81}{x}$

•• In Problems 23 and 24 sketch the graphs of the two stated functions on the same coordinate axis.

23. $f(x) = x^2$ and $f(x) = x^6$

24. $f(x) = x^3$ and $f(x) = x^5$

In Problems 25 and 26 state the degree of the polynomial.

25. $f(x) = 1 - 2x - 5x^2$

26. $f(x) = 3 - 5x$

In Problems 27 to 34 find the vertical asymptotes, if any, and sketch the graph of the stated function.

27. $f(x) = \dfrac{x-2}{x^2-9}$

28. $f(x) = \dfrac{3-x}{x^2-4}$

29. $f(x) = \dfrac{x}{x^2}$

30. $f(x) = \dfrac{1}{x^2+1}$

31. $f(x) = \dfrac{1}{x(x^2-1)}$

32. $f(x) = \dfrac{x-1}{x(x^2-4)}$

33. $f(x) = \dfrac{2x}{(3+2x)(x-1)}$

34. $f(x) = \dfrac{x}{(1+2x)(x-2)}$

In Problems 35 to 38 sketch the graph of the stated function.

35. $f(x) = \begin{cases} x & \text{if } x \geq 1 \\ x^2 + 1 & \text{if } x \leq 0 \end{cases}$

36. $f(x) = \begin{cases} 2x - 1 & \text{if } x \geq 3 \\ x^3 & \text{if } x \leq 1 \end{cases}$

37. $f(x) = \begin{cases} 1 - 2x - 5x^2 & \text{if } x \leq 1 \\ x & \text{if } x > 1 \end{cases}$

38. $f(x) = \begin{cases} -x^2 & \text{if } x \geq 0 \\ x^2 & \text{if } x < 0 \end{cases}$

39. Suppose the profit per item is given by the formula $P(x) = 20(3 - x)(x - 25)$, where x is the number of items that are produced. Draw a graph of $P(x)$ to determine the number of items that should be produced to yield the largest profit per item.

40. A store offers this special sale: if 10 compact disks (CDs) are purchased at the full price of $12, then additional CDs can be purchased at half price. There is a limit of 25 CDs per person. Express the cost of the CDs as a function of the number purchased and draw the graph.

●●● In the programming language BASIC there is a function INT that produces the integral part of a number. For instance, $INT(3.14) = 3$. In Problems 41 and 42 sketch the graph of the given function.

41. $f(x) = INT(x)$ for $x \geq 0$

42. $f(x) = INT(x^2)$ for $x \geq 0$

43. The curve in Figure 9 shows the relation between the implied stock price and the D/A ratio for Santa Clara Industries. What percentage yields the maximal D/A ratio?

44. Figure 10 shows two curves portraying stock market activity during two dramatic months in 1986.
    (a) What was the maximum value of the Dow Jones Industrial average during this period?
    (b) During which week did the Dow Jones Industrial average tumble precipitously?

45. An article in the *Wall Street Journal* analyzed the federal income tax that was signed into law on October 22, 1986.* Consider this statement as an example of the tax law: "If the income lies in the range from $0 through $29,300 (in-

FIGURE 9 Relation Between Implied Stock Price and the D/A Ratio for Santa Clara Industries. (Adapted from B. Campsey and E. Brigham, *Introduction to Financial Management*, Hinsdale, IL, The Dryden Press, 1985, p. 542.)

*Alan Murray, "Winners? Losers? Estimates Show How Impact of Tax Proposal Varies," *Wall Street Journal*, May 9, 1986, p. 25.

Stock Market Indexes

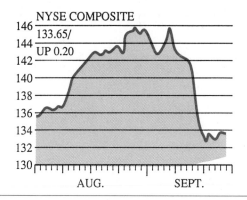

AUG.   SEPT.

FIGURE 10 Stock market activity during two dramatic months in 1986. (Dow–Jones Index.)

clusive) then the tax is 15 percent of the income; if the income exceeds $29,300 then the tax equals $4350 plus 27 percent of the portion of the income above $29,300.''
(a) Write this statement as a function $y = f(x)$, where x is the income and y is the corresponding tax.
(b) Sketch the graph of $f(x)$.

(c) Evaluate $f(x)$ for $x = 29,300$.
(d) Evaluate $f(x)$ for $x = 29,301$.
(e) What do you conclude from parts (c) and (d)?

46. Public health officials must often determine the threshold level at which effects on health begin. The threshold level for certain environmental issues occurs when the introduction of one more unit of pollution becomes a detriment to public health. One study[†] made use of the health function $H(x)$ defined by

$$H(x) = \begin{cases} 0 & \text{if } x < 5 \\ (x - 5)^2 & \text{if } x \geq 5 \end{cases}$$

where x denotes the number of units of pollution.
(a) Sketch the graph of this health function.
(b) At what value of x does the threshold level occur?

47. Human beings have raced over a large span of times and distances. Take women's swimming, for instance. Races have been contested from a scant 50 meters to the 80-kilometer (km) swim across the English Channel. It turns out that when world records for many athletic events are recorded for various distances at any given time, it is possible to derive a power function that relates the time to the distance. The power function for women's swimming up to 1980, with the time t in minutes and the distance x in kilometers (km), is:[‡]

$$t = 10.578x^{1.03256}$$

Use this equation to calculate the time of the world record for (a) 1500 meters and (b) 80 km.

†Louis A. Cox, Jr., ''A New Measure of Attributable Risk for Public Health Applications,'' *Management Science*, July 1985, pp. 800–813.

‡Peter S. Riegel, ''Athletic Records and Human Endurance,'' *American Scientist*, May-June 1981, pp. 285–290.

1–5 THE ALGEBRA OF FUNCTIONS

The four familiar operations on real numbers are addition, subtraction, multiplication, and division. In this section we define these operations on functions and introduce a fifth operation, the composition of functions. We then discuss additional types of functions that are defined in terms of these five operations.

Operations on Functions

Consider the polynomial functions f and g defined by

$$f(x) = 2x^2 - 3x - 7 \qquad \text{and} \qquad g(x) = 2x + 3$$

The **sum** of these functions, denoted by $f + g$, is defined by the formula

$$\begin{aligned}
(f + g)(x) &= f(x) + g(x) \\
&= (2x^2 - 3x - 7) + (2x + 3) \\
&= 2x^2 - x - 4
\end{aligned}$$

The **difference** $f - g$ is defined similarly.

$$\begin{aligned}
(f - g)(x) &= (2x^2 - 3x - 7) - (2x + 3) \\
&= 2x^2 - 5x - 10
\end{aligned}$$

The **product** $f \cdot g$ is defined by carrying out the multiplication of the two polynomials.

$$\begin{aligned}
(f \cdot g)(x) &= f(x) \cdot g(x) \\
&= (2x^2 - 3x - 7) \cdot (2x + 3) \\
&= (2x^2 - 3x - 7)(2x) + (2x^2 - 3x - 7)(3) \\
&= (4x^3 - 6x^2 - 14x) + (6x^2 - 9x - 21) \\
&= 4x^3 - 23x - 21
\end{aligned}$$

The **quotient** f/g is defined by

$$\left(\frac{f}{g}\right)(x) = \frac{f(x)}{g(x)} = \frac{2x^2 - 3x - 7}{2x + 3}$$

Sometimes the quotient can be left in this form. At other times the division of polynomials must be carried out.

$$\begin{array}{r}
x - 3 \\
2x + 3 \overline{\smash{)}2x^2 - 3x - 7} \\
\underline{2x^2 + 3x} \\
-6x - 7 \\
\underline{-6x - 9} \\
2
\end{array}$$

Then

$$\left(\frac{f}{g}\right)(x) = x - 3 + \frac{2}{2x + 3}.$$

The operations of $+$, $-$, and $\cdot$ are defined whenever both $f(x)$ and $g(x)$ are defined, so the domain of each one is the intersection of the domain of f and the domain of g. The quotient f/g is somewhat different. It is defined whenever both $f(x)$ and $g(x)$ are defined except where $g(x) = 0$. In the example above, the domain of f/g consists of all real numbers except $-\frac{3}{2}$ because $g(-\frac{3}{2}) = 0$. The display gives the general definitions.

_____ **Definition** _____

If f and g are any functions for which $f(x)$ and $g(x)$ are defined, then

$$(f + g)(x) = f(x) + g(x) \qquad \text{(sum)}$$
$$(f - g)(x) = f(x) - g(x) \qquad \text{(difference)}$$
$$(f \cdot g)(x) = f(x) \cdot g(x) \qquad \text{(product)}$$
$$\frac{f}{g}(x) = \frac{f(x)}{g(x)} \qquad \text{if } g(x) \neq 0 \quad \text{(quotient)}$$

_____

The four operations $f + g$, $f - g$, $f \cdot g$, and f/g are referred to collectively as the **rational operations.**

EXAMPLE 1

Let f and g be functions defined by

$$f(x) = \frac{x}{x - 1} \qquad \text{and} \qquad g(x) = \frac{x - 3}{x + 1}$$

Problem

Find $f + g$, $f \cdot g$, and f/g. State their domains.

SOLUTION By definition $f + g$ is defined by

$$(f + g)(x) = \frac{x}{x - 1} + \frac{x - 3}{x + 1}$$

The proper form of the sum is obtained by carrying out the addition, which involves finding the least common denominator.

$$(f + g)(x) = \frac{x(x + 1) + (x - 1)(x - 3)}{(x - 1)(x + 1)}$$
$$= \frac{x^2 + x + x^2 - 4x + 3}{x^2 - 1}$$
$$= \frac{2x^2 - 3x + 3}{x^2 - 1}$$

The product $f \cdot g$ is obtained by multiplying the numerators and denominators.

$$(f \cdot g)(x) = \frac{x(x - 3)}{(x - 1)(x + 1)}$$

Often it is convenient to write this expression in a more compact form.

$$(f \cdot g)(x) = \frac{x^2 - 3x}{x^2 - 1}$$

The quotient is obtained from the "invert and multiply" rule.

$$\left(\frac{f}{g}\right)(x) = \frac{x}{x-1} \div \frac{x-3}{x+1} = \frac{x}{x-1} \cdot \frac{x+1}{x-3}$$

$$= \frac{x^2 + x}{x^2 - 4x + 3}$$

The domain of f consists of all real numbers except 1 since $f(1)$ is not defined. Similarly, $g(-1)$ is not defined. Therefore the domain of $f + g$ and $f \cdot g$ is the set of all real numbers except 1 and -1. Since $g(3) = 0$, the domain of f/g is the set of all real numbers except 1, -1, and 3.

If a number a lies in the domain of $f + g$, we write $(f + g)(a) = f(a) + g(a)$. The same notation applies to the other three rational operations on two functions. Example 2 shows how to find such images.

EXAMPLE 2   Let f and g be functions defined by

$$f(x) = \frac{x}{x-1} \quad \text{and} \quad g(x) = \frac{3x-2}{x+1}$$

Problem

Compute $(f + g)(\frac{2}{3})$, $(f \cdot g)(\frac{2}{3})$, and $(f/g)(\frac{2}{3})$, if possible.

SOLUTION First $f(\frac{2}{3}) = -2$ and $g(\frac{2}{3}) = 0$. By definition, $(f + g)(\frac{2}{3}) = f(\frac{2}{3}) + g(\frac{2}{3})$. Since $f(\frac{2}{3}) = -2$ and $g(\frac{2}{3}) = 0$, it follows that $(f + g)(\frac{2}{3}) = -2 + 0 = -2$. Similarly, $(f \cdot g)(\frac{2}{3}) = f(\frac{2}{3}) \cdot g(\frac{2}{3}) = (-2) \cdot 0 = 0$. However, $(f/g)(\frac{2}{3})$ is not defined because $g(\frac{2}{3}) = 0$.

Composition of Functions

Consider once again the polynomial functions

$$f(x) = 2x^2 - 3x - 7 \quad \text{and} \quad g(x) = 2x + 3$$

Regard each function as an "input-output machine" such that each input value comes from the domain and the corresponding output lies in the range. Now put the machines in a line with g in front of f, so the output of g becomes the input of f. For example, take $x = 1$. The output of g is $g(1) = 5$. This value becomes the input of f, and since $f(5) = 28$, its output is 28. Altogether the sequence of "machines" has produced an image of 28 from the original input of $x = 1$. This is shown in Figure 1, where the dashed curve represents the result of the two functions.

The operation of performing one function f on the image of another function g is called *composition* and is denoted by $f \circ g$. It is defined by

$$(f \circ g)(x) = f(g(x))$$

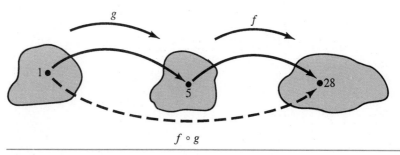

FIGURE 1 The composition of functions.

For the polynomials f and g above,

$$\begin{aligned}
(f \circ g)(x) &= f(g(x)) = f(2x + 3) \\
&= 2(2x + 3)^2 - 3(2x + 3) - 7 \\
&= 2(4x^2 + 12x + 9) - 6x - 9 - 7 \\
&= 8x^2 + 18x + 2
\end{aligned}$$

The display provides the formal definition.

Definition

If f and g are any functions, then the **composition** of f and g is the function $f \circ g$ defined by

$$(f \circ g)(x) = f(g(x))$$

The domain of $f \circ g$ is the set of all real numbers x in the domain of g whose image $g(x)$ lies in the domain of f.

If a is a number in the domain of $f \circ g$, then $(f \circ g)(a) = f(g(a))$. For $f(x) = 2x^2 - 3x - 7$, $g(x) = 2x + 3$, and $a = 1$, we get $(f \circ g)(1) = f(g(1)) = f(5) = 28$. However, $(f \cdot g)(1) = f(1) \cdot g(1) = (-8) \cdot 5 = -40$, so $(f \circ g)(x) \neq (f \cdot g)(x)$.

> **WARNING:** Do not confuse $f \circ g$ with $f \cdot g$. The composition of two functions is entirely different from the product of two functions.

EXAMPLE 3

Let f and g be the functions defined by

$$f(x) = \frac{1}{x + 1} \qquad \text{and} \qquad g(x) = \sqrt{x}$$

Problem

Find $f \circ g$ and $g \circ f$.

SOLUTION

$$(f \circ g)(x) = f(g(x))$$
$$= f(\sqrt{x})$$
$$= \frac{1}{\sqrt{x} + 1}$$

Similarly

$$(g \circ f)(x) = g(f(x)) = g\left(\frac{1}{x + 1}\right) = \sqrt{\frac{1}{x + 1}} = \frac{1}{\sqrt{x + 1}}$$

When forming the composition of two functions the order of the functions is important. Notice, for instance, that for the functions f and g in Example 3, $(f \circ g)(1) = 1/2$ but $(g \circ f)(1) = 1/\sqrt{2}$, so $(g \circ f)(1) \neq (f \circ g)(1)$. We single this out as a warning.

WARNING: In general $f \circ g \neq g \circ f$

When applying the techniques of calculus, often we find it advantageous to regard a given function as the composition of two functions. For example, consider the function $h(x) = \sqrt{x^2 - 3x + 5}$. Since the polynomial $x^2 - 3x + 5$ is evaluated first, set $g(x) = x^2 - 3x + 5$. Then $h(x) = \sqrt{g(x)}$. Now put $f(x) = \sqrt{x}$. Then $(f \circ g)(x) = f(g(x)) = \sqrt{g(x)} = h(x)$, so $h = f \circ g$. In this way $h(x)$ is written as the composition of the two simpler functions f and g.

EXAMPLE 4 | **Problem**

Write the function

$$k(x) = \frac{(x^2 - 3x + 5)^5 - 1}{(x^2 - 3x + 5)^4 + 1}$$

as a composition of two simpler functions.

SOLUTION The polynomial $x^2 - 3x + 5$ is evaluated first, so set $g(x) = x^2 - 3x + 5$. Then $k(x)$ can be written as

$$k(x) = \frac{[g(x)]^5 - 1}{[g(x)]^4 + 1}$$

Define $f(x)$ by substituting x for $g(x)$ in this expression.

$$f(x) = \frac{x^5 - 1}{x^4 + 1}$$

Then $k = f \circ g$ since

$$(f \circ g)(x) = f(g(x)) = \frac{[g(x)]^5 - 1}{[g(x)]^4 + 1} = k(x)$$

Quality Control

For many companies quality is the most important product. Quality control often depends on the production level; a low production level can cause problems with a worker's concentration and morale, while a high production level can lead to fatigue and nervousness. The usual assembly line offers a prime example.

It is possible to measure quality Q in terms of the production level P, where Q ranges from 0 to 100 (100 being a "perfect" product) and P represents the number of units produced per day. Production, in turn, often depends on the time of day t, measured in hours. This means that ultimately the quality Q can be regarded as a function of the time of day t. The underlying concept is the composition of functions. Example 5 addresses this issue.

EXAMPLE 5

Let the quality of a product be defined by

$$Q(P) = 50 + 8P - 0.4P^2$$

where P is the number of units produced per day, with $0 \leq P \leq 18$. Let the daily production be defined by

$$P(t) = 6t - 0.5t^2$$

where t is the time of day in hours, with $0 \leq t \leq 8$.

Problem

Express the quality Q as a function t.

SOLUTION Write Q as a function of $P(t)$.

$$Q(P(t)) = 50 + 8P(t) - 0.4P(t)^2$$

Substitute the expression for $P(t)$.

$$Q(P(t)) = 50 + 8(6t - 0.5t^2) - 0.4(6t - 0.5t^2)^2$$

Collect similar terms.

$$\begin{aligned} Q(P(t)) &= 50 + 48t - 4t^2 - 0.4(36t^2 - 6t^3 + 0.25t^4) \\ &= 50 + 48t - 18.4t^2 + 2.4t^3 - 0.1t^4 \end{aligned}$$

In this way quality Q is defined as a function of time t.

EXERCISE SET 1–5

- In Problems 1 to 4 find $f + g$, $f \cdot g$, and f/g.

1. $f(x) = \dfrac{x + 2}{x - 1}$ and $g(x) = \dfrac{x}{x + 2}$

2. $f(x) = \dfrac{x - 2}{x + 3}$ and $g(x) = \dfrac{x}{x - 3}$

3. $f(x) = \dfrac{x}{x + 1}$ and $g(x) = \dfrac{1}{x}$

4. $f(x) = \dfrac{x}{x - 1}$ and $g(x) = \dfrac{x}{x + 2}$

5. Find the domain of $f + g$ for the functions f and g defined in Problems 1 and 3.

6. Find the domain of $f - g$ for the functions f and g defined in Problems 2 and 4.

7. Find the domain of f/g for the functions f and g defined in Problems 1 and 3.

8. Find the domain of f/g for the functions f and g defined in Problems 2 and 4.

In Problems 9 to 12 compute (a) $(f + g)(a)$, (b) $(f \cdot g)(a)$, and (c) $(f/g)(a)$ for the functions f and g in the stated problem and the given value of a.

9. f and g in Problem 1, $a = 2$

10. f and g in Problem 2, $a = 4$

11. f and g in Problem 3, $a = -2$

12. f and g in Problem 4, $a = 0$

In Problems 13 to 16 find $f \circ g$ and $g \circ f$.

13. $f(x) = \sqrt{x}$ $g(x) = \dfrac{x}{1 - x}$

14. $f(x) = \sqrt{2x}$ $g(x) = \dfrac{1}{1 - x}$

15. $f(x) = 2x$ $g(x) = \dfrac{x - 2}{x + 2}$

16. $f(x) = \dfrac{1}{x}$ $g(x) = \sqrt{x}$

In Problems 17 to 20 write the function as the composition of two functions.

17. $f(x) = \sqrt{x^3 + 2x - 5}$

18. $f(x) = \sqrt{2x + 3}$

19. $f(x) = \dfrac{1}{(x - 2)^2} + (x - 2)^3$

20. $f(x) = \dfrac{1 + (x + 1)^2}{3 + (x + 1)^3}$

In Problems 21 and 22 the quality of a product is defined by $Q(P)$, where P is the number of units produced per day and the daily production is defined by $P(t)$, with t the time of day in hours. Express Q as a function of t.

21. $Q(P) = 20 + 4P - 0.02P^2$
    $P(t) = 8t - 0.5t^2$

22. $Q(P) = 25 + 2P - 0.05P^2$
    $P(t) = 6t - 0.8t^2$

•• In Problems 23 and 24 let $f(x)$ and $g(x)$ be the func-

tions defined in Problem 1 and $h(x) = \dfrac{1}{x - 2}$.

23. Compute (a) $f + g + h$ (b) $f \cdot g \cdot h$.

24. Compute (a) $(f/g)/h$ (b) $f/(g/h)$
    (c) Is $(f/g)/h = f/(g/h)$?

In Problems 25 and 26 let $f(x)$ and $g(x)$ be the functions defined in Problem 2 and $h(x) = \dfrac{1}{x + 3}$.

25. Compute (a) $f + g + h$ (b) $f \cdot g \cdot h$.

26. Compute (a) $(f/g)/h$ (b) $f/(g/h)$
    (c) Is $(f/g)/h = f/(g/h)$?

27. Evaluate $(f \cdot g \cdot h)(0)$ for the functions in Problem 23.

28. Evaluate $(f \cdot g \cdot h)(0)$, where $f(0)$ and $g(0)$ are defined and $h(0) = 0$.

In Problems 29 to 32 compute $f \circ g$

29. $f(x) = \dfrac{1}{\sqrt{x}}$ $g(x) = \dfrac{1}{x^2}$

30. $f(x) = 1 - x$ $g(x) = \dfrac{1}{x^3}$

31. $f(x) = \dfrac{x}{x + 1}$ $g(x) = \dfrac{x + 1}{x}$

32. $f(x) = \sqrt{1 - x}$ $g(x) = 1 - x^2$

33. Let $f(x)$ and $g(x)$ be the functions defined in Problem 31. Is it true that $(f \circ g)(1) = (g \circ f)(1)$?

34. Let $f(x)$ and $g(x)$ be the functions defined in Problem 30. Is it true that $(f \circ g)(-1) = (g \circ f)(-1)$?

In Problems 35 and 36 write the given function as the composition of two functions.

35. $f(x) = \sqrt[3]{(x - 1)^2}$

36. $f(x) = \dfrac{5}{(2 - x)^3}$

In Problems 37 and 38 let f and g be the functions defined in Example 1 in the text.

37. Compute $(f + g)(a)$, where a is a given value of x.

38. Compute $(f + g)(x + h)$.

In Problems 39 and 40 let f and g be the functions defined in Example 3 in the text.

39. Compute $(f + g)(a)$, where a is a given value of x.

40. Compute $(f + g)(x + h)$.

41. For the functions f and g in Problem 1, is it true that $f \cdot g = f \circ g$?

42. For the functions f and g in Problem 3, is it true that $f \cdot g = f \circ g$?

43. Suppose the number of hours required to stock x percent of a shipment is $S(x) = g(x)/h(x)$, where $g(x) = 40x$ and $h(x) = 200 - x$.
    (a) What is the domain of the function $S(x)$?
    (b) For what values of x does $S(x)$ have a practical interpretation in this context?
    (c) How many hours were required to stock 40% of the shipment?
    (d) How many hours were required to stock the entire shipment?
    (e) What percentage of the shipment was stocked after 20 hours?

44. Suppose the cost of conducting a political survey of x percent of the eligible voters in a precinct is $C(x) = g(x)/h(x)$ dollars, where $g(x) = 105x$ and $h(x) = 110 - x$.
    (a) What is the domain of the function $C(x)$?
    (b) For what values of x does $C(x)$ have a practical interpretation in this context?
    (c) How much did it cost to poll 40% of the precinct?
    (d) How much did it cost to poll the entire precinct?
    (e) What percentage of the precinct could be polled with a fund of $500?

45. This problem refers to the profit function $P(x)$ on the sales of x units of an item: $P(x) = R(x) - C(x)$, where $R(x) = 480x - 12x^2 - 2100$ is the revenue function and $C(x) = 96x + 204$ is the cost function.
    (a) Draw the graphs of $R(x)$ and $C(x)$ on the same coordinate axis.

(b) Use the graphs in part (a) to determine the numbers of units of the item that should be sold so the profit is a positive amount.

●●● In Problems 46 and 47 let f, g, and h be the functions

$$f(x) = \frac{1}{x} \quad g(x) = \sqrt{x} \quad h(x) = x + 1$$

46. Compute $\dfrac{f \circ g}{h}$.

47. (a) Compute $(f \circ g) \circ h$.
    (b) Compute $f \circ (g \circ h)$.
    (c) Compute $f \circ g \circ h$.

48. Figure 2 traces Japanese net purchases of foreign stocks and bonds from 1984 through 1987.* Let $B(x)$ denote the purchases of bonds and $S(x)$ denote the purchases of stocks in year x. Write an equation for $B(x) + S(x)$ during the years 1984–1987.

49. The coding of messages using a computer involves the composition of three functions.† The

FIGURE 2

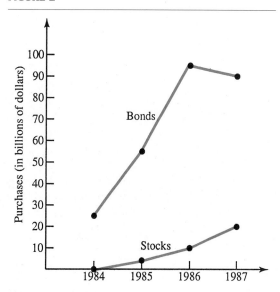

*Mark Beauchamp and John Hines, "We'll Send You VCRs—You Send Us Stocks," *Forbes Magazine,* August 10, 1987, pp. 60–62.

†James Reagan, "Get the Message? Cryptographs, Mathematics, and Computers," *The Mathematics Teacher,* October 1986, pp. 547–553.

TABLE 1

letter	A	B	C	D	⋯	L	⋯	X	Y	Z
value	65	66	67	68	⋯	76	⋯	88	89	90

first function, $h(x)$, assigns a number, called the ASCII value, to each letter in the alphabet. (ASCII stands for the American Standard Code for Information Interchange.) It is partially defined in Table 1.

The second function, $g(x)$, is called the encoding function. Let $g(x)$ be defined by

$$g(x) = \begin{cases} x + 10 & \text{if } 65 \le x \le 80 \\ x - 16 & \text{if } 81 \le x \le 90 \end{cases}$$

The third function, $f(x)$, uses Table 1 to assign a letter to each ASCII value x, where $65 \le x \le 90$. The code function $c(x)$, which replaces each letter in the alphabet by another letter, is defined by $c = f \circ g \circ h$. For instance, $c(L) = (f \circ g \circ h)(L) = f(g(h(L))) = f(g(76)) = f(86) = V$, so the letter L is always replaced in the code by the letter V. Find the code for each letter in the word FUN.

1–6 SOLVING POLYNOMIAL EQUATIONS

Many applications of mathematics require a knowledge of points where the graph of a function f crosses the x-axis. Algebraically this is equivalent to solving the equation $f(x) = 0$.

We begin this section by reviewing quadratic equations and the quadratic formula. Then we apply the method of factoring to solving polynomial equations. We end by using these procedures to find the points of intersection of two graphs.

The Quadratic Formula

The graph of a function f crosses the x-axis at $x = a$ if and only if $f(a) = 0$. Geometrically the point $(a, 0)$ is an x-intercept of the graph. Algebraically the value $x = a$ is called a **solution** (or a **root**) **of the equation** $f(x) = 0$. Therefore solutions of the equation $f(x) = 0$ yield x-intercepts of the graph of f. A **polynomial equation** is an equation $f(x) = 0$, where f is a polynomial function. To "solve a polynomial equation" means to find all of its solutions. Recall that the **degree of a polynomial equation** is the highest power of the polynomial.

A **quadratic equation** is a polynomial equation of degree 2. The general form of a quadratic equation is

$$ax^2 + bx + c = 0$$

where a, b, and c are real numbers with $a \ne 0$. We have seen that the graph of a quadratic function $f(x) = ax^2 + bx + c$ is a parabola, so it has either zero, one, or two x-intercepts. Equivalently, a quadratic equation has either zero, one, or two solutions. The solution of a quadratic equation can be computed from the **quadratic formula.** The quadratic formula also shows when a solution does not exist.

Quadratic Formula

If $ax^2 + bx + c = 0$ and $a \neq 0$, then

$$x = \frac{-b + \sqrt{b^2 - 4ac}}{2a}$$

or

$$x = \frac{-b - \sqrt{b^2 - 4ac}}{2a}$$

The number $D = b^2 - 4ac$, which is called the **discriminant,** determines whether there are zero, one, or two solutions. If $D > 0$, then the quadratic equation has two distinct solutions.

$$x = \frac{-b + \sqrt{D}}{2a} \qquad \text{or} \qquad x = \frac{-b - \sqrt{D}}{2a}$$

If $D = 0$, there is one solution $x = -b/2a$. If $D < 0$, there is no solution because the square root of a negative number is not a real number. Example 1 shows how the three possibilities arise.

EXAMPLE 1 | **Problem**

For each function $f(x)$, solve the quadratic equation $f(x) = 0$ and sketch the graph of $f(x)$.
(a) $f(x) = 2x^2 - x - 10$
(b) $f(x) = x^2 + 4x + 4$
(c) $f(x) = x^2 + 4x + 5$

SOLUTION (a) To use the quadratic formula, set $a = 2$, $b = -1$, and $c = -10$. Then $D = b^2 - 4ac = 81$. Since $D > 0$, the two solutions of the equation $f(x) = 0$ are

$$x = \frac{-(-1) + \sqrt{81}}{2(2)} = \frac{5}{2} \qquad x = \frac{-(-1) - \sqrt{81}}{2(2)} = -2$$

Therefore the graph of $f(x)$ has x-intercepts at the points $(-2, 0)$ and $(\frac{5}{2}, 0)$. Since the graph of a quadratic equation is a parabola that opens upward or downward, it suffices to plot a point (x, y) where $-2 < x < (\frac{5}{2})$. Since $f(0) = -10$, the point $(0, -10)$ lies on the graph so that the parabola opens upward. It is sketched in Figure 1a.
(b) To use the quadratic formula, set $a = 1$, $b = 4$, and $c = 4$. Then $D = 4^2 - 4(1)(4) = 0$, so there is one solution of the equation $f(x) = 0$: $x = -b/2a = -\frac{4}{2} = -2$. This means that the graph of $f(x)$ has one x-intercept, the point $(-2, 0)$, which must be the vertex of the parabola. Since $f(0) = 4$, the point

(0, 4) lies on the graph, so the parabola opens upward. It is sketched in Figure 1b.

(c) Set $a = 1$, $b = 4$, and $c = 5$. Then $D = 4^2 - 4(1)(5) = -4$, so there is no solution of the equation $f(x) = 0$. This means that the graph of $f(x)$ has no x-intercepts. The only means at our disposal for sketching the graph is the plotting method using a table like the one here. The resulting graph is sketched in Figure 1c.

(a)

(b)

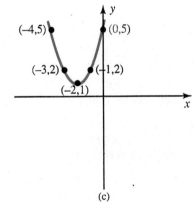

(c)

FIGURE 1

Example 2 illustrates how to use the quadratic formula to determine the domain of a rational function whose denominator is a quadratic function.

EXAMPLE 2   **Problem**

Determine the domain of the function

$$f(x) = \frac{x}{12x^2 - 11x + 2}$$

SOLUTION The domain consists of all real numbers except where the denominator is 0. Set $12x^2 - 11x + 2 = 0$. Since $D = (-11)^2 - 4(12)(2) = 25$, the values of x where $12x^2 - 11x + 2 = 0$ are

$$x = \frac{-b + \sqrt{D}}{2a} = \frac{11 + 5}{24} = \frac{2}{3}$$

or

$$x = \frac{-b - \sqrt{D}}{2a} = \frac{11 - 5}{24} = \frac{1}{4}.$$

Therefore the domain of f is all real numbers except $\frac{2}{3}$ and $\frac{1}{4}$.

Factoring

A number r is a root of a polynomial equation $f(x) = 0$ if and only if $f(x) = (x - r)g(x)$ for some polynomial $g(x)$. Each of the functions $(x - r)$ and $g(x)$ is called a **factor** of $f(x)$. The method of decomposing a polynomial into a product of polynomials of lesser degree is called **factoring.**

Consider the general quadratic equation $ax^2 + bx + c = 0$. Since $a \neq 0$, we can divide both sides of the equation by a and get $x^2 + Bx + C = 0$, where $B = b/a$ and $C = c/a$. If r and s are roots of $x^2 + Bx + C = 0$, then

$$x^2 + Bx + C = (x - r)(x - s)$$

Expand the right side of the equation.

$$(x - r)(x - s) = x^2 - (r + s)x + rs$$

Therefore

$$x^2 + Bx + C = x^2 - (r + s)x + rs$$

Set the coefficients equal: $B = -(r + s)$, $C = rs$. The sum of the roots equals $-B$ (since $r + s = -B$) and the product of the roots equals the constant term C. For instance, if $f(x) = x^2 - x - 2$, then $B = -1$ and $C = -2$, so $r + s = 1$ and $rs = -2$. We are searching for two numbers whose sum is 1 and product is -2. The solution by trial and error is $r = 2$ and $s = -1$. Therefore the linear factors of $x^2 - x - 2$ are $(x - 2)$ and $(x + 1)$, so

$$x^2 - x - 2 = (x - 2)(x + 1)$$

Example 3 uses factoring to determine the domain and x-intercepts of a rational function.

EXAMPLE 3

Consider the function

$$f(x) = \frac{x^2 - 4}{x^2 - 2x - 8}$$

Problem

(a) Find the domain of f. (b) Find all x-intercepts of the graph of f.

SOLUTION (a) The domain of f consists of all real numbers except those where the denominator is 0. Factor the denominator.

$$x^2 - 2x - 8 = (x - 4)(x + 2)$$

The roots are $x = 4$ or $x = -2$. Therefore the domain of f consists of all real numbers except 4 and -2.

(b) The x-intercepts of the graph of f occur where the numerator is 0 and the denominator is not 0. Set $x^2 - 4 = 0$. Since $x^2 - 4 = (x + 2)(x - 2)$, the solutions are $x = 2$ or $x = -2$. However, -2 does not lie in the domain of f. Thus the only possible x-intercept occurs at $x = 2$. Since $x^2 - 2x - 8 \neq 0$ when $x = 2$, $f(x)$ has one x-intercept at $(2, 0)$.

The next two examples show other instances where the solution of a polynomial equation is required. Example 4 makes use of the fact that if a and b are numbers with $ab = 0$, then either $a = 0$ or $b = 0$. This fact extends to functions, in which case if $f(x)$ and $g(x)$ are functions with $f(x)g(x) = 0$, then either $f(x) = 0$ or $g(x) = 0$. Therefore if a polynomial has been factored, its roots are found by setting each of the factors equal to 0.

EXAMPLE 4

Problem

Solve $f(x) = 0$, where

$$f(x) = (2x + 3)(x^2 + 3)(x^2 - 3)$$

SOLUTION The solutions to $f(x) = 0$ are found by setting each of the factors equal to 0. The solution to the linear equation $2x + 3 = 0$ is $x = -\frac{3}{2}$. The equation $x^2 = -3$ has no real solutions, so the factor $(x^2 + 3)$ adds no solutions to $f(x) = 0$. However, $x^2 - 3 = (x + \sqrt{3})(x - \sqrt{3})$, so $\sqrt{3}$ and $-\sqrt{3}$ are solutions to $x^2 - 3 = 0$, and hence are solutions to $f(x) = 0$. Altogether the solutions to $f(x) = 0$ are $-\frac{3}{2}$, $\sqrt{3}$, and $-\sqrt{3}$.

Example 5 makes repeated use of one of the basic rules of algebra.

$$a^2 - b^2 = (a + b)(a - b)$$

In particular, $x^2 - 1 = (x + 1)(x - 1)$.

EXAMPLE 5

Problem

Find the x-intercepts of the function

$$f(x) = x^4 - 1$$

SOLUTION The x-intercepts are found by solving $f(x) = 0$. First factor $f(x)$ by making the substitution $y = x^2$.

$$f(x) = x^4 - 1 = y^2 - 1 = (y + 1)(y - 1)$$

Next substitute for y.

$$f(x) = (x^2 + 1)(x^2 - 1)$$

Then factor $x^2 - 1$.

$$f(x) = (x^2 + 1)(x + 1)(x - 1)$$

The roots of $f(x) = 0$ are found by setting each factor equal to 0: $x = 1$, $x = -1$. These are the x-intercepts of $f(x)$.

Points of Intersection

A point at which two graphs meet is called a **point of intersection** of the graphs. We find the intersection of the graphs of two polynomial functions by a method quite different from the one used for two linear functions.

Let f and g be any two functions. The intersection of their graphs is obtained by solving the equation $f(x) = g(x)$, or, equivalently, by solving $f(x) - g(x) = 0$. If there are no solutions to this equation, then the graphs have no points in common. Otherwise each solution x, along with the corresponding value of $y = f(x)$ or $y = g(x)$, leads to a point of intersection (x, y).

For example, let $f(x) = 3x$ and $g(x) = 4 - x^2$. Then

$$f(x) - g(x) = 0$$
$$3x - (4 - x^2) = 0$$
$$x^2 + 3x - 4 = 0$$
$$(x + 4)(x - 1) = 0$$

The roots are $x = -4$ and $x = 1$. The corresponding values are $f(-4) = g(-4) = -12$ and $f(1) = g(1) = 3$. Each root leads to a point of intersection, so the points of intersection of $f(x)$ and $g(x)$ are $(-4, -12)$ and $(1, 3)$. The graphs are sketched in Figure 2.

An application of this concept comes from a field called **break-even analysis.** We have already introduced the cost function $C(x)$, where x denotes the number of items being produced. The revenue R is also a function of x, called the **revenue function** and denoted by $R(x)$. The **profit function** $P(x)$ is then defined by

$$P(x) = R(x) - C(x)$$

The **break-even points** occur when cost equals revenue: $R(x) = C(x)$. Much of the information can be conveyed visually by drawing the graphs of $R(x)$ and $C(x)$ on the same coordinate system and finding the points of intersection.

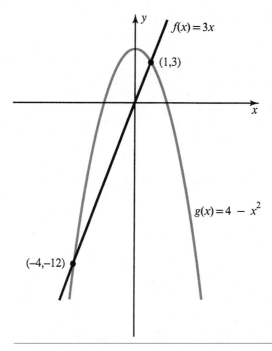

FIGURE 2

EXAMPLE 6

A manufacturer produces compact disks. The variable cost per disk is $2 and the fixed cost is $12. If x disks are manufactured, the revenue function is given by $R(x) = x(10 - x)$.

Problem

What are the break-even points?

SOLUTION The cost function is

$C(x) = $ (variable cost per disk)$x + ($ fixed cost$)$
$C(x) = 2x + 12$

The revenue function is

$R(x) = x(10 - x)$
$R(x) = 10x - x^2$

The break-even points occur when cost equals revenue.

$$C(x) = R(x)$$
$$2x + 12 = 10x - x^2$$
$$x^2 - 8x + 12 = 0$$

The possible solutions to this quadratic equation are $x = 2$ or $x = 6$, so the break-even points are (2, 16) and (6, 24). (See Figure 3.)

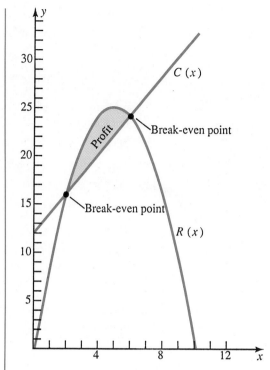

FIGURE 3

It is informative to view Figure 3 for a deeper understanding of break-even analysis. The functions $C(x)$ and $R(x)$ are sketched on the same coordinate axis. If either two disks or six disks are produced, the manufacturer will break even. A profit is realized if the number of disks produced lies between these numbers. The profit function is

$$P(x) = R(x) - C(x) = x^2 + 8x - 12$$

What production level do you think will maximize profit?

EXERCISE SET 1–6

• In Problems 1 to 6 solve the quadratic equation $f(x) = 0$ and sketch the graph of $f(x)$.

1. $f(x) = x^2 + 2x - 8$

2. $f(x) = x^2 - 4x + 4$

3. $f(x) = x^2 + 6x + 9$

4. $f(x) = x^2 - 2x + 5$

5. $f(x) = x^2 + 4$

6. $f(x) = x^2 - 2x - 8$

In Problems 7 to 10 determine the domain and x-intercepts of each function.

7. $f(x) = \dfrac{x - 2}{2x^2 - x - 3}$

8. $f(x) = \dfrac{x - 3}{x^2 + x - 2}$

9. $f(x) = \dfrac{x + 2}{4x^2 + 5x - 6}$

10. $f(x) = \dfrac{x - 3}{5x^2 - 13x - 6}$

In Problems 11 and 12 solve $f(x) = 0$ for the given $f(x)$.

11. $f(x) = (2x - 3)(x^2 - 4)(x^2 + 3)$

12. $f(x) = x(x^2 - 1)(x^2 + 1)$

In Problems 13 to 16 find the x-intercepts of each function.

13. $f(x) = x^4 - 16$ 14. $f(x) = x^4 - 81$

15. $f(x) = (x^2 + 3)(x - 4)(x + 3)$

16. $f(x) = (x^2 + 4)(x + 7)(x + 3)$

In Problems 17 to 20 a cost function and a revenue function are given. Find the break-even points.

17. $R(x) = x(10 - x)$ $C(x) = x + 8$

18. $R(x) = x(10 - x)$ $C(x) = 18 - x$

19. $R(x) = x(8 - x)$ $C(x) = 14 - x$

20. $R(x) = x(8 - x)$ $C(x) = x + 6$

•• In Problems 21 to 24 solve the equation $f(x) = 0$ and sketch the graph of $f(x)$.

21. $f(x) = \dfrac{x - 2}{4x^2 - 12x + 9}$

22. $f(x) = \dfrac{x - 3}{9x^2 + 6x - 1}$

23. $f(x) = \dfrac{x + 1}{x^2 + 1}$ 24. $f(x) = \dfrac{x - 1}{5x^2 + 6}$

In Problems 25 to 28 determine the domain and x-intercepts of each function.

25. $f(x) = \dfrac{x - 2}{x^3 - 4x^2 + 4x}$

26. $f(x) = \dfrac{x - 3}{x^3 + 2x^2 + x}$

27. $f(x) = \dfrac{x^2 - x + 2}{x^3 - 4x^2 + 4x}$

28. $f(x) = \dfrac{x^2 - x - 6}{x^3 - 6x^2 + 9x}$

In Problems 29 to 34 find where the graphs of the functions intersect.

29. $f(x) = x + 2$, $g(x) = x^2$

30. $f(x) = 6x - 5$, $g(x) = x^2$

31. $f(x) = x^3$, $g(x) = x^2$

32. $f(x) = x^2$, $g(x) = \sqrt{x}$

33. $f(x) = x^2 + 4$, $g(x) = 22 - x^2$

34. $f(x) = x^3$, $g(x) = x$

In Problems 35 to 38 an object is thrown vertically upward with height $h(t) = 28t - 4t^2$ (in feet) at time t (in seconds). Determine when the object will reach the stated height.

35. 40 feet   36. 48 feet

37. 49 feet   38. 0 feet

39. Let the revenue function on the sales of x units of an item be $R(x) = 480x - 12x^2 - 2100$ and the cost function be $C(x) = 96x + 204$. Determine the break-even points.

••• Problems 40 to 42 are taken from a lecture that was given by Robert L. Devaney to a sectional meeting of the Mathematical Association of America.* It was accompanied by an exhilarating collection of computer graphics that illustrate the area of mathematics called dynamical systems.

40. Enter any positive number into a calculator. Tap the square-root key. Tap it again. Continue tapping the key about 15 or 20 more times.
    (a) What do you observe?
    (b) Repeat this procedure for several more positive numbers, including numbers between 0 and 1. What general conclusion can be drawn?

*Robert L. Devaney, "Computer Graphics Experiments in Complex Dynamical Systems," Eastern Pennsylvania and Delaware Section of the Mathematical Association of America, Franklin & Marshall College, November 21, 1987.

41. Find all points of intersection of the graphs of the functions $f(x) = \sqrt{x}$ and $g(x) = x$ for $x > 0$.

42. What does Problem 41 have to do with Problem 40?

43. Figure 4 illustrates the dramatic shift in the American public's eating habits from red meats to poultry. It has been drawn from data supplied by the U. S. Department of Agriculture.†
    (a) Interpret the point of intersection of the poultry curve and the pork curve.
    (b) Interpret the point of intersection of the poultry curve and the beef curve.

FIGURE 4

†Drawn from data extracted from Neill Boroski, "Poultry's Prospects Far From Paltry," *The Philadelphia Inquirer,* June 15, 1987, pp. C1 and C6.

CASE STUDY   Powerlifting*

Powerlifting is one of the world's youngest sports, having attained international status in 1968 when the strong local programs from the United States and Great Britain joined together with the developing programs in Canada, Japan, and the Scandinavian countries. It consists of three types of lifts: the squat (the bar is set on the back of the neck, and the lifter does a deep knee bend, returning to a standing position); the bench press (the lifter lies on a bench and pushes the weight away from his chest until his arms are locked); and the deadlift (the weight is lifted waist high from the floor, with the arms kept straight). A lifter's total is the sum of the highest weight lifted in each of the three categories.

This Case Study describes the sport, discusses its brief history, and introduces a sixth-degree polynomial which plays a crucial role in determining the winner of a competition.

The winner of a powerlifting meet is not necessarily the competitor who has lifted the most weight. In fact, there are many winners. Like other sports in

*This Case Study was inspired by Joseph A. Gallian's note "Handicapping Bodyweights," which appeared in the August–September 1986 issue of the *American Mathematical Monthly,* pp. 583–584. We are indebted to Lyle Schwartz and Linc Gotshalk for supplying details. (Linc is the legendary powerlifting coach at Temple University.)

which a competitor's weight is one of the crucial determining factors, powerlifting competitions are divided into several weight classes, and there is a champion in each weight class. However, most meets crown a "Champion of Champions." What criteria should be used to decide this distinction?

Let us consider two lifters, Amos Jones and Bob Brown. Amos weighs 167 pounds (lb) so he competes in the 181-lb class. Bob is a behemoth, who, at 272 lb, competes in the 275-lb class. (There is a super heavyweight class beyond this one.) In a particular meet Amos squatted 710 lb, bench pressed 450 lb, and deadlifted 640 lb, while Bob's best three lifts were 850, 570, and 790 lb, respectively. Therefore Amos's total was 1800 (since 710 + 450 + 640 = 1800) and Bob's total was 2210. Each lifter was a weight class champion in the meet. Which one should win the Champion of Champions award?

Mathematics to the rescue!

The Schwartz formula   There are several ways to approach the question of who is stronger, Amos or Bob. If only the totals are taken into account, then Bob wins by 410 lb. Yet Amos's total is about 11 times his bodyweight, while Bob's total is about 8 times his bodyweight.

The International Powerlifting Federation (IPF) sought a handicapping scheme that would compensate for the widely differing body weights of the lifters. In 1976 the IPF chose a scheme that was developed by Lyle Schwartz to fit the empirical data that were available at the time. It has had the effect of taking into account not only body weights and muscular cross-sectional areas, but also body size versus fixed barbell plate, bar diameter, and muscle-ligament attachment leverage.

Dr. Schwartz had developed the scheme in 1967–1968 while he was a Professor of Materials Science at Northwestern University. (Today he is the Director of the Institute for Materials Science and Engineering at the National Bureau of Standards.) His method was published first in the magazine *Muscular Development* in 1970 and adopted by the U.S. Powerlifting Federation in 1975. The method uses a function of the lifter's bodyweight to produce a number that is then multiplied by the total number of pounds lifted to give the lifter's score. The competitor with the highest score wins the Champion of Champions trophy.

The *Schwartz formula* is the following intimidating sixth-degree polynomial function.

$$S(x) = 6.31926 - 1.189995 \cdot 10^{-1} x + 1.052494 \cdot 10^{-3} \cdot x^2$$
$$- 4.850446 \cdot 10^{-6} \cdot x^3 + 1.132899 \cdot 10^{-8} \cdot x^4$$
$$- 1.037119 \cdot 10^{-11} \cdot x^5 - 6.348189 \cdot 10^{-16} \cdot x^6$$

Here x is the bodyweight in pounds. The domain of $S(x)$ is $114 \le x \le 275$, so the Schwartz formula only applies to weights in this interval.

Let us evaluate the Schwartz formula for Amos, who weighs 167 lb. The details for computing $S(167)$ are shown in Table 1. They reinforce the fact that all seven terms in $S(x)$ must be included. Since the IPF rounds all numbers to four decimal places, we see that $S(167) = 0.6593$.

TABLE 1

Term	Value with $x = 167$	Cumulative Total
6.319260	6.319260	
$-1.189995 \cdot 10^{-1}x$	-19.872917	-13.553657
$1.052494 \cdot 10^{-3}x^2$	29.353005	15.799348
$-4.850446 \cdot 10^{-6}x^3$	-22.590773	-6.791425
$1.132899 \cdot 10^{-8}x^4$	8.811647	2.020222
$-1.037119 \cdot 10^{-11}x^5$	-1.347134	0.673088
$-6.348189 \cdot 10^{-16}x^6$	-0.013770	0.659318

The **Schwartz formula total** (denoted SFT) is equal to $S(x)$ when $114 \leq x \leq 275$. Other methods are used to determine the SFT for lifters who weigh less than 114 lb or more than 275 lb. (See Exercises 9 and 10.) The SFT for Amos is 0.6593. Similar computations show that the SFT for Bob is $S(272) = 0.5232$.

Each lifter's score is the product of the total weight lifted and the SFT. For Amos, the score is $1800 \cdot 0.6593 = 1186.74$. For Bob, the score is $2210 \cdot 0.5232 = 1156.272$. Therefore Amos is declared the Champion of Champions. Their statistics are summarized in Table 2.

TABLE 2

	Amos	Bob
Bodyweight x	167	272
$S(x)$	0.6593	0.5232
Total	1800	2210
Score	1186.7	1156.3

EXAMPLE 1

Jack London won the Pennsylvania state teenage powerlifting championship in 1987. He squatted 701 lb, bench pressed 451 lb, and deadlifted 585 lb. He weighs a mere 275 lb.

Problem

What is his score?

SOLUTION London's total is $701 + 451 + 585 = 1737$. The Schwartz formula total is $S(275) = 0.5214$. Therefore London's score is $1737 \cdot 0.5214 = 905.67$.

It is not necessary for a powerlifting meet director to employ someone with a calculator to evaluate $S(x)$ for each lifter. Instead, the director uses a table like the one in Figure 1, which also contains the Malone formula total (MFT) for female lifters and the Schwartz master's formula (SMF) for lifters of age 40 or more. However, neither the MFT nor the SMF is defined by a polynomial function, so they will not be described here.

Schwartz/Malone Formula (in pounds)

Updated (4/84) Schwartz Formula for men, Malone Formula For Women with bodyweights in pounds. To determine the "Best Lifter," multiply each lifter's coefficient (to the right of each body weight listed) by his or her total. The resulting factor is his Schwartz or her Malone formula total (SFT/MFT). The lifter with the highest SFT/MFT is considered the "Best Lifter." The weight class winner with the highest (SFT/MFT) is the Champion of Champions.

BWT	Schwartz	Malone	BWT	Schwartz	Malone	BWT	Schwartz	Malone	BWT	Schwartz	Malone	BWT	Schwartz	Malone
90	1.2803	1.1756	123	.8783	.9110	155	.7004	.7565	187	.6077	.6595	219	.5556	.6008
91	1.2627	1.1645	124	.8706	.9086	156	.6967	.7520	188	.6056	.6566	220	.5545	.5993
92	1.2455	1.1557	125	.8630	.9019	157	.6930	.7490	189	.6036	.6543	221	.5535	.5981
93	1.2287	1.1450	126	.8556	.8980	158	.6893	.7453	190	.6014	.6521	222	.5524	.5965
94	1.2124	1.1365	127	.8483	.8902	159	.6857	.7431	191	.5994	.6492	223	.5514	.5953
95	1.1965	1.1261	128	.8412	.8851	160	.6822	.7387	192	.5978	.6464	224	.5504	.5938
96	1.1809	1.1180	129	.8343	.8788	161	.6787	.7358	193	.5954	.6442	225	.5494	.5926
97	1.1657	1.1079	130	.8276	.8738	162	.6753	.7322	194	.5935	.6415	226	.5485	.5911
98	1.1509	1.0980	131	.8210	.8676	163	.6720	.7293	195	.5916	.6387	227	.5476	.5896
99	1.1365	1.0903	132	.8146	.8628	164	.6688	.7258	196	.5897	.6366	228	.5467	.5884
100	1.1223	1.0807	133	.8083	.8568	165	.6656	.7230	197	.5879	.6339	229	.5458	.5869
101	1.1086	1.0732	134	.8022	.8508	166	.6624	.7196	198	.5861	.6317	230	.5449	.5858
102	1.0952	1.0657	135	.7961	.8462	167	.6593	.7168	199	.5843	.6300	231	.5441	.5843
103	1.0821	1.0566	136	.7903	.8401	168	.6563	.7134	200	.5826	.6286	232	.5433	.5831
104	1.0693	1.0494	137	.7846	.8358	169	.6533	.7107	201	.5809	.6269	233	.5426	.5817
105	1.0569	1.0405	138	.7790	.8302	170	.6504	.7074	202	.5792	.6256	234	.5418	.5805
106	1.0448	1.0336	139	.7735	.8257	171	.6475	.7040	203	.5776	.6239	235	.5411	.5791
107	1.0329	1.0250	140	.7682	.8202	172	.6447	.7014	204	.5760	.6226	236	.5405	.5779
108	1.0214	1.0165	141	.7630	.8159	173	.6420	.6981	205	.5744	.6209	237	.5398	.5765
109	1.0101	1.0098	142	.7579	.8105	174	.6392	.6956	206	.5729	.6196	238	.5391	.5754
110	.9991	1.0016	143	.7528	.8052	175	.6365	.6923	207	.5714	.6180	239	.5385	.5740
111	.9884	.9952	144	.7479	.8010	176	.6339	.6898	208	.5700	.6167	240	.5379	.5725
112	.9779	.9872	145	.7432	.7959	177	.6313	.6866	209	.5685	.6151	241	.5373	.5714
113	.9677	.9809	146	.7385	.7918	178	.6288	.6837	210	.5670	.6134	242	.5367	.5700
114	.9578	.9731	147	.7339	.7867	179	.6262	.6810	211	.5657	.6122	243	.5362	.5693
115	.9481	.9670	148	.7294	.7827	180	.6238	.6786	212	.5643	.6109	244	.5357	.5686
116	.9385	.9595	149	.7250	.7769	181	.6214	.6755	213	.5630	.6093	245	.5352	.5681
117	.9293	.9536	150	.7207	.7737	182	.6190	.6731	214	.5617	.6077	246	.5347	.5671
118	.9203	.9462	151	.7165	.7697	183	.6167	.6701	215	.5604	.6064	247	.5342	.5669
119	.9115	.9390	152	.7124	.7666	184	.6144	.6671	216	.5592	.6049	248	.5337	.5662
120	.9029	.9333	153	.7083	.7627	185	.6121	.6639	217	.5580	.6036	249	.5333	.5656
121	.8946	.9263	154	.7044	.7596	186	6099	.6618	218	.5568	.6021	250	.5328	.5649
122	.8863	.9208												

FIGURE 1   Schwartz/Malone Formula. (Courtesy of the United States Powerlifting Federation.)

Schwartz Formula (Continued)

BWT	Schwartz	BWT	Schwartz	BWT	Schwartz	BWT	Schwartz
251	.5325	279	.5192	307	.5043	335	.4909
252	.5320	280	.5186	308	.5037	336	.4905
253	.5316	281	.5180	309	.5032	337	.4901
254	.5312	282	.5175	310	.5027	338	.4896
255	.5308	283	.5169	311	.5022	339	.4891
256	.5304	284	.5164	312	.5017	340	.4887
257	.5300	285	.5158	313	.5013	341	.4883
258	.5296	286	.5154	314	.5007	342	.4878
259	.5292	287	.5147	315	.5002	343	.4874
260	.5289	288	.5142	316	.4998	344	.4870
261	.5284	289	.5137	317	.4992	345	.4866
262	.5281	290	.5132	318	.4988	346	.4862
263	.5276	291	.5126	319	.4982	347	.4858
264	.5273	292	.5121	320	.4978	348	.4854
265	.5268	293	.5115	321	.4973	349	.4850
266	.5263	294	.5109	322	.4968	350	.4845
267	.5259	295	.5104	323	.4964	351	.4841
268	.5254	296	.5098	324	.4959	352	.4837
269	.5248	297	.5094	325	.4955	353	.4833
270	.5243	298	.5088	326	.4950	354	.4829
271	.5239	299	.5083	327	.4946	355	.4825
272	.5232	300	.5077	328	.4941	356	.4821
273	.5227	301	.5072	329	.4937	357	.4817
274	.5220	302	.5067	330	.4932	358	.4813
275	.5214	303	.5062	331	.4928	359	.4809
276	.5208	304	.5057	332	.4924	360	.4805
277	.5203	305	.5053	333	.4919	361	.4801
278	.5197	306	.5047	334	.4914	362	.4796

Schwartz Master's Formula*

Age	SMF	Age	SMF
40	1.000	61	1.700
41	1.003	62	1.755
42	1.009	63	1.810
43	1.018	64	1.865
44	1.031	65	1.920
45	1.048	66	1.970
46	1.069	67	2.010
47	1.092	68	2.030
48	1.117	69	2.048
49	1.144	70	2.062
50	1.173	71	2.070
51	1.204	72	2.076
52	1.239	73	2.080
53	1.281	74	2.082
54	1.330	75	2.083
55	1.380	76	2.084
56	1.430	77	2.085
57	1.480	78	2.086
58	1.535	79	2.087
59	1.590	80	2.088
60	1.645		

*Use after the regular formula has been used.

Case Study Exercises

1. Use the Schwartz formula $S(x)$ to determine the Schwartz formula total for a 200-pound (lb) male powerlifter.

2. Use the Schwartz formula $S(x)$ to determine the Schwartz formula total for a 125-lb male powerlifter.

3. Use Figure 1 to verify your answer to Problem 1.

4. Use Figure 1 to verify your answer to Problem 2.

5. Suppose Bob had lost 4 lb and lifted 10 lb less. Would he have won the Champion of Champions award?

Problems 6 to 8 require the use of Figure 1.

6. What is the score of a female powerlifter who weighs 140 lb and lifts 950 lb?

7. Suppose that a 120-lb female powerlifter lifts 640 lb. If she gains 6 lb, how much more weight will she have to lift to maintain the same score?

8. What is the score of a 63-year-old person who lifts 580 lb and weighs 180 lb?

Problems 9 and 10 define the Schwartz Formula in terms of body weight x (in kilograms) for certain male behemoths. Sketch the graph of each formula.

9.
$$S(x) = \begin{cases} 0.5208 - 0.0012(x - 125) \\ \qquad \text{if } 125 < x \le 135 \\ \\ 0.5088 - 0.0011(x - 135) \\ \qquad \text{if } 135 < x \le 145 \end{cases}$$

10.
$$S(x) = \begin{cases} 0.4978 - 0.0010(x - 145) \\ \qquad \text{if } 145 < x \le 155 \\ \\ 0.4878 - 0.0009(x - 155) \\ \qquad \text{if } 155 < x \le 165 \end{cases}$$

Problems 11 to 13 refer to performances at the 1988 National Collegiate Powerlifting Championships.

11. There were four session champions chosen at this meet. For the two lifters listed here, determine their score based on the Schwartz formula. (a) Sheridan Suttles of Middle Tennessee State College, who weighed 129 lb, squatted 440 lb, bench pressed 303 lb, and deadlifted 518 lb. (b) Ty Stapleton Jr. of the University of Oklahoma, who weighed 164 lb, squatted 573 lb, bench pressed 413 lb, and deadlifted 562 lb.

12. Jack London of Temple University weighed 289 lb at this competition, so he entered the superheavyweight class. He squatted 771 lb, bench pressed 458 lb, and deadlifted 661 lb. (a) Use Figure 1 to determine his score. (b) Is this score better than his score in the competition described in Example 1?

13. The Champion of Champions was based on the results of Sheridan Suttles, Ty Stapleton Jr., and Jack London. Who won this prestigious award?

CHAPTER REVIEW

Key Terms

1–1 **The Cartesian Coordinate System**

Real Line	Cartesian Coordinate System
Origin	Quadrants
Unit Distance	Graph
x-Axis; y-Axis	The Plotting Method
x-Coordinate; y-Coordinate	Linear Equation
Coordinates	

1–2 **Lines**

General Form	Parallel Lines
x-Intercept	Horizontal Lines
y-Intercept	Vertical Lines
Slope	Point-Slope Form
Slope-Intercept Form	

1–3 **Linear and Quadratic Functions**

Function	Quadratic Function
Image	Vertex
Preimage	Vertical Line Test
Plotting Method	Cost Function
Linear Function	Fixed Cost
Domain	Variable Cost
Range	Marginal Cost
Function Defined at a Number	

1–4 **Polynomial and Other Functions**

Parabola	Hyperbola
Power Function	Vertical Asymptote
Polynomial Function	Rational Function
Degree of a Polynomial	

1–5 **The Algebra of Functions**

Sum of Functions: $f + g$	Quotient of Functions: f/g
Difference of Functions: $f - g$	Rational Operations
Product of Functions: $f \cdot g$	Composition of Functions: $f \circ g$

1–6 **Solving Polynomial Equations**

Solution of an Equation	Factor
Root of an Equation	Factoring
Polynomial Equation	Point of Intersection
Degree of a Polynomial Equation	Break-Even Analysis
Quadratic Equation	Revenue Function
Quadratic Formula	Profit Function
Discriminant	Break-Even Point

Summary of Important Concepts

Quadratic Formula

If $ax^2 + bx + c = 0$ and $a \neq 0$, then

$$x = \frac{-b + \sqrt{b^2 - 4ac}}{2a}$$

or

$$x = \frac{-b - \sqrt{b^2 - 4ac}}{2a}$$

REVIEW PROBLEMS

Problems 1 to 3 refer to the function $f(x) = x^2 - 4x - 6$.

1. Compute $f(3)$, $f(2)$, $f(1)$, $f(0)$, and $f(-2)$.

2. Draw the graph of f.

3. Evaluate $f(x + h) - f(x)$.

In Problems 4 and 5 determine the domain of each function.

4. $f(x) = \sqrt{2x - 8}$ **5.** $f(x) = \dfrac{1}{1 - x}$

In Problems 6 and 7 determine whether each curve represents the graph of a function.

6. 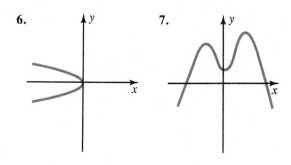   **7.**

8. Use the plotting method to draw the graph of the function $f(x) = 2x^3 - 3x^2 - 12x + 1$.

9. Find the vertical asymptotes of the graph of the function $f(x) = \dfrac{x}{(2 - x)(x^2 - 9)}$

10. Draw the graph of the function

$$f(x) = \begin{cases} x^2 + 4 & \text{if } x \geq 1 \\ 3x + 2 & \text{if } x < 1 \end{cases}$$

In Problems 11 to 13 let

$$f(x) = \frac{x + 2}{x - 1} \quad \text{and} \quad g(x) = \frac{x}{x + 2}$$

11. Find (a) $f + g$ (b) $f \cdot g$ (c) f/g.

12. State the domains of (a) $f + g$ (b) $f \cdot g$ (c) f/g.

13. Evaluate (a) $(f + g)(2)$ (b) $(f \cdot g)(2)$ (c) $(f/g)(2)$

14. Find $f \circ g$ and $g \circ f$ for $f(x) = \sqrt{\dfrac{3}{x}}$ and $g(x) = x^2$.

15. Write $f(x)$ as the composition of two functions. $f(x) = x + 2 - \sqrt{x + 2}$

16. Express Q as a function of t, where

$$Q(P) = 40 - 0.1P^2 \quad \text{and} \quad P(t) = 5t - 0.2t^2$$

In Problems 17 to 19 find the x-intercepts of each polynomial function.

17. $f(x) = x^2 + 4x + 5$

18. $f(x) = 6x^2 - 13x + 6$

19. $f(x) = 3(x - 1)(x^2 + 4)(x^2 - 4)$

20. State the domain of the function

$$f(x) = \frac{x - 3}{4x^2 - 12x + 9}$$

21. Determine the domain and x-intercepts of the function

$$f(x) = \frac{x^2 - 1}{x^2 + x - 2}$$

22. Find all points where the graphs of $f(x) = x^4$ and $g(x) = x^2$ intersect.

2 The Derivative

CHAPTER OVERVIEW   In this chapter we introduce the part of calculus called differentiation. The basic notion is the limit of a function. In Section 2–1 we motivate limits by considering the velocity of a moving object, the rate of change of a function, and lines that are tangent to the graph of a function. The concept is applied to supply and demand curves. In Section 2–2 we develop limits intuitively in terms of "closer and closer" and "approaching." We also introduce limits at infinity.

The derivative is introduced in Section 2–3. This section relies on the properties of functions developed in Chapter 1 and the notion of a limit introduced in Sections 2–1 and 2–2. The derivatives of several functions are computed from the definition, and the definition is applied to tangent lines and rates of change. We also introduce and contrast continuous functions and differentiable functions.

2–1 RATES OF CHANGE AND TANGENTS

There are two broad categories of applications of differential calculus: the rate of change and optimization. In this section the rate of change of a function is introduced by means of the velocity of an object that moves in a straight line. Then the tangent line to the graph of a function is presented by analogy with the rate of change. The tangent forms the basis for many optimization problems. Also discussed is the relationship between average rates of change and secant lines, as well as the relationship between the instantaneous rate of change and the tangent line.

Average Rate of Change

The **velocity** of an object that moves in a straight line is the rate of change of its distance with respect to time. The direction in which the object is traveling must also be taken into account, with a plus sign indicating one direction and a minus sign the opposite direction. We begin by computing the velocity in a specific setting.

Let $d(t)$ be the distance (in miles) that a car travels due east in time t (in

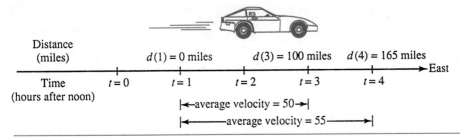

Distance
(miles) $d(1) = 0$ miles $d(3) = 100$ miles $d(4) = 165$ miles
                                                                                          →East
Time $t = 0$ $t = 1$ $t = 2$ $t = 3$ $t = 4$
(hours after noon)

|←average velocity = 50→|

|←————average velocity = 55————→|

FIGURE 1

hours after noon). If the car starts at 1 P.M., then $d = 0$ at $t = 1$, so that $d(1) = 0$. If the car travels 100 miles (mi) by 3 P.M., then $d(3) = 100$. The average velocity over the interval of time from 1 to 3 P.M. is 50 miles per hour (mph) since the car traveled 100 mi in 2 hours. This is shown in Figure 1.

Suppose that the car traveled 165 mi by 4 P.M., so that $d(4) = 165$. Then the average velocity over the interval from $t = 1$ to $t = 4$ is 55 mph since

$$\frac{165 - 0}{4 - 1} = 55$$

In general if a and b denote points in time with $a < b$, then the average velocity over the interval from $t = a$ to $t = b$ is given by the formula

$$\frac{d(b) - d(a)}{b - a}$$

This definition generalizes to the average rate of change of an arbitrary function.

_____ **Definition** _____

The **average rate of change** of the function $y = f(x)$ over the interval from $x = a$ to $x = b$, with $a < b$, is

$$\frac{f(b) - f(a)}{b - a}$$

_____

EXAMPLE 1   Consider the function $f(x) = 3x - 1$.

Problem

(a) What is the average rate of change of $f(x)$ over the interval from $x = -1$ to $x = 4$? (b) What is the average rate of change of $f(x)$ from $x = 1$ to $x = 2$?

SOLUTION (a) First $f(4) = 11$ and $f(-1) = -4$. By definition the average rate of change over the interval from $x = -1$ to $x = 4$ is

$$\frac{f(4) - f(-1)}{4 - (-1)} = \frac{11 - (-4)}{5} = \frac{15}{5} = 3$$

(b) The average rate of change over the interval from $x = 1$ to $x = 2$ is

$$\frac{f(2) - f(1)}{2 - 1} = \frac{5 - 2}{1} = 3$$

In Example 1 the average rate of change of $f(x) = 3x - 1$ over each interval is equal to 3. The exercises will show that the average rate of change of a linear function $f(x) = mx + b$ over any interval is always equal to the slope m. The exercises will also show that linear functions are the only functions whose average rates of change are constant over all intervals. Example 2 indicates that nonlinear functions can have different rates of change over different intervals.

EXAMPLE 2 | **Problem**

Compute the average rate of change of $f(x) = x^3$ over the intervals (a) from 1 to 3 and (b) from 1 to 1.5.

SOLUTION (a) Since $f(3) = 27$ and $f(1) = 1$, the average rate of change over the interval from $x = 1$ to $x = 3$ is

$$\frac{f(3) - f(1)}{3 - 1} = \frac{27 - 1}{2} = 13$$

(b) The average rate of change over the interval from $x = 1$ to $x = 1.5$ is

$$\frac{f(1.5) - f(1)}{1.5 - 1} = \frac{3.375 - 1}{0.5} = 4.75$$

Instantaneous Rate of Change

Table 1 lists the average rates of change of the function $f(x) = x^3$ over the intervals from $x = 1$ to $x = 1 + h$ for various values of h. Example 2 covered $h = 2$ and $h = 0.5$.

Table 2 shows what happens to the average rate of change of $f(x) = x^3$ as the values of h approach 0 for negative values of h.

TABLE 1

h	Interval	Average Rate of Change
2	from 1 to 3	13
1	from 1 to 2	7
0.5	from 1 to 1.5	4.75
0.1	from 1 to 1.1	3.31
0.01	from 1 to 1.01	3.0301
0.001	from 1 to 1.001	3.003001

TABLE 2

h	Interval	Average Rate of Change
−0.1	from 0.9 to 1	2.71
−0.01	from 0.99 to 1	2.9701
−0.001	from 0.999 to 1	2.997001

We can define the instantaneous rate of change of the function at the value $x = 1$ by viewing the numbers in the column for the average rates of change in both Tables 1 and 2. As h approaches 0 from either direction (positive or negative values), the average rates of change approach 3. The number 3 is called the *instantaneous rate of change* of $f(x)$ at $x = 1$, or just the *rate of change* of $f(x)$ at $x = 1$. The general definition is given here.

_____ **Definition** _____

The **instantaneous rate of change** of the function $y = f(x)$ at a value of x is the number (if it exists) approached by the average rates of change from x to $x + h$ as h approaches 0 and $h \neq 0$.

EXAMPLE 3 | **Problem**

Find the instantaneous rate of change of the function $f(x) = 5 - x^2$ at $x = 2$.

SOLUTION Table 3 shows the average rates of change of $f(x) = 5 - x^2$ over various intervals for $h > 0$ and $h < 0$. The top line has $h = 0.1$, so the average rate of change over the interval from 2 to 2.1 is

$$\frac{f(2.1) - f(2)}{2.1 - 2} = \frac{(5 - 4.41) - (5 - 4)}{0.1} = \frac{-0.41}{0.1} = -4.1$$

As h gets closer and closer to 0, the average rates of change get closer and closer to -4. Therefore the instantaneous rate of change at $x = 2$ is equal to -4.

TABLE 3

h	Interval	Average Rate of Change
0.1	from 2 to 2.1	−4.1
0.01	from 2 to 2.01	−4.01
0.001	from 2 to 2.001	−4.001
−0.1	from 1.9 to 2	−3.9
−0.01	from 1.99 to 2	−3.99
−0.001	from 1.999 to 2	−3.999

Secant and Tangent Lines

Thus far we have used algebra to define the instantaneous rate of change in terms of the average rates of change. There is also a geometric analog to this idea.

In general a line that passes through two points on a curve is called a **secant line.** (Recall that the secant line of a circle passes through two points on the circle.) Consider the function $f(x) = x^2$, whose graph is a parabola. The line through the points (2, 4) and (3, 9) is a secant line. (See Figure 2.) By definition the slope of this secant line is equal to

$$m = \frac{9 - 4}{3 - 2} = 5$$

Notice that the average rate of change of $f(x) = x^2$ over the interval from $x = 2$ to $x = 3$ is defined similarly.

$$\frac{f(3) - f(2)}{3 - 2} = \frac{9 - 4}{1} = 5$$

These equalities illustrate the fact that *the slope of the secant line corresponds to the average rate of change*.

What geometric entity corresponds to the instantaneous rate of change? To answer this question, look at Figure 3, which shows the secant lines of the graph of $f(x) = 5 - x^2$ from $x = 2$ to $x = 2 + h$ for values of h that approach 0 and $h \neq 0$. Observe that the secant lines get closer and closer to one particular line. This line is called a **tangent line.** (Recall that the tangent to a circle is a line that touches the circle at precisely one point.) Because of this, we define the slope of the line tangent to a given point on a curve to be the number (if it exists) that the slopes of the secant lines approach. From our earlier discussion it follows that *the slope of the line tangent to a point corresponds to the instantaneous rate of change at that point*.

The equation of the tangent line through a given point on the graph of a function can be derived once the slope is known. Example 4 makes use of Example 3 to illustrate the method that is used to derive the equation of the tangent line.

FIGURE 2

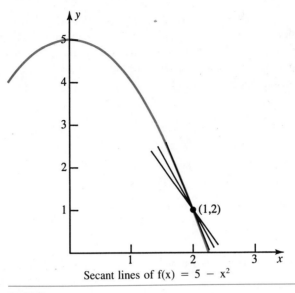

Secant lines of f(x) = 5 − x²

FIGURE 3

EXAMPLE 4 | **Problem**

Find the slope-intercept form of the line that is tangent to the graph of $f(x) = 5 - x^2$ at $x = 2$.

SOLUTION Example 3 showed that the instantaneous rate of change of $f(x)$ at

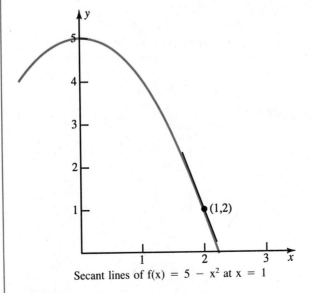

Secant lines of f(x) = 5 − x² at x = 1

FIGURE 4

$x = 2$ is equal to -4, so the slope of the tangent line is -4. Since $f(2) = 1$, the tangent line passes through the point $(2, 1)$. Therefore the equation of the tangent line in point-slope form is $y - 1 = -4(x - 2)$. The slope-intercept form is $y = -4x + 9$. (See Figure 4.)

The **slope of a function** $f(x)$ at a point on the graph of $f(x)$ is defined to be the slope of the tangent line through that point. The procedure for computing the slope of a function $f(x)$ at any value of x involves three steps:

---

1. Form the **difference quotient.**

$$\frac{f(x + h) - f(x)}{(x + h) - x} = \frac{f(x + h) - f(x)}{h}$$

2. Simplify the difference quotient, using the fact that $h \neq 0$.
3. The slope of $f(x)$ is the limit, if it exists, as h approaches 0 and $h \neq 0$.

EXAMPLE 5

Problem

Find the slope of $f(x) = x^3$ at an arbitrary value of x.

SOLUTION It is instructive to refer to Figure 5 while carrying out the three steps. By step 1 the difference quotient is

$$\frac{f(x + h) - f(x)}{h} = \frac{(x + h)^3 - x^3}{h}$$

By step 2

$$\frac{f(x + h) - f(x)}{h} = \frac{(x^3 + 3x^2h + 3xh^2 + h^3) - x^3}{h}$$

$$= \frac{h(3x^2 + 3xh + h^2)}{h}$$

The factor h can be canceled since $h \neq 0$, so

$$\frac{f(x + h) - f(x)}{h} = 3x^2 + 3xh + h^2$$

By step 3 let h approach 0. The terms $3xh$ and h^2 also approach 0, so $3x^2 + 3xh + h^2$ approaches $3x^2$. Thus the slope of $f(x) = x^3$ at any value of x is $3x^2$.

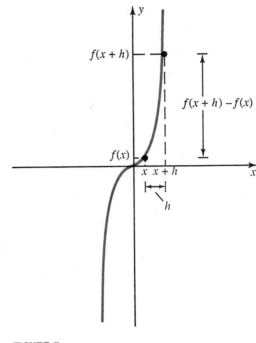

FIGURE 5

The tangent to a curve measures the steepness of the curve. An intuitive way of visualizing this is to put a roller coaster car, with a long board attached to the hubs of the wheels, on the curve. The car is the point; the board is the tangent. (See Figure 6.) As the car moves along the curve from left to right, the board reflects the steepness of the climb and the descent. As the car ascends, the board points upward so the tangent is a positive number. When the car is on the top of the hill, the board is horizontal so the tangent is 0. As the car descends, the board points downward so the tangent is a negative number. We will develop these ideas in greater detail in Chapter 4.

FIGURE 6 The board attached to the roller coaster car approximates the tangent line.

Supply and Demand

The ideas in this section can be applied to the part of economics that deals with supply and demand curves. The **demand** for a product is the number $D(x)$ of items that consumers are willing and able to buy at a given price per unit x. For example, suppose the demand for a new toy called "Cubic" is $D(x) = 5 - x^2$, where x is in dollars per Cubic and $D(x)$ is in thousands of Cubics per week. Notice that the demand slackens as the price rises. For instance, the average rate of change of the demand if the price of each Cubic is raised from \$1 to \$2 is

$$\frac{D(2) - D(1)}{2 - 1} = \frac{1 - 4}{1} = -3$$

This means that the demand will decrease by 3 ($= 3000$ Cubics per week) if the price of the Cubics is raised from \$1 to \$2. The instantaneous rate of change of the demand at any price x is called the **marginal demand.** Example 6 computes the marginal demand of this function.

EXAMPLE 6

Problem

If the demand function is $D(x) = 5 - x^2$, what is the marginal demand?

SOLUTION The marginal demand can be obtained from the three-step procedure. By step 1 the difference quotient is

$$\frac{D(x + h) - D(x)}{h} = \frac{[5 - (x + h)^2] - (5 - x^2)}{h}$$

By step 2

$$\frac{D(x + h) - D(x)}{h} = \frac{5 - x^2 - 2xh - h^2 - (5 - x^2)}{h}$$

$$= \frac{h(-2x - h)}{h}$$

The factor h can be canceled since $h \neq 0$, so

$$\frac{D(x + h) - D(x)}{h} = -2x - h$$

By step 3, as h gets close to 0, the term $-2x - h$ gets close to $-2x$. Thus the marginal demand at an arbitrary price x is equal to $-2x$. This means that the demand for Cubics decreases at a rate that is double the price of each Cubic.

EXERCISE SET 2–1

• In Problems 1 to 8 compute the average rate of change of the function $f(x)$ over the stated intervals.

1. $f(x) = 2x - 3$
   (a) from 1 to 2   (b) from -1 to 4

2. $f(x) = 4x - 3$
   (a) from 1 to 2   (b) from -1 to 4

3. $f(x) = 3 - 4x$
   (a) from 0 to 5   (b) from -3 to -2

4. $f(x) = 3 - 2x$
   (a) from 0 to 5   (b) from -3 to -2

5. $f(x) = x^3$
   (a) from 0 to 1   (b) from 0 to 0.1

6. $f(x) = x^3$
   (a) from -1 to 0 (b) from -0.1 to 0

7. $f(x) = x^4$
   (a) from 0 to 1   (b) from 0 to 0.1

8. $f(x) = x^4$
   (a) from -1 to 0 (b) from -0.1 to 0

In Problems 9 to 12 use the method shown in Example 3 to compute the instantaneous rate of change of the function $f(x)$ at the stated value of x.

9. $f(x) = x^2 - 5$ at $x = 1$

10. $f(x) = x^2 - 5$ at $x = -2$

11. $f(x) = 10 - 2x^2$ at $x = -2$

12. $f(x) = 10 - 2x^2$ at $x = 1$

In Problems 13 to 16 find the slope-intercept form of the line tangent to the graph of the function in the stated problem above.

13. Problem 9   14. Problem 10

15. Problem 11   16. Problem 12

In Problems 17 to 20 find the marginal demand at an arbitrary price x, where the demand function is given.

17. $D(x) = 10 - 2x^2$ 18. $D(x) = x^2$

19. $D(x) = x^2 + 2$ 20. $D(x) = 1 - 2x^2$

•• In Problems 21 to 26 compute the average rate of change of the function $f(x)$ over the stated intervals.

21. $f(x) = x^2 - 5$ (a) from 2 to 2.1 (b) from 2 to 2.01 (c) from 2 to 2.001

22. $f(x) = 10 - x^2$ (a) from 2 to 2.1 (b) from 2 to 2.01 (c) from 2 to 2.001

23. $f(x) = 2x^2 - 4x - 3$ (a) from 3 to 3.1 (b) from 3 to 3.01 (c) from 3 to 3.001

24. $f(x) = 3x^2 + 2x - 29$ (a) from 3 to 3.1 (b) from 3 to 3.01 (c) from 3 to 3.001

25. $f(x) = \sqrt{x}$ (a) from 4 to 4.1 (b) from 4 to 4.01 (c) from 4 to 4.001

26. $f(x) = 1/x$ (a) from 0.5 to 0.51 (b) from 0.5 to 0.501 (c) from 0.5 to 0.5001

In Problems 27 and 28 compute the instantaneous rate of change of the function $f(x)$ at the stated value of x.

27. $f(x) = x^3 + x + 1$ at $x = 1$

28. $f(x) = x^3 + x^2 + 1$ at $x = 1$

In Problems 29 and 30 find the slope-intercept form of the line tangent to the graph of the function in the preceding stated problem.

29. Problem 27   30. Problem 28

In Problems 31 to 34 find the slope of the line tangent to the graph of the function $f(x)$ at an arbitrary value of x.

31. $f(x) = 2x^2 - 4x - 3$

32. $f(x) = 3x^2 + 2x - 29$

33. $f(x) = x^3 + x + 1$

34. $f(x) = x^3 + x^2 + 1$

In Problems 35 to 38 the height of a ball thrown into the air is given by $d(t)$. Find the slope of the line tangent to the graph of the function $d(t)$ at an arbitrary value of t.

35. $d(t) = -16t^2 + 192t$

36. $d(t) = 70t - 16t^2$

37. $d(t) = -16t^2 + 128t + 64$

38. $d(t) = -16t^2 + 32t + 64$

••• Problems 39 and 40 refer to Figure 7, which shows the personal consumption expenditures (in billions of dollars) in the United States on durable goods over the first part of the 1980s.*

39. Estimate the average rate of change of the expenditures.
    (a) from 1980 to 1984
    (b) from 1982 to 1984

*Source: Bureau of Economic Analysis, U.S. Commerce Department.

FIGURE 7

FIGURE 8

40. Estimate the instantaneous rate of change of the expenditures in 1981.

Problems 41 and 42 refer to Figure 8, which shows the U.S. Federal Budget Deficit as a percentage of the Gross National Product (GNP) for the first half of the 1980's.†

41. Estimate the average rate of change of the percentage of the GNP
    (a) from 1981 to 1983
    (b) from 1984 to 1985

42. Estimate the instantaneous rate of change of the percentage of the GNP in 1985.

43. An article by Alan Murray in the *Wall Street Journal* used the fictitious Business Corporation to illustrate the tax law of 1986.‡ The tax law permitted Business Corporation to depreciate its new $6.3 million building over a period of 31.5 years by the "straightline method." What is the rate of depreciation? (The rate of depreciation is defined as the slope of the straight line where the x-axis is time in years and the y-axis is the cost in millions of dollars. The rate is expressed as a percent.)

Problems 44 and 45 refer to an editorial by Paul Craig Roberts, who holds the William E. Simon chair of political economy at the Center for Strategic and International Studies in Washington.§ The editorial compared the economy under presidents Carter and Reagan during two 58-month periods using these facts: under President Carter the unemployment rate fell 27% and the CPI (consumer price index) rose 48%, while under President Reagan the unemployment rate fell 45% and the CPI rose 17%.

44. (a) If graphs for the unemployment rate under each administration are sketched with years on the x-axis and unemployment rates on

†*Source:* From *The World Almanac and Book of Facts,* New York, Newspaper Enterprise Association Inc., 1986, p. 102.

‡Alan Murray, "Individuals' Top Rate Would Plunge to 28 Percent; Tax-Break Curbs Offset Benefits to Wealthy," *Wall Street Journal,* August 18, 1986, p. 6.

§Paul Craig Roberts, "While Critics Carp, the Supply-Side Revolution Sweeps On," *Business Week,* April 25, 1988, p. 15.

the y-axis, are the slopes of the functions (which are defined by the graphs) positive or negative?

(b) Which administration had a better record regarding unemployment?

45. (a) If graphs for the CPI under each administration are sketched with years on the x-axis and the CPI on the y-axis, are the slopes positive or negative?

(b) Which administration had a better record regarding the CPI?

46. Let m, a, b, and c be real numbers and $f(x) = mx + b$.

(a) Compute the average rate of change of $f(x)$ over the interval from a to c.

(b) Compute the instantaneous rate of change of $f(x)$ at $x = a$.

47. Let a, b, and c be real numbers. Compute the slope of the line tangent to the graph of $f(x) = ax^2 + bx + c$ at an arbitrary value of x.

48. Prove that if $f(x)$ is a function having the property that the average rate of change is the same number over all intervals, then $f(x) = mx + b$ for some m and b.

49. Prove that if $f(x)$ is a function having the property that the slope of the tangent line at each point on the graph of $f(x)$ is the same number, then $f(x) = mx + b$ for some m and b.

2–2 LIMITS

In Section 2–1 we discussed a function $f(x)$ approaching a number L as x approaches a given number. The number L is called the limit, and the process of finding limits forms the basis for rates of change and tangents to the graphs of functions. The present section introduces three approaches to limits. The first approach, which builds on Section 2–1, might be called an "arithmetic approach" because it relies on a table in much the same way that the plotting method does. The geometric approach refers to the use of a graph to evaluate limits. Finally the "algebraic approach" uses the full range of algebraic properties of numbers to evaluate the limits of a wide variety of functions.

Arithmetic Approach

In Section 2–1 we derived the instantaneous rate of change of the function $f(x) = x^3$ at $x = 1$ from a table whose essential ingredients are reproduced here (rounded to three places).

h	−0.1	−0.01	−0.001	0.001	0.01	0.1
Average rate	2.71	2.970	2.997	3.003	3.030	3.31

The average rates of change approach 3 as h approaches 0. This leads to an intuitive definition of the limit of a function.

—————————————————————— **Definition** ——————————————————————

If there is a unique number L with the property that $f(x)$ approaches L as x approaches the number a and $x \neq a$, then L is called the **limit** of $f(x)$ as x approaches a. In symbols

$$\lim_{x \to a} f(x) = L$$

The notation $\lim_{x \to a} f(x)$ is read "the limit of f of x as x approaches a." If the limit L exists, then it is the only number with the defining property.

To illustrate the definition, consider the function

$$f(x) = x^3 - 3x^2 + 5$$

for values of x that are *near* 1, but are *not equal to* 1. Construct Table 1 with values of x that approach 1 in the top row and the corresponding values of $f(x)$ below them (rounded to three decimal places). It is important to observe that the values of x approach 1 from two directions: from the left (numbers that are less than 1) and from the right (numbers that are greater than 1).

TABLE 1

	approach 1 from the left $\longrightarrow$				$\longleftarrow$ approach 1 from the right		
x	0.9	0.99	0.999	1	1.001	1.01	1.1
$f(x)$	3.299	3.030	3.003		2.997	2.970	2.701

It follows from Table 1 that $f(x)$ approaches 3 as x approaches 1 from both directions, so

$$\lim_{x \to 1} f(x) = 3$$

It is possible to obtain this limit directly by evaluating $f(1)$ since $f(1) = 3$. However, such a substitution is not always possible, as the next illustration demonstrates. Let

$$f(x) = \frac{x^2 - 8x + 15}{3 - x}$$

What is the behavior of $f(x)$ when x assumes values that are *near* 3 but *not equal to* 3? To answer the question, construct Table 2. Notice that $f(x)$ is not defined at $x = 3$ because both the numerator and denominator are 0.

TABLE 2

	approach 3 from the left →				← approach 3 from the right		
x	2.9	2.99	2.999	3	3.001	3.01	3.1
$f(x)$	2.1	2.01	2.001		1.999	1.99	1.9

It follows from Table 2 that $f(x)$ approaches 2 as x approaches 3 from both directions, so

$$\lim_{x \to 3} f(x) = 2$$

EXAMPLE 1

Problem

Compute $\lim_{x \to 1} f(x)$ for $f(x) = \dfrac{4x^2 - 4x}{x - 1}$.

SOLUTION Construct a table with values of x approaching 1 from both sides in the top row. Compute the corresponding values of $f(x)$ and place them in the bottom row.

	approach 1 from the left →				← approach 1 from the right		
x	0.9	0.99	0.999	1	1.001	1.01	1.1
$f(x)$	3.6	3.96	3.996		4.004	4.04	4.4

The table shows that $f(x)$ approaches 4 as x approaches 1 from both sides, so

$$\lim_{x \to 1} f(x) = 4$$

In Example 1 the function is defined for values of x that are near 1 but are not equal to 1, yet the limit exists at $x = 1$ (and is equal to 4). Example 2 examines an entirely different situation.

EXAMPLE 2

Problem

Compute $\lim_{x \to 2} f(x)$ where $f(x) = \dfrac{x + 1}{x - 2}$.

SOLUTION The appropriate table is shown here.

	approach 2 from the left →				← approach 2 from the right		
x	1.9	1.99	1.999	2	2.001	2.01	2.1
$f(x)$	-29	-299	-2999		3001	301	31

The table shows that as x approaches 2 from the left, the values of $f(x)$ get smaller and smaller, and as x approaches 2 from the right, the values of $f(x)$ get larger and larger. Therefore $f(x)$ does not have a limit when x approaches 2.

In several places we have previously added the parenthetical expression "if it exists." Example 2 presents the first occurrence of a **limit that does not exist,** meaning that there is no fixed number that the values of $f(x)$ approach as x approaches 2. The standard expression for this is that $f(x)$ **decreases without bound** to the left of $x = 2$ and **increases without bound** to the right of $x = 2$. When a function either increases or decreases without bound near a value $x = a$, we write

$$\lim_{x \to a} f(x) \quad \text{does not exist}$$

We will see other ways in which a limit might fail to exist.

Geometric Approach

Sometimes it is possible to derive the limit of a function from its graph. For instance, consider again the function

$$f(x) = \frac{x^2 - 8x + 15}{3 - x}$$

Its graph is the line with the "hole" in it drawn in Figure 1. The open circle

FIGURE 1

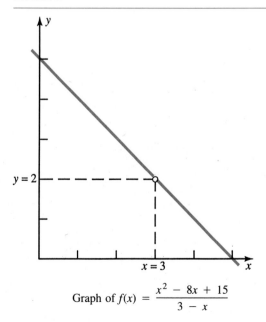

Graph of $f(x) = \dfrac{x^2 - 8x + 15}{3 - x}$

indicates that $f(3)$ is not defined. However, even though the function is not defined at $x = 3$, the limit is defined there because the functional values of $f(x)$ get close to 2 as x gets close to 3. The geometric interpretation of the limit is the height of the graph when the values of x get close to 3.

Figure 2 shows that the graph of the function from Example 2, $f(x) = \dfrac{x + 1}{x - 2}$, has a vertical asymptote at $x = 2$. As x approaches 2 from the left, $f(x)$ decreases without bound, and as x approaches 2 from the right, $f(x)$ increases without bound. Therefore

$$\lim_{x \to 2} f(x) \quad \text{does not exist}$$

In Example 1 the limit of $f(x)$ exists, even though $f(x)$ is not defined at that limiting value. Next we provide an example of a function whose limit does not exist even though the function is defined there. The example deals with the graph of the "quality points function," which many universities use to define grade point average (GPA). If a student's test average is 90 or above, the student gets an A and four quality points are awarded. The chart shows the other values.

Test average	≥ 90	80s	70s	60s	<60
Quality points	4	3	2	1	0

FIGURE 2

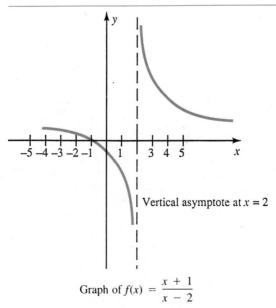

Graph of $f(x) = \dfrac{x + 1}{x - 2}$

Vertical asymptote at $x = 2$

EXAMPLE 3

Let $Q(x)$ denote the quality points function whose graph is drawn in Figure 3. An open circle means that the function is not defined at that end point.

Problem

Evaluate $\lim\limits_{x \to 80} Q(x)$

SOLUTION The notation means to find the quality points when the test average is close to 80. Note that $Q(80) = 3$ since an average of 80 yields three quality points. In fact, for any number x that is greater than 80 but close to 80, $Q(x) = 3$. However, for values of x that are close to 80 but less than 80, $Q(x) = 2$.

As x gets closer to 80, we cannot choose whether $Q(x)$ gets closer to 3 or 2. Since the limit must be a unique number, we conclude that the limit does not exist.

FIGURE 3   Graph of the quality points function $Q(x)$.

Algebraic Approach

We have evaluated limits by arithmetic and geometric methods. The most powerful method, however, is to use algebraic simplifications based on the following properties of limits. These properties can be derived from the definition of the limit of a function.

Properties of Limits

Let $f(x)$ and $g(x)$ be functions for which $\lim\limits_{x \to a} f(x)$ and $\lim\limits_{x \to a} g(x)$ exist. Then

1. $\lim\limits_{x \to a} f(x) = \lim\limits_{x \to a} g(x)$ if $f(x) = g(x)$ for all $x \neq a$

2. $\lim\limits_{x \to a} cf(x) = c \lim\limits_{x \to a} f(x)$ for any real number c

3. $\lim\limits_{x \to a} [f(x) \pm g(x)] = \lim\limits_{x \to a} f(x) \pm \lim\limits_{x \to a} g(x)$

4. $\lim\limits_{x \to a} [f(x)g(x)] = \lim\limits_{x \to a} f(x) \lim\limits_{x \to a} g(x)$

5. $\lim\limits_{x \to a} \dfrac{f(x)}{g(x)} = \dfrac{\lim\limits_{x \to a} f(x)}{\lim\limits_{x \to a} g(x)}$,   provided that $\lim\limits_{x \to a} g(x) \neq 0$

6. $\lim\limits_{x \to a} [f(x)]^r = \left[\lim\limits_{x \to a} f(x) \right]^r$ for any real number r

(where x must be restricted to avoid even roots of negative numbers)

The next example makes use of two properties that can be proved from the definition of a limit.

$$\lim_{x \to a} x = a \quad \text{and} \quad \lim_{x \to a} c = c$$

EXAMPLE 4   **Problem**

Evaluate $\lim\limits_{x \to 0.5} (2x^2 - x - 4)$

SOLUTION It follows from the stated properties that

$$\lim_{x \to 0.5} x = 0.5 \quad \text{and} \quad \lim_{x \to 0.5} 4 = 4$$

By properties 2 and 4

$$\lim_{x \to 0.5} 2x^2 = 2 \left(\lim_{x \to 0.5} x^2 \right)$$

$$= 2 \left(\lim_{x \to 0.5} x \right) \left(\lim_{x \to 0.5} x \right)$$

$$= 2(0.5)(0.5)$$

$$= 0.5$$

Using property 3

$$\lim_{x \to 0.5} (2x^2 - x - 4) = \lim_{x \to 0.5} 2x^2 - \lim_{x \to 0.5} x - \lim_{x \to 0.5} 4$$

$$= 0.5 - 0.5 - 4$$

$$= -4$$

Property 1 is a powerful tool for evaluating many of the limits that appear in calculus. For instance, let

$$f(x) = \frac{x^2 - 8x + 15}{3 - x}$$

We have seen that $\lim\limits_{x \to 3} f(x) = 2$. The algebraic approach is to reduce $f(x)$ to a function $g(x)$ such that $g(3)$ is defined and $g(x) = f(x)$ for all x except $x = 3$. For this function

$$\frac{x^2 - 8x + 15}{3 - x} = \frac{(x - 3)(x - 5)}{-(x - 3)} = 5 - x$$

The common factor $(x - 3)$ can be canceled because $x \neq 3$. Set $g(x) = 5 - x$. Then $g(x) = f(x)$ for all x except $x = 3$. Since $g(3) = 2$, it follows that $\lim_{x \to 3} f(x) = g(3)$, so

$$\lim_{x \to 3} \frac{x^2 - 8x + 15}{3 - x} = \lim_{x \to 3} (5 - x) = 2$$

Examples 5, 6, and 7 exploit property 1.

EXAMPLE 5 | **Problem**

Evaluate $\lim_{x \to 2} \dfrac{x^2 - 2x}{x - 2}$.

SOLUTION $\lim_{x \to 2} \dfrac{x^2 - 2x}{x - 2} = \lim_{x \to 2} \dfrac{x(x - 2)}{x - 2} = \lim_{x \to 2} x = 2$

EXAMPLE 6 | **Problem**

Evaluate $\lim_{x \to 3} \dfrac{3 - x}{x^2 - 6x + 9}$.

SOLUTION $\lim_{x \to 3} \dfrac{3 - x}{x^2 - 6x + 9} = \lim_{x \to 3} \dfrac{-(x - 3)}{(x - 3)^2}$

$$= \lim_{x \to 3} \frac{-1}{x - 3}$$

$$= \lim_{x \to 3} \frac{1}{3 - x}$$

In general the last step is not necessary. We carried it out, however, because we evaluated the limit geometrically from Figure 2. Since this limit does not exist, the stated limit does not exist either.

Example 7 shows that it is sometimes necessary to rationalize the numerator in order to divide out the factor that results in an expression of the form $\dfrac{0}{0}$.

EXAMPLE 7 | **Problem**

Evaluate $\lim_{x \to 9} \dfrac{\sqrt{x} - 3}{x - 9}$

SOLUTION $\lim\limits_{x \to 9} \dfrac{\sqrt{x} - 3}{x - 9} = \lim\limits_{x \to 9} \dfrac{(\sqrt{x} - 3)(\sqrt{x} + 3)}{(x - 9)(\sqrt{x} + 3)}$

$\qquad\qquad = \lim\limits_{x \to 9} \dfrac{x - 9}{(x - 9)(\sqrt{x} + 3)}$

$\qquad\qquad = \lim\limits_{x \to 9} \dfrac{1}{\sqrt{x} + 3} = \dfrac{1}{6}$

Limits at Infinity

Consider the function $f(x) = \dfrac{1 + 2x}{x}$. Table 3 shows that as x assumes larger and larger values in the positive direction, the corresponding values of $f(x)$ approach the number 2. This is called taking a **limit at infinity.** The notation is

$$\lim\limits_{x \to \infty} f(x) = 2$$

It is read "the limit of $f(x)$ as x approaches infinity is 2," where the symbol for infinity is ∞.

TABLE 3

x	10	100	1000	10000
y	2.1	2.01	2.001	2.0001

Table 4 shows that for values of x that get smaller and smaller in the negative direction, the corresponding values of $f(x)$ also approach the number 2, so

$$\lim\limits_{x \to -\infty} f(x) = 2$$

This is read "the limit of $f(x)$ as x approaches minus infinity is 2."

TABLE 4

x	-10	-100	-1000	-10000
y	1.9	1.99	1.999	1.9999

The display contains the general definitions.

_____ **Definition** _____

$\lim\limits_{x \to \infty} f(x) = L$ if there is a unique number L such that the values of $f(x)$ approach L as x assumes larger and larger values. $\lim\limits_{x \to -\infty} f(x) = L$ if there is a unique number L such that the values of $f(x)$ approach L as x assumes smaller and smaller values. If either $\lim\limits_{x \to \infty} f(x) = L$ or $\lim\limits_{x \to -\infty} f(x) = L$, then the horizontal line $y = L$ is a **horizontal asymptote** of $f(x)$.

Horizontal asymptotes of rational functions can be found by dividing the numerator and the denominator by x^n, where n is the larger of the degrees of the polynomials in the numerator and the denominator. Example 8 illustrates the method.

EXAMPLE 8 | Let

$$f(x) = \frac{5x^2 - 1}{2x^2 - x + 5}$$

Problem

(a) Evaluate $\lim\limits_{x \to \infty} f(x)$ (b) Find all horizontal asymptotes of $f(x)$.

SOLUTION (a) Both the numerator and denominator are polynomials of degree 2. Divide each one by x^2.

$$f(x) = \frac{5x^2 - 1}{2x^2 - x + 5} = \frac{\dfrac{5x^2 - 1}{x^2}}{\dfrac{2x^2 - x + 5}{x^2}} = \frac{\dfrac{5x^2}{x^2} - \dfrac{1}{x^2}}{\dfrac{2x^2}{x^2} - \dfrac{x}{x^2} + \dfrac{5}{x^2}}$$

$$= \frac{5 - \dfrac{1}{x^2}}{2 - \dfrac{1}{x} + \dfrac{5}{x^2}}$$

As the values of x get larger and larger, $\dfrac{1}{x^2}$ approaches 0. Therefore the

numerator $5 - \dfrac{1}{x^2}$ approaches 5. Similarly the denominator $2 - \dfrac{1}{x} + \dfrac{5}{x^2}$

approaches 2. Thus the desired limit is $\frac{5}{2}$.

(b) Part (a) shows that $\lim\limits_{x \to \infty} f(x) = \frac{5}{2}$. Similarly $\lim\limits_{x \to -\infty} f(x) = \frac{5}{2}$. Therefore the line $y = \frac{5}{2}$ is the only horizontal asymptote of $f(x)$.

EXERCISE SET 2–2

● In Problems 1 to 8 compute the given limits by constructing tables.

1. $\lim\limits_{x \to 1} f(x)$ for $f(x) = \dfrac{3x^2 - 3x}{x - 1}$

2. $\lim\limits_{x \to 1} f(x)$ for $f(x) = \dfrac{2x^2 - 2x}{x - 1}$

3. $\lim\limits_{x \to 3} f(x)$ for $f(x) = \dfrac{2x^2 - 6x}{x - 3}$

4. $\lim\limits_{x \to 3} f(x)$ for $f(x) = \dfrac{3x^2 - 9x}{x - 3}$

5. $\lim\limits_{x \to 2} f(x)$ for $f(x) = \dfrac{x + 1000}{x - 2}$

6. $\lim\limits_{x \to 2} f(x)$ for $f(x) = \dfrac{x + 2}{x - 2}$

7. $\lim\limits_{x \to 3} f(x)$ for $f(x) = \dfrac{3 + x}{3 - x}$

8. $\lim\limits_{x \to 3} f(x)$ for $f(x) = \dfrac{x + 1000}{3 - x}$

In Problems 9 to 12 evaluate the limit, if it exists, of the quality points function $Q(x)$.

9. $\lim\limits_{x \to 70} Q(x)$ 10. $\lim\limits_{x \to 90} Q(x)$

11. $\lim\limits_{x \to 81} Q(x)$ 12. $\lim\limits_{x \to 79} Q(x)$

In Problems 13 to 26 evaluate each limit, if it exists, by algebraic methods.

13. $\lim\limits_{x \to 2} x^2 - 2x$ 14. $\lim\limits_{x \to 4} x^2 - 4x$

15. $\lim\limits_{x \to 1} 1 - x$ 16. $\lim\limits_{x \to 2} 2 - x$

17. $\lim\limits_{x \to 2} \dfrac{x^2 - 2x}{x - 2}$ 18. $\lim\limits_{x \to 4} \dfrac{x^2 - 4x}{x - 4}$

19. $\lim\limits_{x \to 1} \dfrac{1 - x}{x^2 - 2x + 1}$ 20. $\lim\limits_{x \to 2} \dfrac{2 - x}{x^2 - 4x + 4}$

21. $\lim\limits_{x \to -2} \dfrac{x + 2}{x^2 + 4x + 4}$

22. $\lim\limits_{x \to -1} \dfrac{x + 1}{x^2 + 2x + 1}$

23. $\lim\limits_{x \to 1} \dfrac{\sqrt{x} - 1}{x - 1}$ 24. $\lim\limits_{x \to 4} \dfrac{\sqrt{x} - 2}{x - 2}$

25. $\lim\limits_{x \to 9} \dfrac{3 - \sqrt{x}}{9 - x}$ 26. $\lim\limits_{x \to 1} \dfrac{1 - \sqrt{x}}{1 - x}$

In Problems 27 to 32 evaluate $\lim\limits_{x \to \infty} f(x)$ for each function.

27. $f(x) = \dfrac{5}{x^2}$ 28. $f(x) = \dfrac{2}{5x^3}$

29. $f(x) = \dfrac{1 - 4x}{2 + x}$ 30. $f(x) = \dfrac{9 + 2x}{1 - 3x}$

31. $f(x) = \dfrac{2x^2 + 3x - 5}{6x^2 - 5}$

32. $f(x) = \dfrac{2x^2 + 12}{5x^3 - 6}$

•• In Problems 33 to 36 compute each limit, if it exists, by constructing tables.

33. $\lim\limits_{x \to 1} f(x)$ for $f(x) = \dfrac{3x^3 - 3x^2}{x^2 - x}$

34. $\lim\limits_{x \to 1} f(x)$ for $f(x) = \dfrac{2x^3 - 2x^2}{x^2 - x}$

35. $\lim\limits_{x \to 2} f(x)$ for $f(x) = \dfrac{x - 2}{x + 2}$

36. $\lim\limits_{x \to 3} f(x)$ for $f(x) = \dfrac{3 - x}{3 + x}$

In Problems 37 to 40 compute each limit, if it exists, by the geometric approach.

37. $\lim\limits_{x \to 0} |x|$ 38. $\lim\limits_{x \to 0} \dfrac{1}{x^2}$

39. $\lim\limits_{x \to 0} f(x)$, where $f(x) = \begin{cases} -x & \text{if } x < 0 \\ 3x & \text{if } x \geq 0 \end{cases}$

40. $\lim\limits_{x \to 1} f(x)$, where $f(x) = \begin{cases} x + 1 & \text{if } x < 1 \\ x & \text{if } x \geq 1 \end{cases}$

In Problems 41 to 52 compute each limit, if it exists, by any method.

41. $\lim\limits_{x \to 1} \dfrac{x^2 - 9}{x - 3}$ 42. $\lim\limits_{x \to 1} \dfrac{x^2 - 4}{x - 2}$

43. $\lim\limits_{x \to -2} \dfrac{x^2 + 3x + 2}{x^2 - 4}$

44. $\lim\limits_{x \to 1} \dfrac{x^2 + 3x + 2}{x^2 - 1}$

45. $\lim\limits_{x \to 3} \dfrac{x^3 - 9x}{x - 3}$ 46. $\lim\limits_{x \to -3} \dfrac{x^3 - 9x}{x + 3}$

47. $\lim\limits_{x \to 2} \dfrac{x^3 - 4x}{x - 2}$ 48. $\lim\limits_{x \to 0} \dfrac{x^2 - 3x}{x}$

49. $\lim\limits_{x \to -2} \dfrac{x^2 + 4x + 4}{x + 2}$

50. $\lim\limits_{x \to -1} \dfrac{x^2 + 2x + 1}{x + 1}$

51. $\lim\limits_{x \to 1} \dfrac{x - 1}{\sqrt{x} - 1}$

52. $\lim\limits_{x \to 4} \dfrac{x - 2}{\sqrt{x} - 2}$

In Problems 53 to 56 evaluate $\lim\limits_{x \to -\infty} f(x)$ for each function.

53. $f(x) = \dfrac{4x + 3}{2x - 6}$

54. $f(x) = \dfrac{x - 3}{5x + 1}$

55. $f(x) = \dfrac{3x^2 - 4x - 5}{x^2 + 5}$

56. $f(x) = \dfrac{7x^2 + 12}{3x^2 - 6}$

In Problems 57 to 60 evaluate $\lim\limits_{x \to \infty} f(x)$ for each function.

57. $f(x) = \dfrac{2 - 3x}{x^3}$

58. $f(x) = \dfrac{x^3}{4 + 5x}$

59. $f(x) = \dfrac{x^2 + 6}{x - 1}$

60. $f(x) = \dfrac{9 + 2x}{x^2 - 3}$

Problems 61 to 64 depict situations where the limit of a function is determined by examining values close to the limiting value.

61. A small midwestern newspaper services 20 small towns within a 20-mile radius of the city where the newspaper is published. A reporter who is charged with recording the daily temperature table notices that 19 of the towns report a noon temperature of 90°F, but the remaining town is missing. What number do you suppose the reporter filled in?

62. The table below lists the population of a bac- terium culture, where t is measured in hours and $P(t)$ in millions of bacteria. The measurement is missing at $t = 5$. What value should be in- serted?

t	4.6	4.7	4.8	4.9	5	5.1	5.2
$P(t)$	12	14	18	26		74	138

63. A rocket is designed to explode precisely 5 sec- onds after it has been fired into the air. An instrument measures the distance the rocket travels up to the point of the explosion, but it cannot measure the distance at the point of explosion. The table gives the distance $s(t)$ traveled in miles after t seconds. How far did the rocket travel before the explosion?

t	4.9	4.99	4.999
$s(t)$	199	199.9	199.99

64. A laboratory tests a new adhesive to see how many pounds it will hold for 1 minute. The exact weight cannot be measured because if the weight falls, it was too heavy, and if the weight does not fall, it was too light. The laboratory com- piled the data in the table. What is the maximum weight that the adhesive can hold?

	Held			Did not hold		
Weight (lb)	198	199	199.9	200.1	201	202
Time (min)	60	60	60	59.8	59.5	59

●●● A study by K. Rohrbach proposed a new accounting strategy for auditing called *monetary unit acceptance sampling*.* Figure 4 describes the so-called power functions of the four different statistical distributions which he used. (The nominal power function is based on the binomial distribution.) Problems 65 and 66 refer to Figure 4.

*Kermit J. Rohrbach, "Monetary Unit Acceptance Sampling," *Journal of Accounting Research*, Spring 1986, 127–150.

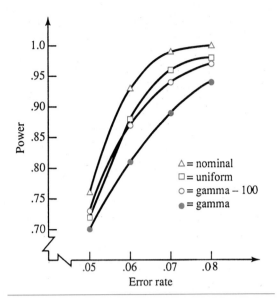

FIGURE 4

FIGURE 5

65. Complete the following table for the gamma − 100 power function $f(x)$ and use it to evaluate $\lim_{x \to .08} f(x)$.

x	.05	.06	.07	.08
$f(x)$				

66. Repeat Problem 65 for the uniform power function.

67. In an article concerning various demands and price structures of finished goods inventories, F. Arcelus and G. Srinivasan† introduced Figure 5 to show how increases in the price elasticity of demand G affect the markup rate m. Use Figure 5 to evaluate $\lim_{-G \to \infty} f(x)$ for each of these functions.

   (a) $f(x) = m_1(x)$
   (b) $f(x) = m_2(x)$

†F. J. Arcelus and G. Srinivasan, "Inventory Policies Under Various Optimizing Criteria and Variable Markup Rates," *Management Science*, June 1987, pp. 756–762.

2−3 DEFINITION OF THE DERIVATIVE

Section 2−1 provided the motivation and Section 2−2 laid the foundation for the definition of the derivative. Here we define the derivative, show how to compute it, and apply it to graphing techniques and rates of change. Then we discuss continuous and differentiable functions.

Procedure

The methods of computing the instantaneous rate of change of a function $f(x)$ at a value $x = a$ and the line tangent to the graph of $f(x)$ at the point $(a, f(a))$ involve the same procedure: form a difference quotient and evaluate a limit as the denominator approaches 0. This motivates the following fundamental definition.

Definition

The **derivative of a function** $f(x)$ is a function, denoted by $f'(x)$, defined by

$$f'(x) = \lim_{h \to 0} \frac{f(x + h) - f(x)}{h}$$

The domain of $f'(x)$ consists of all x in the domain of $f(x)$ for which the limit exists.

The method of computing $f'(x)$ is the same method that was introduced in Section 2–1 to compute the slope of the line tangent to the graph of $f(x)$. The crucial step in evaluating the limit is to perform algebraic simplifications that enable the denominator h to be canceled.

EXAMPLE 1

Problem

Compute $f'(x)$ for $f(x) = x^3 - 2x$.

SOLUTION The difference quotient is

$$\frac{f(x + h) - f(x)}{h} = \frac{[(x + h)^3 - 2(x + h)] - (x^3 - 2x)}{h}$$

$$= \frac{x^3 + 3x^2h + 3xh^2 + h^3 - 2x - 2h - x^3 + 2x}{h}$$

$$= \frac{h(3x^2 + 3xh + h^2 - 2)}{h}$$

Cancel the common factor h.

$$\frac{f(x + h) - f(x)}{h} = 3x^2 + 3xh + h^2 - 2$$

Then

$$f'(x) = \lim_{h \to 0} (3x^2 + 3xh + h^2 - 2) = 3x^2 - 2$$

The display contains the important fact that the derivative of a function can be interpreted as a rate of change.

The instantaneous rate of change of a function $f(x)$ at a value $x = a$ is the derivative $f'(a)$, provided that $f'(a)$ exists.

EXAMPLE 2 | **Problem**

Find the instantaneous rate of change of

$$f(x) = \sqrt{x} \quad \text{at } x = 9.$$

SOLUTION First find $f'(x)$. The difference-quotient is

$$\frac{f(x + h) - f(x)}{h} = \frac{\sqrt{x + h} - \sqrt{x}}{h}$$

The algebraic steps that enable h to be canceled are

$$\frac{\sqrt{x + h} - \sqrt{x}}{h} = \frac{(\sqrt{x + h} - \sqrt{x})}{h} \cdot \frac{(\sqrt{x + h} + \sqrt{x})}{(\sqrt{x + h} + \sqrt{x})}$$

$$= \frac{(x + h) - x}{h(\sqrt{x + h} + \sqrt{x})} = \frac{1}{\sqrt{x + h} + \sqrt{x}}$$

Then

$$f'(x) = \lim_{h \to 0} \frac{1}{\sqrt{x + h} + \sqrt{x}} = \frac{1}{\sqrt{x} + \sqrt{x}} = \frac{1}{2\sqrt{x}}$$

Therefore the instantaneous rate of change at $x = 9$ is

$$f'(9) = \frac{1}{2\sqrt{9}} = \frac{1}{6}$$

Tangents

In Section 2–1 the slope of the line tangent to the graph of a function $f(x)$ was shown to correspond to the instantaneous rate of change of $f(x)$. This means that if the derivative $f'(x)$ can be evaluated at $x = a$, then $f'(a)$ is the slope of the line tangent to that point. The equation of the tangent line can then be written because one point $(a, f(a))$ and the slope $f'(a)$ are known.

> An equation of the line tangent to the graph of a function $f(x)$ at the point $(a, f(a))$ is
>
> $$y - f(a) = f'(a)(x - a)$$
>
> provided that $f'(a)$ exists.

EXAMPLE 3 | **Problem**

Find the slope-intercept form of the line that is tangent to the graph of $f(x) = 2/x$ at $x = -3$.

SOLUTION The approach is to find $f'(x)$ and to evaluate it at $x = -3$.

$$\frac{f(x + h) - f(x)}{h} = \frac{\dfrac{2}{x + h} - \dfrac{2}{x}}{h}$$

$$= \frac{\dfrac{2x - 2(x + h)}{(x + h)x}}{h}$$

$$= \frac{2x - 2(x + h)}{(x + h)x} \cdot \frac{1}{h}$$

$$= \frac{-2h}{(x + h)xh} = \frac{-2}{(x + h)x}$$

Then

$$f'(x) = \lim_{h \to 0} \frac{-2}{(x + h)x} = \frac{-2}{(x)x} = \frac{-2}{x^2}$$

Therefore $f'(-3) = \dfrac{-2}{(-3)^2} = \dfrac{-2}{9}$. The equation of the tangent line is

$$y - f(-3) = f'(-3)(x + 3)$$

$$y - \left(\frac{2}{-3}\right) = \left(\frac{-2}{9}\right)(x + 3)$$

$$y = \frac{-2}{9}x - \frac{4}{3}$$

Velocity

In Section 2–1 the instantaneous velocity of an object moving in a straight line was shown to correspond to the instantaneous rate of change. But the instantaneous rate of change is the derivative. Therefore the derivative can also be interpreted as the velocity of a moving object at a given time.

If $d(t)$ is the distance function of a moving object, then the velocity $v(t)$ is the derivative, provided that $d'(t)$ exists. In symbols, $v(t) = d'(t)$.

If a ball is thrown into the air, it will reach a high point before returning to earth. The ball's velocity is a positive number on the way up, and a negative number on the way down. At the instant when the ball reaches its zenith, the velocity is 0. This observation is crucial for solving problems like Example 4.

EXAMPLE 4 | The height of a ball thrown from ground level into the air with an initial velocity of 128 feet per second is given by the equation

$$d(t) = -16t^2 + 128t$$

where d is measured in feet and t in seconds.

Problem

(a) When will the ball reach its maximum height? (b) What is the maximum height?

SOLUTION (a) The velocity at any time t is given by

$$d'(t) = \lim_{h \to 0} \frac{d(t + h) - d(t)}{h}$$

$$= \lim_{h \to 0} \frac{[-16(t + h)^2 + 128(t + h)] - (-16t^2 + 128t)}{h}$$

$$= \lim_{h \to 0} \frac{-16t^2 - 32th - 16h^2 + 128t + 128h + 16t^2 - 128t}{h}$$

$$= \lim_{h \to 0} \frac{h(-32t - 16h + 128)}{h}$$

$$= \lim_{h \to 0} (-32t - 16h + 128)$$

$$= -32t + 128$$

The ball will reach its maximum height when its velocity is 0, so $d'(t) = 0$. Then $-32t + 128 = 0$, so $t = 4$. Therefore the ball will reach its maximum height after 4 seconds. (b) By part (a) the maximum height occurs at $t = 4$. Since $d(4) = -16 \cdot 4^2 + 128 \cdot 4 = 256$, the maximum height is 256 feet.

Marginal Cost

Marginal cost functions were introduced in Chapter 1. The *marginal cost* of producing an item is the rate of change in cost when that item is produced. It is an approximation to the change in cost for the next item. For instance, let the cost (in dollars) of producing x liters of distilled water be

$$C(x) = x^2 - 2x + 6 \quad 1 \le x \le 6$$

The additional cost of producing the third liter of distilled water is $3, the difference between the cost of producing 3 liters and the cost of producing 2 liters, since $C(3) - C(2)/(3 - 2) = 9 - 6 = 3$. This can be seen in Figure 1 as the slope of the secant line from $(2, 6)$ to $(3, 9)$.

However, a liter is an arbitrary unit. If deciliters are used instead, the interval from 2 to 3 liters becomes 20 to 30 deciliters. Then the marginal cost of producing

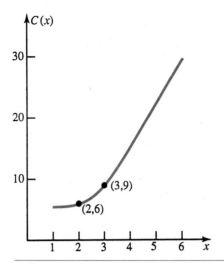

FIGURE 1

the third liter becomes $C(3) - C(2.9)/(3 - 2.9)$ because the interval runs from 29 to 30 deciliters. If centiliters are used, the marginal cost is $C(3) - C(2.99)/(3 - 2.99)$, while if milliliters are used, the marginal cost is $C(3) - C(2.999)/(3 - 2.999)$. This limit motivates the following definition.

_____ **Definition** _____

If $C(x)$ is the cost of producing x items of a product, the **marginal cost** is $C'(x)$. The marginal cost of producing the nth item is $C'(n)$.

_____

EXAMPLE 5

Problem

What is the marginal cost of producing the third item if the cost function is $C(x) = x^2 - 2x + 6$?

SOLUTION The aim is to compute $C'(x)$ and then to substitute $x = 3$.

$$C'(x) = \lim_{h \to 0} \frac{[(x + h)^2 - 2(x + h) + 6] - (x^2 - 2x + 6)}{h}$$

$$= \lim_{h \to 0} \frac{x^2 + 2xh + h^2 - 2x - 2h + 6 - x^2 + 2x - 6}{h}$$

$$= \lim_{h \to 0} \frac{2xh + h^2 - 2h}{h} = \lim_{h \to 0} \frac{h(2x + h - 2)}{h}$$

$$= \lim_{h \to 0} (2x + h - 2)$$

$$= 2x - 2$$

When $x = 3$, $C'(x) = 4$ so the marginal cost of producing the third item is $4.

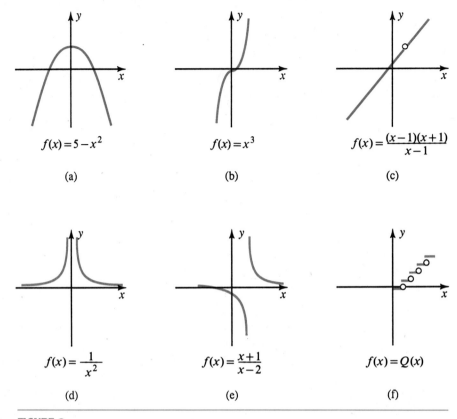

$f(x) = 5 - x^2$

(a)

$f(x) = x^3$

(b)

$f(x) = \dfrac{(x-1)(x+1)}{x-1}$

(c)

$f(x) = \dfrac{1}{x^2}$

(d)

$f(x) = \dfrac{x+1}{x-2}$

(e)

$f(x) = Q(x)$

(f)

FIGURE 2

Continuous and Differentiable Functions

Figure 2 portrays the graphs of various functions. The graphs in parts (a) and (b) can be sketched without the pencil leaving the paper, so the functions that define them are called *continuous*. The other four graphs require the pencil to be lifted from the paper, so they are said to be *discontinuous* at each value of x where the pencil is lifted. The function

$$f(x) = \frac{(x-1)(x+1)}{x-1}$$

in part (c) is discontinuous at $x = 1$ because the pencil is lifted at $x = 1$. The hole in the graph is indicated by the open circle. Discontinuities can also occur at gaps like those in parts (d) and (e) and at jumps like the one in part (f).

The display contains the precise definitions.

―――――――――――――――――――― **Definition** ――――――――――――――――――――

A function $f(x)$ is **continuous at a value** $x = a$ if

1. $f(x)$ is defined at $x = a$,

2. $\lim\limits_{x \to a} f(x)$ exists, and

3. $\lim\limits_{x \to a} f(x) = f(a)$.

If any one of these conditions is not satisfied, then $f(x)$ is **discontinuous** at $x = a$. $f(x)$ is **continuous in an interval** if $f(x)$ is continuous at every point in that interval. $f(x)$ is a **continuous function** if $f(x)$ is continuous at every real number.

Although the points of discontinuity of a function can be found by sketching its graph, general statements can be made about some special kinds of functions without appealing to their graphs. For instance, *a polynomial function is continuous at every value of x,* as can be seen in parts (a) and (b) in Figure 2. More generally, *a rational function* $\dfrac{f(x)}{g(x)}$ *is continuous at all values of x for which* $g(x) \neq 0$. This can be seen in parts (c), (d), and (e) in Figure 2.

EXAMPLE 6 | **Problem**

Determine the points where each function is discontinuous.

(a) $F(x) = \dfrac{x}{|x|}$ (b) $G(x) = \dfrac{x + 2}{x^2 + 2x}$

SOLUTION (a) The graph of $F(x)$ is sketched in Figure 3. The graph shows that $F(x)$ is discontinuous at $x = 0$ and is continuous at every other value of x.
(b) $G(x)$ is a rational function, so its continuity can be determined without sketching the graph. The points of discontinuity occur where $x^2 + 2x = 0$. Since $x^2 + 2x = x(x + 2)$, $G(x)$ is discontinuous at $x = 0$ and at $x = -2$.

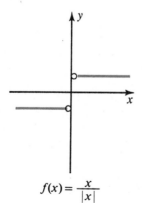

$f(x) = \dfrac{x}{|x|}$ **FIGURE 3**

A function $f(x)$ is **differentiable at a value** $x = a$ if $f'(a)$ exists and is **differentiable** if $f'(x)$ exists at all real numbers. Although this definition might not seem to carry any geometric connotation, the graph of a differentiable function can be viewed as a "smooth" curve. Historically there was no need to distinguish between continuous functions and differentiable functions until the difference became vital in applications during the first part of the 19th century, when it was proved that *if a function is differentiable, then it is also continuous.* However, *a continuous function need not be differentiable,* the classic example being $f(x) = |x|$, which is continuous at all values of x but not differentiable at $x = 0$. Figure 4 shows that the graph can be sketched without the pencil leaving the paper, but it is not a smooth curve at $x = 0$, so it is not differentiable.

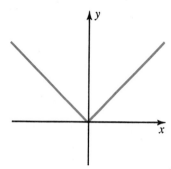

$f(x) = |x|$ is continuous at $x=0$
but not differentiable at $x=0$

FIGURE 4

EXERCISE SET 2–3

• In Problems 1 to 10 compute the derivative of the function.

1. $f(x) = x^2 + 3$ 2. $f(x) = x^2 - 5$

3. $f(x) = \sqrt{2x}$ 4. $f(x) = -\sqrt{x}$

5. $f(x) = 1 - \sqrt{x}$ 6. $f(x) = 1 + \sqrt{x}$

7. $f(x) = 1/x$ 8. $f(x) = 3/x$

9. $f(x) = 1 + 1/x$ 10. $f(x) = 1 - 1/x$

In Problems 11 to 14 find the equation of the tangent line to the function at the stated value of x.

11. $f(x) = x^2 - 2x + 6$ at $x = 0$

12. $f(x) = x^2 - 2x + 6$ at $x = 1$

13. $f(x) = x^2 + x - 5$ at $x = -1$

14. $f(x) = x^2 + x - 5$ at $x = 1$

In Problems 15 to 18 the height of a ball thrown from ground level into the air is given by $d(t)$, where d is measured in feet and t in seconds. Find (a) the amount of time it takes for the ball to reach its maximum height and (b) the maximum height.

15. $d(t) = -16t^2 + 192t$

16. $d(t) = -16t^2 + 48t$

17. $d(t) = 70t - 16t^2$ 18. $d(t) = 96t - 16t^2$

In Problems 19 to 22 determine the marginal cost of producing the third item for the given cost function $C(x)$.

19. $C(x) = x^2 + x + 2$

20. $C(x) = x^2 - x + 5$

21. $C(x) = x^2 - 1$ 22. $C(x) = x^2 + 4$

In Problems 23 to 26 determine the points where each function is discontinuous.

23. $f(x) = \dfrac{2}{|x|}$ 24. $f(x) = \dfrac{|x|}{x}$

25. $f(x) = \dfrac{x - 3}{x^2 - 3x}$ 26. $f(x) = \dfrac{x^2 + 3x}{x + 3}$

•• In Problems 27 to 30 compute the derivative of the function.

27. $f(x) = 1/\sqrt{x}$ 28. $f(x) = x^3$

29. $f(x) = x^{-2}$ 30. $f(x) = x^{-3}$

In Problems 31 to 36 find the equation of the tangent line to the function at the stated value of x.

31. $f(x) = 1/x$ at $x = 2$

32. $f(x) = 6/x$ at $x = -2$

33. $f(x) = 1/x^2$ at $x = -2$

34. $f(x) = \sqrt{x}$ at $x = 4$

35. $f(x) = 3x - 5$ at $x = 0$

36. $f(x) = 4 - 2x$ at $x = 1$

In Problems 37 and 38 the height of a ball thrown from the top of a 64-foot building into the air is given by $d(t)$, where d is measured in feet and t in seconds. Find (a) the amount of time it takes for the ball to reach its maximum height and (b) the maximum height.

37. $d(t) = -16t^2 + 128t + 64$

38. $d(t) = -16t^2 + 32t + 64$

39. For how many seconds will the ball in Problem 15 be in the air?

40. For how many seconds will the ball in Problem 18 be in the air?

In Problems 41 to 44 determine if the given function is continuous at $x = 2$.

41. $f(x) = x - 2$ 42. $f(x) = x^2 - 2$

43. $f(x) = \dfrac{1}{x - 2}$ 44. $f(x) = \dfrac{x - 2}{x^2 - 4}$

In Problems 45 to 48 draw the graph of the function and determine whether it is continuous or differentiable.

45. $f(x) = \sqrt{|x|}$

46. $f(x) = \begin{cases} x + 6 & \text{if } x \geq 3 \\ x^2 & \text{if } x < 3 \end{cases}$

47. $f(x) = \dfrac{x^2 - 1}{x - 1}$

48. $f(x) = \begin{cases} x^2 & \text{if } x \leq 0 \\ -x^2 & \text{if } x > 0 \end{cases}$

••• In Problems 49 and 50 let a, b, and c be real numbers. Find the derivatives of the stated function.

49. $f(x) = ax^2 + bx + c$

50. $f(x) = ax + b$

In the context of income taxes, the *marginal tax rate* is the derivative of the tax function as a percent. Also, *tax brackets* are intervals of income where the derivative is constant. Problems 51 and 52 refer to the following tax function $T(x)$, where x represents a family's income in dollars.*

$$T(x) = \begin{cases} .15x & \text{if } 0 \leq x \leq 29{,}750 \\ .28(x - 29{,}750) + 4463 & \text{if } 29{,}750 < x \leq 71{,}900 \\ .33(x - 71{,}900) + 16{,}265 & \text{if } 71{,}900 < x \leq 227{,}250 \\ .28(x - 227{,}250) + 67{,}531 & \text{if } x > 227{,}250 \end{cases}$$

51. What is the marginal tax rate for a family whose income is
(a) $25,000 (b) $50,000
(c) $100,000 (d) $500,000

52. What interval (or intervals) represents each of these tax brackets?
(a) 15 percent (b) 28 percent
(c) 33 percent

The authoritative *Automobile Magazine* declared the Porsche 959 sports car to be the car of the future.†

---

*Alan Murray, "Individuals' Top Rate Would Plunge to 28 Percent; Tax-Break Curbs Offset Benefits to Wealthy," *Wall Street Journal*, August 18, 1986, p. 6.

†Mel Nichols, "Review: Porsche 959," *Automobile Magazine*, November 1987, pp. 60–71.

Problems 53 and 54 refer to Figure 5, which gives some indication of the Porsche's astounding power and technology by listing the units for power on the left and the units for torque on the right.

53. (a) At what rpm is the derivative of the power function equal to 0?
   (b) What is the power at that point?

54. (a) At what rpm is the derivative of the torque function equal to 0?
   (b) What is the torque at that point?

55. Figure 6 shows the rate of change of certain economic indicators for 1987. Let $f(x)$ denote the function defined by the graph.
   (a) Interpret the meaning of the expression "percent change" in mathematical terms.
   (b) What month had highest value of the derivative $f'(x)$?
   (c) What month had lowest value of the derivative $f'(x)$?

FIGURE 5

FIGURE 6   *Source:* Commerce Department.

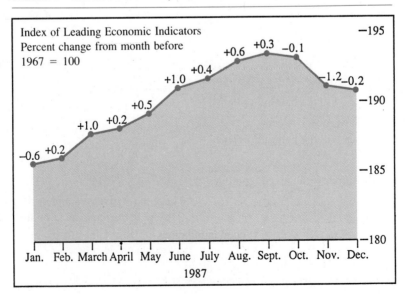

CHAPTER REVIEW

Key Terms

2–1 **Rate of Change and Tangents**

Velocity	Tangent Line
Average Rate of Change	Slope of a Function
Instantaneous Rate of Change	Difference Quotient
Rate of Change	Demand
Secant Line	Marginal Demand

2–2 **Limits**

Limit	Increases Without Bound
Limit Does Not Exist	Limit at Infinity
Decreases Without Bound	Horizontal Asymptote

2–3 **Definition of the Derivative**

Derivative	Continuous in an Interval
Marginal Cost	Continuous Function
Continuous at a Value	Differentiable at a Value
Discontinuous Function	Differentiable Function

Summary of Important Concepts

Formula

$$f'(x) = \lim_{h \to 0} \frac{f(x + h) - f(x)}{h}$$

Interpretations of the derivative

1. Instantaneous rate of change.
2. Tangent line.
3. Velocity of a moving object.

Steps in computing the derivative of $f(x)$

1. Form the difference quotient $\dfrac{f(x + h) - f(x)}{h}$.
2. Perform algebraic manipulations to factor h from the numerator of the difference quotient.
3. Cancel the h factors.
4. Take the limit as h approaches 0.

REVIEW PROBLEMS

1. Compute the average rate of change of the function $f(x) = x^2 + 1$ over the stated intervals.
   (a) from 1 to 2   (b) from 1 to 1.1

2. Compute the instantaneous rate of change of the function

   $$f(x) = 2x^2 + 1 \quad \text{at } x = -1.$$

3. Find the slope of the tangent line to the graph of the function $f(x) = x^2 - x + 4$ at an arbitrary value of x.

4. Compute the limit by constructing a table.

   $$\lim_{x \to 3} f(x) \quad \text{for } f(x) = \frac{x^2 - x - 6}{3 - x}$$

   In Problems 5 and 6 evaluate the limit, if it exists, by algebraic methods.

5. $\lim\limits_{x \to 1} \dfrac{2x^2 - 2x}{x - 1}$

6. $\lim\limits_{x \to 4} \dfrac{2 - \sqrt{x}}{4 - x}$

   In Problems 7 to 10 compute each limit, if it exists, by any method.

7. $\lim\limits_{x \to 1} \dfrac{x^2 + 2x - 3}{x^2 + x - 2}$

8. $\lim\limits_{x \to 0} \dfrac{1}{x}$

9. $\lim\limits_{x \to 0} x^2 + x$

10. $\lim\limits_{x \to 1} \dfrac{x - 1}{(x - 1)^2}$

    In Problems 11 and 12 compute the derivative of the function from the definition of the derivative.

11. $f(x) = x^2 - 3x - 7$

12. $f(x) = \dfrac{1}{x - 2}$

13. Find the equation of the tangent line to the function $f(x) = x^2 - 3x - 7$ at the value $x = 4$.

14. The height of a ball thrown from the top of a 64-foot building into the air is given by $d(t) = -16t^2 + 16t + 64$, where d is measured in feet and t in seconds. Find
    (a) the amount of time it takes for the ball to reach its maximum height.
    (b) the maximum height.

3 Differentiation Rules

CHAPTER OVERVIEW Calculus can be applied to a wide variety of real-world problems. Many examples will be given in this chapter and following ones. In each case the application will be described by a function. The problem is solved by analyzing the derivative of the function because the derivative gives information about the original function. Thus we will need to calculate the derivative of many types of functions. This chapter considers various rules for calculating derivatives.

The first section deals with elementary rules that govern functions that can be expressed as the sum of simpler functions or the product of a constant times a function. In the next section more complicated functions are handled, those that can be expressed as the product or quotient of other functions. In the third section the chain rule is presented. The chain rule governs functions that are raised to a power. The fourth section deals with higher derivatives.

CASE STUDY PREVIEW

The primary treatments for cancer are surgery and radiotherapy. More than half of all tumors can be treated by radiation therapy. The treatment consists of several beams directed at various parts of the tumor. Each beam has a specific intensity depending on the mass of the tumor and the amount of healthy cells the beam will affect. There are many variables that affect the design of an individual patient's treatment. Medical researchers are using mathematical optimization techniques to maximize the amount of cancer cells killed while also minimizing the number of healthy cells affected. The case study shows how the derivative plays an essential role in balancing these two objectives.

3-1 A FEW ELEMENTARY RULES

We have calculated the derivative of several functions using the definition of the derivative. The definition is cumbersome, so in this section some elementary differentiation rules are developed to simplify the task of finding derivatives.

Marginal Product

Figure 1 shows a typical production function from economics. It measures the output of a firm at a given level of input. The input is usually labor or land. For example, the production function given by Samuelson* can be represented by the equation

$$f(x) = -0.16x^2 + 1.6x$$

Here $f(x)$ is the total product, or amount of output, of the firm with a labor force of size x, where x is measured in hundreds of workers and $f(x)$ is measured in units of $10,000. An important economic concept derived from the production function is **marginal product.** It is defined as the rate of increase of production per unit of input. Thus marginal product is the derivative of the production function. In this section we study how to find the derivative of functions like $f(x)$ above. At the end of the section we show how marginal product is a key ingredient that economists use to determine a fair wage.

Other Notation for the Derivative

There are several notations for expressing the derivative of $y = f(x)$. The three most common notations are

$$f'(x) \quad D_x f(x) \quad \frac{df}{dx}$$

*Paul A. Samuelson, *Economics,* New York, McGraw-Hill Book Co., 1987.

FIGURE 1   *Source:* Paul A. Samuelson, *Economics,* New York, McGraw-Hill Book Co., 1987.

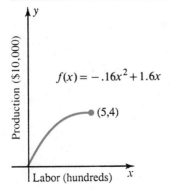

$$f(x) = -.16x^2 + 1.6x$$

$(5,4)$

Production ($10,000)

Labor (hundreds)

The latter can also be written as

$$\frac{dy}{dx} \qquad \text{or} \qquad \frac{d}{dx} f(x)$$

Each notation has an advantage in a particular situation so that no one notation is preferred over the others in every context. An advantage to the $f'(x)$ notation is that it is more convenient to express the derivative when evaluated at a particular value. For example, it is easier to write $f'(3)$ with this notation than the others. Using the $\frac{dy}{dx}$ and $D_x f(x)$ notations, we express $f'(3)$ as

$$D_x f(x)\big|_{x=3} \qquad \frac{d}{dx} f(x)\bigg|_{x=3}$$

The D_x notation is more convenient to use in formulas and when expressing the derivative of more complicated functions in terms of the derivative of simpler functions.

Constant Functions

Consider the constant function $f(x) = c$. Its graph is a horizontal straight line. (See Figure 2.) The slope of such a line is 0. Since the derivative measures the slope of the tangent line and the tangent to a line is the line itself, the derivative of a constant function is 0. This can be remembered as "the derivative of a constant is 0."

Constant Rule

If $f(x) = c$ for c a real number, then $f'(x) = 0$.
This is also expressed as

$$D_x(c) = 0 \qquad \text{for any constant } c$$

FIGURE 2

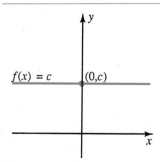

The constant function $f(x)=c$

This rule is actually a theorem whose proof is straightforward.

Proof of Rule D1

If $f(x) = c$ for some real number c, then

$$f'(x) = \lim_{h \to 0} \frac{f(x + h) - f(x)}{h} = \lim_{h \to 0} \frac{c - c}{h}$$

$$= \lim_{h \to 0} \frac{0}{h} = 0$$

Hence $f'(x) = 0$ for every real number x.

Power Rule

The following derivatives are calculated from the previous chapter:

$$D_x(x) = 1 \quad D_x(x^2) = 2x \quad D_x(x^3) = 3x^2 \quad D_x(x^{1/2}) = \tfrac{1}{2} x^{-1/2}$$

Each of these functions has the form $y = x^r$, where r is 1, 2, 3 and $\tfrac{1}{2}$, respectively. There is a pattern to the derivatives. It suggests an important rule called the *power rule*.

Power Rule

$$D_x(x^r) = rx^{r-1} \qquad \text{for any real number } r$$

The exercises indicate how to prove this theorem for r a positive integer. The proof for other real numbers is beyond the scope of this text. However, we emphasize that the power rule works for all real numbers as illustrated by Example 1.

EXAMPLE 1

Problem

Find $f'(x)$ for (a) $f(x) = x^{-5}$ and (b) $f(x) = \dfrac{1}{\sqrt{x}}$.

SOLUTION (a) Use the power rule with $r = -5$.

$$f'(x) = D_x[x^{-5}]$$
$$= -5x^{-5-1}$$
$$= -5x^{-6}$$

(b) Express the function as $f(x) = x^{-1/2}$. Then use the power rule with $r = -\tfrac{1}{2}$.

$$f'(x) = D_x[x^{-1/2}]$$
$$= (-\tfrac{1}{2})x^{(-1/2)-1}$$
$$= (-\tfrac{1}{2})x^{-3/2}$$

Constant Times a Function

From Section 2–3 we compute $D_x[4x^2] = 8x = 4 \cdot 2x = 4 \cdot D_x[x^2]$. The general rule that governs this derivative is "the derivative of a constant times a function is the constant times the derivative of the function."

Constant Times a Function Rule

$$D_x[cf(x)] = cD_x[f(x)] \qquad \text{for any real number } c$$

Proof of the Constant Times a Function Rule

If $g(x) = cf(x)$ for some constant c, then

$$g'(x) = \lim_{h \to 0} \frac{g(x + h) - g(x)}{h} = \lim_{h \to 0} \frac{cf(x + h) - cf(x)}{h}$$

$$= \lim_{h \to 0} \frac{c[f(x + h) - f(x)]}{h}$$

$$= \lim_{h \to 0} c \frac{f(x + h) - f(x)}{h}$$

$$= c \lim_{h \to 0} \frac{f(x + h) - f(x)}{h}$$

$$= cf'(x)$$

EXAMPLE 2

Problem

Find $f'(x)$ for (a) $f(x) = -3x$ and (b) $f(x) = 2x^5$.

SOLUTION (a) Use the constant times a function rule with $c = -3$ and $f(x) = x$.

$$f'(x) = D_x[-3x] = -3D_x[x]$$
$$= -3 \cdot 1 = -3$$

(b)

$$f'(x) = D_x[2x^5]$$
$$= 2D_x[x^5]$$
$$= 2 \cdot 5x^4 = 10x^4$$

Derivative of a Sum

The next formula states that "the derivative of a sum is the sum of the derivatives." The rule states that this is true for the sum of two functions, but it is also true for the sum of any finite number of functions.

Sum Rule

$$D_x[f(x) + g(x)] = D_x[f(x)] + D_x[g(x)]$$

Since the difference of two functions, $f(x) - g(x)$, can be regarded as the sum, $f(x) + [-g(x)]$, the derivative of the difference of two functions is the difference of their derivatives.

The exercises will indicate how to prove the sum rule.

EXAMPLE 3

Problem

Find $f'(x)$ for (a) $f(x) = 4x - 3/x$ and (b) $f(x) = 8 + 2x^{10} + 3x^{-5} - 7x^{5/3}$.

SOLUTION (a)

$$\begin{aligned}
f'(x) &= D_x(4x - 3x^{-1}) \\
&= D_x(4x) + D_x(-3x^{-1}) \\
&= 4D_x(x) - 3D_x(x^{-1}) = 4 \cdot 1 - 3(-1)x^{-2} \\
&= 4 + \frac{3}{x^2}
\end{aligned}$$

(b)

$$\begin{aligned}
f'(x) &= D_x(8 + 2x^{10} + 3x^{-5} - 7x^{5/3}) \\
&= D_x(8) + D_x(2x^{10}) + D_x(3x^{-5}) + D_x(-7x^{5/3}) \\
&= 0 + 2 \cdot 10x^9 + 3(-5)x^{-6} - 7\left(\frac{5}{3}\right)x^{(5/3 - 1)} \\
&= 20x^9 - 15x^{-6} - \left(\frac{35}{3}\right)x^{2/3}
\end{aligned}$$

The next example computes the equation of the tangent line to a curve.

EXAMPLE 4

Problem

Find the equation of the tangent line to the graph of the function $f(x) = x^2 - 5x + 3$ at $x = 1$, that is, at the point $(1, -1)$.

SOLUTION The slope of the tangent line is measured by $f'(1)$. Since $f'(x) = 2x - 5$, $f'(1) = -3$. Using the point-slope form of the equation of the tangent line yields

$$y - (-1) = -3(x - 1)$$

Simplifying gives the equation $y = -3x + 2$.

Applications

The derivative measures instantaneous rate of change. If an object is dropped so that gravity is the only force acting on it, then the relationship between the distance traveled, d, and time, t, is $d(t) = -16t^2$, where d is measured in feet and t in seconds. A negative sign is in the expression because the object is falling and hence is moving in the negative direction. The average velocity is the average rate of change of $d(t)$ with respect to t. The instantaneous velocity measures how fast the object is falling at that instant. It is measured by $d'(t)$. Thus the instantaneous velocity is $v(t) = d'(t) = -32t$. Similarly the instantaneous acceleration of the object is the instantaneous rate of change of its velocity with respect to t. Thus the object's instantaneous acceleration is $a(t) = v'(t) = -32$.

EXAMPLE 5 | An object is thrown upward and its position is given by $d(t) = -16t^2 + 64t$.

Problem

Find (a) the velocity and acceleration of the object, (b) the time at which it reaches its maximum height, and (c) how many feet it travels upward.

SOLUTION (a) The velocity is $v(t) = d'(t) = -32t + 64$ and the acceleration is $a(t) = v'(t) = -32$.
(b) The object will reach its maximum height when $v(t) = 0$ because it stops at that instant. Letting $v(t) = 0$ yields $-32t - 64 = 0$, which implies $t = 2$ seconds.
(c) The object travels upward until it stops, at which time it starts downward. Thus it travels upward from $t = 0$ to $t = 2$ seconds. Hence the object travels $d(2)$ feet upward.

$$d(2) = -16(2)^2 + 64(2) = 64$$

The object travels 64 feet upward.

Marginal Product Revisited

The production function from Samuelson given at the beginning of the section is

$$f(x) = -0.16x^2 + 1.6x$$

The marginal product of the firm is the derivative $f'(x)$.

$$f'(x) = -0.32x + 1.6$$

The marginal product measures the rate of change of output as input changes. It makes no sense to let x take on values less than 0, and it is assumed that the largest number of labor units available to the firm is $x = 5$. This function is valid only for values of x in the interval $[0, 5]$. It is a quadratic function whose graph is given in Figure 1. It is typical of production functions because as x increases $f'(x)$ decreases. For example, $f'(0) = 1.6$ and $f'(1) = 1.28$. Another way to see this is to note that the marginal product is a linear function with a negative slope. Thus as x increases $f'(x)$ decreases. Economists refer to this as the "law of diminishing returns." It means that as more labor is used, more output is generated, but the rate of increase of output, the marginal product, decreases.

Economists use marginal product to determine fair wages. The fundamental principle is that *wages* = *marginal product*. This assumption is derived from the fact that if an extra worker were hired, the worker would increase the firm's productivity only by an amount equal to that worker's marginal product. For instance, consider a particular segment of the work force, so that all of these workers are paid the same wage. If there are 100 workers, thus $x = 1$, then the last one hired, the 100th worker hired, would increase the firm's productivity by the marginal product $f'(1)$. Therefore, in theory, all workers would receive the same fair wage, $f'(1)$, the marginal product of the last worker hired.

EXAMPLE 6

Consider the firm with the production function

$$f(x) = -0.16x^2 + 1.6x$$

Suppose the firm has 100 employees with the same wage.

Problem

Find the theoretical fair wage of the employees.

SOLUTION The theoretical fair wage is the marginal product at $x = 1$. It is $f'(1)$ where $f'(x) = -0.32x + 1.6$. Thus $f'(1) = -0.32(1) + 1.6 = 1.28$. Each worker is paid $12,800.

Are workers always paid their marginal product? Economist Robert H. Frank answers this question in an article appearing in the *American Economic Review*. His conclusion is that some workers, such as salespersons, do command wages equal to their marginal product, but others, such as college professors, are paid less. Exercise 41 deals with some of the data from this study.

EXERCISE SET 3–1

● In Problems 1 to 26 find $f'(x)$.

1. $f(x) = x^3 + 5$
2. $f(x) = x^6 - 2$
3. $f(x) = 2x^4 - 7$
4. $f(x) = 3x^5 + 1$
5. $f(x) = 5x^{-3} + 5x$
6. $f(x) = 4x^{-5} - x^3 + 1$
7. $f(x) = 2x^{1/3} + 3x^{1/2} - 4x$
8. $f(x) = 7x^{1/5} - x^{2/5} + 2x$

9. $f(x) = 2x^{-1/3} + 7x^{-1/4} - 5$
10. $f(x) = 6x^{-1/6} - 3x^{-3/7} - 6x$
11. $f(x) = x^{0.3} + 3x^{0.4}$
12. $f(x) = 1.2x^{-1.5} - 3.1x^{-2.2}$
13. $f(x) = 1/x^3 + 3x^4$
14. $f(x) = 4x^{0.5/5} - 3x^{2.2/7}$
15. $f(x) = 1/x^{1.5} + 4/x^{1/4} + 4x^{2.4}$

16. $f(x) = 4/x^{2.5} - 2/x^{2/5} - 2x^{2/5}$

17. $f(x) = 2/x^{-3.5} + 4/x^{1.4} + 4x^{-1.3}$

18. $f(x) = 1/x^{-2.5} - 2x^{-3/5} - x^{-2.7/3}$

19. $f(x) = 0.25/x^{-0.5} + 0.48/x^{0.4}$

20. $f(x) = 1.5x^{-0.5} - 2.75x^{-0.05}$

•• 21. $f(x) = 2\sqrt{x} - 2x$

22. $f(x) = \sqrt{x}/9 + 9/x$

23. $f(x) = 5\sqrt{x} - 2/\sqrt{x}$

24. $f(x) = \sqrt{9/x} + 2/\sqrt{7x}$

25. $f(x) = 18/\sqrt{3x} + 2/\sqrt[3]{4x}$

26. $f(x) = 8/\sqrt[4]{3x} + \frac{2}{3}\sqrt[4]{x}$

In Problems 27 to 30 find the equation of the tangent line to $y = f(x)$ at the given value of x.

27. $f(x) = x^2 + 2x$ at $x = 1$

28. $f(x) = x^3 - 4x$ at $x = 2$

29. $f(x) = x^{-2} + 2x^2$ at $x = 1$

30. $f(x) = x^{-3/2} - x$ at $x = 1$

31. An object is thrown upward, and its position is given by $d(t) = -16t^2 + 32t$. Find (a) the velocity and acceleration of the object, (b) the time at which it reaches its maximum height, and (c) how many feet it travels upward.

32. An object is thrown upward, and its position is given by $d(t) = -16t^2 + 50t + 6$. Find (a) the velocity and acceleration of the object, (b) the time at which it reaches its maximum height, and (c) how many feet it travels upward.

33. A firm determines that the relationship of the quantity x that the public is willing to purchase at price p is given by

$$p(x) = 5000 - 300x^{-1}$$

Find the rate of change of $p(x)$ when $x = 5$.

34. The profit P in thousands of dollars from the sales volume of x records is found to be

$$P(x) = 0.2x^2 + 10x - 100$$

Find the marginal profit when $x = 10$.

35. The cost C in thousands of dollars to produce x million batteries is determined to be

$$C(x) = 10 + 20x - 0.1x^2$$

Find the marginal cost when $x = 50$.

36. Business analysts call a product a "fad" if its total sales increase rapidly and then level off after a short period of time. Let S represent total sales in thousands of units and let x represent time in weeks. A function that represents the behavior of a fad is

$$S(x) = 100 - 50x^{-1}$$

Show that the rate of change of $S(x)$ with respect to x is decreasing as x increases by computing $S'(x)$ for (a) $x = 1$, (b) $x = 5$, and (c) $x = 10$.

••• 37. The binomial theorem states that for any real numbers p and q and any positive integer n

$$(p + q)^n = p^n + np^{n-1}q + n(n - 1)/2p^{n-2}q^2$$
$$+ \cdots + npq^{n-1} + q^n$$

Use the binomial theorem to prove Rule D2 for positive integers.

38. Use the definition of the derivative to prove the sum rule.

39. In an article illustrating the benefits of quality control to executives, A. W. Whitton, Jr.,* Senior Corporate Quality Consultant for Abbott Laboratories, uses Figure 3 to demonstrate how to reduce total quality cost. Locate point a on the graph of $y = TQC(x)$. It represents the lowest point on the TQC curve because its y-coordinate is the smallest of all points on the graph. The slope of the tangent line at A is 0. From the graph show that the following equations are true:

$$TQC(x) = FC(x) + AP(x) + PC(x)$$
$$TQC'(a) = FC'(a) + AP'(a) + PC'(a)$$

*A. W. Whitton, Jr., "Methods for Selling Total Quality Cost Systems," *American Society of Quality Control—Technical Conference Transactions*, 1972, pp. 365–371.

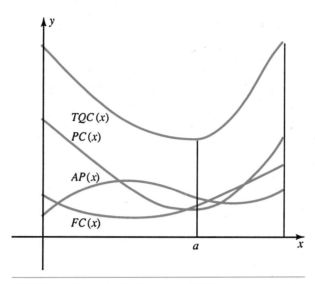

FIGURE 3   *Source:* A. W. Whitton, Jr., "Methods for Selling Total Quality Cost Systems," *American Society of Quality Control — Technical Conference Transactions,* 1972, pp. 365–371.

40. In the study of optics the subjective brightness a viewer senses is compared with the actual brightness from the source.† If x measures the actual brightness, then the subjective brightness, S, is a function of x. The functional relationship is usually described by a power function $S(x) = kx^n$, where k and n are constants depending on the light intensity and the type of light the viewer is coming from before the experiment. The *subjective sensitivity* is defined as the rate of change of the subjective brightness. Thus subjective sensitivity is $S'(x)$. Suppose for those in relative darkness before an experiment that the subjective brightness is given by $S_1(x) = 0.001x^{1/4}$ and for those in light the subjective brightness $S_2(x) = 0.002x^{1/2}$. At what value of x are the subjective sensitivities equal?

41. In an article studying whether certain segments of the work force are paid their theoretical fair wage, defined by economists as the marginal product of the last worker hired, Robert H. Frank computes the following production function for real estate sales agents, where $f(x)$ represents the earnings for a sales force of size x.‡

$$f(x) = -0.024x^2 + 0.575x$$

Compute the marginal product function.

†Robert Sekuler et al., "Structural Modeling of Spatial Vision," *Vision Research,* Vol. 24, 1984, pp. 689–700, and Heywood M. Petry et al., "Spatial Contrast Sensitivity of the Tree Shrew," *Vision Research,* Vol. 24, 1984, pp. 1037–1042.

‡Robert H. Frank, "Are Workers Paid Their Marginal Products?" *American Economic Review,* Vol. 74, No. 4, 1984, pp. 549–571.

3–2 PRODUCT AND QUOTIENT RULES

The formulas developed in the previous section are used to compute the derivatives of polynomial functions and functions that are sums of terms of the form ax^r. Some functions do not fit one of these forms. Examples of functions whose derivatives cannot be found directly by these rules are

$$f(x) = x^2(5x + 11)$$

$$h(x) = \frac{x^3}{(4x + 1)}$$

We develop two formulas in this section. The first handles functions that are the product of simpler functions, like $f(x)$, and the second is applied to quotients, like $h(x)$.

The Product Rule

At first it might seem that the derivative of a product is the product of the derivatives, similar to the derivative of a sum. A quick inspection shows that this is not the case. Let $f(x) = x$ and $g(x) = x^2$. Then $f(x) \cdot g(x) = x^3$, so the derivative of the product is $3x^2$. But the product of the derivatives is $f'(x) \cdot g'(x) = 1 \cdot 2x = 2x$. Thus the derivative of the product is *not* the product of the derivatives. The formula for the derivative of a product is more complicated.

The Product Rule

If $f(x) = g(x) \cdot k(x)$, then

$$f'(x) = g(x)k'(x) + k(x)g'(x)$$

The **product rule** states that "the derivative of a product is the first function times the derivative of the second plus the second times the derivative of the first." We prove this theorem at the end of the section.

EXAMPLE 1

Problem

Use the product rule to find $D_x x^2(5x + 11)$.

SOLUTION Let $f(x) = x^2(5x + 11)$ with $g(x) = x^2$ and $k(x) = 5x + 11$. Then $f(x) = g(x) \cdot k(x)$. Since $g'(x) = 2x$ and $k'(x) = 5$, we have

$$f'(x) = g(x)k'(x) + k(x)g'(x) = x^2(5) + (5x + 11)(2x)$$
$$= 5x^2 + 10x^2 + 22x = 15x^2 + 22x$$

The function in Example 1 could be expressed as $f(x) = 5x^3 + 11x^2$ by multiplying through by x^2. Using the rules of the previous section, we have $f'(x) = 15x^2 + 22x$, which is the same derivative computed by the product rule. In all of the examples of this section the functions that are products can be expanded in this way, but in the next section some functions require the product rule to compute their derivatives.

For example, care is needed when computing the derivative of functions that are raised to a power. Consider $f(x) = (x^2 + 1)^2$. It might seem from the power rule that $f'(x) = 2(x^2 + 1)$. But by using the product rule with $g(x) = k(x) = x^2 + 1$, we get

$$f'(x) = (x^2 + 1)2x + (x^2 + 1)2x = 4x(x^2 + 1)$$

In the next section we will learn how to handle power functions that cannot be expanded like this function.

Let us look at one more example of the product rule.

EXAMPLE 2 | **Problem**

Find $D_x(x^2 - 1)(x^3 + x)$.

SOLUTION To apply the product rule let $f(x) = (x^2 - 1)(x^3 + x)$ with $g(x) = x^2 - 1$ and $k(x) = x^3 + x$. Since $g'(x) = 2x$ and $k'(x) = 3x^2 + 1$, we have

$$
\begin{aligned}
f'(x) &= g(x)k'(x) + k(x)g'(x) \\
&= (x^2 - 1)(3x^2 + 1) + (x^3 + x)(2x) \\
&= 3x^4 - 3x^2 + x^2 - 1 + 2x^4 + 2x^2 \\
&= 5x^4 - 1
\end{aligned}
$$

The Quotient Rule

Sometimes functions can be expressed as the quotient of simpler functions, such as $h(x) = x^3/(4x + 1)$. To find the derivative of such functions use the following rule.

The Quotient Rule

If $f(x) = \dfrac{g(x)}{k(x)}$, then

$$f'(x) = \frac{k(x)g'(x) - g(x)k'(x)}{[k(x)]^2}$$

The numerator in the quotient rule is similar to the expression in the product rule, but there is a significant difference. The two terms cannot be interchanged as they can in the product rule because the first term has a positive sign and the second has a negative sign. Examples 3 and 4 illustrate how to apply the quotient rule.

EXAMPLE 3

Problem

Find $D_x \dfrac{x^3}{4x + 1}$

SOLUTION Let $f(x) = x^3/(4x + 1)$ with $g(x) = x^3$ and $k(x) = 4x + 1$. Then $f(x) = g(x)/k(x)$. Since $g'(x) = 3x^2$ and $k'(x) = 4$, we have

$$f'(x) = \frac{k(x)g'(x) - g(x)k'(x)}{[k(x)]^2}$$

$$= \frac{(4x + 1)(3x^2) - x^3(4)}{(4x + 1)^2}$$

$$= \frac{8x^3 + 3x^2}{(4x + 1)^2}$$

EXAMPLE 4

Problem

Find $D_x \dfrac{x^3}{x^2 + 5}$

SOLUTION

$$D_x \frac{x^3}{x^2 + 5} = \frac{(x^2 + 5)D_x x^3 - x^3 D_x(x^2 + 5)}{(x^2 + 5)^2}$$

$$= \frac{(x^2 + 5)3x^2 - x^3(2x)}{(x^2 + 5)^2}$$

$$= \frac{3x^4 + 15x^2 - 2x^4}{(x^2 + 5)^2}$$

$$= \frac{x^4 + 15x^2}{(x^2 + 5)^2}$$

An Application: The Lerner Index

There are many methods to measure a particular firm's power and influence in the marketplace. One measure, called the **Lerner index,** named after economist A. P. Lerner of England and Michigan State University, depends on the difference between price P and marginal cost, expressed as a proportion of P.* It is defined as

$$\text{Lerner index} = \frac{P - MC}{P}$$

The Lerner index is sometimes referred to as a measure of a firm's monopoly power. If an industry has perfect competition, then $P = MC$, which means that the price

*Milton H. Spenser, *Contemporary Macroeconomics,* 6th ed., New York, Worth Publishers, 1986.

each firm establishes for its product is equal to the rate of change of cost, marginal cost. Under perfect competition the Lerner index is 0, meaning the firm has no monopoly power. If a firm has a large amount of monopoly power, it can set a price that is much larger than marginal cost, so the Lerner index will increase. The closer the index is to 1 (its limiting value), the more monopoly power the firm has. Once a firm calculates its price and cost functions it is interested in the rate of change of the Lerner index L, measured by L'.

EXAMPLE 5

A firm determines that its price $P(x)$ and cost $C(x)$ per level of production x are

$$P(x) = x^2 + 4$$
$$C(x) = 3x + 10$$

Problem

Compute the firm's Lerner index $L(x)$ and find the rate of change of $L(x)$.

SOLUTION First calculate the marginal cost function, $MC(x) = C'(x)$.

$$MC(x) = C'(x) = 3$$

Substituting this equation along with the equation of $P(x)$ into the definition of $L(x)$ yields

$$L(x) = \frac{P - MC}{P} = \frac{x^2 + 4 - 3}{x^2 + 4} = \frac{x^2 + 1}{x^2 + 4}$$

Applying the quotient rule gives

$$L'(x) = \frac{2x(x^2 + 4) - (x^2 + 1)(2x)}{(x^2 + 4)^2}$$

$$= \frac{2x^3 + 8x - (2x^3 + 2x)}{(x^2 + 4)^2}$$

$$= \frac{6x}{(x^2 + 4)^2}$$

Proof of the Product Rule

The product rule is not so readily apparent as the derivative of a sum. In fact, intuition leads to an incorrect formula; the derivative of a product looks like it should be the product of the derivatives. Why intuition fails is answered in the proof of the theorem. The proof uses an algebraic ploy—adding and subtracting the same quantity to and from an expression. This results in adding 0 to the expression, so that it remains the same. This is done so the expression can be factored in a certain way. This is why the formula contains two terms, each containing one derivative.

Proof of the Product Rule

If $f(x) = g(x) \cdot k(x)$, then $f(x + h) = g(x + h) \cdot k(x + h)$. From the definition of the derivative of $f(x)$

$$f'(x) = \lim_{h \to 0} \frac{f(x + h) - f(x)}{h}$$

$$= \lim_{h \to 0} \frac{g(x + h)k(x + h) - g(x)k(x)}{h}$$

Here we add and subtract $g(x + h)k(x)$ in the numerator.

$$f'(x) = \lim_{h \to 0} \frac{g(x + h)k(x + h) - g(x + h)k(x) + g(x + h)k(x) - g(x)k(x)}{h}$$

$$= \lim_{h \to 0} \frac{g(x + h)[k(x + h) - k(x)] + k(x)[g(x + h) - g(x)]}{h}$$

$$= \lim_{h \to 0} g(x + h) \left[\frac{k(x + h) - k(x)}{h} \right] + \lim_{h \to 0} k(x) \left[\frac{g(x + h) - g(x)}{h} \right]$$

$$= \lim_{h \to 0} g(x + h) \lim_{h \to 0} \frac{k(x + h) - k(x)}{h} + \lim_{h \to 0} k(x) \lim_{h \to 0} \frac{g(x + h) - g(x)}{h}$$

$$= g(x)k'(x) + k(x)g'(x)$$

provided that $g'(x)$ and $k'(x)$ exist.

EXERCISE SET 3–2

- In Problems 1 to 18 find $f'(x)$.

1. $f(x) = x^4(x + 5)$ 2. $f(x) = x^6(x - 2)$

3. $f(x) = 2x^3(3x - 7)$

4. $f(x) = 3x^5(2x + 1)$

5. $f(x) = (x^{-3} + 5)(3x^2 - 17)$

6. $f(x) = (x^{-5} - 9)(12x^3 + 1)$

7. $f(x) = (x^{1/3} + x)(3x^{1/2} - 4x)$

8. $f(x) = (x^{1/5} - x)(x^{3/5} + 2x)$

9. $f(x) = (x^{-1/3} + 7x^{-1/4})(4x^2 - 5)$

10. $f(x) = (2x^{-1/6} - x^{-3/7} + 2)(x^{1/2} - 6)$

11. $f(x) = (2x + 1)^2$ 12. $f(x) = (4x - 3)^2$

13. $f(x) = (x^3 + 1)^2$ 14. $f(x) = (x^4 - 2)^2$

15. $f(x) = x/(2x + 1)$

16. $f(x) = 2x/(5x - 4)$

17. $f(x) = (x^3 + 5)/3x^2$

18. $f(x) = (x^5 - 11)/x^3$

In Problems 19 to 22 find the equation of the tangent line to $y = f(x)$ at the given point

19. $f(x) = (x^{1/2} - 4)^2$, $(1, 9)$

20. $f(x) = (x^{4/5} + 1)^3$, $(1, 8)$

21. $f(x) = (x^3 - 2)/(x^2 - 1)$, $(0, 2)$

22. $f(x) = (x^{1/2} - 1)/(x^{1/2} + 8)$, $(1, 0)$

In Problems 23 and 24 a firm determines that its price $P(x)$ and cost $C(x)$ per level of production x are the given functions. Compute the firm's Lerner index $L(x)$ and find the rate of change of $L(x)$.

23. $P(x) = 0.5x^2 + 5$, $C(x) = 2x + 7$

24. $P(x) = 0.22x^2 + 4.1, \quad C(x) = 4x + 3$

•• In Problems 25 to 32 find $f'(x)$.

25. $f(x) = x^2(x^3 + 15)(x^2 - 17)$

26. $f(x) = x^{1/2}(x^{-5} - 9)(12x^3 + 1)$

27. $f(x) = x^{-1}(x^{1/3} + 2)(x^2 - 3)$

28. $f(x) = x^{-1/2}(x^{1.5} - 1)(2x^3 - 3)$

29. $f(x) = x^2(x + 5)/(x^2 - 7)$

30. $f(x) = x^{1/2}(x^2 - 1)/(x^3 + 1)$

31. $f(x) = (x^2 + 3)/x(x^2 - 1)$

32. $f(x) = (x^3 - 2)/x^2(x^{-1} + 1)$

33. The total cost of producing x amount of books is given by $C(x) = 100 + 23x(1 - 2x^2)$. Find the marginal cost.

The *average cost per good* is defined to be the total cost to produce x amount of goods divided by x. If $C(x)$ measures the total cost to produce x amount of goods and $AC(x)$ is the average cost per good, then $AC(x)$ is given by

$$AC(x) = \frac{C(x)}{x}$$

In Problems 34 and 35 find the marginal average cost, $AC'(x)$.

34. $C(x) = 10 + 25x - 2x^2$

35. $C(x) = 100 + 3.5x - 0.4x^2$

••• 36. Use the definition of the derivative and an argument similar to the proof of the product rule to prove Rule D6, the quotient rule.

37. Show why the Lerner index $L = \dfrac{(P - MC)}{P}$, for price P and marginal cost MC, is never more than 1.

38. With the breakup of AT&T, the transition from monopoly to competition necessitated the use of different tools for business decisions. In an article* describing how GTE Telephone Operations decided on prices for various products, one model centered on the Styleline phone. Of key interest is the total quantity sold q after x years, where q is measured in millions of units. The data can be approximated by

$$q(x) = 0.13x(x - 10)^2$$

GTE was concerned with finding when $q(x)$ would attain its maximum rate of increase. In the next chapter we will show how this is done by using $q'(x)$. Compute $q'(x)$.

39. The Atomic Energy Commission has demanded extensive quality control in nuclear power plants, especially since the Three Mile Island and Chernobyl incidents. In an article describing the significance of AEC regulations, even for "minor" equipment, John B. Silverwood,† Reliability and Assurance Manager for United Engineers and Constructors, outlines various procedures that are necessary in performance testing. In one example he describes the efficiency rating of a centrifugal pump by using a graph (Figure 1) whose equation is approximated by

$$f(x) = \left(\frac{x}{4}\right)\left(5 - \frac{x}{32}\right)$$

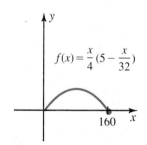

FIGURE 1   *Source:* John B. Silverwood, "Objective Quality Evidence of Product Integrity," *American Society of Quality Control — Technical Conference Transactions,* 1972, pp. 276–283.

*Arvind G. Jadhav, "Market Simulation for Telecommunications Services," *Forecasting Public Utilities,* New York, North-Holland Publishing Co., pp. 165–186.

†John B. Silverwood, "Objective Quality Evidence of Product Integrity," *American Society of Quality Control—Technical Conference Transactions,* 1972, pp. 276–283.

$f(x)$ measures the efficiency, in percent, at which the pump is operating when x gallons are being pumped per minute. The function is defined in $[0, 100]$. From the graph it can be seen that as x increases, the efficiency increases for a while and then it starts to decrease. At that point the tangent line has slope 0. Find the point where $f(x)$ changes from increasing to decreasing by finding the value of x that solves $f'(x) = 0$.

40. In a paper‡ studying the effects of private pond fishing on preexisting common property fishing,

researchers centered their attention on the crawfish market. The cost function for wild crawfish, those caught in common property waters, can be approximated by

$$C(x) = \frac{7.12}{13.7x - x^2}$$

Here x is the quantity of crawfish supplied to the market, measured in millions of pounds, and C is measured in millions of dollars. Find the marginal cost $C'(x)$.

‡Frederick W. Bell, "Competition from Fish Farming in Influencing Rent Dissipation: The Crawfish Fishery," *American Journal of Agricultural Economics,* Vol. 68, pp. 95–101.

3–3 THE CHAIN RULE

Sometimes more complicated functions can be differentiated by separating the function into two parts, an "inner" and "outer" part, and then computing the derivative of each. The rule that allows us to find a derivative this way is called the chain rule. It has several forms, each governing a different type of function. In this section we study the form of the chain rule called the generalized power rule, so-called because it deals with the derivative of functions raised to a power. In general the chain rule is used to find the derivative of a composite function of the form $f(g(x))$. With this notation, $g(x)$ is the "inner" function and $f(x)$ is the "outer" function. At the end of the section the chain rule is studied in this more general setting so that it can be applied later to other types of functions.

The Generalized Power Rule

The differentiation rules developed so far do not allow us to find the derivative of more complicated functions, such as

$$f(x) = (x^3 + x)^5 \quad s(x) = (x^2 + 1)^{-1} \quad h(x) = (3x^2 + x)^{1/2}$$

Each of these functions is an expression in x raised to a power. The general form of such a function is

$$f(x) = g(x)^n$$

This type of function is called a **general power function.** In each case $g(x)$ is referred to as the "inside part" of the function or the inside function. For example, by letting $g(x) = x^3 + x$ and $n = 5$, we can form the previous function $f(x) = (x^3 + x)^5$. Similarly $s(x)$ has $g(x) = x^2 + 1$ and $n = -1$, while $h(x)$ has $g(x) = 3x^2 + x$ and $n = \frac{1}{2}$.

Let us find the derivative of a specific power function.

EXAMPLE 1

Problem

Use the product rule to find the derivative of (a) $f(x) = (x^3 + x)^2$, and (b) $g(x) = (x^3 + x)^3$

SOLUTION (a) Consider $f(x)$ as the product of two functions, each being $x^3 + x$. That is, $f(x) = (x^3 + x)(x^3 + x)$. By the product rule

$$f'(x) = (x^3 + x)D_x(x^3 + x) + (x^3 + x)D_x(x^3 + x)$$
$$= 2(x^3 + x)D_x(x^3 + x)$$
$$= 2(x^3 + x)(3x^2 + 1)$$

(b) Consider $g(x)$ as the product of the two functions $f(x) = (x^3 + x)^2$ from (a) and $x^3 + x$. Thus

$$g(x) = (x^3 + x)^2(x^3 + x)$$

Use the product rule, with $f'(x) = 2(x^3 + x)(3x^2 + 1)$ from (a).

$$g'(x) = (x^3 + x)^2 D_x(x^3 + x) + (x^3 + x)D_x(x^3 + x)^2$$
$$= (x^3 + x)^2(3x^2 + 1) + (x^3 + x)[(2)(x^3 + x)(3x^2 + 1)]$$

Factor out $3x^2 + 1$ and get

$$g'(x) = [(x^3 + x)^2 + 2(x^3 + x)^2](3x^2 + 1)$$
$$= 3(x^3 + x)^2(3x^2 + 1)$$

Recall the power rule:

$$D_x x^n = nx^{n-1}$$

It governs only functions of the form $f(x) = x^n$. In Example 1, when finding $f'(x)$, it is tempting to use just the power rule, in which case we would get $2(x^3 + x)$. While this expression is part of the derivative, it is not precisely the derivative. It varies by a factor of $3x^2 + 1$, which is the derivative of the inside function, $x^3 + x$. The generalized power rule requires that the derivative of a general power function contain the derivative of the inside part of the function as a factor.

   Example 1 motivates the chain rule for general power functions, called the **generalized power rule.**

The Generalized Power Rule

$$D_x[g(x)]^n = ng(x)^{n-1}D_x[g(x)]$$

Notice that the power rule is a special case of the generalized power rule when $g(x) = x$. Let us apply the rule to the function in Example 1(a). The function is $f(x) = (x^3 + x)^2$ where $g(x) = x^3 + x$ and $n = 2$. Then

$$f'(x) = 2(x^3 + x)D_x(x^3 + x)$$
$$= 2(x^3 + x)(3x^2 + 1)$$

The next three examples illustrate the generalized power rule. Notice that these functions cannot be differentiated by the product rule.

EXAMPLE 2

Problem

Find $D_x f(x)$ where $f(x) = (x^4 + 2x)^{1/2}$.

SOLUTION In the generalized power rule we have $g(x) = x^4 + 2x$ and $n = \frac{1}{2}$. Therefore

$$D_x(x^4 + 2x)^{1/2} = (\tfrac{1}{2})(x^4 + 2x)^{(1/2)-1}D_x(x^4 + 2x)$$
$$= (\tfrac{1}{2})(x^4 + 2x)^{-1/2}(4x^3 + 2)$$
$$= (2x^3 + 1)(x^4 + 2x)^{-1/2}$$

EXAMPLE 3

Problem

Find $f'(x)$ where $f(x) = (x^3 + 6x^2)^{-4}$.

SOLUTION In the generalized power rule we have $g(x) = x^3 + 6x^2$ and $n = -4$. Therefore

$$f'(x) = -4(x^3 + 6x^2)^{-4-1}D_x(x^3 + 6x^2)$$
$$= -4(x^3 + 6x^2)^{-5}(3x^2 + 12x)$$
$$= -12x(x + 4)(x^3 + 6x^2)^{-5}$$

EXAMPLE 4

Problem

Find $f'(x)$ where $f(x) = (x^4 + 12x + 4)^{-1/4}$.

SOLUTION Let $g(x) = x^4 + 12x + 4$ and $n = -\frac{1}{4}$. Then

$$f'(x) = (-\tfrac{1}{4})(x^4 + 12x + 4)^{-5/4}D_x(x^4 + 12x + 4)$$
$$= (-\tfrac{1}{4})(x^4 + 12x + 4)^{-5/4}(4x^3 + 12)$$
$$= -(x^3 + 3)(x^4 + 12x + 4)^{-5/4}$$

The next example demonstrates how to use the generalized power rule in conjunction with the product rule.

EXAMPLE 5

Problem

Compute $f'(x)$ where $f(x) = x^3(x^2 + 5)^{-1}$.

SOLUTION Use the product rule, with the first function being x^3 and the second function being $(x^2 + 5)^{-1}$.

$$
\begin{aligned}
f'(x) &= x^3 D_x(x^2 + 5)^{-1} + (x^2 + 5)^{-1} D_x x^3 \\
&= x^3(-1)(x^2 + 5)^{-2}(2x) + (x^2 + 5)^{-1}(3x^2) \\
&= -2x^4(x^2 + 5)^{-2} + 3x^2(x^2 + 5)^{-1} \\
&= (x^2 + 5)^{-2}[-2x^4 + 3x^2(x^2 + 5)] \\
&= (x^2 + 5)^{-2}(-2x^4 + 3x^4 + 15x^2) \\
&= (x^4 + 15x^2)(x^2 + 5)^{-2} \\
&= x^2(x^2 + 15)(x^2 + 5)^{-2}
\end{aligned}
$$

Notice that the quotient rule could have been used in Example 5 by expressing $f(x)$ as a quotient, $x^3/(x^2 + 5)$.

The General Form of the Chain Rule

The generalized power rule is a special case of the **chain rule.** The general form of the chain rule is stated in terms of the composition of functions.

The Chain Rule

If $f(x) = h(g(x))$, then $f'(x) = h'(g(x))g'(x)$.

To illustrate, suppose $f(x) = (x^2 + 1)^3$. Let $h(x) = x^3$ and $g(x) = x^2 + 1$ so that $f(x) = h(g(x))$. To find $f'(x)$ by the general form of the chain rule, we use $h'(x) = 3x^2$ and $g'(x) = 2x$. Then $h'(g(x)) = 3[g(x)]^2 = 3(x^2 + 1)^2$ so that

$$
\begin{aligned}
f'(x) &= h'(g(x))g'(x) = 3(x^2 + 1)^2(2x) \\
&= 6x(x^2 + 1)^2
\end{aligned}
$$

The chain rule can be expressed in another form, which is easy to memorize. This form uses the $\dfrac{dy}{dx}$ notation of the derivative. Suppose $y = f(g(x))$. Make the substitution of variable $u = g(x)$ so that $y = f(u)$. Then y can be expressed as a function of u or of x. With this notation, we write the derivatives as $g'(x) = \dfrac{du}{dx}$, $f'(u) = \dfrac{dy}{du}$, and $f'(x) = \dfrac{dy}{dx}$. Then the chain rule states

$$
\frac{dy}{dx} = \frac{dy}{du}\frac{du}{dx}
$$

In the illustration above, $y = f(x) = (x^2 + 1)^3$ and $u = g(x) = x^2 + 1$ so that $y = f(u) = u^3$. Then $\dfrac{dy}{du} = 3u^2$ and $\dfrac{du}{dx} = 2x$. Therefore

$$\frac{dy}{dx} = \frac{dy}{du}\frac{du}{dx} = 3u^2(2x) = 3(x^2 + 1)^2(2x)$$

$$= 6x(x^2 + 1)^2$$

This agrees with the preceding expression for $f'(x)$.

The generalized power rule is a special case of the chain rule. We now derive the special case, the generalized power rule, from the more general theorem, the chain rule.

Derivation of the Generalized Power Rule from the Chain Rule

Consider the power function $f(x) = g(x)^n$. Let $h(x) = x^n$. Then $f(x) = h(g(x))$ and $h'(x) = nx^{n-1}$, so $h'(g(x)) = ng(x)^{n-1}$. From the general form of the chain rule we have

$$f'(x) = h'(g(x))g'(x) = ng(x)^{n-1}g'(x)$$

This is the statement of the generalized power rule.

An Application

There are many situations in business where one variable depends upon another, and the second variable depends on a third. An example is the important business tool, quality control. Quality $Q(x)$ is measured in terms of level of production x, and x is measured in terms of time t. Q is measured on a scale from 0 to 100 (with 100 being a product with no flaws), x is measured in units of 100, and t is measured in hours. The next example shows how the rate of change of Q can be computed in two ways, by expressing Q in terms of t directly and by using the chain rule.

EXAMPLE 6

A company determines that

$$Q(x) = 90 - 0.4(x - 10)^2 = 50 + 8x - 0.4x^2$$

where $Q(x)$ is defined for $0 < x < 18$. The graph of $Q(x)$ is given in Figure 1. Production level, measured in hours, depends on the time of day, t, and is given by

$$x(t) = 18 - 0.5(t - 6)^2 = 6t - 0.5t^2$$

where $x(t)$ is defined for $0 \le t \le 8$. This means that at time t the factory is producing goods at a rate such that if that rate were held constant all day $100x(t)$ units would be produced. The graph of $x(t)$ is given in Figure 2.

Problem

Find D_tQ by expressing Q as a function of t and differentiating, and also by using the chain rule. Show that the two expressions are equal.

SOLUTION Since Q is a function of x and x is a function of t, Q is a function of t; namely

$$Q(t) = Q(x(t)) = Q(6t - 0.5t^2)$$
$$= 50 + 8(6t - 0.5t^2) - 0.4(6t - 0.5t^2)^2$$

To find $Q'(t)$ directly from this formula, differentiate with respect to t.

$$Q'(t) = 8(6 - t) - 0.8(6t - 0.5t^2)(6 - t)$$
$$= 48 - 36.8t + 7.2t^2 - 0.4t^3$$

Using the chain rule to find $Q'(t)$ yields

$$Q'(t) = Q'(x(t))x'(t)$$
$$= D_x[Q(x)]D_t[x(t)]$$
$$= D_x[50 + 8x - 0.4x^2]D_t[6t - 0.5t^2]$$
$$= [8 - 0.8x][6 - t]$$

Now substitute the expression $x(t) = 6t - 0.5t^2$ for x.

$$Q'(t) = [8 - 0.8(6t - 0.5t^2)](6 - t)$$
$$= (8 - 4.8t + 0.4t^2)(6 - t)$$
$$= 48 - 36.8t + 7.2t^2 - 0.4t^3$$

Thus the two expressions for $Q'(t)$ are equal.

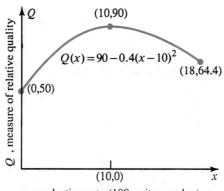

x, production rate (100 units per day)   **FIGURE 1**

FIGURE 2

$$Q(t) = 50 + 8(6t - \tfrac{1}{2}t^2) - 0.4(6t - \tfrac{1}{2}t^2)^2$$

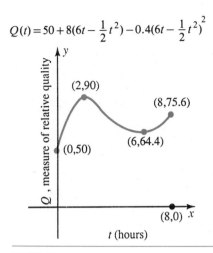

FIGURE 3

The graph of $Q(t)$ is given in Figure 3. In Section 4–2 we will see how to obtain this graph. The quality is lowest at the start of the shift when it is 50; that is, $Q(0) = 50$. The graph of Q then gradually increases until it reaches its largest value, 90, when $t = 2$, meaning that two hours into the shift quality is at its peak. At this point $Q'(t)$ is 0 since the rate of change of Q goes from positive when Q is increasing to negative when it is decreasing. Then Q decreases to the value 64.4 at $t = 6$ hours, at which time it then starts to increase again. Thus when $t = 6$, $Q'(t) = 0$. We will discuss these concepts further in the next chapter.

An Application: The Lerner Index

The measure of a firm's monopoly power, called the Lerner index L, was presented in the previous section. It is defined by

$$L = \frac{P - MC}{P}$$

P is the price of the firm's product and MC is the marginal cost. If MC is fixed, then L is a function of P. If P is a function of the units produced x, then L is also a function of x. In the previous section we expressed L as a function of x and then computed $L'(x)$. The next example shows that the chain rule can also be used to find $L'(x)$. The price and cost functions are from Example 5 in the previous section.

EXAMPLE 7 | A firm determines that its price $P(x)$ and cost $C(x)$ per level of production x are

$$P(x) = x^2 + 4$$
$$C(x) = 3x + 10$$
$$MC(x) = C'(x) = 3$$

Problem

Use the chain rule to compute $L'(x)$.

SOLUTION Since MC is the constant 3, we can express L as a function of P.

$$L(P) = \frac{P - MC}{P} = \frac{P - 3}{P} = 1 - 3P^{-1}$$

Then we have

$$\frac{dL}{dP} = 3P^{-2}$$

Also,

$$\frac{dP}{dx} = 2x$$

Applying the chain rule yields

$$\frac{dL}{dx} = \frac{dL}{dP}\frac{dP}{dx} = (3P^{-2})(2x) = 3(x^2 + 4)^{-2}(2x)$$

$$= \frac{6x}{(x^2 + 4)^2}$$

This is the same expression for the derivative that was computed in the last section.

EXERCISE SET 3–3

● In Problems 1 to 10 find $f'(x)$ by using the generalized power rule.

1. $f(x) = (5x + 2)^3$
2. $f(x) = (15x + 2)^4$
3. $f(x) = (x^3 + 4)^3$
4. $f(x) = (5x^2 + 2)^6$
5. $f(x) = (x^3 + 3x^2)^3$
6. $f(x) = (2x^4 + 3x^3)^6$
7. $f(x) = (x^{3/2} + 4)^{1/3}$
8. $f(x) = (x^{3/5} - 2)^{2/3}$
9. $f(x) = (x^{1/2} + 2x^{-1})^{-6}$
10. $f(x) = (5x^{1/3} + x^{-2})^{-1/2}$

In Problems 11 to 16 find $f'(x)$ using the generalized power rule in conjunction with the product rule or the quotient rule.

11. $f(x) = x(5x^2 + 2)^2$

12. $f(x) = 3x(3x^2 + 5)^4$
13. $f(x) = x^2(x^3 + 2x)^2$
14. $f(x) = 3x^2(x^3 + 3x)^3$
15. $f(x) = x^2/(x^3 - x)^3$
16. $f(x) = 4x^4(x^2 - 2x)^{1/2}$

In Problems 17 to 20 find the equation of the tangent line at the given value of x.

17. $f(x) = (x^2 + 2)^3$ at $x = 0$
18. $f(x) = (x^3 - 1)^2$ at $x = 0$
19. $f(x) = (x^3 + 4x)^{-1}$ at $x = 1$
20. $f(x) = (3x^2 - x - 1)^{-2}$ at $x = 0$

A company determines that its quality control $Q(x)$, measured in terms of level of production x, measured in terms of time t, is given by the functions in Problems 21 and 22. Find D_tQ by expressing Q as a function of t and differentiating and also by using the chain rule. Show that the two expressions are equal.

21. $Q(x) = 95 - 0.2(x - 5)^2$,
    $x(t) = 10 - 0.2(t - 6)^2$

22. $Q(x) = 100 - 0.3(x - 8)^2$,
    $x(t) = 10 - 0.1(t - 6)^2$

A company determines that its price $P(x)$ and cost $C(x)$ per level of production x are given by the functions in Problems 23 and 24. Find the Lerner index $L(x)$ and $L'(x)$.

23. $P(x) = x^2 + 10$, $C(x) = 5x + 20$

24. $P(x) = 100 + 15x - 2x^2$,
    $C(x) = 4x + 25$

•• In Problems 25 to 34 find $f'(x)$ by using the chain rule together with the rules in the previous section.

25. $f(x) = x^2(x^3 + 2x^2)^4$

26. $f(x) = 3x^4(2x^4 + 3x^3)^6$

27. $f(x) = 2x^{1/2}(x^{3/2} + 4)^{-3}$

28. $f(x) = 4x^{-2}(x^{5/2} + 2x^{-1})^{1/6}$

29. $f(x) = (x^2 + 1)^2(x^3 + 1)^3$

30. $f(x) = (x^2 - 3)^{-1}(x^3 - 1)^2$

31. $f(x) = (x^2 + 1)^{1/2}(x^3 + 3)^{1/3}$

32. $f(x) = (x^{1/2} + 3)^{1/3}(x^2 + 3x)^{-1}$

33. $f(x) = 2x^{1/2}/(x^2 + 2x^{-2})^3$

34. $f(x) = (x^4 + 1)^2/(3x^4 + 3)^{1/3}$

In Problems 35 to 38 find $f'(x)$ by using the chain rule and also by expressing the function as a quotient and using the quotient rule.

35. $f(x) = x(x^2 + 1)^{-1}$ 36. $f(x) = x^2(x^3 - 1)^{-1}$

37. $f(x) = (x - 2)(x^2 + 5)^{-1}$

38. $f(x) = (x + 3)^2(4x - 5)^{-1}$

39. The total profit $p(x)$ for x amount of goods sold is

$$p(x) = 500(200 - x)^2$$

Find the marginal profit.

40. The strength $S(x)$ of an individual's reaction to the quantity x of a drug administered into the bloodstream is given by

$$S(x) = 5x(20 - x/2)^{1/2}$$

Find the rate of change of $S(x)$ with respect to x when $x = 8$.

••• 41. The chain rule can be used in conjunction with the product rule to derive the quotient rule. Suppose $f(x)$ is a quotient function $f(x) = \dfrac{g(x)}{h(x)}$. Express $f(x)$ as the product function $f(x) = g(x)h(x)^{-1}$. First apply the product rule and then the chain rule for the derivative of $h(x)^{-1}$. Factor $h(x)^{-2}$ from each term to get the quotient rule.

42. Franchised businesses account for over 38% of all retail sales in the United States and originate more than 12% of the gross national product. In an article analyzing methods to optimize franchise contractual agreements, the researchers* derive the following function that relates franchisee's profits P in terms of the quantity Q of goods sold by the equation

$$P = aQ - bQ^2 - F$$

where F is the initial lump-sum franchise fee paid by the franchisee, and a and b are constants whose values depend on the type of industry to which the model is being applied. It is also shown that Q is a function of R where R is the royalty rate charged by the parent company to the franchisee

$$Q(R) = c - dR$$

The constants c and d also depend on the particular industry. Q is measured in thousands of units and P in thousands of dollars. Typical values for the constants in the fast food industry

---

*Roger D. Blair and David L. Kaserman, "Optimal Franchising," *Southern Economics Journal*, Vol. 49, 1983, pp. 495–505.

are $a = 0.35$, $b = 0.11$, $c = 0.1$, and Q is measured in 1000 units. Find the marginal profit with respect to R via the chain rule and then express P as a function of R and find the marginal profit with respect to R directly. Show that the two expressions are equal.

43. Many variables affect quality control of production. In describing different quality control procedures for various products, Bartlett and Provost† used as an example the net weight of a #303 can of whole kernel corn. This type of can should weigh 16 ounces. They show that the acceptance of a lot by a retailer from the producer depends on the average net weight of all cans in the lot, and the average net weight depends on the size of the sample of cans selected from the lot to be inspected. Let P represent the likelihood that a lot will be accepted by the retailer, W the average net weight of all the cans in the lot, and x the sample size. The relationships between the variables can be expressed as follows:

$$P(W) = 0.007W^2 - 0.33W + 100$$
$$W(x) = 16 - 0.41x^{-1/2}$$

Express P as a function of x and find $P'(x)$.

†Richard P. Bartlett and Lloyd P. Provost, "Tolerances in Standards and Specifications," *Quality Progress*, 1973, pp. 14–18.

3–4   HIGHER DERIVATIVES AND OTHER NOTATION

Just as the derivative $f'(x)$ of a function gives information about $f(x)$, so too the derivative of the derivative gives information about $f'(x)$. The derivative of the derivative is called the "second derivative." This section shows how to compute the second derivative and higher derivatives, as well as discusses a few additional points about alternate notation.

The Second Derivative

The derivative $f'(x)$ of a function is another function that measures the slope of the tangent line of $f(x)$. In the same way the derivative of $f'(x)$ is another function, called the **second derivative** of $f(x)$, and is denoted by $f''(x)$. Using alternate notation, $D_x(D_xf(x)) = f''(x)$, which says that the derivative of the derivative is the second derivative. The first example illustrates how to compute the second derivative.

EXAMPLE 1 | **Problem.**

Find $f''(x)$ for (a) $f(x) = 2x^3 + 5x$ and (b) $f(x) = \sqrt{x} - 1/x$

SOLUTION (a) $f'(x) = 6x^2 + 5$ and so

$$f''(x) = 12x$$

(b) First write the function as $f(x) = x^{1/2} - x^{-1}$. Then

$$f'(x) = (\tfrac{1}{2})x^{-1/2} - (-x^{-2})$$
$$= (\tfrac{1}{2})x^{-1/2} + x^{-2}$$
$$f''(x) = -(\tfrac{1}{4})x^{-3/2} - 2x^{-3}$$

The second derivative provides important information about $f(x)$. In particular, it helps determine what the shape of the curve of $f(x)$ is in the vicinity of a given point. This topic will be amplified in the next chapter.

Higher Derivatives

In some applications it is necessary to compute higher derivatives. For example, if $f(x) = x^3$, then $f'(x) = 3x^2$, $f''(x) = 6x$ and the derivative of $f''(x)$, called the **third derivative,** denoted by $f'''(x)$, is the derivative of $6x$. Hence $f'''(x) = 6$. For even **higher derivatives** the notation changes. We write $f^{(n)}(x)$ for $n > 3$ to represent higher derivatives after the third one. In this example $f^{(4)}(x) = 0$ as do all higher derivatives.

EXAMPLE 2 | **Problem**

Find $f^{(n)}(x)$ for $f(x) = x^{-1}$.

SOLUTION Find the first, second, and third derivatives to determine a pattern.

$$f'(x) = -x^{-2}$$
$$f''(x) = -(-2)x^{-3} = 2x^{-3}$$
$$f'''(x) = 2(-3)x^{-4} = -6x^{-4}$$

Successive derivatives change sign so a factor $(-1)^n$ is necessary. The absolute value of the coefficient of each derivative has the form $1 \cdot 2 \cdot 3 \cdot \ldots \cdot n = n!$. The power of x in each derivative is negative and its absolute value is one more than the number of the derivative. For example, the third derivative has the power -4. Hence the nth derivative has the factor x to the power $-(n + 1)$. Therefore the nth derivative is

$$f^{(n)}(x) = (-1)^n n! x^{-(n+1)}$$

In particular, $f^{(4)}(x) = (-1)^4 4! x^{-5} = 24x^{-5}$ and $f^{(5)}(x) = -5! x^{-6} = -120x^{-6}$.

Other Notation

There are several notations for expressing higher derivatives of $y = f(x)$. If the function is $y = f(x)$, then the second derivative is expressed by each of these.

$$f''(x) \qquad \frac{d^2y}{dx^2} \quad \frac{d^2f}{dx^2} \quad \frac{d^2}{dx^2} f(x) \qquad D_x^2 f(x)$$

These notations might seem to contain powers, but they are merely contractions of more cumbersome notation. Referring to the second derivative as the "derivative of the derivative," the D_x notation is

$$f''(x) = D_x[D_x f(x)]$$

This is shortened to $D_x^2[f(x)]$. The contraction is more obviously preferred when trying to write the third and higher derivatives with this notation. The third derivative is written $D_x^3 f(x)$ instead of $D_x[D_x(D_x f(x))]$. In a similar way the notation $\dfrac{d^2 y}{dx^2}$ is a shortened form of $\dfrac{d}{dx}\dfrac{dy}{dx}$. Similarly the notation for the third derivative is $\dfrac{d^3 y}{dx^3}$.

The second derivative of $y = f(x)$ at $x = 3$ is expressed as $f''(3)$. Using the $\dfrac{dy}{dx}$ and $D_x f(x)$ notations, $f''(3)$ is expressed as

$$\left. \frac{d^2 y}{dx^2} \right|_{x=3} \qquad \left. D_x^2[f(x)] \right|_{x=3}$$

The next example provides practice with these different notations.

EXAMPLE 3 | **Problem**

If $f(x) = x^3 - 2x^2 + 5$, find (a) $\left. \dfrac{d^2 y}{dx^2} \right|_{x=3}$ and (b) $\left. D_x^2 f(x) \right|_{x=4}$

SOLUTION $f'(x) = 3x^2 - 4x$ and $f''(x) = 6x - 4$. Therefore
(a)

$$\left. \frac{d^2 y}{dx^2} \right|_{x=3} = 6(3) - 4 = 14$$

(b)

$$\left. D_x^2 f(x) \right|_{x=4} = 6(4) - 4 = 20$$

Other Variables

In some applications it is convenient to use more appropriate variables than x and y, such as t and v when discussing time and velocity. When the variable is not x, the notation for the derivative is also changed. For instance, if the function is given as $v(t) = t^4$, then we write $v'(t) = D_t v = \dfrac{dv}{dt} = 4t^3$. The name of the variable does not affect the meaning of the derivative. Thus if $f(x) = 2x$ and $p(s) = 2s$, then $f'(x) = 2$ and $p'(s) = 2$ mean the same thing as far as the derivative is concerned—the slope of the tangent line at any value of the variable is 2. This illustrates that any letter can be used as a variable.

EXAMPLE 4 | **Problem**

Compute the derivative of each function: (a) $G(u) = (3u + 5)^{1/2}$ and (b) $z(y) = y^{-3} - y$.

SOLUTION (a)

$$G'(u) = (\tfrac{1}{2})(3u + 5)^{-1/2} \cdot 3$$
$$= (\tfrac{3}{2})(3u + 5)^{-1/2}$$

(b)

$$D_y z(y) = -3y^{-4} - 1$$

One of the most important applications of calculus is in the movement of objects. Many modern inventions are based on this application, from radar screens to television broadcasts. In Section 3–1 we considered an object moving in a straight line given by the function $y = f(t)$ at time t. Its velocity, $v(t)$, is given by $v(t) = f'(t)$. The rate of change of velocity is defined to be *acceleration*, $a(t)$. Thus if $f(t)$ is the position function, then $f'(t) = v(t)$ is the velocity and $f''(t) = a(t)$ is the acceleration. Examples 5 and 6 illustrate how to compute acceleration.

EXAMPLE 5

An object is moving in a straight line with its position at time t given by $f(t) = t^3 - 12t^2 + 36t + 1$.

Problem

Find the value of t where (a) its velocity is 0 and (b) its acceleration is 0.

SOLUTION Find the first and second derivatives.

$$\begin{aligned} f'(t) &= 3t^2 - 24t + 36 = 3(t^2 - 8t + 12) \\ &= 3(t - 6)(t - 2) \\ f''(t) &= 6t - 24 = 6(t - 4) \end{aligned}$$

(a) Set $v(t) = f'(t) = 3(t - 6)(t - 2) = 0$. Then $v(t) = 0$ when $t = 6$ and when $t = 2$.
(b) Set $a(t) = 6(t - 4) = 0$. Then $a(t) = 0$ when $t = 4$.

An object comes to rest when $v(t) = 0$. So the object in Example 5 comes to rest at $t = 2$ and $t = 6$. In the next chapter we will learn that when the acceleration is 0, the object's speed changes from slowing down to speeding up.

When studying the velocity and acceleration of an object that is in motion in a vertical direction, usually the positive direction is upward and the negative direction is downward. This convention is used in the next example.

EXAMPLE 6

A ball is thrown straight upward. The ball's vertical distance $d(t)$ feet from the ground after t seconds is measured by the function

$$d(t) = -16t^2 + 128t + 6$$

Problem

(a) Find the velocity after $t = 1$ and $t = 6$ seconds.
(b) Find the acceleration when $t = 1$ second.
(c) At what value of t is the velocity equal to -32 ft/sec?
(d) When is the acceleration equal to -32 ft/sec^2?

(e) As far as the motion of the ball is concerned, what does it mean to say that the acceleration is a constant function?

SOLUTION (a) First find the velocity function $v(t) = d'(t)$.

$$v(t) = d'(t) = -32t + 128$$

Evaluate $v(t)$ at $t = 1$ and $t = 6$.

$$v(1) = -32 + 128 = 96$$
$$v(6) = -192 + 128 = -64$$

When $t = 1$ second the velocity of the ball is 96 ft/sec. When $t = 6$ seconds the velocity is -64 ft/sec. The negative sign means that the ball is in a downward direction; that is, it is falling.

(b) Compute the acceleration function $a(t) = d''(t)$.

$$a(t) = d''(t) = -32$$

Evaluate $a(t)$ at $t = 1$.

$$a(1) = -32$$

When $t = 1$ second the acceleration is -32 ft/sec^2.

(c) Let $v(t)$ equal -32 and solve for t.

$$v(t) = -32$$
$$-32t + 128 = -32$$
$$-32t = -32 - 128 = -160$$
$$t = 5$$

When $t = 5$ seconds the velocity is -32 ft/sec.

(d) The acceleration function is the constant function $a(t) = -32$. This means that $a(t)$ is -32 ft/sec^2 when evaluated at every value of t.

(e) The second derivative measures the rate of change of the first derivative. Therefore the acceleration measures the rate of change of the velocity. When the ball is thrown in the air the only force acting on it is the force of gravity. It is a constant acceleration of -32 ft/sec^2. Gravity causes the velocity to decrease at this constant rate.

The second derivative is used in business to measure the rate of change of various quantities. For example, the derivative of cost is marginal cost. The rate of change of marginal cost is the derivative of marginal cost, which is the second derivative of cost.

EXAMPLE 7   A firm determines that its cost $C(x)$ of producing x units is given by the function

$$C(x) = 24 - 5x + 0.4x^2 - 0.01x^3$$

Problem

Find the derivative of the marginal cost function.

SOLUTION The marginal cost function $MC(x)$ is

$$MC(x) = C'(x) = -5 + 0.8x - 0.03x^2$$

The derivative of $MC(x)$ is

$$MC'(x) = C''(x) = 0.8 - 0.06x$$

In the next chapter we will see that $MC'(x)$ gives information about $C(x)$.

In an example showing how firms such as fast food chains use price and demand to determine maximum profit, McCarty* uses the second derivative of a profit function. The next example shows how to compute the second derivative of this profit function.

EXAMPLE 8

The profit $P(Q)$ of a firm producing a quantity of Q goods is estimated to be

$$P(Q) = -50 + 390Q - 70.8Q^2$$

Problem

Find the second derivative of $P(Q)$.

SOLUTION Compute $P'(Q)$ and then $P''(Q)$.

$$P'(Q) = 390 - 141.6Q$$
$$P''(Q) = -141.6$$

*Marilu Hurt McCarty, *Managerial Economics*, Glenview, IL, Scott, Foresman and Co., 1986, pp. 350–352.

EXERCISE SET 3–4

● In Problems 1 to 14 find $f''(x)$.

1. $f(x) = x^2 + 1$

2. $f(x) = x^3 - 2$

3. $f(x) = 2x^4 - 7$

4. $f(x) = 3x^5 + 1$

5. $f(x) = 4x^{-1} + 5$

6. $f(x) = 2x^{-2} - 1$

7. $f(x) = x^{-3} + 2$

8. $f(x) = 3x^{-4} - 4$

9. $f(x) = x^4 - x^2$

10. $f(x) = x^5 + 3x^2$

11. $f(x) = 2x^{-1/3} + 7x^{-1/4} - 5$

12. $f(x) = 6x^{-1/6} - 3x^{-3/7} - 6x$

13. $f(x) = x^{0.3} + 3x^{0.4}$

14. $f(x) = 1.2x^{-1.5} - 3.1x^{-2.2}$

In Problems 15 to 18 compute $D_x^2 f(x)\Big|_{x=2}$.

15. $f(x) = x^2 + 2x^3$

16. $f(x) = 2x^2 + x^4 - 5$

17. $f(x) = x^{-2} - 2$

18. $f(x) = 2x^{-3} + x^{-4}$

In Problems 19 to 22 compute the second derivative.

19. $g(t) = t^2 + 2t^3$

20. $s(u) = 2u^2 + u^3 - 5$

21. $F(g) = g^{1/2} - 2g$

22. $A(r) = 3r^{-1/3} + r^{-1}$

In Problems 23 to 26 an object is moving in a straight line with its position at time t given by $f(t)$. Find the value of t where (a) its velocity is 0 and (b) its acceleration is 0.

23. $f(t) = t^2 - 2t + 3$

24. $f(t) = 2t^2 - 4t + 5$

25. $f(t) = t^3 - 2t^2 + t + 5$

26. $f(t) = 3t^3 - t^2 + 4t$

In Problems 27 and 28 a ball is thrown straight upward. The function $d(t)$ is the distance the ball is from the ground after t seconds. Find the acceleration function.

27. $d(t) = -32t^2 + 132t + 10$

28. $d(t) = -32t^2 + 164t + 5$

In Problems 29 and 30 the cost $C(x)$ of a firm producing x units is given. Find the derivative of the marginal cost function.

29. $C(x) = 120 - 22x + 0.9x^2 - 0.003x^3$

30. $C(x) = 233 - 32x + 0.5x^2 - 0.04x^3 + 0.001x^4$

In Problems 31 and 32 the profit $P(x)$ of a firm producing x units is given. Find the second derivative of the profit function.

31. $P(x) = -134 + x(122 - 0.4x)^2$

32. $P(x) = -533 + x(455 - 3.5x)^2$

•• In Problems 33 to 38 find $f''(x)$.

33. $f(x) = (x^2 + 1)^2$

34. $f(x) = (x^3 + 4)^2$

35. $f(x) = (2x^2 + 1)^{-1}$

36. $f(x) = (x^3 + 4)^{-2}$

37. $f(x) = (x^2 + 1)^{1/2}$

38. $f(x) = (x^3 + 4)^{2/3}$

In Problems 39 to 42 find the values of x for which $f''(x) = 0$.

39. $f(x) = x^4 - 4x^3$

40. $f(x) = x^4 + 2x^3$

41. $f(x) = x^4 - 6x^2$

42. $f(x) = x^5 + 10x^3$

••• In Problems 43 to 46 find $f^{(3)}(x)$ and $f^{(4)}(x)$.

43. $f(x) = x^4 + 3x$

44. $f(x) = 2x^4 - x^3 + x$

45. $f(x) = 2x^{-1} + x^{-2}$

46. $f(x) = (1 + 2x)^{-1}$

In Problems 47 and 48 find $f^{(n)}(x)$.

47. $f(x) = x^{-2}$

48. $f(x) = (1 - x)^{1}$

49. When a vendor supplies a business with a large amount of a product, the recipient often selects a small sample to test the quality. If the sample has an acceptable quality level, then the entire lot is accepted. There are two costs involved: the cost of using too many defectives when the lot was accepted but should have been rejected, and the cost of rejecting the lot. The breakeven point is defined to be that fraction of defectives where the two costs are equal. If the fraction of defectives in the sample is greater than the breakeven point, then the lot is rejected. Edgar Dawes,* Director of Quality for the Dictaphone Corporation, illustrates how the breakeven point can be calculated by using the second derivative. In one particular sampling plan he defines the operating characteristic curve $y = f(x)$, where f measures the fraction of times the lot should be accepted when x defectives are found. Thus $f(0) = 1$ implies that when no defectives are found the lot should always be accepted. The function is approximated by

$$f(x) = 960x^3 - 180x^2 + 1$$

The function is defined in $[0, 0.16]$. In the next chapter we will see that the breakeven point

---

*Edgar W. Dawes, "Optimizing Attribute Sampling Costs—A Case Study," *American Society of Quality Control—Technical Conference Transactions*, 1972, pp. 181–190.

occurs when $f''(x) = 0$. Determine the break-even point for this sampling plan.

50. In an article studying local public goods, such as police and fire protection, Hayes[†] used empirical data from 90 Illinois municipalities. The cost $C(x)$ of these services can be estimated by the function

$$C(x) = 0.5x^{-0.7}$$

Find the derivative of the marginal cost function.

[†]K. Hayes, "Local Public Good Demands and Demographic Effects," *Applied Economics*, Vol. 18, 1986, pp. 1039–1045.

CASE STUDY   Radiotherapy Treatment Design[*]

This case study describes a new type of weapon used in the treatment of cancer. It discusses the mathematical models developed by David Sonderman and Philip Abrahamson to help cancer therapists devise a comprehensive treatment for their patients.

Cancer therapy   Cancer is second only to heart disease as a cause of human mortality. One-fourth of all deaths of Americans are caused by cancer. After surgery the most important treatment of cancer is radiation therapy. More than half of all tumors can be treated by radiation therapy, which consists of directing ionizing radiation from external beams through a patient's body to the cancer cells. However, the radiation damages healthy cells along with cancerous tissue. Therefore it is necessary to develop a specific strategy of intensity and direction of the radiation for each individual patient in order to minimize the effect of the treatment on healthy cells. The process of selecting the intensity and direction of the beams is called *radiotherapy treatment design*.

It is common practice to develop a treatment design by using a trial-and-error process. The designer studies the patient's anatomy and proposes a treatment by using a series of several beams with various intensities and directions. Computer software is then used to calculate the effect of the radiation at specific points in the patient's body. If the designer is not satisfied with the predicted outcome of the plan, the design is modified or an entirely different treatment is proposed. The new treatment design is then evaluated, and the procedure continues until the predicted outcome is acceptable. This process is time-consuming and offers little guarantee that the design finally implemented will satisfy all the important constraints of the treatment, especially eliminating the cancer cells and minimizing the damage to healthy tissue.

Today many researchers are developing treatment design techniques that use mathematical optimization methods. Mathematical programming and calculus optimization techniques are replacing the trial-and-error process to produce better

[*]This case study is based on David Sonderman and Philip Abrahamson, "Radiotherapy Treatment Design Using Mathematical Programming Models," *Operations Research*, Vol. 33, No. 4, 1985, pp. 705–725.

treatment designs in much less time. These new rigorous techniques also ensure that all important constraints are satisfied.

Ionizing radiation   Radiation from external beams is administered by sophisticated machines that accurately direct radiation that can vary in energy from several thousand electron volts to more than 40 million electron volts. Radiation dosage is measured in units called rads. While different patients require different dose levels, a common range is 3000 to 7000 rads. The human body cannot withstand such a large dose in one treatment so it is delivered in several treatments of smaller doses over a span of three to six weeks. However, the treatment design is based only on the total dose and its distribution throughout the body. Once the treatment design is formulated, the patient's requirements are determined to decide how often treatments will be given and what fractions of the total dose will be administered at various intervals.

Malignant tissue is slightly more sensitive to radiation than healthy tissue, so the radiation dose at each point of the treatment region must be large enough to kill the cancerous cells but small enough to spare the healthy cells. Tumor cells are usually microscopically interspersed among healthy cells. If the total dose given at a particular point of the tumor is too low, then cancerous tissue will survive and the disease will recur. Too large a dose can cause serious side effects. Therefore the treatment design must include a homogeneous dose distribution over the tumor region.

The total integral dose versus homogeneity   The sum of the individual doses over a finite set of grid points throughout the treatment region is called the *total integral dose* and is denoted by A. The term *homogeneity* is defined to be the percentage of the tumor region that is covered by the radiation beams. If the only concern was to kill the cancerous tissue, then the homogeneity would be set at 100%. But as the homogeneity increases so does the number of healthy cells that are killed. This means that the total integral dose, A, is a function of the homogeneity, u. As u increases so does A, so A is an increasing function of u.

For almost all patients there is a very gradual increase in A when u ranges between 80 and 85%. However, there is a dramatic increase in A when u approaches 95%. Therefore most treatment designers set u between 80 and 95%. The rule of thumb when designing a treatment is to increase u as much as possible until the increase in A is too great. This means that the rate of increase of A with respect to u, which is measured by the derivative, $A'(u)$, must not be greater than a specific value, usually set at $\frac{2}{3}$ or 0.67. That is, u can be increased as long as $A'(u)$ is less than 0.67.

A specific case   To illustrate their methods, Sonderman and Abrahamson present a specific case of a patient suffering from lung cancer. After studying the patient's anatomy, they used their linear programming techniques to devise a treatment based on maximizing u and minimizing A, subject to various constraints, including the direction and intensity of the beams, the location of the tumor, and the ability of the patient's anatomy to withstand a certain level of radiation. The final step is to determine the proper value of u. Their model

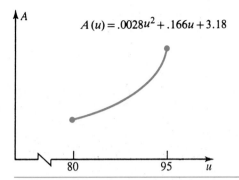

$$A(u) = .0028u^2 + .166u + 3.18$$

FIGURE 1 *Source:* David Sonderman and Philip Abrahamson, *Operations Research*, Vol. 33, No. 4, 1985, pp. 705–725.

produced the relationship between A and u, graphed in Figure 1, where the domain of A is [80, 100]; that is, u ranges from 80 to 100%. The graph can be approximated in the interval [80, 95] by the function

$$A(u) = 0.0028u^2 + 0.166u + 3.18$$

Figure 1 shows that there is a gradual increase in A when u increases from 80 to 85% but when u is larger than 95% A shows a dramatic rise. This case is typical. To find the value of u for the treatment, first find the derivative of A.

$$A'(u) = 0.0056u + 0.166$$

To demonstrate the behavior of the rate of change of A, we calculate some values of A': $A'(80) = 0.614$, $A'(85) = 0.642$, $A'(90) = 0.670$, and $A'(95) = 0.698$. The rule of thumb shows that the designer should set u at a value of about 90%. At this level of homogeneity the total integral dose of radiation to healthy cells is $A(90) = 40.8$.

The article closes with a comparison of the standard practice method of design treatment versus the mathematical optimization method. The standard practice design would either have yielded a smaller value of u than 90% to produce the same value of A, or, if the same value of u was used, the value of A would have been 49.6. Thus the mathematical model produced a savings of about 22% of A for the same homogeneity.

Case Study Exercises

Suppose a particular patient's profile yielded the following function of A with respect to u. Find the value of u such that $A' = \frac{2}{3}$.

1. $A(u) = 0.003u^2 + 0.18u + 3.5$

2. $A(u) = 0.003u^2 + 0.15u + 3.1$

3. $A(u) = 0.002u^2 + 0.33u + 2.5$

4. $A(u) = 0.002u^2 + 0.302u + 1.1$

CHAPTER REVIEW

Key Terms

3–1 A Few Elementary Rules

Marginal Product   Derivative of a Sum

Derivative of a Constant   Derivative of a Constant Times a Function

3–2 Product and Quotient Rules

Product Rule   Lerner Index

Quotient Rule

3–3 The Chain Rule

General Power Function   Chain Rule

Generalized Power Rule

3–4 Higher Derivatives

Second Derivative   Higher Derivatives

Third Derivative

Summary of Important Concepts

Formulas

$D_x(c) = 0$ for any constant c

$D_x(x^r) = rx^{r-1}$

$D_x[cf(x)] = cD_x[f(x)]$ for any constant c

$D_x[f(x) + g(x)] = D_x[f(x)] + D_x[g(x)]$

If $f(x) = g(x)k(x)$ then
$f'(x) = g(x)k'(x) + k(x)g'(x)$

If $f(x) = \dfrac{g(x)}{k(x)}$ then

$$f'(x) = \frac{k(x)g'(x) - g(x)k'(x)}{[k(x)]^2}$$

$D_x[g(x)]^n = ng(x)^{n-1} D_x[g(x)]$

REVIEW PROBLEMS

In Problems 1 to 4 find $f'(x)$.

1. $f(x) = 4x^2 + x^3 + 1$

2. $f(x) = 5x^4 - 2x^{-2}$

3. $f(x) = 4x^{-5} - x^{1/3} + 1$

4. $f(x) = \dfrac{5}{x^2} - \dfrac{2}{x^{1/2}}$

5. Find the equation of the tangent line to
$f(x) = x^{-3/2} - x$ at $x = 1$

In Problems 6 to 16 find $f'(x)$.

6. $f(x) = (x^4 - 7)(11x^3 + 1)$

7. $f(x) = \dfrac{x^3}{(x^2 + 2)}$

8. $f(x) = \dfrac{(x^3 - 2)}{(x^2 + 2x)}$

9. $f(x) = \dfrac{x^{1/3}}{(x^{4/5} + 2)}$

10. $f(x) = (x^3 + 2)^3$

11. $f(x) = (2x^3 + x)^4$

12. $f(x) = (2 - 3x^{-1})^{-2}$

13. $f(x) = 3x^2 + 3(x^3 + 4x)^3$

14. $f(x) = x^2(x^3 + 2x)^3$

15. $f(x) = (x^2 + 1)^2(x^3 + 1)^3$

16. $f(x) = \dfrac{x^2}{(x^2 + 3x)^2}$

17. Find the equation of the tangent line of
$f(x) = (x^3 + 3)^2$ at $x = 1$.

In Problems 18 and 19 find $f''(x)$.

18. $f(x) = 6x^{1/2} - 3x^{-3} - 6x$

19. $f(x) = (x^3 + 4)^{-2}$

20. Find the values of x for which $f''(x) = 0$ where
$f(x) = x^3 - 4x^2$

21. Find $f''(x)$ for $f(x) = x^{-3}$.

4 Applications of the Derivative

CHAPTER OVERVIEW

When a real-world application uses rate of change, the natural tool used is the derivative. This chapter shows how the derivative measures various properties about functions that are used in applications. These include rate of inflation, maximum profit, the most efficient route, minimizing cost, and minimizing traffic noise.

The first section deals with the effect the derivative has in determining where a function is increasing and where it is decreasing. Section 4–2 covers relative extrema, that is, the point on a graph that is higher or lower than any other point on the graph in the vicinity. The derivative is used to determine where a function has a relative maximum and where it has a relative minimum. The second derivative also gives useful information about the function. It helps to determine where a function is concave up or down, which means, intuitively, where the function is bending up or bending down.

The next section applies the ideas on relative extrema to problems where an absolute extremum is sought. Section 4–5 deals with problems in which two variables are changing. These are called related rates problems and they occur often in applications.

CASE STUDY PREVIEW

One of the most important considerations of highway operation is safety. Obtaining optimum flow of traffic is also desired, but frequently these two goals work at odds with each other. The case study tells the story of how highway administrators in Houston used calculus to minimize acceleration noise, which is defined as the disturbance of a vehicle's speed from uniform speed. Acceleration noise is a good measure of the threat of hazardous conditions. So minimizing acceleration noise produces safer traffic conditions. Researchers found that the optimum speed for minimizing acceleration noise is very close to the speed that maximizes traffic volume.

4–1 INCREASING AND DECREASING FUNCTIONS

One of the most important applications of calculus centers on when a function is increasing and when it is decreasing. A firm wants to know when profit is increasing and when cost is decreasing. A political party studies the latest polls to find out if its percentage of the vote is increasing or decreasing. In this section we study when graphs of functions are increasing and decreasing. The definition of increasing and decreasing functions has a firm geometrical foundation. However, it is also necessary to have an algebraic means for determining where a function is increasing and decreasing. The primary tool is the derivative of the function.

An Application: Laffer's Curve

Let us start by studying a concept from business that depends on when a function is increasing or decreasing.

Supply-side economics states that governments should stimulate production by giving incentives to business in order to increase economic growth. Other economists believe the best way to increase economic growth is to increase demand by giving incentives to consumers in the way of tax benefits. At the heart of the matter is the concept of *marginal tax rates,* the taxes paid on the last increment of income. Should the government increase or decrease marginal tax rates? Economist Arthur Laffer answers the question with a graph, called the *Laffer curve.** It measures a tax revenue function, $T(x)$, where x is the marginal tax rate measured in percent. A typical Laffer curve is given in Figure 1, where $T(x) = 100x - x^2$. For this curve, tax revenue is highest when $x = 50\%$. From $x = 0$ to $x = 50$, $T(x)$ is increasing. This means that if the marginal tax rate were at a value from 0 to 50%, an increase in x that is not beyond 50% would generate larger tax revenue. But if x is greater than 50%, an increase in the marginal tax rate would generate less tax revenue. It is this concept of when a function is increasing and decreasing that we study in this section.

Increasing and Decreasing Functions

Consider the graph in Figure 2. Intuitively, the graph is increasing from point A to point B because it goes up as x increases from left to right. The curve is decreasing

*Milton H. Spenser, *Contemporary Economics,* 6th ed., New York, Worth Publishers, 1986, p. 293.

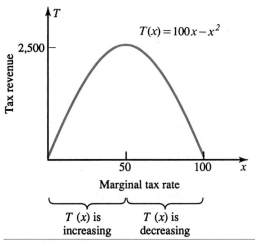

FIGURE 1 *Source:* Milton H. Spenser, *Contemporary Economics,* 6th ed., New York, Worth Publishers, 1986, p. 293.

from B to C because it goes down from left to right. Thus the property of an increasing or decreasing function implies the convention that x always moves from left to right. From the point of view of the x-axis, the curve in Figure 2 is increasing from $x = a$ to $x = b$ and decreasing from $x = b$ to $x = c$.

_____ Definition _____

A function $f(x)$ is said to be **increasing** from $x = a$ to $x = b$ if for any x_1 and x_2 in the interval (a, b)

$$\text{when } x_1 < x_2, \quad \text{then } f(x_1) < f(x_2)$$

A function $f(x)$ is **decreasing** from $x = a$ to $x = b$ if for any x_1 and x_2 in (a, b)

$$\text{when } x_1 < x_2, \quad \text{then } f(x_1) > f(x_2)$$

FIGURE 2

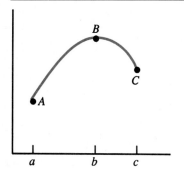

A function $f(x)$ is said to be *increasing at* $x = a$ if $f(x)$ is increasing in an interval containing a, and $f(x)$ is said to be *decreasing at* $x = a$ if $f(x)$ is decreasing in an interval containing a.

EXAMPLE 1

Problem

Consider $f(x) = x^2$ whose graph is in Figure 3. Find the intervals where $f(x)$ is increasing and where $f(x)$ is decreasing.

SOLUTION Because $f(x)$ is going down from left to right for all $x < 0$, f is decreasing in the interval $(-\infty, 0)$. For any x_1 and x_2 in $(-\infty, 0)$, if $x_1 < x_2$, then $f(x_1) > f(x_2)$ because $x_1{}^2 > x_2{}^2$. For example, $-2 < -1$ and $f(-2) = (-2)^2 = 4 > 1 = f(-1)$. Similarly $f(x)$ is increasing in $(0, \infty)$.

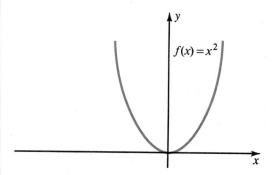

FIGURE 3

A Test for Increasing and Decreasing Functions

The primary algebraic tool for determining the intervals where a function is increasing and decreasing is the derivative. For example, if $f(x) = x^2$, then $f'(x) = 2x$. This says that the slope of the tangent line of $f(x)$ is negative for all $x < 0$ and the slope of the tangent line is positive for all $x > 0$. Coupling this fact with Example 1 means that the function $f(x) = x^2$ is increasing when $f'(x) > 0$ and decreasing when $f'(x) < 0$. This is true in general. If the derivative is positive for every value in an interval, then the function is increasing in that interval. A negative derivative means the function is decreasing.

Test for Increasing and Decreasing Functions

If $f(x)$ is differentiable in an interval (a, b), then

1. $f(x)$ is increasing in (a, b) if $f'(x) > 0$ for all $x \in (a, b)$
2. $f(x)$ is decreasing in (a, b) if $f'(x) < 0$ for all $x \in (a, b)$

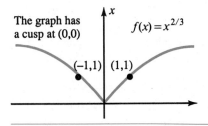

The graph has a cusp at (0,0)

$f(x) = x^{2/3}$

(–1,1) (1,1)

FIGURE 4

If $f'(x) = 0$ at $x = a$, then the slope of the tangent line of $y = f(x)$ at $x = a$ is 0. The tangent line is horizontal at such a point. These are the points that often divide the intervals where $f(x)$ changes from increasing to decreasing.

The function's derivative can also change sign at points where $f'(x)$ does not exist. For example, the derivative of $f(x) = x^{2/3}$ is $f'(x) = \frac{2}{3}x^{-1/3}$, and so $f'(0)$ does not exist. Since $f'(x) < 0$ for all x in $(-\infty, 0)$, $f(x)$ is decreasing to the left of $x = 0$; and since $f'(x) > 0$ in $(0, \infty)$, $f(x)$ is increasing to the right of $x = 0$. Therefore, at $x = 0$, the function's derivative changes sign and the graph changes from decreasing to increasing. The graph of $f(x) = x^{2/3}$ is given in Figure 4. The graph has a "cusp" at $x = 0$.

This suggests how to determine the intervals where $f(x)$ is increasing and decreasing. First find the values $x = a$ where $f'(a) = 0$ or $f'(a)$ does not exist (but $f(a)$ does exist). These values of x are called the **critical values** of $f(x)$. The critical values divide the domain of the function into distinct intervals where $f(x)$ is either increasing or decreasing. To determine the sign of $f'(x)$ in an interval, it is enough to select one value in the interval, say, $x = c$, and then to find $f'(c)$. If $f'(c) > 0$, then $f'(x)$ is positive for all values in the interval. Thus $f(x)$ is increasing in that interval. We treat $f'(c) < 0$ in a similar manner.

Procedure for Determining Where $f(x)$ Is Increasing and Decreasing

1. Find the critical values of $f(x)$, that is, the values $x = a$ where $f'(a) = 0$ or $f'(a)$ does not exist (but $f(a)$ does exist).
2. Determine the distinct intervals where $f(x)$ is either increasing or decreasing using the values in step 1 as end points.
3. For each interval select a value c in the interval and find $f'(c)$.
   i. If $f'(c) > 0$, then $f'(x) > 0$ for all x in that interval.
   ii. If $f'(c) < 0$, then $f'(x) < 0$ for all x in that interval.
4. Apply the test for increasing and decreasing functions.

EXAMPLE 2 | **Problem**

Find the intervals where $f(x) = x^2 - 4x + 3$ is increasing and decreasing.

SOLUTION Let us follow the steps in the procedure.

1. Compute $f'(x) = 2x - 4$. Solve $f'(x) = 2x - 4 = 0$. This implies $x = 2$. Thus $f'(2) = 0$. There are no values where $f'(x)$ does not exist.
2. This divides the real line into $(-\infty, 2)$ and $(2, \infty)$.
3. Select $0 \in (-\infty, 2)$ and $3 \in (2, \infty)$. (Remember, any suitable values can be chosen.)
    a. for $x = 0$, $f'(0) = -4$, so $f'(x) < 0$ in $(-\infty, 2)$
    b. for $x = 3$, $f'(3) = 2$, so $f'(x) > 0$ in $(2, \infty)$
4. a. $f(x)$ is decreasing in $(-\infty, 2)$ because $f'(x) < 0$ in $(-\infty, 2)$.
    b. $f(x)$ is increasing in $(2, \infty)$ because $f'(x) > 0$ in $(2, \infty)$.

This information is recorded in the chart below.

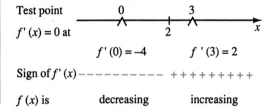

The graph of $f(x) = x^2 - 4x + 3$ is given in Figure 5. Example 2 shows where $f(x)$ is increasing and decreasing. Also note that there is a horizontal tangent line at $x = 2$ because $f'(2) = 0$. From these facts the graph can be obtained by plotting just a few points. This demonstrates that knowing where a function is increasing and decreasing is a useful tool in graphing the function. In the next few sections we discuss other properties of the graphs of functions.
    Example 3 looks at a more complicated function.

FIGURE 5

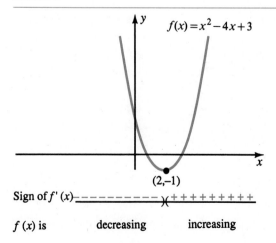

EXAMPLE 3

Problem

Find the intervals where $f(x) = x^3 - 6x^2 + 9x$ is increasing and decreasing. Locate the points where the graph has a horizontal tangent line. Graph the function.

SOLUTION

1. Solve $f'(x) = 0$ where

   $$f'(x) = 3x^2 - 12x + 9 = 3(x - 1)(x - 3)$$

   Then $3(x - 1)(x - 3) = 0$ implies $x = 1$ or $x = 3$. Therefore $f'(1) = f'(3) = 0$. There are no values where $f'(x)$ does not exist.
2. This divides the real line into the three intervals $(-\infty, 1)$, $(1, 3)$, and $(3, \infty)$.
3. Select $0 \in (-\infty, 1)$, $2 \in (1, 3)$, and $4 \in (3, \infty)$.
   a. for $x = 0$, $f'(0) = 9$, so $f'(x) > 0$ in $(-\infty, 1)$
   b. for $x = 2$, $f'(2) = -3$, so $f'(x) < 0$ in $(1, 3)$
   c. for $x = 4$, $f'(4) = 9$, so $f'(x) > 0$ in $(3, \infty)$
4. a. $f(x)$ is increasing in $(-\infty, 1)$ because $f'(x) > 0$ in $(-\infty, 1)$.
   b. $f(x)$ is decreasing in $(1, 3)$ because $f'(x) < 0$ in $(1, 3)$.
   c. $f(x)$ is increasing in $(3, \infty)$ because $f'(x) > 0$ in $(3, \infty)$.

$f(x)$ has a horizontal tangent line when $f'(x) = 0$. Since $f'(1) = f'(3) = 0$, $f(x)$ has a horizontal tangent line at $x = 1, 3$, that is, at the points $(1, 4)$ and $(3, 0)$.

This information is recorded in the following chart.

Test point	0		2		4	
$f'(x) = 0$ at		1		3		x
	$f'(0) = 9$		$f'(2) = -3$		$f'(4) = 9$	
Sign of $f'(x)$	$+ + + + + + + + +$		$- - - - - - - - -$		$+ + + + + + + + +$	
$f(x)$ is	increasing		decreasing		increasing	

The graph is in Figure 6.

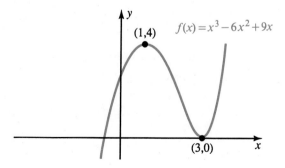

$$f(x) = x^3 - 6x^2 + 9x$$

(1,4)

(3,0)

Sign of $f'(x)$	$+ + + + + + + +$	$- - - - - -$	$+ + + + + +$	
$f(x)$ is	increasing	decreasing	increasing	**FIGURE 6**

EXAMPLE 4 | **Problem**

Let $f(x) = x + \dfrac{4}{x}$. Find the intervals where $f(x)$ is increasing and where $f(x)$ is decreasing. Locate the points where the graph has a horizontal tangent line. Graph the function.

SOLUTION

1. Set $f'(x) = 1 - 4/x^2$ equal to 0. Then $1 = 4/x^2$, so $x^2 = 4$. Hence $x = \pm 2$ are critical values. Also, $f'(x)$ does not exist at $x = 0$, but $x = 0$ is not a critical value because $f(0)$ does not exist.
2. This divides the real line into the four intervals $(-\infty, -2)$, $(-2, 0)$, $(0, 2)$, and $(2, \infty)$.
3. Select $-3 \in (-\infty, -2)$, $-1 \in (-2, 0)$, $1 \in (0, 2)$, and $3 \in (2, \infty)$.
   a. for $x = -3$, $f'(-3) = \frac{5}{9}$, so $f'(x) > 0$ in $(-\infty, -2)$
   b. for $x = -1$, $f'(-1) = -3$, so $f'(x) < 0$ in $(-2, 0)$
   c. for $x = 1$, $f'(1) = -3$, so $f'(x) < 0$ in $(0, 2)$
   d. for $x = 3$, $f'(3) = \frac{5}{9}$, so $f'(x) > 0$ in $(2, \infty)$
4. a. $f(x)$ is increasing in $(-\infty, -2)$ because $f'(x) > 0$ in $(-\infty, -2)$
   b. $f(x)$ is decreasing in $(-2, 0)$ because $f'(x) < 0$ in $(-2, 0)$
   c. $f(x)$ is decreasing in $(0, 2)$ because $f'(x) < 0$ in $(0, 2)$
   d. $f(x)$ is increasing in $(2, \infty)$ because $f'(x) > 0$ in $(2, \infty)$

$f(x)$ has a horizontal tangent line when $f'(x) = 0$. Since $f'(-2) = f'(2) = 0$, $f(x)$ has a horizontal tangent line at $x = -2$ or $x = 2$; that is, at the points $(-2, -4)$ and $(2, 4)$. The graph approaches the y-axis asymptotically both to the left and right of 0. This information is recorded in the chart below.

The graph is in Figure 7.

An Application from Economics

The revenue of a firm depends on many factors, one of which is the quantity q of goods produced. A typical revenue function is dome-shaped as in Figure 8. When q is 0, that is, no goods are produced, the revenue $R(q)$ is 0. As q increases, so does $R(q)$ until it reaches a peak where maximum revenue occurs. In Figure 8 this occurs at $q = 5$. Then the values of $R(q)$ decrease for values of q greater than 5. There are several reasons that this decrease in $R(q)$ takes place. One reason is that the firm might not be able to sell such a large amount of goods. The next example computes the intervals where the revenue function in Figure 1 is increasing and decreasing. This function is defined only in the interval $(0, 10)$.

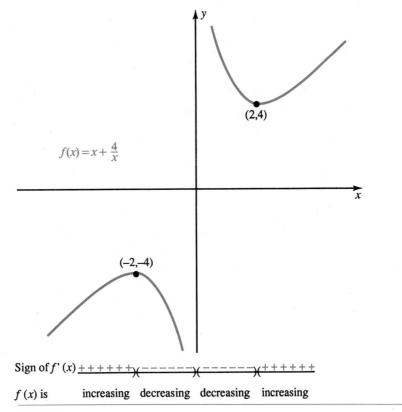

$$f(x) = x + \frac{4}{x}$$

(2,4)

(−2,−4)

Sign of $f'(x)$ ++++++ −−−−−− −−−−−− ++++++

$f(x)$ is   increasing decreasing decreasing increasing

FIGURE 7

FIGURE 8 *Source:* Based on Paul A. Samuelson, *Economics*, 10th ed., New York, McGraw-Hill Book Co., 1985, p. 479.

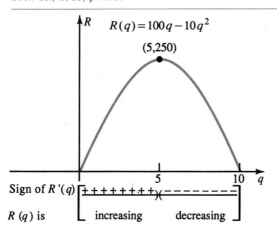

$$R(q) = 100q - 10q^2$$

(5,250)

Sign of $R'(q)$ +++++++++ −−−−−−−−−

$R(q)$ is   increasing   decreasing

EXAMPLE 5 | **Problem**

Determine the intervals where $R(q)$ is increasing and decreasing.

$$R(q) = 100q - 10q^2, \qquad q \geq 0$$

SOLUTION Compute $R'(q)$.

$$R'(q) = 100 - 20q$$

1. Solve $R'(q) = 0$. Then $100 - 20q = 0$ implies $q = 5$. There are no values where $R'(q)$ does not exist.
2. This divides the domain into $[0, 5)$ and $(5, 10]$.
3. Select $1 \in [0, 5)$ and $6 \in (5, 10]$.
   a. for $q = 1$, $R'(1) = 80$, so $R'(q) > 0$ in $[0, 5)$.
   b. for $q = 6$, $R'(6) = -20$, so $R'(q) < 0$ in $(5, 10]$.
4. a. $R(q)$ is increasing in $[0, 5)$ because $R'(q) > 0$ in $[0, 5)$.
   b. $R(q)$ is decreasing in $(5, 10]$ because $R'(q) < 0$ in $(5, 10]$.

EXERCISE SET 4–1

● In Problems 1 to 28 find intervals where $f(x)$ is increasing and decreasing. Find the point(s) where $f(x)$ has a horizontal tangent line. Graph the function.

1. $f(x) = x^2 + 5$

2. $f(x) = 3x^2 - 2$

3. $f(x) = 7 - 2x^2$

4. $f(x) = 1 - 3x^2$

5. $f(x) = x^2 - 9$

6. $f(x) = 4x^2 - 25$

7. $f(x) = x(x - 4)$

8. $f(x) = x(x - 8)$

9. $f(x) = x^2 + 4x - 1$

10. $f(x) = x^2 - 8x + 3$

11. $f(x) = x^3 + 3x^2 - 9x$

12. $f(x) = 2x^3 + 3x^2 - 12x + 1$

13. $f(x) = x^3 - 3x^2 - 9x$

14. $f(x) = 2x^3 + 3x^2 - 36x + 3$

15. $f(x) = -x^3 + 3x^2 + 9x$

16. $f(x) = -x^3 + 6x^2$

17. $f(x) = x + 9/x$

18. $f(x) = 4x + 25/x$

19. $f(x) = (x^2 + 12)/x$

20. $f(x) = (4x^2 + 9)/x$

●● 21. $f(x) = x(x^2 - 4)$

22. $f(x) = x(9 - x^2)$

23. $f(x) = x^2(x - 2)$

24. $f(x) = x^2(3 - x)$

25. $f(x) = x/(x + 1)$

26. $f(x) = x/(x - 4)$

27. $f(x) = (x - 1)/(x + 1)$

28. $f(x) = (x + 3)/(x - 5)$

29. The average cost $AC(x)$ to produce the xth item of production, where $AC(x)$ is measured in dollars and x is measured in hundreds of items, is given by

$$AC(x) = 100 + 48x - 7x^2 + \frac{x^3}{3}$$

If $AC(x)$ is defined for $0 < x < 10$, find the intervals where $AC(x)$ is increasing and where it is decreasing.

30. The amount of glucose $G(t)$ in a patient's bloodstream, measured in tenths of a percent, is measured at time t in minutes during a stress test. The function is

$$G(t) = 12t - 2.1t^2 + 0.1t^3,$$
where $0 < t < 12$

Find the intervals where t is increasing and decreasing.

In Problems 31 to 34 use the information given to make a sketch of the graph of $f(x)$, assuming that $f(x)$ is continuous for all real numbers.

31. Values where $f'(x) = 0$

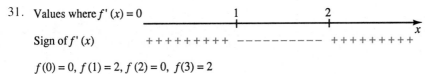

Sign of $f'(x)$

$f(0) = 0$, $f(1) = 2$, $f(2) = 0$, $f(3) = 2$

32. Values where $f'(x) = 0$

Sign of $f'(x)$

$f(-2) = 1$, $f(-1) = 0$, $f(3) = 1$, $f(4) = 0$

33. Values where $f'(x) = 0$

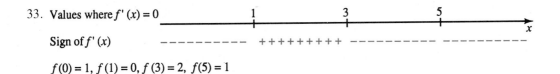

Sign of $f'(x)$

$f(0) = 1$, $f(1) = 0$, $f(3) = 2$, $f(5) = 1$

34. Values where $f'(x) = 0$

Sign of $f'(x)$

$f(0) = -1$, $f(3) = 2$, $f(6) = 4$,

••• In Problems 35 to 42 find the intervals where $f(x)$ is increasing and where $f(x)$ is decreasing. Graph the functions.

35. $f(x) = x^2/(x - 2)$
36. $f(x) = x^2/(1 - 2x)$
37. $f(x) = x^2(x^2 - 4)$
38. $f(x) = x^2(9 - x^2)$
39. $f(x) = (x^2 - 1)(x^2 - 4)$
40. $f(x) = (x^2 - 3)(9 - x^2)$
41. $f(x) = (x - 5)^2(x + 1)^2$
42. $f(x) = (x + 6)^2(x - 4)^2$

43. In a paper studying the Environmental Protection Agency's motor vehicle emission testing procedures, researchers measured the risk, r, of an incorrect decision on a part versus the percentage, x, of actual deterioration of the part.* That is, laboratory standards were developed for the deterioration level of each part of the emis-sion system. When a part reached the deterioration level where it should fail the EPA test, x was assigned the value 1. For $x < 1$, the part should pass, and it should fail if $x > 1$. (The units of the values assigned to r were arbitrary, but the higher the value the greater the risk, meaning the greater the error.) Field agents were then given the parts to test. There were two types of failure: the part should be passed and the agent failed it, and the part should fail and the agent passed it. The graph in Figure 9 was given measuring the risk r, versus the deterioration x. Explain why the curve is decreasing for $x < 1$ and increasing for $x > 1$.

44. In the theory of traffic science a "bottleneck" is defined as a section of roadway with a flow capacity less than the road ahead. When the volume reaches the capacity of the bottleneck,

*George Miller et al., "Application of Sequential Testing to Motor Vehicle Emission Certification," *Decision Sciences*, Vol. 31, 1985, pp. 249–263.

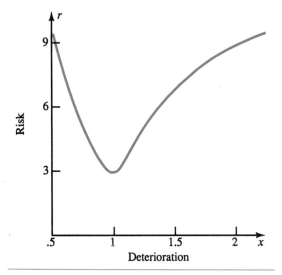

FIGURE 9   *Source:* George Miller et al., "Application of Sequential Testing to Motor Vehicle Emission Certification," *Decision Sciences,* Vol. 31, 1985, pp. 249–263.

the velocity in the bottleneck is much less than that ahead of the bottleneck. A further increase in volume creates a queue in back of the bottleneck. In an article describing why congestion lasts longer than the interval in which demand exceeds capacity, a graph measuring the volume during rush-hour traffic on a large city freeway is studied.[†] The data are approximated by the function $f(t) = -25t^3 + 75t + 150$, where $f(t)$ is the volume of automobiles passing through the bottleneck at time t, and where $t = 0$ represents 7:00 A.M. and $t = 1$ represents 7:30

A.M. The function is defined in the interval [0, 2], that is, from 7:00 A.M. until 8:00 A.M.

(a) Find where $f(t)$ is increasing and where $f(t)$ is decreasing.
(b) The capacity of the bottleneck is $f(t) = 170$. From the graph estimate when demand exceeds capacity.

45. In an article[‡] appearing in the *Journal of Quality Progress* two administrators in the U.S. Department of Agriculture defined "tolerance" as any limit or allowance that is specified by a contract to measure quality. One example referred to the common practice of produce shippers, such as egg packers, of specifying that "no more than 20% of the eggs may be below A quality." The primary question addressed in the article is when should a shipment with this specification be rejected? The function used is $y = f(x)$, where f measures the percent of times the shipment was accepted with $x\%$ of the eggs below A quality. The function is approximated by

$$f(x) = \frac{x^3}{135} - \frac{x^2}{3} + 100$$

(a) Find where is $f(x)$ is increasing and decreasing.
(b) From the definition of $f(x)$, why is it reasonable to assume $f(x)$ is decreasing there?
(c) Find $f(20)$.
(d) This function models the data only in the interval [0, 30]. Give a reason for this.

[†]Karl Moskowitz and Leonard Newman, "Notes on Freeway Capacity," *Highway Resources Board Record,* Vol. 27, 1963.

[‡]Richard P. Bartlett and Lloyd P. Provost, "Tolerances in Standards and Specifications," *Journal of Quality Progress,* 1973, pp. 14–18.

4–2 RELATIVE EXTREMA: THE FIRST DERIVATIVE TEST

A common type of rate of change problem requires that a variable be maximized or minimized. A business wants to maximize profit; a physician needs to know the maximum amount of a drug that can be administered safely; a physiologist studies the minimum rate of stimulus needed to achieve a given response from an organism. The derivative plays a fundamental role in the solution of such problems.

Relative Extrema

Consider the graph in Figure 1, which we use to describe the geometric meaning of extrema. Point A is the highest point on the curve because its second coordinate, $f(a)$, is greater than all others. We call $f(a)$ an *absolute maximum*. Similarly B is the lowest point of the graph so $f(b)$ is called an *absolute minimum*. Point C is not the absolute highest point, but it is higher than any of the points in its vicinity. For this reason $f(c)$ is called a *relative maximum*. Similarly $f(d)$ is a *relative minimum*. The term **extremum** refers to either a **maximum** or **minimum.** We also say that $f(x)$ has a relative maximum at $x = c$ and $f(x)$ has a relative minimum at $x = d$. This section deals with relative extrema and the next section considers absolute extrema.

──────────────── **The Definition of Relative Extrema** ────────────────

1. $f(a)$ is a **relative maximum** of $f(x)$ if there is an interval (c, d) containing a such that

   $$f(x) \le f(a) \qquad \text{for all } x \text{ in } (c, d)$$

2. $f(a)$ is a **relative minimum** of $f(x)$ if there is an interval (c, d) containing a such that

   $$f(x) \ge f(a) \qquad \text{for all } x \text{ in } (c, d)$$

The first example ties together the definition and the geometric meaning of extrema. It considers a function that was graphed in Example 3 of the previous section.

FIGURE 1

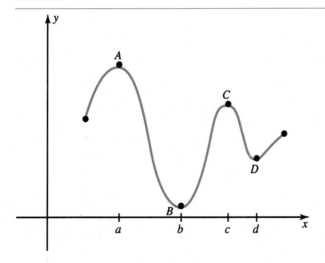

EXAMPLE 1 | **Problem**

Find the extrema of $f(x) = x^3 - 6x^2 + 9x$ by looking at the graph in Figure 2. Then show that the definition is satisfied for each extremum.

SOLUTION The graph has a "peak" at $(1, 4)$ and a "valley" at $(3, 0)$. Thus the graph has a relative maximum of $f(1) = 4$ when $x = 1$ and a relative minimum of $f(3) = 0$ when $x = 3$. To see that $f(1)$ satisfies the definition of a relative maximum, we need an interval containing 1 such that the functional values of all other points in the interval are less than $f(1) = 4$. From Figure 2 it is clear that $f(x) < 4$ for all $x < 3$, so the interval can be chosen to be $(-\infty, 3)$. Any smaller interval containing 1 will also work. (Can a larger interval be selected?) Similarly the largest interval for the relative minimum $f(3) = 0$ is $(0, \infty)$ because the $f(x) > 0$ for all $x > 0$ except $x = 3$.

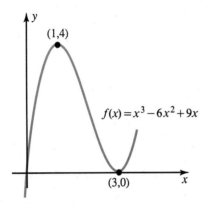

$f(x) = x^3 - 6x^2 + 9x$

FIGURE 2

At a relative extremum a curve either has a horizontal tangent line or the derivative does not exist. For instance, in Example 1, $f(x) = x^3 - 6x^2 + 9x$ has a horizontal tangent line at $x = 1$ and $x = 3$. The graph in Figure 2 has a horizontal tangent line at $(1, 4)$ and $(3, 0)$. In the previous section it was pointed out that the graph of a function has a horizontal tangent line when $f'(x) = 0$. In Example 2, $f'(x) = 3x^2 - 12x + 9 = 3(x - 1)(x - 3)$. By solving $f'(x) = 0$, we get the values $x = 1, 3$ where the graph has a horizontal tangent line. This shows that a relative extremum can occur at a value of $f(x)$ where $f'(x) = 0$.

The other type of value $x = a$ that sometimes yields a relative extremum is where $f'(a)$ does not exist but $f(a)$ does exist. Consider the function $f(x) = x^{2/3}$ whose graph is shown in Figure 3. From the graph we see a relative minimum at $x = 0$, and $f'(x) = \frac{2}{3}x^{-1/3}$ does not exist at $x = 0$. Thus a relative extremum can also occur at a value where $f'(x)$ does not exist.

Recall that the critical values of $f(x)$ are the values $x = a$ in the domain of $f(x)$ where $f'(a) = 0$ or where $f'(a)$ does not exist. This shows that to locate relative extrema, we need to find the critical values of a function.

But a word of caution is in order. It is not true that a relative extremum occurs

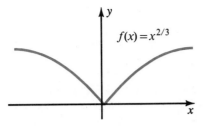

FIGURE 3

at every critical value. Consider the function $f(x) = x^3$ whose graph is in Figure 4. To find the critical value(s), set $f'(x) = 3x^2$ equal to 0. This yields the only critical value $x = 0$, since the derivative exists at all values of x. The graph shows that the function has no relative extremum at $x = 0$; in fact the graph has no relative extremum anywhere.

Let us summarize these results.

> A **relative extremum** of the function $f(x)$ will occur at a **critical value,** that is, a value $x = a$ in the domain such that $f'(a) = 0$ or $f'(a)$ does not exist. Not all critical values yield relative extrema.

The First Derivative Test

The relative maxima and minima of $f(x)$ are computed by finding the critical values and then applying a test to see whether the extremum is (1) a relative maximum, (2) a relative minimum, or (3) neither a maximum nor a minimum. This test involves the derivative of $f(x)$ and uses the properties of $f(x)$ studied in the previous section.

Consider again $f(x) = x^3 - 6x^2 + 9x$. As shown in Figure 2 the graph has a relative maximum at $x = 1$ and a relative minimum at $x = 3$. Also in Figure 2,

FIGURE 4

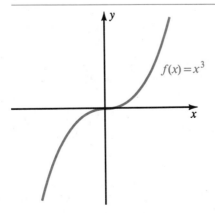

note that the graph is increasing to the left of $x = 1$ and decreasing to the right of $x = 1$. This is why the critical value $x = 1$ is a relative maximum. Similarly the critical value $x = 3$ is a relative minimum because the graph is decreasing to the left and increasing to the right of $x = 3$.

Recall from the previous section that the intervals where the function is increasing and decreasing are found by finding where the derivative is greater than and less than 0. This applies to all functions. It is called the **first derivative test.**

The First Derivative Test

To find the relative extrema of $f(x)$
1. Find the critical values of $f(x)$ and the intervals where $f(x)$ is increasing and where $f(x)$ is decreasing.
2. If $x = a$ is a critical value of $f(x)$, then
   i. $f(a)$ is a relative maximum if $f(x)$ is increasing to the left of a and decreasing to the right of a
   ii. $f(a)$ is a relative minimum if $f(x)$ is decreasing to the left of a and increasing to the right of a
   iii. $f(a)$ is not a relative extremum if $f(x)$ is
       (a) decreasing to the left and to the right of a
       (b) increasing to the left and to the right of a

This means that for a critical value $x = a$, $f(a)$ is a relative maximum if $f'(x)$ changes from positive to the left of $x = a$ to negative to the right of $x = a$; $f(a)$ is a relative minimum if $f'(x)$ changes from negative to the left of $x = a$ to positive to the right of $x = a$; $f(a)$ is not a relative extremum if $f'(x)$ does not change sign.

EXAMPLE 2 | **Problem**

Find the relative extrema of $f(x) = (x - 2)^2(x - 5)$. Sketch the graph of the function.

SOLUTION We follow the steps in the first derivative test.

1. Find $f'(x)$.

$$f'(x) = 2(x - 2)(x - 5) + (x - 2)^2$$
$$= (x - 2)(2x - 10 + x - 2)$$
$$= 3(x - 2)(x - 4)$$

Solve $f'(x) = 3(x - 2)(x - 4) = 0$. This yields $x = 2, 4$. These are the only critical values.
Solve $f'(x) > 0$ and $f'(x) < 0$. The intervals are $(-\infty, 2)$, $(2, 4)$, and $(4, \infty)$. Choose test points, say, 0, 3, and 5. Since $f'(0) = 24$, $f'(3) = -3$ and $f'(5) = 9$, $f(x)$ is increasing in $(-\infty, 2)$ and $(4, \infty)$, while $f(x)$ is decreasing in $(2, 4)$. Record these facts in a chart as follows.

Test point	0		3		5	

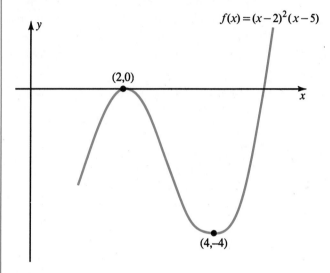

2. $f(2) = 0$ is a relative maximum because $f(x)$ is increasing to the left and decreasing to the right of $x = 2$.

$f(4) = -4$ is a relative minimum because $f(x)$ is decreasing to the left and increasing to the right of $x = 4$.

From this information and by plotting the points $(2, 0)$ and $(4, -4)$, we can sketch the graph as shown in Figure 5.

FIGURE 5

EXAMPLE 3

Problem

Find the relative extrema, and sketch the graph of the function

$$f(x) = x^4 - 8x^3 + 16x^2$$

SOLUTION

1. Find $f'(x)$.

$$f'(x) = 4x^3 - 24x^2 + 32x$$
$$= 4x(x^2 - 6x + 8)$$
$$= 4x(x - 2)(x - 4)$$

Solve $f'(x) = 0$. This yields $x = 0, 2, 4$. These are the only critical values.

Solve $f'(x) > 0$ and $f'(x) < 0$. The intervals are $(-\infty, 0)$, $(0, 2)$, $(2, 4)$, and $(4, \infty)$. Choose test points, say, -1, 1, 3, and 5. Evaluate $f'(x)$ at each value and record the results in the following chart.

Test point	-1		1		3		5	
Critical value		0		2		4		
	$f'(-1)$		$f'(1)$		$f'(3)$		$f'(5)$	
	$= -60$		$= 12$		$= -12$		$= 60$	

Sign of $f'(x)$ $-----------$ $+++++++++$ $----------$ $+++++++++$

f is   decreasing   increasing   decreasing   increasing

2. $f(0) = 0$ is a relative minimum because f is decreasing to the left and increasing to the right of $x = 0$.

   $f(2) = 16$ is a relative maximum because f is increasing to the left and decreasing to the right of $x = 2$.

   $f(4) = 0$ is a relative minimum because f is decreasing to the left and increasing to the right of $x = 4$.

From this information and by plotting the points $(0, 0)$, $(2, 16)$, and $(4, 0)$, we can sketch the graph as shown in Figure 6.

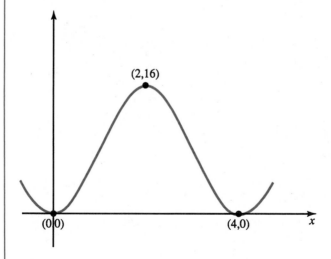

FIGURE 6

The next example shows how to identify when a function has a critical value that does not correspond to an extremum and when a function has an extremum at a point where the derivative does not exist.

EXAMPLE 4

Problem

Find the relative extrema, and sketch the graph of each function.
(a) $g(x) = 2 + x^3$, and (b) $f(x) = x - 3x^{1/3}$.

SOLUTION
(a) Compute

$$g'(x) = 3x^2$$

1. Solve $g'(x) = 0$. This yields $x = 0$. Thus the only critical value is $x = 0$. Find the intervals where $g(x)$ is increasing and decreasing by arranging the necessary information in the following chart.

2. $g(0) = 2$ is neither a relative maximum nor a relative minimum because $g(x)$ is increasing to the left and to the right of $x = 0$.

From this information and by plotting the point $(0, 2)$, we can sketch the graph as shown in Figure 7.

FIGURE 7

(b) Compute

$$f'(x) = 1 - x^{-2/3} = 1 - \frac{1}{x^{2/3}}$$

1. Solve $f'(x) = 0$. This yields $x = \pm 1$. Also, $f'(x)$ does not exist when $x = 0$, but $f(0) = 0$. Thus $x = 0$ is a critical value. The function may change from increasing to decreasing at $x = 0$, so it is included in the analysis. The critical values are $x = -1$, $x = 0$, and $x = 1$. Find the intervals where $f(x)$ is increasing and decreasing by arranging the necessary information in the following chart.

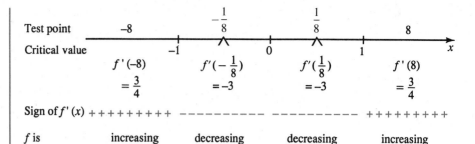

Sign of $f'(x)$ ++++++++++ ---------- ---------- ++++++++++

f is   increasing   decreasing   decreasing   increasing

2. $f(-1) = 2$ is a relative maximum because f is increasing to the left and decreasing to the right of $x = -1$.

   $f(0) = 0$ is neither a relative maximum nor a relative minimum because f is decreasing to the left and to the right of $x = 0$.

   $f(1) = -2$ is a relative minimum because f is decreasing to the left and increasing to the right of $x = 1$.

From this information and by plotting the points $(-1, 2)$, $(0, 0)$, and $(1, -2)$, we can sketch the graph as shown in Figure 8. The graph has a vertical tangent at $(0, 0)$, implying that the derivative does not exist at $x = 0$.

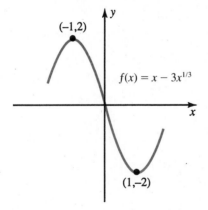

FIGURE 8

An Application: Maximizing Profit

The ultimate goal of most businesses is to maximize profit. A fundamental principle of economics is that a firm will maximize its profit when marginal revenue equals marginal cost.* The first derivative test explains why this is so. The profit $P(x)$ that a firm makes when it sells x units of goods is defined as the difference between the revenue $R(x)$ and cost $C(x)$ functions.

$$P(x) = R(x) - C(x)$$

*Paul A. Samuelson, *Economics,* New York, McGraw-Hill Book Co., 1986, p. 491.

To maximize $P(x)$, solve $P'(x) = 0$. From the definition of $P(x)$ we have

$$P'(x) = R'(x) - C'(x)$$

Solve $P'(x) = 0$. This yields $R'(x) - C'(x) = 0$. Rearranging the latter equation gives

$$R'(x) = C'(x)$$

This states that marginal revenue $R'(x)$ equals marginal cost $C'(x)$. That is, to maximize profit, find the value of x where marginal revenue equals marginal cost.

The next example is an economic model of a grocery chain.† The model uses the first derivative test to maximize the profit function of a particular product.

EXAMPLE 5

A firm determines that the revenue and cost functions for the volume of sales x for a particular product are

$$R(x) = 20x - 0.1x^2, \qquad x \geq 0$$
$$C(x) = 50 + 9x + 0.1x^2, \qquad x \geq 0$$

Problem

(a) Find the profit function $P(x)$ and the volume of sales x that maximizes profit.
(b) Show that the value of x that maximizes $P(x)$ in part (a) is also the solution of the equation $R'(x) = C'(x)$.

SOLUTION (a) From the definition of the profit function

$$\begin{aligned} P(x) &= R(x) - C(x) \\ &= 20x - 0.1x^2 - (50 + 9x + 0.1x^2) \\ &= -50 + 11x - 0.2x^2 \end{aligned}$$

Solve $P'(x) = 0$.

$$P'(x) = 11 - 0.4x = 0$$

$$x = \frac{11}{0.4} = 27.5$$

The first derivative test shows that $P(x)$ has a relative maximum of $P(27.5) = 101.25$. The graph of $P(x)$ is given in Figure 9.
(b) Solve marginal revenue equal to marginal cost.

$$\begin{aligned} R'(x) &= C'(x) \\ 20 - 0.2x &= 9 + 0.2x \\ -0.4x &= -11 \\ x &= 27.5 \end{aligned}$$

This states that maximum profit occurs at that volume of sales where marginal revenue equals marginal cost.

†Marilu Hurt McCarty, *Managerial Economics*, Glenview, IL, Scott, Foresman and Co., 1986, pp. 28–29.

FIGURE 9   *Source:* Marilu Hurt McCarty, *Managerial Economics,* Glenview, IL, Scott, Foresman and Co., 1986, pp. 28–29.

EXERCISE SET 4–2

● In Problems 1 to 30 find the relative extrema and sketch the graph of the functions.

1. $f(x) = x^2 - 4$

2. $f(x) = 2x^2 - 5$

3. $f(x) = 2 - 3x^2$

4. $f(x) = 3 - 4x^2$

5. $f(x) = x^2 - 4x$

6. $f(x) = 3x^2 - 6x$

7. $f(x) = x^2(x + 1)$

8. $f(x) = x^2(x - 3)$

9. $f(x) = x^3 - 6x^2 - 1$

10. $f(x) = x^3 - 9x^2 + 4$

11. $f(x) = x^3 + 3x^2 - 9x$

12. $f(x) = 2x^3 + 3x^2 - 12x + 1$

13. $f(x) = x^3 - 3x^2 - 9x$

14. $f(x) = 2x^3 + 3x^2 - 36x + 3$

15. $f(x) = x^4 - 2x^2$

16. $f(x) = -x^4 + 6x^2$

17. $f(x) = -x^4 + 8x^2$

18. $f(x) = -x^4 + 12x^2 + 1$

19. $f(x) = x + 9/x$

20. $f(x) = 4x + 25/x$

21. $f(x) = (x^2 + 12)/x$

22. $f(x) = (4x^2 + 9)/x$

●● 23. $f(x) = x^3(x - 4)$

24. $f(x) = x^3(8 - x)$

25. $f(x) = x^2(x^2 - 2)$

26. $f(x) = x^2(4 - x^2)$

27. $f(x) = x^2(x^2 + 2)$

28. $f(x) = x^2(4 + x^2)$

29. $f(x) = (x - 1)/(x + 1)$

30. $f(x) = (x + 3)/(x - 5)$

In Problems 31 to 34 find the relative extrema, and sketch the graph of the function from the given information. Assume $f(x)$ is continuous for all real numbers.

31 **Critical values**

Sign of $f'(x)$

$f(0) = 0,\ f(3) = 2$

32 **Critical values**

Sign of $f'(x)$

$f(-2) = 1,\ f(0) = 2$

33. **Critical values**

Sign of $f'(x)$

$f(-1) = 1,\ f(2) = 3,\ f(5) = 4$

34. **Critical values**

Sign of $f'(x)$

$f(2) = -1,\ f(3) = 2,\ f(4) = 4,$

35. The profit $P(x)$ in thousands of dollars for selling x television sets in hundreds of sets is

$$P(x) = x^2(x - 8)^2 \qquad \text{for } 0 < x < 6$$

Find the number of television sets that maximizes profit.

36. Psychologists define learning rate $L(x)$ to be the percent of learning that an individual exhibits in x minutes while solving an intricate puzzle. For a particular puzzle the average $L(x)$ is given by

$$L(x) = 100x^2 - 20x^3 + x^4 \text{ for } 0 < x < 12$$

Find where $L(x)$ is a maximum.

●●● In Problems 37 to 44 find the relative extrema, and sketch the graph of the functions.

37. $f(x) = x^3(x^2 - 5)$

38. $f(x) = x^2(10 - x^3)$

39. $f(x) = (x^2 - 1)(x^2 - 4)$

40. $f(x) = (x^2 - 3)(9 - x^2)$

41. $f(x) = (x - 5)^2(x + 1)^2$

42. $f(x) = (x + 6)^2(x - 4)^2$

43. $f(x) = (x - 5)^{2/3}$

44. $f(x) = (x + 6)^{2/3}$

45. Nursing is the largest cost item in most hospitals. In a paper* examining decision support models for budgeting nursing work-force requirements in a hospital, a key function is the cost function $C(x)$, where x is the number of nursing hours per week, measured in 1000 hours, and $C(x)$ is the cost of nursing, measured in 100,000 dollars. Several models are developed and applied to data from a large metropolitan hospital in the Sunbelt. For one model the cost function can be described by

$$C(x) = 0.115x^2 - 3.046x + 28.686$$

Use the first derivative test to find the value of x that minimizes C.

*Edward P. Kao and Maurice Queyranne, "Budgeting Costs of Nursing in a Hospital," *Management Science*, Vol. 31, No. 5, pp. 608–621.

46. A national sample of married couples was surveyed to untangle the complex relationship between a person's actual pay and the feeling of being underpaid.[†] The data show that as income rises, people shift from comparing their income to that which is needed to "get along" to that which is required to "get ahead." One illustrative function measures the actual pay, x, of husbands versus the average amount, $f(x)$, they felt that they were underpaid. The data are approximated by the function

$$f(x) = 0.018x^2 - 0.757x + 9.047$$

This means that, on the average, husbands surveyed earning salary x (in thousands of dollars) felt they were underpaid by $f(x)$ thousands of dollars.

(a) Find the salary where the underpayment is a minimum. The authors refer to this salary as the "optimum pay."

(b) Find the amount of underpayment at the optimum pay.

(c) Graph the function and use it to describe what the authors refer to as a shift in the feelings of an income needed to "get along" versus to "get ahead."

[†]John Mirowsky, "The Psycho-Economics of Feeling Underpaid: Distributive Justice and the Earnings of Husbands and Wives," *American Journal of Sociology*, Vol. 92, No. 6, 1987, pp. 1404–1434.

4–3 CONCAVITY AND THE SECOND DERIVATIVE TEST

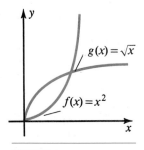

FIGURE 1

Consider the graphs of the two functions in Figure 1. Each function is increasing in $(0, \infty)$, yet the *rate* of increase is different. The graph of $f(x) = x^2$ is increasing at a much greater rate than that of $g(x) = \sqrt{x}$. To illustrate this, let each function represent the position of a car moving in a straight line, so that at time x the car is at position y. Then the derivative of each function represents its velocity. Each function is increasing, but when x increases from 0 to 4, $f(x)$ moves from position $f(0) = 0$ to $f(4) = 16$, while $g(x)$ moves from $g(0) = 0$ to $g(4) = 2$. This is because the derivative of $f(x)$ is increasing, while the derivative of $g(x)$ is decreasing. In this section we study the rate of increase of the derivative of a function. We learn that the second derivative of the function provides valuable information about the graph of the function.

FIGURE 2

(a)   (b)

The tangent line crosses the graph at the point of inflection

(a)

$f''(a)$ does not exist

(b)

Each graph changes concavity at $x = a$

FIGURE 3

Concavity

The difference between the two graphs in Figure 1 can be described geometrically by saying that the graph of $f(x) = x^2$ lies above its tangent line at each point in $(0, \infty)$, while the graph of $g(x) = \sqrt{x}$ lies below its tangent line at each point in $(0, \infty)$. This is because $f'(x) = 2x$ is increasing while $g'(x) = \dfrac{1}{2\sqrt{x}}$ is decreasing on $(0, \infty)$. In Figure 2 two graphs are given along with three tangent lines for each: the curve in (a) lies on or above each tangent line and it is said to "open up," while the curve in (b) lies on or below each tangent line and it is said to "open down."

_____ **Definition of Concavity** _____

The graph of a function $y = f(x)$ is **concave up** *on the interval (a, b) if $f'(x)$ is* increasing on (a, b) and is **concave down** *on (a, b) if $f'(x)$ is decreasing on (a, b).*

A function is defined to be *concave up at $x = a$* if the graph of the function is concave up in some interval containing a and is *concave down at $x = a$* if the graph of the function is concave down in some interval containing a. Geometrically, a function is concave up at $x = a$ *if its graph close to a* lies above its tangent line at a, and it is concave down at $x = a$ if its graph close to a lies below its tangent line at a.

A point on the graph where the graph changes concavity is called a **point of inflection.** Thus at a point of inflection the graph changes from concave up to concave down, or vice versa. At a point of inflection the graph crosses its tangent line or $f''(x)$ does not exist. See Figure 3. Example 1 illustrates these definitions.

EXAMPLE 1 | **Problem**

For the graph in Figure 4 locate the intervals where $f(x)$ is concave up and where $f(x)$ is concave down, and find the points of inflection.

SOLUTION $f(x)$ is concave up in $(0, 2)$ and in $(5, 8)$, while it is concave down in $(2, 5)$. There are points of inflection at $x = 2$ and $x = 5$.

It is tempting to try to draw a connection between when a function is concave up and when the function is increasing. But a look at Figure 4 shows that a function can be both increasing and decreasing in an interval where it is concave up: $f(x)$ is decreasing in $(0, 1)$ and increasing in $(1, 2)$, yet $f(x)$ is concave up in $(0, 2)$. Similarly, a function can be increasing or decreasing in an interval where it is concave down. The chart on the next page shows the four possibilities.

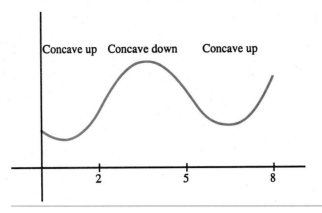

FIGURE 4

f(x) is Increasing or Decreasing	f(x) is Concave Up or Concave Down	Graph of f(x)
A. Increasing	concave up	
B. Increasing	concave down	
C. Decreasing	concave up	
D. Decreasing	concave down	

FIGURE 5

Slope of the tangent line is increasing

(a)

Slope of the tangent line is decreasing

(b)

In Figure 5 two functions are graphed along with three tangent lines for each function. The function in (a) is concave up and the function in (b) is concave down. Start at $x = a$ on the left of each graph and move to $x = b$; that is, let x increase from a to b. What can be said about the slopes of the tangent lines? In (a) the slopes of the tangent lines increase; they are negative on the left part of the graph, then increase to 0 and then become positive. In (b) the slopes decrease as they start out positive, decrease to 0 and then become negative. This indicates that the derivative increases for a function that is concave up and the derivative decreases for a function that is concave down.

This statement can be coupled with the result from Section 4–1 about increasing and decreasing functions to yield a test for concavity. To derive the test, we combine the following two results. Statement I is from the preceding discussion and statement II is from Section 4–1.

I. $f(x)$ is concave up in (a, b) if $f'(x)$ is increasing on (a, b)

II. $f(x)$ is increasing in (a, b) if $f'(x) > 0$ on (a, b)

To tie the two together, replace $f(x)$ by $f'(x)$ in the second statement. We then get

III. $f'(x)$ is increasing in (a, b) if $f''(x) > 0$ on (a, b)

Combining I and III yields the test for concavity.

Procedure for Determining Intervals of Concavity

Assume that the first and second derivatives of $f(x)$ exist for all values of x in the interval (a, b).
1. $f(x)$ is concave up in (a, b) if $f''(x) > 0$ for all x in (a, b)
2. $f(x)$ is concave down in (a, b) if $f''(x) < 0$ for all x in (a, b).

EXAMPLE 2 | **Problem**

Find the intervals of concavity and the point(s) of inflection of $f(x) = x^3 + 3x^2$.

SOLUTION To apply the procedure, first calculate $f'(x)$ and $f''(x)$: $f'(x) = 3x^2 + 6x$ and $f''(x) = 6x + 6 = 6(x + 1)$. Thus $f''(-1) = 0$. Next find where $f''(x)$ is greater than and less than 0. Choose test points to the left and right of $x = -1$, say, $x = -2$ and $x = 0$. This gives $f''(-2) = 6(-1) = -6 < 0$ and $f''(0) = 6 > 0$. Record these facts in the following chart.

From the procedure $f(x)$ is concave up in $(-1, \infty)$ and concave down in $(-\infty, -1)$. Since $f(x)$ changes from concave down to concave up at $x = -1$, $f(x)$ has a point of inflection at $(-1, 2)$. See Figure 6.

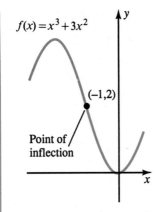

$f(x) = x^3 + 3x^2$

(−1,2)

Point of inflection

FIGURE 6

If $(a, f(a))$ is a point of inflection, $f''(x)$ changes from positive on one side of $x = a$ to negative on the other. For instance, in Example 2, $f(x)$ has a point of inflection at $(-1, 2)$ and $f''(x)$ is negative to the left of $x = -1$ and positive to the right of $x = -1$. What happens to $f''(x)$ at the point of inflection? In Example 2, $f''(-1) = 0$. But there is another possibility. In Figure 7 the graph of $f(x) = x^{1/3}$ is given. It has a point of inflection at $x = 0$ and $f''(0)$ does not exist. Thus there are two possibilities for $f''(x)$ at a point of inflection.

Second Derivative at Point of Inflection

If $f(x)$ has a point of inflection at $(a, f(a))$, then either $f''(a) = 0$ or $f''(a)$ does not exist.

FIGURE 7

$f(x) = x^{1/3}$

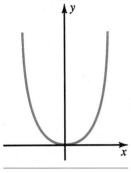

FIGURE 8

It is very important to recognize what this result does not say. To find where a function has a point of inflection, we also need to compute the values where $f''(x) = 0$ and where $f''(x)$ does not exist. Recall that it is not enough to compute critical values when trying to locate relative extrema; more information is required about the behavior around the critical value, which is the essence of the first derivative test. So, too, to determine if a value where $f''(x) = 0$ or $f''(x)$ does not exist yields a point of inflection, we need to find where the function is concave up and concave down. For example, $f(x) = x^4$ has $f''(x) = 12x^2$, and $f''(x) = 0$ when $x = 0$. But since $f''(x) > 0$ for all x to the left and to the right of $x = 0$, f is concave up both to the left and to the right of $x = 0$. Thus f does not have a point of inflection at $x = 0$ even though $f''(0) = 0$. See Figure 8.

The Second Derivative Test

The two graphs in Figure 9 illustrate the connection between concavity and relative extrema. Each graph has a horizontal tangent at $x = a$, so $f'(a) = 0$ and the function has a critical value at $x = a$. The graph in (a) is concave down and $f(x)$ has a relative maximum at $x = a$. The graph in (b) is concave up and $f(x)$ has a relative minimum at $x = a$. To summarize, let $x = a$ be a critical value of $f(x)$.

 I. If $f(x)$ is concave down at $x = a$, then $f(x)$ has a relative maximum at $x = a$.

 II. If $f(x)$ is concave up at $x = a$, then $f(x)$ has a relative minimum at $x = a$.

FIGURE 9

$f'(a) = 0$
$f(x)$ is concave down
$f(a)$ is a relative maximum

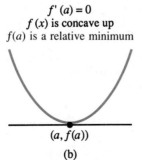

$(a, f(a))$

(a)

$f'(a) = 0$
$f(x)$ is concave up
$f(a)$ is a relative minimum

$(a, f(a))$

(b)

Therefore concavity can be used to determine whether a critical value yields a relative maximum or a relative minimum. Since the second derivative measures concavity, the second derivative can be used to determine whether a critical value is a relative maximum or a relative minimum. Statements I and II can be refined by using the test for concavity from the previous section.

Let $x = a$ be a critical value of $f(x)$.

III. If $f''(a) < 0$, then $f(x)$ is concave down at $x = a$, and thus $f(x)$ has a relative maximum at $x = a$.

IV. If $f''(a) > 0$, then $f(x)$ is concave up at $x = a$, and thus $f(x)$ has a relative minimum at $x = a$.

This is the essence of the second derivative test, which, like the first derivative test, determines whether a critical value is a relative maximum or a relative minimum.

The Second Derivative Test

Suppose $f(x)$ is differentiable and $f''(x)$ exists on an interval containing a. Let $x = a$ be a critical value.
1. $f(a)$ is a relative maximum if $f''(x) < 0$ at $x = a$.
2. $f(a)$ is a relative minimum if $f''(x) > 0$ at $x = a$.
3. If $f''(a) = 0$, the test gives no conclusion about a relative extremum at $x = a$; that is, $f(a)$ may be a relative maximum, a relative minimum, or neither.

EXAMPLE 3

Problem

Use the second derivative test to find the relative extrema for
(a) $f(x) = 5x^2 - 10x$ and (b) $f(x) = x^3 - 3x^2$.

SOLUTION (a) Compute $f'(x) = 10x - 10 = 10(x - 1)$. Solve $f'(x) = 0$. The only critical value is $x = 1$. To apply the second derivative test, we compute the second derivative, $f''(x) = 10$. This is a constant function, so

$$f''(1) = 10 > 0$$

Since $f''(1) > 0$, apply part 2 of the second derivative test and conclude that $f(x)$ has a relative minimum at $x = 1$. The relative minimum is $f(1) = -5$.
(b) Compute $f'(x) = 3x^2 - 6x$. Solve $f'(x) = 3x^2 - 6x = 3x(x - 2) = 0$. So $x = 0, 2$. The critical values are $x = 0$ and $x = 2$. Compute $f''(x) = 6x - 6$. Substitute each critical value into f''. This yields

$$f''(0) = 0 - 6 = -6 < 0 \qquad f''(2) = 12 - 6 = 6 > 0$$

Part 1 of the second derivative test states that $f(x)$ has a relative maximum at $x = 0$ because $f''(0) < 0$; part 2 states that $f(x)$ has a relative minimum at $x = 2$ because $f''(2) > 0$. The relative maximum is $f(0) = 0$ and the relative minimum is $f(2) = -4$.

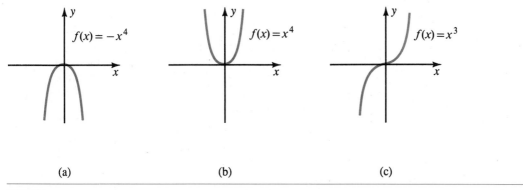

(a)   (b)   (c)

FIGURE 10

Part 3 of the second derivative test states that the test fails when $f''(a) = 0$ since no conclusion can be drawn about relative extrema. Figure 10 shows why this is the case. Even though each of the three functions graphed in Figure 10 has $f''(0) = 0$, their behavior at $x = 0$ is different. In (a), $f(x) = -x^4$ has a relative maximum at $x = 0$; in (b), $f(x) = x^4$ has a relative minimum at $x = 0$; in (c), $f(x) = x^3$ has neither a relative maximum nor a relative minimum because it has a point of inflection at $x = 0$. This means that if $f''(a) = 0$ for a critical value $x = a$, then the first derivative test must be applied to $x = a$. Also, if $f''(x)$ does not exist, the second derivative test cannot be used so the first derivative test must be applied.

Curve Sketching

The ideas presented thus far in this chapter can be combined to form a handy procedure for sketching graphs. The summary of these various techniques is a process called **curve sketching.**

Curve-Sketching Techniques

1. Compute $f'(x)$.
   a. Find the critical values by solving $f'(x) = 0$ and finding where $f'(x)$ does not exist.
   b. Find the intervals where $f(x)$ is increasing and decreasing by solving $f'(x)$ greater than and less than 0.
2. Compute $f''(x)$.
   a. Find the possible points of inflection by solving $f''(x) = 0$ and locating where $f''(x)$ does not exist.
   b. Find the intervals where $f(x)$ is concave up and concave down by setting $f''(x)$ greater than and less than 0. Use this information to locate the points of inflection.
3. Use the first or second derivative test to locate the relative maxima and the relative minima.

4. Plot the relative extrema and the points of inflection. Keep in mind where the function is increasing, decreasing, concave up, and concave down, and plot several additional points that help to get an accurate sketch. For instance, plot the x- and y-intercepts if they exist and are easy to calculate; if the function is defined on a closed or half-open interval, plot the points corresponding to the end points.
5. Draw a sketch of the curve from this information.

Example 4 shows how to use the curve-sketching techniques by applying them to the functions in Example 3.

EXAMPLE 4

Problem

Sketch the graph of the functions (a) $f(x) = 5x^2 - 10x$, and (b) $f(x) = x^3 - 3x^2$.

SOLUTION We follow the steps in the procedure even though they do not have to be done precisely in the given order.

(a)

1. $f'(x) = 10x - 10 = 10(x - 1) = 0$ implies $x = 1$.
   a. Since $f'(1) = 0$, the only critical value is $x = 1$.
   b. $f'(x) > 0$ in $(1, \infty)$ and $f'(x) < 0$ in $(-\infty, 1)$, so $f(x)$ is increasing in $(1, \infty)$ and decreasing in $(-\infty, 1)$.
2. $f''(x) = 10$
   a. Since $f''(x)$ is always greater than 0, $f(x)$ is concave up in $(-\infty, \infty)$.
   b. $f''(x)$ is never equal to 0, so there is no point of inflection.

This information is recorded in the following chart.

Critical value		1	x
Sign of $f'(x)$	$-\,-\,-\,-\,-\,-\,-\,-\,-$	$+\,+\,+\,+\,+\,+\,+\,+\,+\,+$	
$f(x)$ is	decreasing	increasing	
Sign of $f''(x)$	$+\,+\,+\,+\,+\,+\,+\,+\,+\,+\,+\,+\,+\,+\,+\,+\,+\,+\,+\,+$		
$f(x)$ is	concave up		

3. The second derivative test shows that $f(x)$ has a relative minimum at $x = 1$ because $f''(1) = 10 > 0$.
4. Plot $(1, -5)$, where $f(x)$ has its relative minimum. To find the y-intercept, let $x = 0$ and get $f(0) = 0$. Since $f(x) = 0$ at $x = 0$ and $x = 2$, plot the x-intercepts $(0, 0)$ and $(2, 0)$.
5. The graph is in Figure 11.

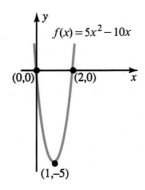

FIGURE 11

(b)

1. $f'(x) = 3x^2 - 6x = 3x(x - 2)$
   a. Since $f'(0) = f'(2) = 0$, the critical values occur at $x = 0$ and $x = 2$.
   b. $f'(x) > 0$ in $(-\infty, 0)$ and $(2, \infty)$; $f'(x) < 0$ in $(0, 2)$. Thus $f(x)$ is increasing in $(-\infty, 0)$ and $(2, \infty)$ and $f(x)$ is decreasing in $(0, 2)$.
2. $f''(x) = 6x - 6 = 6(x - 1)$
   a. Since $f''(x) > 0$ in $(1, \infty)$ and $f''(x) < 0$ in $(-\infty, 1)$, $f(x)$ is concave up in $(1, \infty)$ and concave down in $(-\infty, 1)$.
   b. $f''(1) = 0$ and $f(x)$ changes concavity at $x = 1$, so there is a point of inflection at $(1, -2)$.

This information is recorded in the following chart.

Possible point of inflection		1	
Critical values	0		2
Sign of $f'(x)$	$+ + + + + + + + + +$	$- - - - - - - - - -$	$+ + + + + + + + +$
$f(x)$ is	increasing	decreasing	increasing
Sign of $f''(x)$	$- - - - - - - - - - - - - - - -$		$+ + + + + + + + + + + + + +$
$f(x)$ is	concave down		concave up

3. The second derivative test shows that $f(x)$ has a relative maximum at $x = 0$ because $f''(0) = -6 < 0$ and $f(x)$ has a relative minimum at $x = 2$ because $f''(2) = 6 > 0$.
4. Plot $(0, 0)$ and $(2, -4)$, where $f(x)$ has its relative extrema. Plot the point of inflection, $(1, -2)$. To find the y-intercept, let $x = 0$ and get $f(0) = 0$, but $(0, 0)$ is already plotted. Since $f(x) = 0$ at $x = 0$ and $x = 3$, plot the additional x-intercept $(3, 0)$.
5. The graph is in Figure 12.

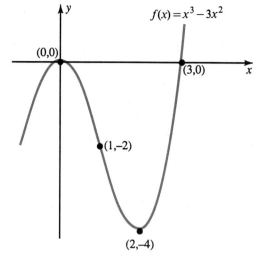

FIGURE 12

An Application: Public Subsidy

In some situations it is not satisfactory to analyze production decisions merely on the basis of cost-effectiveness or profit optimization. Certain not-for-profit institutions provide services to a community at prices below production costs. Total revenue from sales may be insufficient to pay total costs and a subsidy may be required. Examples of such institutions are public transit companies, low-cost higher education institutions, and cultural organizations. To illustrate production decisions when a public subsidy is required, economist Marilu McCarty used as an example a mass transit system in a congested urban area.* The cost $C(x)$ and revenue $R(x)$ functions were estimated for the daily demand x, where $C(x)$ and $R(x)$ are measured in thousands of dollars and x is the number of riders in thousands. If the system expected to make a profit, then it would seek to maximize $R(x) - C(x)$. But the system is designed to absorb a loss. The objective is then to minimize the *economic loss EL(x)*, defined by

$$EL(x) = C(x) - R(x)$$

The goal is to determine the demand x that minimizes $EL(x)$. This minimum value of $EL(x)$ is the subsidy paid by the government to the institution. The next example shows how to solve this problem for cost and revenue functions that are similar to those of the mass transit system. Problem 48 of the exercises deals with the actual data.

EXAMPLE 5   The cost and revenue functions for the mass transit system are

$$C(x) = 2000 + 140x - 30x^2 + x^3$$
$$R(x) = 20x - 9x^2$$

*Marilu Hurt McCarty, *Managerial Economics,* Glenview, IL, Scott, Foresman and Co., 1986, pp. 479–486.

Problem

Determine the minimum economic loss.

SOLUTION The economic loss function is defined by $EL(x) = C(x) - R(x)$. It is

$$EL(x) = 2000 + 120x - 21x^2 + x^3$$

To find the minimum, use the second derivative test. Compute $EL'(x)$ and $EL''(x)$.

$$EL'(x) = 120 - 42x + 3x^2 = 3(40 - 14x + x^2)$$
$$EL''(x) = -42 + 6x$$

To solve $EL'(x) = 0$, either factor the quadratic or use the quadratic formula.

$$EL'(x) = 3(x - 10)(x - 4) = 0$$

This implies that the critical values are $x = 4$ and $x = 10$. Substituting these values into $EL''(x)$ yields

$$EL''(4) = -18 \quad \text{and} \quad EL''(10) = 18$$

By using the second derivative test, we conclude that the relative minimum of $EL(x)$ occurs at $x = 10$. It is $EL(10) = 2100$. The function is graphed in Figure 13. Notice that there are functional values of $EL(x)$ that are smaller than $EL(10)$; in fact $EL(0) = 2000$. But these values of x are too small to be practical. That is, the transit system would not be properly serving the public if it accommodated so few riders. This means that an implicit assumption in the problem is that the transit system will service a certain minimum number of riders. In the next section we study problems that have natural boundaries like this.

FIGURE 13 *Source:* Marilu Hurt McCarty, *Managerial Economics,* Glenview, IL, Scott, Foresman and Co., 1986, pp. 479–496.

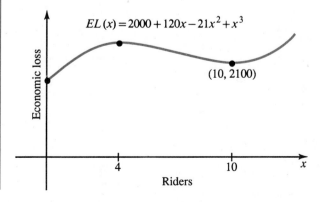

EXERCISE SET 4–3

• In Problems 1 to 6 find the largest intervals where the function is concave up and concave down. Find the points of inflection. Assume that the functions are defined on the whole real line.

1.

2.

3.

4.

5.

6.

In Problems 7 to 12 find the intervals where the function is concave up and concave down. Find any point of inflection.

7. $f(x) = x^2 - 4$

8. $f(x) = 2x^2 - 5$

9. $f(x) = x^3 - 6x^2 - 2$

10. $f(x) = x^3 - 9x^2$

11. $f(x) = x^3 + 3x^2 - 9x$

12. $f(x) = 2x^3 + 3x^2 - 12x + 1$

In Problems 13 to 18 apply the second derivative test to find the relative extrema.

13. $f(x) = x^3 - 3x^2 - 9x$

14. $f(x) = 2x^3 + 3x^2 - 36x + 3$

15. $f(x) = x^4 - 2x^2$

16. $f(x) = -x^4 + 6x^2$

17. $f(x) = -x^4 + 8x^2$

18. $f(x) = -x^4 + 12x^2 + 1$

In Problems 19 to 36 use the curve-sketching techniques to graph the function.

19. $f(x) = x^3 + 6x^2$ 20. $f(x) = x^3 - x^2$

21. $f(x) = x^3 - 3x^2 - 9x$

22. $f(x) = 2x^3 + 3x^2 - 36x + 3$

23. $f(x) = x^4 - 2x^2$ 24. $f(x) = -x^4 + 6x^2$

25. $f(x) = 64x - 2x^4$ 26. $f(x) = 2x - 3x^{2/3}$

•• 27. $f(x) = x + 9/x$ 28. $f(x) = 4x + 25/x$

29. $f(x) = (x^2 + 12)/x$ 30. $f(x) = (4x^2 + 9)/x$

31. $f(x) = x^2(x^2 - 2)$ 32. $f(x) = x^2(4 - x^2)$

33. $f(x) = x^2(x^2 + 2)$ 34. $f(x) = x^2(4 + x^2)$

35. $f(x) = (x - 1)/(x + 1)$

36. $f(x) = (x + 3)/(x - 5)$

In Problems 37 to 40 use the information given to sketch the graph of $f(x)$. Assume $f(x)$ is continuous for all real numbers.

37.
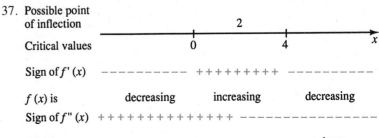

$f(0) = 0, f(2) = 1, f(4) = 2$

38. Possible point
   of inflection

   Critical values

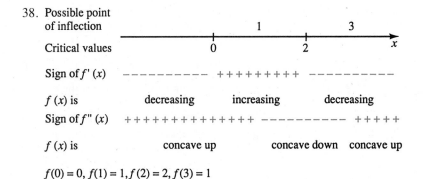

   Sign of $f'(x)$ $----------$ $+++++++++$ $----------$

   $f(x)$ is   decreasing   increasing   decreasing

   Sign of $f''(x)$ $++++++++++++++++$ $----------$ $+++++$

   $f(x)$ is   concave up   concave down   concave up

   $f(0) = 0, f(1) = 1, f(2) = 2, f(3) = 1$

39. Possible points
   of inflection

   Critical values

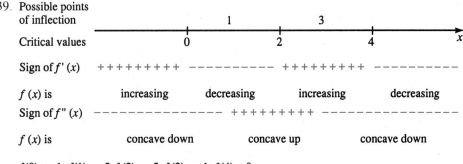

   Sign of $f'(x)$ $+++++++++$ $----------$ $+++++++++$ $----------$

   $f(x)$ is   increasing   decreasing   increasing   decreasing

   Sign of $f''(x)$ $----------------$ $+++++++++$ $------------------$

   $f(x)$ is   concave down   concave up   concave down

   $f(0) = -1, f(1) = -2, f(2) = -3, f(3) = -1, f(4) = 0$

40. Possible points
   of inflection

   Critical values

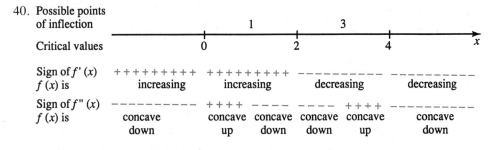

   Sign of $f'(x)$ $++++++++++$ $+++++++++$ $----------$ $----------$
   $f(x)$ is   increasing   increasing   decreasing   decreasing

   Sign of $f''(x)$ $----------$ $++++$ $----$ $----$ $++++$ $----------$
   $f(x)$ is   concave   concave   concave   concave   concave   concave
                   down   up   down   down   up   down

   $f(0) = 0, f(1) = 2, f(2) = 4, f(3) = 2, f(4) = 1$

••• In Problems 41 to 45 use the curve-sketching techniques to graph the function.

41. $f(x) = x^3(x^2 - 15)$ 42. $f(x) = x^2(10 - x^3)$

43. $f(x) = (x^2 - 1)(x^2 - 4)$

44. $f(x) = (x + 6)^2(x - 4)^2$

45. $f(x) = (x + 6)^{2/3}$

46. In 1970 the U.S. government commissioned the Army to conduct a thorough study of the cost of converting to the metric system of measure. The methodology of the Army study could then be applied by various organizations, which would be similarly affected by the change. In the report* the cumulative costs of metrication for the U.S. Army are estimated by the function

$$f(x) = -0.02x^3 + 0.12x^2 + 0.72x$$

*Walter A. Lilius, "The Methodology of the Army Metric Study," *American Society of Quality Control—Technical Conference Transactions*, 1972, pp. 242–249.

where x is measured in units of five-year periods starting with 1972 and $f(x)$ is measured in billions of dollars. For example, $x = 1$ represents 1977 (one five-year period after 1972) and $f(1) = 0.82$ means that the estimated cost of metrication for the Army by 1977 would be $0.82 billion. Find the year when the rate of increase of the budget would be a maximum.

47. Production models in business help a firm decide what is the optimum quantity to produce either to maximize price or profit or to minimize cost. In an article† studying agricultural production functions, with an emphasis on the swine industry, the following function relating the price P (dollars) to the weight x (lb) of an individual animal is given:

$$P(x) = 50.536 + 0.0988x - 0.00024x^2$$

Find the value of x that maximizes the price.

48. For the economic loss function $EL(x)$ given in McCarty‡ for a mass transit system, find the value of x that minimizes $EL(x)$, assuming that the system must accommodate at least $x = 4$ thousand riders:

$$EL(x) = 2000 + 72x - 1.1x^2 + 0.005x^3$$

†Jean-Paul Chavas et al., "Modeling Dynamic Agricultural Production Response: The Case of Swine Production," *American Journal of Agricultural Economics,* Vol. 67, 1987, pp. 636–646.

‡Marilu Hurt McCarty, *Managerial Economics,* Glenview, IL, Scott, Foresman and Co., 1986, pp. 479–486.

4–4 ABSOLUTE EXTREMA AND OPTIMIZATION PROBLEMS

A function may have a relative maximum or a relative minimum without being an absolute maximum or an absolute minimum. In fact most of the functions graphed in the previous sections have functional values that are larger or smaller than the relative extrema. In this section we cover absolute extrema. If $f(a)$ is a relative maximum, then $f(a)$ is greater than or equal to the other functional values $f(x)$ for all x restricted to an interval containing a. If $f(a)$ is an absolute maximum, then $f(a)$ is greater than or equal to all other functional values $f(x)$. A similar statement holds for relative and absolute minima.

There are several possibilities for absolute extrema. Some functions have neither an absolute maximum nor an absolute minimum. If the function is defined on a closed interval the absolute extremum can occur at an end point. And sometimes an absolute extremum occurs at a relative extremum.

Absolute Extrema

Consider the function $y = f(x)$ graphed in Figure 1. It has a relative minimum at $x = b$ and a relative maximum at $x = c$. But neither is an absolute extremum; $f(a)$ is greater than $f(c)$, so $f(c)$ is not an absolute maximum, and $f(d)$ is less than $f(b)$, so $f(b)$ is not an absolute minimum. In addition, $f(a)$ is greater than all other functional values, so it is the absolute maximum while $f(d)$ is the absolute minimum. Sometimes an absolute extremum occurs at an end point of an interval, as does $f(a)$ in Figure 1, and sometimes it does not, like $f(d)$. This leads to the following definition.

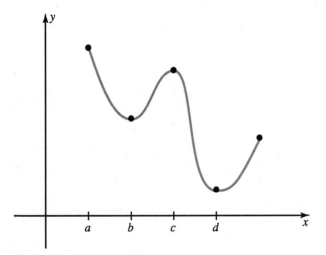

FIGURE 1

_____ **Definition** _____

Let $y = f(x)$ be defined on an interval I, where I can be either an open or closed interval. Suppose a is in I. Then

1. $f(a)$ is the **absolute maximum** _of_ $f(x)$ _on I_ if

    $f(a) \geq f(x)$ for all x in I

2. $f(a)$ is the **absolute minimum** _of_ $f(x)$ _on I_ if

    $f(a) \leq f(x)$ for all x in I

The graph in Figure 1 shows that an absolute extremum may occur at an end point of the interval. Of course, this can happen only when the end point is included in the interval, that is when it is closed or half-open (which means one end point is in the interval and the other is not). If the interval is open, then the absolute extrema cannot occur at an end point because the end points are not part of the interval. In fact the function may not have an absolute maximum or an absolute minimum. On the other hand, if the interval is closed, the end points are candidates for the absolute extrema. If an absolute extremum does not occur at an end point, then it occurs at an interior relative extremum. In more advanced courses the following theorem is proved.

_____ **Theorem** _____

If $f(x)$ is continuous on the closed interval $[a, b]$, then $f(x)$ has an absolute maximum and an absolute minimum on $[a, b]$.

A nonvertical straight line drawn on an open interval is an example of a function that has neither an absolute maximum nor an absolute minimum on an open interval. If the end points are included, however, meaning the interval is closed, the absolute extrema of a nonvertical straight line occur at the end points.

The procedure to compute absolute extrema is the same as that for finding relative extrema, except that the end points must also be considered. That is, absolute extrema occur either at points that are relative extrema or at end points.

The Procedure for Finding Absolute Extrema for $f(x)$ Defined on a Closed Interval $[a, b]$

1. Calculate the critical values, $c_1, c_2, \ldots, c_n$ that lie in the interval $[a, b]$.
2. Calculate $f(a)$, $f(b)$, $f(c_1)$, $f(c_2)$, $\ldots$, $f(c_n)$.
3. The absolute maximum is the largest value of the numbers in step 2, and the absolute minimum is the smallest value of these numbers.

Example 1 shows how to compute absolute extrema once the relative extrema are known. It uses a function that was defined in Example 2 of Section 4–2 where the relative extrema were computed.

EXAMPLE 1   Let the function $f(x) = (x - 2)^2(x - 5)$ be defined on the interval $[\frac{3}{2}, 6]$.

Problem

Find the absolute extrema of $f(x)$.

SOLUTION In Example 2 of Section 4–2, $f(x)$ was found to have a relative maximum of $f(2) = 0$ and a relative minimum of $f(4) = -4$. See Figure 2 for the graph of the function. To see whether the relative extrema are absolute extrema, check the functional values of the end points. The end points occur at $x = \frac{3}{2}$ and $x = 6$. Their functional values are $f(\frac{3}{2}) = -\frac{7}{8}$ and $f(6) = 16$. To find the absolute extrema, find the largest and smallest functional values of the critical values and the end points. We arrange these four functional values in the following table:

	Critical Values		End Points	
x	2	4	$\frac{3}{2}$	6
$f(x)$	0	-4	$-\frac{7}{8}$	16

The largest of the four numbers for $f(x)$ is 16 and the smallest is -4. Therefore the absolute maximum is $f(6) = 16$ and the absolute minimum is $f(4) = -4$. The absolute maximum occurs at the right-hand end-point, while the absolute minimum occurs at an interior relative minimum.

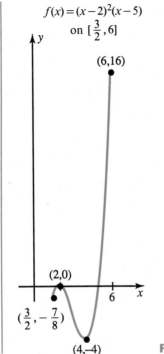

$$f(x) = (x-2)^2(x-5)$$
$$\text{on } [\tfrac{3}{2}, 6]$$

(6,16)

(2,0)

$(\tfrac{3}{2}, -\tfrac{7}{8})$

6

(4,–4)

FIGURE 2

Optimization Problems

Many applications of calculus depend on maximizing or minimizing a certain quantity. Often the variables in such problems have natural boundaries. For example, if x represents the number of items produced by a firm, x must be greater than or equal to 0. If t represents time, measured in hours, during an 8-hour shift, then t is restricted to the interval $[0, 8]$. If an absolute extremum is sought, then the end points must be checked.

An optimization problem is ordinarily a word problem, so it must first be translated into mathematics. Then the relative extrema are computed as well as the functional values of the end points. The extrema are then interpreted in terms of the original problem. Let us demonstrate the method with an example that is made easy because the function is given.

EXAMPLE 2

An object is thrown straight upward so that its height after t seconds is
$f(t) = -16t^2 + 64t$, where f is measured in feet.

Problem

How high will the object travel and after how many seconds will it reach its maximum height?

SOLUTION The object will reach its maximum height when $f(t)$ has its absolute maximum.

Are there natural end points for t? It makes no sense for t to be less than 0, and a negative value for $f(t)$ means the object is below ground level. Solving $f(t) < 0$ yields

$$-16t^2 + 64t = -16t(t - 4) < 0$$

This is true when $t < 0$ or when $t > 4$. Thus the natural restriction for $f(t)$ is $[0, 4]$.

There are no values of t where $f'(t)$ fails to exist, so to find the critical values, solve $f'(t) = 0$. Compute $f'(t) = -32t + 64$. Then $32t = 64$ and $t = 2$. The functional values of the end points are $f(0) = 0 = f(4)$. The absolute maximum occurs at the largest of the three values

$$f(0) = 0, \quad f(2) = 64, \quad f(4) = 0$$

Hence the object reaches its maximum height of $f(2) = 64$ at $t = 2$ seconds. The function is graphed in Figure 3.

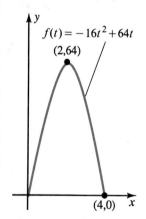

FIGURE 3

The next two examples show how to apply the procedure for finding absolute extrema to optimization problems.

EXAMPLE 3 | A department store wants to build a rectangular outdoor storage facility on the rear wall of its building. An iron fence will enclose the facility on three sides and the wall of the building will make the fourth side. (See Figure 4.)

Problem

Find the dimensions of the storage facility with the largest area that can be built using 200 feet (ft) of fencing.

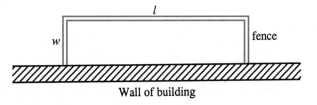

FIGURE 4

SOLUTION From Figure 4 we assign letters to the variable quantities. Let l represent the length and w the width of the storage facility. Then the area, A, is given by the formula

$$A = lw \qquad (1)$$

The variable A is to be maximized since we are asked for the largest storage facility. A must be expressed as a function of l or w alone, so there must be another relationship between w and l. The amount of fencing is fixed at 200 ft, and it is the sum of the three sides, whose lengths are w, w, and l. Therefore

$$200 = 2w + l$$

Solving for l yields

$$1 = 200 - 2w \qquad (2)$$

Substitute equation 2 into equation 1, which results in an expression of A as a function of w.

$$A(w) = w(200 - 2w) = 200w - 2w^2$$

To find the end points, note that it makes no sense for A to be negative. So set $A(w) \geq 0$. This yields

$$w(200 - 2w) \geq 0$$

This is satisfied for w in $[0, 100]$. Thus the end points are $w = 0$ and $w = 100$. To find the critical values, solve $A'(w) = 0$.

$$A'(w) = 200 - 4w = 0$$

The solution is $w = 50$. The end points are $w = 0$ and $w = 100$. The three values that are candidates for the absolute maximum are

$$A(50) = 5000$$
$$A(0) = 0$$
$$A(100) = 0$$

The largest value is $A(50) = 5000$. So the storage facility with the largest area has width 50 ft and length $l = 200 - 2w = 200 - 100 = 100$ ft.

EXAMPLE 4 An apple orchard owner has determined that if 40 trees are planted per acre, the yield is 500 apples per tree. For each tree per acre that is added beyond 40, the yield is reduced by 10 apples per tree.

Problem

Find the number of trees per acre that should be planted to yield the maximum crop.

SOLUTION Let x represent the number of trees planted per acre beyond 40. Let Y represent the total crop per acre. Then Y is the number of trees per acre times the number of apples per tree. For example, if $x = 0$, then the number of trees planted per acre is 40 and yield per tree is 500 apples, so $Y = 40 \cdot 500 = 20{,}000$ apples per acre.

The number of trees per acre is $40 + x$. The yield per tree diminishes by 10 for each tree planted beyond 40 so the yield per tree is $500 - 10x$. Hence the formula for $Y(x)$ is

$$Y(x) = \underset{\substack{\text{number} \\ \text{of trees}}}{(40 + x)} \underset{\substack{\text{number of} \\ \text{apples}}}{(500 - 10x)}$$
$$= 20{,}000 + 100x - 10x^2$$

We must find the value of x that maximizes $Y(x)$. First find the critical values. Solve $Y'(x) = 100 - 20x$ equal to 0. This yields

$$Y'(x) = 100 - 20x = 0$$

Thus $x = 5$ is the only critical value. The natural end points are $x = 0$ and $x = 50$. The three values to check are

$$Y(5) = (45)(450) = 20{,}250$$
$$Y(0) = 20{,}000$$
$$Y(50) = 0$$

Therefore the absolute maximum is $Y(5) = 20{,}250$. Hence the maximum yield occurs when 45 trees are planted per acre.

An Application: Dayton Trucking Company

In the mid–1970s the price of gasoline skyrocketed as OPEC (Organization of Petroleum Exporting Countries) forced up the price of crude oil. This caused great hardship for many industries, especially those whose transportation costs were significant. Winger gives an account of how one company, the Dayton Trucking Co., approached the problem.* The price of gasoline had become a critical variable in their cost evaluation for the first time. Truck performance was analyzed. They wanted to determine the speed that produced maximum fuel economy, that is, the speed at which miles per gallon (mpg) was a maximum. The data were summarized in the following function that measured fuel economy $F(x)$ in mpg for a truck operating at speed x, measured in miles per hour (mph).

$$F(x) = 0.15x - 0.0015x^2$$

*Bernard J. Winger, *Cases in Management Economics,* Columbus, OH, Grid Publishing, 1979, pp. 78–81.

The natural bounds for x were $x = 0$, because a negative value for x makes no sense, and $x = 55$, because the government had just imposed the 55 mph speed limit. The next example solves the problem.

EXAMPLE 5 | **Problem**

Find the absolute maximum of the fuel economy function

$$F(x) = 0.15x - 0.0015x^2$$

where the domain of the function is [0, 55].

SOLUTION Find the critical values by solving $F'(x) = 0$.

$$F'(x) = 0.15 - 0.003x = 0$$
$$x = 50$$

Compare the three values $F(0) = 0$, $F(50) = 3.75$, and $F(55) = 3.7125$. The absolute maximum is $F(50) = 3.75$ mpg. Therefore the speed of 50 mph produced the greatest fuel economy.

EXERCISE SET 4–4

• In Problems 1 to 14 find the absolute extrema, if they exist, for the function defined on the given interval.

1. $f(x) = x^2 + 2x + 1$ on $[-4, 4]$
2. $f(x) = x^2 - 2x + 1$ on $[-4, 4]$
3. $f(x) = 5 - 4x - x^2$ on $[-4, 6]$
4. $f(x) = 8x - 2x^2$ on $[-4, 6]$
5. $f(x) = x^3 + 3x^2 + 1$ on $[-2, 2]$
6. $f(x) = x^3 - 6x^2 + 1$ on $[-1, 5]$
7. $f(x) = x^3 + 3x^2 + 1$ on $[-3, -1]$
8. $f(x) = x^3 - 6x^2 + 1$ on $[-2, 2]$
9. $f(x) = x^4 - 2x^2 + 1$ on $[-2, 2]$
10. $f(x) = x^4 + 4x^3 + 1$ on $[-4, 2]$
11. $f(x) = (x - 2)^2(x - 1)$ on $[-1, 4]$
12. $f(x) = (x + 1)^2(x - 1)$ on $[0, 4]$
13. $f(x) = (x - 3)^2(x - 1)^2$ on $[-1, 4]$
14. $f(x) = (x + 1)^2(x - 1)^2$ on $[0, 4]$

15. An object is thrown straight upward so that its height after t seconds is $f(t) = -16t^2 + 128t$, where f is measured in feet. How high will the object travel and after how many seconds will it reach its maximum height?

16. An object is thrown straight upward so that its height after t seconds is $f(t) = -16t^2 + 256t + 6$, where f is measured in feet. How high will the object travel and after how many seconds will it reach its maximum height?

17. A family wants to build a rectangular deck on the back of their house. It will be enclosed on three sides by wooden fencing with the back of the house being the fourth side. If 40 feet of fencing is to be used, find the dimensions of the deck with the largest area that can be built.

18. An apple orchard owner has determined that if 30 trees are planted per acre, the yield is 600 apples per tree. For each tree per acre that is added beyond 30, the yield is reduced by 15 apples per tree. Find the number of trees per acre that should be planted to yield the maximum crop.

•• 19. Divide the number 300 into two parts whose sum is 300 such that the product of one part and the square of the other is a maximum.

20. Divide the number 120 into two parts whose sum is 120 such that the product of one part and the cube of the other is a maximum.

21. A farmer wants to enclose with a fence a rectangular field that is adjacent to a straight river, where no fencing is required along the river. What are the dimensions of the field if 4000 feet of fencing is to be used and the area of the field is to be a maximum?

22. A manufacturing company offers the following discount schedule of prices: $20 for orders of 500 or fewer, with the cost being reduced 2 cents for each unit above 500. Find the order that will maximize the company's revenue.

23. A person is on an island 10 miles from a straight shore. The person wishes to get to a point 12 miles up the coast. If the person can row at the rate of 2 mph (miles per hour) and walk at the rate of 4 mph, to what point on the shore should be rowed in order to minimize the time?

24. A manufacturing company makes an open box from a 5 by 6-inch rectangular piece of cardboard by cutting out equal squares at each corner and then folding up the sides. What are the dimensions of the box with maximum volume?

25. A manufacturing company wants to package its product in a rectangular box with a square base and a volume of 32 in.^{3}. The cost of the material used for the top and bottom is 3 cents per square inch (in.^{2}), while the cost of the material used for the sides is 6 cents per in.^{2}. What are the dimensions of the box with the minimum cost?

26. A manufacturing company wants to package its product in a closed can in the shape of a right circular cylinder. The company has determined that the volume of the can must be 16 in.^{3}. What are the dimensions of the can with the minimum surface area?

27. A manufacturing company wants to package its product in a rectangular box with a square base and a volume of 32 in.^{3}. The cost of the material used for the top is 1 cent per in.^{2}, the cost of the material used for the bottom is 3 cents per in.^{2}, and the cost of the material used for the sides is 2 cents per in.^{2}. What are the dimensions of the box with the minimum cost?

28. At 1:00 P.M. plane A is 650 miles due west of plane B. If plane A flies at 100 mph due south and plane B flies at 150 mph due west, when will they be nearest each other and how near will they be?

29. U.S. postal regulations state that packages that are sent parcel post must have the sum of the length plus the girth (perimeter of a cross section) no more than 84 in. What are the dimensions of the rectangular package with square ends with the greatest volume that can be sent by parcel post?

30. U.S. postal regulations state that packages that are sent parcel post must have the sum of the length plus the girth (perimeter of a cross section) no more than 84 in. What are the dimensions of the cylindrical package with the greatest volume that can be sent by parcel post?

31. A "Norman window" consists of a rectangle with a semicircular area on top. See the accompanying figure. Find the width of the rectangle if the perimeter of the window is 12 feet and the area is as large as possible.

32. A long rectangular sheet of aluminum is to be made into a gutter by turning up equal sides the length of the strip at right angles. How many inches should be turned up on each side to maximize the volume of the gutter?

33. A rectangular pen is to be fenced in, and then

an additional fence will be constructed in the center of the pen to make two identical rectangular enclosures. If the area is to be 60 square yards, find the dimensions of the pen with the minimum amount of fencing.

●●● 34. Show that among all rectangles having a fixed perimeter, the square has the maximum area.

35. Show that among all rectangles having a fixed area, the square has the minimum perimeter.

36. A principle in economics states that maximum profit is obtained when the marginal revenue, which is the derivative of the revenue function, equals the marginal cost, which is the derivative of the cost function. Explain this principle if $P(x)$ is the profit function, $R(x)$ is the revenue function, and $C(x)$ is the cost function when x goods are produced.

37. A box with no top and a square bottom is to have a volume of 10 in.^{3}. Find the dimensions of the box of least cost if the material of which the sides are to be made costs one-third per square inch as much as the material of which the bottom is to be made.

38. In a study* to increase the productivity of the average personal automobile, Purdue University's Interdisciplinary Engineering Studies researched the feasibility of a plan for members of an enterprise to own and operate jointly a fleet of vehicles. In one model 26 families were monitored by having them keep diaries of when they would use a shared vehicle. By keeping records of cost and availability of the shared vehicles, the researchers generated the following function that relates the number of vehicles, x, to the cost per family per week, $f(x)$:

$$f(x) = 11.8x^3 - 138.2x + 200$$

The function is valid in the interval $[0, 4]$. Find the value of x that minimizes the cost per family per week.

39. There is extensive literature on the optimal governmental regulation of open-access resources, such as fisheries and watersheds. In a study† of market behavior and exploitation of fisheries an important function is the total revenue R produced that depends on the fishing effort x, measured by the number of days fished by standard boats. One model considered assumes the cost $C(x)$ is a linear function. A typical pair of functions for $R(x)$ and $C(x)$ is

$$R(x) = 0.4x - 0.002x^2$$
$$\text{and} \quad C(x) = 0.11x$$

The natural bounds for these functions would be $x \geq 0$ on the left and the natural bound on the right would be that value of x such that $R(x) = C(x)$.
(a) Find the value of x that maximizes $R(x)$.
(b) Define the profit function $P(x) = R(x) - C(x)$ and find the value of x that maximizes $P(x)$.
(c) Graph $R(x)$ and $C(x)$ in the same coordinate system and state why the right-hand bound is a natural boundary for the problem.

40. In an article studying the trade-offs that research and development teams must make, an example from space research was analyzed.‡ NASA was trying to determine whether it was cost-effective to manufacture silicon cells for use in future applications. The price of silicon when the study was done (1981) was $65 per kilogram (kg). NASA set a goal of producing silicon at a price of $20 per kilogram. They could make silicon at this price but its quality was not high enough.

---

*Jeffery K. Cochran and F. T. Sparrow, "Optimal Management of a Shared Fleet with Peak Demands," *Applications of Management Science*, Vol. 4, 1985, pp. 81–105.

†Lee G. Anderson and Dwight R. Lee, "Optimal Governing Instrument, Operation Level, and Enforcement in Natural Resource Regulation: The Case of the Fishery," *American Journal of Agricultural Economics*, Vol. 68, 1986, pp. 678–690.

‡Virendra S. Sherlekar and Burton V. Dean, "Assessment of R & D Risk-Cost Trade-Offs," *Applications of Management Science*, Vol. 2, 1983, pp. 40–51.

If they produced silicon at a higher price, what was the likelihood that it would be high quality? The researchers generated a function that measured the likelihood $P(x)$, measured in percent, that the silicon produced at price x would be high quality. The function is approximated by

$$P(x) = -0.0001x^3 + 0.0045x^2 - 0.25$$

There were two natural bounds for x. It made no sense to let x take on negative values and the price of $28 per kg was set as the highest price they would be willing to pay. So the domain of $P(x)$ is [0, 28]. Find the absolute maximum of $P(x)$ and graph the function.

4–5 IMPLICIT DIFFERENTIATION AND RELATED RATES

Thus far the relationships between the variables have been presented as **explicit functions,** meaning that an equation in the form $y = f(x)$ has been given. Two examples are

$$y = x^2 - 4 \qquad y = (x^3 + 4x)^{1/2}$$

However, often in applications equations are given that do not express y as an explicit function of x. For example, the equation

$$x^2 + y^2 = 1$$

whose graph is a circle of radius 1, is not of the form $y = f(x)$. In this section we study equations that are not expressed as functions. They are called **implicit functions.** We also compute their derivatives. Applications of these types of derivatives, called *related rates,* are also studied.

Implicit Functions

The equation $x^2 + y^2 = 1$ does not express y as a function of x directly because, for every value of x in the interval $(-1, 1)$, there are two corresponding values of y that satisfy the equation. For instance, if $x = 0$, then y can be either 1 or -1; that is, $(0, 1)$ and $(0, -1)$ are on the graph. Geometrically, Figure 1 shows that any vertical line $x = a$ for $-1 < a < 1$ intersects the graph in more than one point, so the graph is not a function.

We say that equations such as $x^2 + y^2 = 1$ define y *implicitly* as a function of x. Intuitively this means that even though the equation does not express y in terms of x directly, it is theoretically possible to solve for y. However, this is not always an easy task. For example, given the equation $xy = 1$, it is easy to solve for y to get $y = 1/x$; but it is much more difficult to solve for y given the equation $y^5 - y^3 + xy = 1$. It may be possible to express the equation as several formulas. For example, $x^2 + y^2 = 1$ can be expressed as the two functions $y = \sqrt{1 - x^2}$ and $y = -\sqrt{1 - x^2}$.

Implicit Differentiation

If an equation expresses y as an implicit function of x, it is possible to compute the derivative of y with respect to x without having to solve for y directly. This

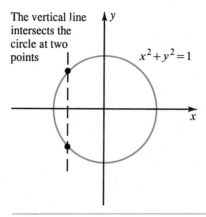

The vertical line intersects the circle at two points

$x^2 + y^2 = 1$

FIGURE 1

technique is called **implicit differentiation.** It assumes that the equation can be expressed in the form $y = f(x)$, and then the derivative of each term of the equation is computed, keeping in mind that y is a function of x. For example, to find y' given the equation $x^2 + y^2 = 1$, find the derivative of each term separately. The derivative of the first term, x^2, is $2x$. However, the derivative of the second term, y^2, is not merely $2y$. The chain rule must be used to compute the derivative of y^2, because it is assumed that $y^2 = [f(x)]^2$. Therefore, using the chain rule, the derivative of y^2 is $2yy'$. The derivative of the constant term, 1, is 0. Putting these steps together and solving for y' yields

$$D_x(x^2 + y^2) = D_x 1$$
$$2x + 2yy' = 0$$
$$2yy' = -2x$$
$$y' = -\frac{2x}{2y}$$
$$y' = -\frac{x}{y}$$

The equation for the derivative contains both x and y, which is usually the case when computing a derivative implicitly. To evaluate the derivative at a particular point both the x and y coordinates must be substituted into the equation.

The important idea to remember when computing implicit derivatives is that the derivative of any term containing y will require the use of the chain rule so that it will have y' as part of its derivative. Also, any term containing a product, such as x^2y^3, will require the product rule as well as the chain rule. For instance, suppose an implicit function contains the term x^2y^3, then the derivative of this term is

$$D_x(x^2y^3) = x^2 D_x(y^3) + y^3 D_x(x^2)$$
$$= x^2 \cdot 3y^2 y' + y^3 \cdot 2x$$
$$= 3x^2 y^2 y' + 2xy^3$$

Example 1 illustrates how to compute the derivative of an implicit function and how to evaluate the derivative at a point that satisfies the implicit function.

EXAMPLE 1 | **Problem**

Consider the equation

$$x^3 + x^2y - y^4 = 1$$

(a) Find y' using implicit differentiation.
(b) Evaluate y' at $(1, 1)$.

SOLUTION (a) Find the derivative of each term.

$$D_x(x^3 + x^2y - y^4) = D_x(1)$$
$$D_x(x^3) + D_x(x^2y) - D_x(y^4) = 0$$
$$D_x(x^3) + x^2D_x(y) + yD_x(x^2) - D_x(y^4) = 0$$
$$3x^2 + x^2{\cdot}y' + y{\cdot}2x - 4y^3y' = 0$$
$$3x^2 + x^2y' + 2xy - 4y^3y' = 0$$

Gather terms containing y'.

$$x^2y' - 4y^3y' = -3x^2 - 2xy$$
$$(x^2 - 4y^3)y' = -3x^2 - 2xy$$
$$y' = (-3x^2 - 2xy)/(x^2 - 4y^3)$$

(b) Substitute $x = 1$ and $y = 1$ into the equation for y' in (a).

$$y' = \frac{-3 - 2}{1 - 4} = \frac{-5}{-3} = \frac{5}{3}$$

Often in applications the two variables in an implicit function both depend on a third variable. For example, suppose x and y are related by the equation $x^2 + y^2 = 2500$ where x and y represent distances, and each is changing with respect to time. If t represents time, then x and y are functions of t. By differentiating the equation relating x and y, we get a relationship in the two derivatives D_tx and D_ty. The chain rule must be used on each term, including x^2. That is, the derivative of x^2 with respect to t is not merely $2x$ but $2xx'$, where x' is D_tx. Example 2 shows how to carry this out.

EXAMPLE 2 | Let x and y be related by the equation

$$x^2 + y^2 = 2500$$

Suppose x and y change with respect to time, t.

Problem

Find an equation relating D_tx and D_ty. Then evaluate the equation for $x = 40$, $y = 30$, and $D_ty = -4$.

SOLUTION Think of the equation as an implicit function of t, where both x and y are implicit functions of t. Differentiate the equation with respect to t by differentiating each term with respect to t. This yields

$$D_t(x^2 + y^2) = D_t 2500$$
$$2x \cdot D_t x + 2y \cdot D_t y = 0$$
$$2x D_t x = -2y D_t y$$
$$D_t x = \frac{-y D_t y}{x}$$

Evaluating the equation for $D_t x$ when $x = 40$, $y = 30$, and $D_t y = -4$ yields

$$D_t x = \frac{-30(-4)}{40}$$

$$= 3$$

Related Rates

In Example 2 the variables x and y are functions of a third variable t. The rate of change of x is related to the rate of change of y by the formula derived in the example by differentiating the original implicit function with respect to t. Such a problem is called a **related rates** problem. The next example illustrates this type of application.

EXAMPLE 3

A tropical storm is 50 miles offshore, and its path is perpendicular to a straight shoreline. It is approaching the shore at the rate of 4 mph. Meteorologists studying the behavior of the storm from a van on the shore want to stay exactly 50 miles from the storm and remain on the shoreline. They start at the point on the shoreline in the path of the storm.

Problem

Determine a formula for the speed that the truck must maintain to remain 50 miles from the storm. Then find the speed of the truck when the storm is 40 miles from the shore.

SOLUTION Usually the first step in a related rates problem is to make a sketch of the problem. From the sketch the variables can be labeled. See Figure 2. Let the distance from the storm to the shore be represented by y and the distance that the van has traveled be represented by x. From the Pythagorean Theorem the relationship between x and y is

$$x^2 + y^2 = 50^2 = 2500$$

The problem asks for $D_t x$, given that $D_t y = -4$, which is negative because y is decreasing. The relationship between $D_t x$ and $D_t y$ is derived by differentiating the formula relating x and y with respect to t. This was done in Example 2. The relationship is

$$D_t x = \frac{-y D_t y}{x}$$

Since $D_t y = -4$, the formula becomes

$$D_t x = \frac{4y}{x}$$

When the storm is 40 mi from the shore; that is, $y = 40$, then from the Pythagorean Theorem $x = 30$. Therefore the speed of the van when $y = 40$ is

$$D_t x = \frac{4(40)}{30} = \frac{16}{3} \text{ mph}$$

FIGURE 2

Example 3 demonstrates the six steps used in solving related rates problems.

The Key Steps in Solving Related Rates Problems

1. Draw a sketch of the problem if possible.
2. Identify all quantities, both constants and variables, and give labels to the variables.
3. Find a formula relating the variables.
4. Using implicit differentiation, find the derivative with respect to time, t, of each side of the formula. Remember that each variable in the formula is a function of t.
5. The problem asks for the rate of a particular variable. Solve the formula derived in step 4 for this derivative.
6. Substitute the given quantities into the formula for the desired derivative.

It is important to perform steps 4 and 5 before doing the substitution called for in step 6. If the substitution is made beforehand, the variables will not appear in the formula when applying implicit differentiation. Thus the formula derived in step 4 will be incorrect. To illustrate, suppose in Example 3 the value $y = 40$ is substituted into the formula $x^2 + y^2 = 2500$ before the derivative is found. This would result in the formula $x^2 + 1600 = 2500$. Using implicit differentiation on this formula results in $2xx' = 0$, so $x' = 0$, which is certainly incorrect.

The next two examples illustrate how to apply the steps used to solve related rates problems.

EXAMPLE 4

A rock is thrown into a still pond and circular ripples move out. The radius of the disturbed region increases at the rate of 3 ft/sec.

Problem

Find the rate at which the surface area of the disturbed region is increasing when the farthest ripple is 20 feet (ft) from the place where the rock struck the pond.

SOLUTION The sketch of the problem is one of concentric circles given in Figure 3, each circle representing a ripple. The farthest ripple is the largest circle. Let the radius of the farthest ripple be represented by r and the area of the disturbed region by A. The formula relating r and A is the area of a circle.

$$A = \pi r^2$$

It is given that $D_t r = 3$ and we are asked to find $D_t A$ when $r = 20$. Next use implicit differentiation to find $D_t A$ in terms of $D_t r$.

$$D_t A = 2\pi r D_t r$$

Substitute into this formula $r = 20$ and $D_t r = 3$.

$$D_t A = 2\pi(20)(3) = 120\pi \approx 376.99 \text{ ft}^2/\text{sec}$$

Thus when the radius is 20 ft the surface area of the disturbed region is increasing at the rate of about 377 ft²/sec.

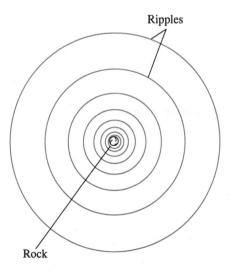

Ripples

Rock   **FIGURE 3**

EXAMPLE 5

An industrial filter in the shape of a cone is 10 cm (centimeters) high and has a radius of 20 cm at its top. A solution is poured through the filter, and residue gathers in the filter at the rate of 2 cm³/min.

Problem

Find the rate at which the height of the residue is increasing when the height is 2 cm.

SOLUTION The sketch is given in Figure 4. Let the height of the residue be represented by h, the radius by r, and the volume by V. The residue gathers in the shape of a cone. The formula for the volume of a cone is

$$V = \left(\frac{1}{3}\right) \pi r^2 h$$

It is given that $D_t V = 2$ and we are asked to find $D_t h$ when $h = 2$. It is possible to use implicit differentiation to find $D_t h$ in terms of $D_t V$ and $D_t r$, but the value of $D_t r$ is unknown. Since V is expressed as a function of two variables, r and h, and only one value of the two derivatives is given, it is necessary to express V as a function of h alone. The relationship between r and h is given by similar triangles. From Figure 4 the triangle of the cone and the triangle of the residue are similar. The ratios of corresponding sides of similar triangles are equal, which yields

$$\frac{r}{20} = \frac{h}{10}$$

Thus $r = 2h$. Substituting this into the formula for V yields

$$V = \left(\frac{1}{3}\right) \pi (2h)^2 h = \left(\frac{4}{3}\right) \pi h^3$$

Applying implicit differentiation to the formula yields

$$D_t V = \left(\frac{4}{3}\right) \pi \cdot 3h^2 D_t h$$
$$= 4\pi h^2 D_t h$$

Solving for $D_t h$ and making the substitution $D_t V = 2$ gives

$$D_t h = \frac{2}{4\pi h^2}$$

This is the formula for $D_t h$. Next substitute the value $h = 2$ to find the rate at which the height of the residue is increasing when the height is 2 cm.

$$D_t h = \frac{2}{4\pi 2^2} = \frac{1}{8\pi} \approx 0.04 \text{ cm/min}$$

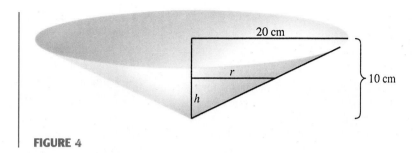

FIGURE 4

EXERCISE SET 4–5

• In Problems 1 to 8 find $D_x y$ by using implicit differentiation.

1. $x^2 + 3y^2 = 4$

2. $x^2 - 5y^3 = 6$

3. $xy + x^2 y = 2$

4. $xy - x^2 y^2 = 4$

5. $x^3 + xy^3 - y^4 = 1$

6. $x^4 - x^2 y^3 + y^5 = 4$

7. $x^2 + xy^3 - y^{-1} = 1$

8. $x^4 - 2x^2 y + y^{-2} = 0$

In Problems 9 to 12 find $D_x y$ evaluated at the given point.

9. $x^2 + y^3 = 0; \quad (1, -1)$

10. $x^3 + y^2 = 2; \quad (1, -1)$

11. $3xy + x^2 y = 3; \quad (1, 0)$

12. $4xy - x^2 y^2 = 3; \quad (1, 1)$

In Problems 13 to 16 x and y are functions of t. Express $D_t y$ in terms of $D_t x$. Then find $D_t y$ when $x = 1$, $y = 1$, and $D_t x = 2$.

13. $x^2 + y^3 = 2$ 14. $x^3 + 4y^3 = 5$

15. $3x - x^2 + y^2 = 3$

16. $x^{1/2} + 3y - y^{1/2} = 3$

17. A tropical storm is 100 miles offshore, and its path is perpendicular to a straight shoreline. It is approaching the shore at the rate of 6 mph. Meteorologists studying the behavior of the storm from a van on the shore want to stay exactly 100 miles from the storm and remain on the shoreline. Determine the speed of the truck when the storm is 60 miles from the shore.

18. A rock is thrown into a still pond and circular ripples move out, the radius of the disturbed region increasing at the rate of 2 ft/sec. Find the rate at which the surface area of the disturbed region is increasing when the farthest ripple is 30 ft from the place where the rock struck the pond.

19. A filter in the shape of an inverted cone is 20 cm high and has a radius of 30 cm at its top. A solution is poured through the filter, and residue gathers in the filter at the rate of 5 cm³/min. Find the rate at which the height of the residue is increasing when the height is 3 cm.

20. A firm determines that output y is related to input x, both measured in hundreds of units, by the formula $y = 3x^{1/3}$. If the demand is increasing at the rate of 20 units per month, at what rate must the input increase to meet this increase in demand if the current input is 2700 units?

•• 21. A ladder 50 ft long is standing against a building. The base of the ladder is moved away from the building at the rate of 3 ft/sec. How fast is the top of the ladder moving down the wall when the base of the ladder is 40 ft from the foot of the building?

22. Water is being pumped into a cylindrical trough at the rate of 10 ft³/min. If the depth of the trough is 12 ft, the width of the top is 12 ft, and the length of the top is 20 ft, at what rate is the height of the water rising when the height is 6 ft?

23. A 6-ft tall person is walking away from a streetlight at the constant rate of 3 ft/sec. If the streetlight is 20 ft high, how fast is the length of the person's shadow increasing when the person is 30 ft from the streetlight?

24. A water tank is 30 meters (m) wide, 60 m long, and 10 m deep at one end and 15 m deep at the other end, with the depth increasing linearly. Water is pumped in at the rate of 5 m³/min. How fast is the water level rising when the water level is 3 m.?

25. A spherical rubber balloon is being inflated with gas at a constant rate of 15 cm³/sec. What is the rate of change of the radius when the volume is 45 cm³?

26. Water is condensing on the surface of a spherical drop of water so that the water surface remains a sphere and the volume is increasing at the rate of 5 in.³/min. At what rate are the surface area and the radius increasing when the radius is 10 in.?

27. A pulley, mounted on a dock, stands 10 ft above the water and a boat is being towed in toward the dock. The rope is attached to the boat at a point 2 ft above the water. If 0.5 ft of rope is being drawn in each second, at what rate is the boat moving when it is 6 ft from the dock?

28. An off-shore oil drill springs a leak. The oil slick spreads in a circular shape, and the radius of the slick is increasing at the rate of 30 ft/hr. Find the rate that the surface area of the oil slick is increasing when the radius is 100 ft.

29. A balloon is rising vertically at the rate of 15 ft/sec. An observer is standing on the ground 900 ft from the point where the balloon left the ground. Find the rate that the distance between the balloon and the observer is changing when the balloon is 1200 ft high.

30. Two roads intersect at right angles. A truck, traveling at the rate of 30 mph, reaches the intersection 5 minutes (min) before a car that is traveling at the rate of 40 mph on the other road. How fast is the distance between the truck and the car changing 5 min after the car reaches the intersection?

••• 31. One airplane, traveling due north at 200 mph, passes a point 1 hour after another airplane, traveling due east at 150 mph, passes the same point. How fast is the distance between the airplanes increasing 1 hour after the first airplane passed the point?

32. A baseball player, running at the speed of 30 ft/sec, is running in a straight line from second to third base. A fielder, standing at a point 100 ft from third base on the left field foul line (which is at right angles to the path of the runner), throws the ball, at the speed of 150 ft/sec, when the runner is 50 ft from third base. What is the rate of change of the distance between the ball and the runner 0.5 sec after the fielder throws the ball?

33. When helium expands, its pressure P is related to its volume V by the formula $P = kV^{-5/3}$. When the volume of a helium-filled balloon is 18 m³ the pressure is 0.3 kg/m². What is the rate of change of the volume if the pressure is increasing at the rate of 0.05 kg/m²/sec?

34. Liquid is being drawn at the rate of 2 m³/hr from a tank whose shape is a hemisphere with radius 10 m. How fast is the height of the liquid level changing 2 hours after the tank was full?

35. In a paper* studying efficient glass-cutting techniques, various cost-saving measures were explored, including minimizing the amount of waste

*Adolf Diegel and Hans Vocker, "Optimal Dimensions of Virgin Stock in Cutting Glass to Order," *Decision Sciences*, Vol. 15, 1984, pp. 260–272.

FIGURE 5 *Source:* Adolf Diegel and Hans Vocker, "Optimal Dimensions of Virgin Stock in Cutting Glass to Order," *Decision Sciences,* Vol. 15, 1984, pp. 260–272.

and the time used in the cutting procedure. Once a firm has an inventory of various sizes of virgin stock, the problem is to determine what sizes of glass should be cut from the different sheets of stock. Solve the following problem, which is representative of those considered in the paper. Two pieces of glass are to be produced by making two cuts in a sheet of stock. A vertical cut will produce one piece and a horizontal cut in the remaining piece of stock will produce the second piece. The piece left over is called the waste and it is to be made square. (See Figure 5.) If the perimeter of the stock plate is 20 in. and the total surface area of two pieces produced by the cuts is 20 in.^{2}, find the dimensions of the stock plate that minimizes the total length of the two cuts.

36. Population structure and regeneration patterns of trees are important aspects of forest management. In a paper studying size and age structure of two forested strands in the western Cascade Range in Oregon, Stewart studied four species of trees growing in various types of habitats.[†] In one part of the study a typical problem involved planting two types of trees, each planted on a different square plot of land. In the first plot seedlings were planted in rows every $1\frac{1}{3}$ meters, and in the other plot the seedlings were planted in rows every 4 meters. Find the minimum area of each plot that would be needed to plant 320 rows of trees.

[†]Glenn H. Stewart, "Population Dynamics of a Montane Conifer Forest, Western Cascade Range, Oregon, USA," *Ecology,* Vol. 67, 1986, pp. 534–544.

CASE STUDY   Minimizing Traffic Noise

The private car is by far the most popular means of transportation. However, for many people the car represents more than convenient transportation; their dreams and egos are front seat passengers. It is not surprising that the automobile population rate far outdistances its human counterpart. The role of highway transportation in our lives is profound. Indeed, a significant measure of a country's "development" is its highway system.

The Department of Transportation used to be an agency whose sole responsibility was the structure of roads. Most of our highways were built with static road design as the only critical issue. No concern was given to dynamic factors such as traffic flow, ramp control, merging, and congestion. But today traffic controllers must consider a multitude of factors to minimize traffic snarls and accidents. Estimating traffic demand and capacity has become as much an art as it is a science.

One of the most important considerations of highway operation is acceleration noise control. In the past the sole criterion for improving highway operations was to maximize volume throughout. Little consideration was given to the quality of highway conditions. For example, it is standard practice to reduce highway bottlenecks by ramp control techniques, such as using stoplights to control entrance to the roadway. But nowadays highway administrators give equal concern to the qualitative changes in traffic due to these controls. For instance, a particular control might have reduced overall travel time, but perhaps it created hazardous locations such as rapid decelerations far from the bottlenecks. Measurement of acceleration noise is a valuable tool to identify any hazardous locations. Noise control not only provides more comfortable travel, but it also helps maintain a safer and smoother operation.

One of the first and most extensive traffic flow research projects* was conducted by the Texas Highway Department and the U.S. Bureau of Public Roads in 1963, called the "Gulf Freeway Project." Various surveillance and control devices, including television and noise monitors, were installed along the Gulf Freeway in Houston. The freeway was divided into 56 sections, and each was monitored for several characteristics, including noise level.

Acceleration noise is defined as the disturbance of a vehicle's speed from a uniform speed. It is thus a measure of the deviation from a smooth ride and so it measures the smoothness of traffic. The frequency of violent acceleration and deceleration affects acceleration noise and reflects potentially hazardous conditions.

The primary goal of the Gulf Freeway study was to determine the means to ensure efficient traffic flow. This meant not only maximizing volume but also minimizing the threat of hazardous conditions by minimizing acceleration noise. What range of speeds would best satisfy both of these criteria?

Acceleration noise is a function of speed. Figure 1 is a representation of the relationship of acceleration noise, a, to the average speed of traffic, u, for 1 of the 56 sections of the freeway. The curve that represents the data was given as

$$a = 1.693 - 0.00284u^2 + 0.000045u^3$$

where a is measured in ft/sec^2 and u is measured in miles per hour(mph). This function is valid only in the interval [10, 60]. To see why 10 is a natural endpoint, refer to Figure 1: when $u = 0$, the vehicle is at rest, and $a = 0$. As the vehicle starts to accelerate quickly to merge, acceleration noise increases rapidly. It is represented by the dashed line. As the vehicle enters the traffic pattern, acceleration noise decreases.

*Donald R. Drew, *Traffic Flow Theory and Control*, New York, McGraw-Hill Book Co., 1968.

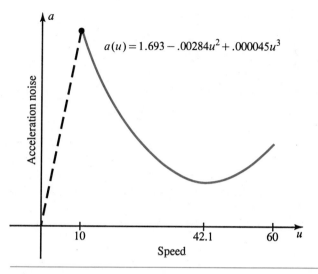

$$a(u) = 1.693 - .00284u^2 + .000045u^3$$

FIGURE 1   *Source:* Donald R. Drew, *Traffic Flow Theory and Control,* New York, McGraw-Hill Book Co., 1968.

The speed that yields minimum acceleration noise is found by calculating the derivative, a', and setting $a' = 0$. The derivative is computed as follows:

$$a' = -2 \cdot 0.00284u + 3 \cdot 0.000045u^2$$
$$= -0.00568u + 0.000135u^2$$
$$= u(-0.00568 + 0.000135u)$$

Setting $a' = 0$ yields

$$u(-0.00568 + 0.000135u) = 0$$

This means that either $u = 0$ or $-0.00568 + 0.000135u = 0$. But 0 is not in the domain of the function so the only critical value is the latter one. To find the critical value, we solve for u.

$$-0.00568 + 0.000135u = 0$$
$$0.000135u = 0.00568$$
$$u \approx 42.1$$

The second derivative is $a'' = 0.000135$ and $a''(42.1) > 0$, so a has a minimum at $u = 42.1$. Thus the speed that yields the minimum amount of acceleration noise is 42.1 mph.

In another study the speed that yields the maximum flow of volume was found to be 35 mph. This allowed the highway administrators to establish a target speed interval of 35 to 42 mph. This ensured that while the freeway's average speed fluctuated within this interval, they were either maximizing volume, if the speed was close to 35 mph, or minimizing acceleration noise, if it was close to 42 mph.

Case Study Exercises

In Problems 1 to 4 the acceleration noise levels of several other sections of the Gulf Freeway are given. Find the speed at which the acceleration noise is a minimum.

1. $a = 1.572 - 0.00345u^2 + 0.000078u^3$

2. $a = 1.916 - 0.00402u^2 + 0.000088u^3$

3. $a = 2.006 - 0.00501u^2 + 0.000131u^3$

4. $a = 2.118 - 0.00298u^2 + 0.000073u^3$

5. If an acceleration noise test was done on a highway passing through a large city, would you expect the level of acceleration noise to be greater at 1:00 A.M. or at 5:00 P.M. and why?

CHAPTER REVIEW

Key Terms

4–1   Increasing and Decreasing Functions
Increasing Function   Decreasing Function
Critical Value

4–2   Relative Extrema: The First Derivative Test
Extrema   Relative Minimum
Maximum   Relative Extremum
Minimum   Critical Value
Relative Maximum   First Derivative Test

4–3   Concavity and the Second Derivative Test
Concave Up   Point of Inflection
Concave Down   Curve Sketching

4–4   Absolute Extrema and Optimization Problems
Absolute Maximum   Absolute Extremum
Absolute Minimum

4–5   Implicit Differentiation and Related Rates
Explicit Function   Implicit Differentiation
Implicit Function   Related Rates

Summary of Important Concepts

First Derivative Test: If $f(a)$ is a critical value of $f(x)$, then

1. $f(a)$ is a relative maximum if $f(x)$ is increasing to the left of a and decreasing to the right of a;

2. $f(a)$ is a relative minimum if $f(x)$ is decreasing to the left of a and increasing to the right of a.

The Second Derivative Test: If a is a critical value of $f(x)$, then

1. $f(a)$ is a relative maximum if $f''(x) < 0$ at $x = a$;
2. $f(a)$ is a relative minimum if $f''(x) > 0$ at $x = a$;
3. the test fails if $f''(a) = 0$.

REVIEW PROBLEMS

In Problems 1 to 4 find intervals where $f(x)$ is increasing and decreasing. Find the point(s) where $f(x)$ has a horizontal tangent line. Sketch the graph of the function.

1. $f(x) = 9x^2 - 25$ **2.** $f(x) = x^2(4 - x)$

3. $f(x) = x/(3 - x)$ **4.** $f(x) = x^2/(4 - x)$

In Problems 5 to 8 find the relative extrema, and sketch the graph of the function.

5. $f(x) = x^2 + 6x - 10$

6. $f(x) = x^2(x - 4)$

7. $f(x) = x^2(4 - x^2)$ **8.** $f(x) = 3x^{2/3} - 2x$

In Problems 9 and 10 use the second derivative test to find the relative extrema.

9. $f(x) = 3x^4 - 4x^3$ **10.** $f(x) = x^4 - 2x^2$

In Problems 11 and 12 find the intervals where the function is concave up and concave down, and sketch the graph.

11. $f(x) = 2x^3 - x^4$ **12.** $f(x) = x^4 - 6x^2$

In Problems 13 to 15 find the absolute extrema, if they exist.

13. $f(x) = x^3 - 3x^2 + 4$ on $[-2, 2]$

14. $f(x) = (x + 1)^2(x - 1)^2$ on $[0, 4]$

15. $f(x) = x - x^{1/3}$ on $[-1, 8]$

16. An open box is to be constructed from a 10 by 12 inch rectangular piece of cardboard by cutting out equal squares at each corner and then folding up the sides. What are the dimensions of the box with maximum volume?

In Problems 17 and 18 find D_xy by using implicit differentiation.

17. $x^2 + xy + y^2 = 1$

18. $x^3 - 4x^2y + y^{-2} = 1$

19. A water tank is 10 meters (m) wide, 40 m long, and 20 m deep at one end and 30 m deep at the other end, with the depth increasing linearly. Water is pumped in at the rate of 5 m^3/min. How fast is the water level rising when the water level is 3 m.?

5 Exponential and Logarithmic Functions

CHAPTER OVERVIEW   Perhaps the most useful and universal functions in the application of calculus to the real world are exponential and logarithmic functions. They are used in such diverse areas as the growth of a population, the decay of radioactive material, the growth of investment, and the rate of learning. The first section defines exponential functions and investigates some of their properties. Some graphs of simple types of exponential functions are sketched. The second section does the same for logarithmic functions. Sections 5–3 and 5–4 cover the derivatives of these functions and show how to use graphing techniques to draw the graphs of more complicated functions.

CASE STUDY PREVIEW   **The Frustration of Dieting**

Losing weight can be a trying experience. Excuses abound for our inability to diet successfully. This Case Study describes the mathematical foundation for weight control. It offers the frustrated dieter a bit of evidence for the case against short-term dieting.

CASE STUDY PREVIEW   Sears, Clothespins, and Unnatural Logarithms

Sears Roebuck and Co. started its mail-order business at the turn of the 20th century with the same kind of catalogs that are used today. These catalogs are still a major concern for Sears, but they also serve several purposes for economists. The Case Study uses logarithmic functions to explore the U.S. economy and lifestyle by examining automobiles, the Consumer Price Index, and clothespins.

CASE STUDY PREVIEW   Denver Fire Department Deployment Policy Analysis

In 1972 the mayor and fire chief of Denver agreed that a study of the fire department was in order in the face of mounting costs. The challenge was to minimize cost while keeping the same level of fire protection. A research team from the University of Colorado was·commissioned to solve the problem. An exponential model was used to predict future trends by means of simulation methods.

5–1   EXPONENTIAL FUNCTIONS

In this section we introduce a new kind of function, called an exponential function, in which the exponent is a variable instead of a constant. An exponential function describes an object that "grows exponentially," meaning that it either grows or shrinks at an increasingly fast rate. We draw the graphs of exponential functions, study some of their properties, and define *the* exponential function involving the number e. The material is applied to an area of psychology that deals with learning curves.

Graphs

The graph of the quadratic function $f(x) = x^2$ is a parabola. It is quite different from the graph of the function $f(x) = 2^x$, which is called an exponential function because the exponent is a variable. Table 1 lists some points on the graph of $f(x) = 2^x$. These points are joined by the curve in Figure 1. In completing the table, remember that negative exponents refer to reciprocals, so that, for instance,

$$2^{-3} = \frac{1}{2^3} = \frac{1}{8}.$$

Table 1 lists only integral exponents, but exponential functions are defined for all real numbers. Rational exponents refer to roots. For instance, $2^{1/2} = \sqrt{2}$,

TABLE 1

x	-3	-2	-1	0	1	2	3
2^x	$\frac{1}{8}$	$\frac{1}{4}$	$\frac{1}{2}$	1	2	4	8

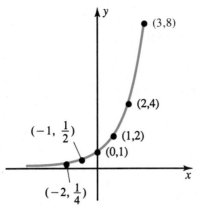

Graph of $f(x) = 2^x$

FIGURE 1

$2^{5/3} = \sqrt[3]{2^5} = (\sqrt[3]{2})^5$, and $2^{-3/4} = \dfrac{1}{2^{3/4}} = \dfrac{1}{\sqrt[4]{2^3}} = \dfrac{1}{\sqrt[4]{8}}$. The meaning of an exponential function at an irrational exponent, such as $2^{\sqrt{2}}$, is beyond the scope of this book. However, it can be viewed intuitively from the graph in Figure 1 or derived from a calculator experiment by examining the limit of the numbers $2^{1.4}$, $2^{1.41}$, $2^{1.412}$, ...

_____ Definition _____

If b is a positive number with $b \ne 1$, then a function $f(x) = b^x$ is called an **exponential function.** The number b is the **base** of $f(x)$.

The restriction $b \ne 1$ prevents the graph of $f(x) = b^x$ from being the horizontal line $f(x) = 1$. The restriction $b > 0$ ensures that the domain of an exponential function is the set of all real numbers, since otherwise there would be some values of x for which b^x is not a real number. For example, if $b = -1$ and $x = \frac{1}{2}$, then $(-1)^{1/2} = \sqrt{-1}$ is not a real number.

EXAMPLE 1

Problem

Sketch the graph of $f(x) = 3^x$.

SOLUTION Table 2 lists some points on the graph, which is sketched in Figure 2.

TABLE 2

x	-3	-2	-1	0	1	2	3
3^x	$\frac{1}{27}$	$\frac{1}{9}$	$\frac{1}{3}$	1	3	9	27

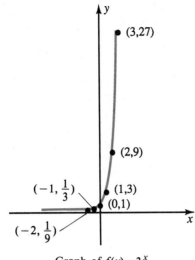

Graph of $f(x) = 3^x$

FIGURE 2

The graphs of $f(x) = 2^x$ and $f(x) = 3^x$ are typical of the graphs of those exponential functions $f(x) = b^x$ with $b > 1$. The graphs pass through the point $(0, 1)$ and increase over the entire domain. As x gets larger and larger the graphs increase without bound, which is expressed compactly by writing $\lim_{x \to \infty} b^x = \infty$. In addition, $\lim_{x \to -\infty} b^x = 0$, so the x-axis is a horizontal asymptote of the graph of $f(x)$.

Example 2 examines a typical exponential function $f(x) = b^x$ for $0 < b < 1$.

EXAMPLE 2

Problem

Sketch the graph of $f(x) = \left(\dfrac{1}{2}\right)^x$.

SOLUTION Write $\left(\dfrac{1}{2}\right)^x$ in the form $\left(\dfrac{1}{2}\right)^x = \dfrac{1}{2^x}$. Then construct Table 3. The graph is drawn in Figure 3.

TABLE 3

x	-3	-2	-1	0	1	2	3
$(\frac{1}{2})^x$	8	4	2	1	$\frac{1}{2}$	$\frac{1}{4}$	$\frac{1}{8}$

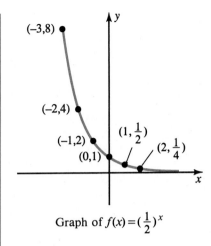

Graph of $f(x) = (\frac{1}{2})^x$

FIGURE 3

Properties

The graph of an exponential function $f(x) = b^x$ shows that if $b^r = b^s$, then $r = s$. This is the first of several properties of exponential functions that we list for reference. The other properties should be familiar from a previous study of algebra.

1. If $b^r = b^s$, then $r = s$
2. $b^r \cdot b^s = b^{r+s}$
3. $\dfrac{b^r}{b^s} = b^r \cdot b^{-s} = b^{r-s}$
4. $(b^r)^s = b^{rs}$

EXAMPLE 3

Problem

Find all values of x for which (a) $2^x = 64$ and (b) $9^{(x+3)} = 3^{(3x-9)}$.

SOLUTION (a) Since $2^6 = 64$, one value of x for which $2^x = 64$ is $x = 6$. According to property 1, this is the only value.
(b) Write $9^{(x+3)} = (3^2)^{(x+3)}$. By property 4, $(3^2)^{(x+3)} = 3^{2(x+3)}$. Thus $9^{(x+3)} = 3^{(2x+6)}$. It was given that $9^{(x+3)} = 3^{(3x-9)}$, so $3^{(2x+6)} = 3^{(3x-9)}$. By property 1, $2x + 6 = 3x - 9$. Therefore $x = 15$. This answer can be verified using the properties because $3^{(45-9)} = 3^{36} = (3^2)^{18} = 9^{18} = 9^{(15+3)}$.

THE Exponential Function

Consider the function defined by

$$f(x) = \left(1 + \frac{1}{x}\right)^x$$

Then $f(1) = 2$ and $f(2) = \left(\frac{3}{2}\right)^2 = 2.25$. A calculator with a $\boxed{y^x}$ key was used to compute the additional values of $f(x)$ in Table 4. The numbers are rounded to five decimal places. Table 4 indicates that $f(x)$ approaches $2.71828 \ldots$ as x gets larger and larger. The exact number that $f(x)$ approaches is denoted by the letter e.

—————————————————— Definition ——————————————————

$$e = \lim_{x \to \infty} \left(1 + \frac{1}{x}\right)^x$$

Since e lies between 2 and 3, the graph of the function $f(x) = e^x$ is sandwiched between the graphs of $f(x) = 2^x$ and $f(x) = 3^x$, as shown in Figure 4. Although the number e might seem to be contrived, the function $f(x) = e^x$ occurs so often that it is called *the* exponential function. A table in the Appendix lists the values of e^x for various values of x. Many calculators have an $\boxed{e^x}$ key. An approximation of e^k can be obtained by the sequence $\boxed{k}\ \boxed{e^x}$, although a color-coded key sometimes has to be pressed before tapping the $\boxed{e^x}$ key.

TABLE 4

x	$f(x)$
1	$(2/1)^1 = 2$
2	$(3/2)^2 = 2.25$
3	$(4/3)^3 \approx 2.37037$
4	$(5/4)^4 \approx 2.44141$
5	$(6/5)^5 \approx 2.48832$
6	$(7/6)^6 \approx 2.52163$
7	$(8/7)^7 \approx 2.54650$
8	$(9/8)^8 \approx 2.56578$
9	$(10/9)^9 \approx 2.58117$
10	≈ 2.59374
100	≈ 2.70481
1,000	≈ 2.71692
10,000	≈ 2.71815
100,000	≈ 2.71827
1,000,000	≈ 2.71828
10,000,000	≈ 2.71828

Graphs of three
exponential functions

FIGURE 4

_____ Definition _____

The function $f(x) = e^x$ is called **the exponential function.** The domain consists
of all real numbers and the range is the set of positive numbers.

Example 2 examined the graph of the function $f(x) = \left(\dfrac{1}{2}\right)^x$, which is typical
of all exponential functions of the form $f(x) = b^x$ for $0 < b < 1$. Example 4
examines the function $f(x) = e^{-x}$. It is important to remember that

$$e^{-x} = \left(\frac{1}{e}\right)^x = \frac{1}{e^x}$$

EXAMPLE 4 | **Problem**

Sketch the graph of $f(x) = e^{-x}$.

SOLUTION Construct Table 5 by using a calculator or by selecting the
appropriate values for x from the table in the Appendix. The graph of
$f(x) = e^{-x}$ is sketched by joining these points with the smooth curve shown in
Figure 5.

TABLE 5

x	-2	-1	0	1	2
e^{-x}	7.38906	2.71828	1	0.36788	0.13534

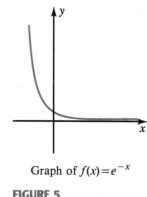

Graph of $f(x) = e^{-x}$

FIGURE 5

The graph of $f(x) = e^{-x}$ is similar to the graph of $f(x) = \left(\dfrac{1}{2}\right)^x$. To see this, compare Figure 5 with Figure 3. Notice that both curves pass through the point $(0, 1)$, decrease for all values of x, have the x-axis as a horizontal asymptote, and increase without bound as the values of x get smaller and smaller. The last two properties can be written $\lim\limits_{x \to \infty} e^{-x} = 0$ and $\lim\limits_{x \to -\infty} e^{-x} = \infty$.

Learning Curves

We have all experienced the range of emotions involved in learning a new skill, be it a physical activity like jumping rope or a mental discipline like differentiating functions. At first we are overwhelmed by the seeming complexity of the new task, but soon progress begins and improvement proceeds apace. However, later we reach a point at which further progress requires increasingly more time and dedication.

Psychologists introduced the name "learning curve" to deal with such situations in which a person's ability to learn a new skill is a function of the time devoted to learning it. A **learning curve** is the graph of an exponential function of the form $f(t) = a - be^{-ct}$, where a, b, and c are positive constants. Example 5 presents a typical setting for a learning curve.

EXAMPLE 5

The rate at which a person can learn to type is given by $f(t) = 70 - 60e^{-0.1t}$, where t is in weeks and $f(t)$ is in words per minute (wpm).

Problem

How many words per minute will the person ultimately be able to type?

SOLUTION We derive the answer from the learning curve. The domain of $f(t)$ is restricted to $t \geq 0$ due to physical considerations. Without any training $(t = 0)$, the person can type $f(0) = 70 - 60 = 10$ wpm. After one week the person can type $f(1) = 70 - 60e^{-0.1} = 70 - 54.3 = 15.7$ wpm. Table 6 lists additional time periods. (All wpm are rounded to one decimal place.)

The learning curve is the graph that is obtained from this table. It is sketched in Figure 6 and shows that $\lim\limits_{t\to\infty} f(t) = 70$, so as time goes by, the person's speed approaches 70 wpm.

TABLE 6

t	0	1	2	4	10	20	30	40
$f(t)$	10	15.7	20.9	29.8	47.9	61.9	67	68.9

FIGURE 6

EXERCISE SET 5–1

• In Problems 1 to 8 sketch the graph of the stated function by the plotting method.

1. $f(x) = 5^x$
2. $f(x) = 4^x$
3. $f(x) = 1.5^x$
4. $f(x) = 2.5^x$
5. $f(x) = (1/3)^x$
6. $f(x) = (1/4)^x$
7. $f(x) = (0.2)^x$
8. $f(x) = (0.1)^x$

In Problems 9 to 14 find all values of x that satisfy the given equality.

9. $2^x = 32$
10. $2^x = 128$
11. $5^x = 625$
12. $7^x = 16,807$
13. $9^{x+6} = 3^{20-2x}$
14. $9^{x+3} = 3^{2x+10}$

In Problems 15 to 18 sketch the graph of the given exponential function.

15. $f(x) = e^{x+2}$
16. $f(x) = e^{x-2}$

17. $f(x) = e^{2-x}$
18. $f(x) = e^{-2x}$

In Problems 19 and 20 $f(t)$ represents the rate at which a person can learn to type, where t is in weeks and $f(t)$ is in words per minute (wpm). Find (a) the number of wpm that the person will be able to type after 50 weeks and (b) the number of wpm that the person will ultimately be able to type.

19. $f(t) = 60 - 55e^{-0.1t}$
20. $f(t) = 45 - 40e^{-0.01t}$

In Problems 21 and 22 $f(t)$ represents the number of bricks that an apprentice bricklayer can lay per hour, where t is in weeks. Find (a) the number of bricks that the apprentice will be able to lay per hour after 10 weeks and (b) the number of bricks that the apprentice will ultimately be able to lay.

21. $f(t) = 28 - 27e^{-0.5t}$

22. $f(t) = 45 - 40e^{-0.2t}$

•• In Problems 23 to 30 sketch the graph of the stated function by the plotting method.

23. $f(x) = 1.1^x$ 24. $f(x) = 0.9^x$

25. $f(x) = 6 - 3e^{-x}$

26. $f(x) = 15 - 20e^{-x}$

27. $f(x) = \dfrac{5}{1 + 3e^{-4x}}$

28. $f(x) = \dfrac{6}{1 + 2e^{-3x}}$

29. $f(x) = \dfrac{e^x - e^{-x}}{2}$ 30. $f(x) = \dfrac{e^x + e^{-x}}{2}$

In Problems 31 to 34 find all values of x that satisfy the given equality.

31. $2^{2x+4} = 4^{8-x}$ 32. $2^{2x-3} = 8^{x-4}$

33. $2^x = 6^x$ 34. $3^x = 2^x$

35. Sketch the graphs of the exponential functions $f(x)$ and $g(x)$ on the same coordinate axis, where

$$f(x) = (e^x)^2 \quad \text{and} \quad g(x) = e^{(x^2)}$$

36. Sketch the graph of the exponential function

$$f(x) = e^{-x^2}$$

In Problems 37 and 38 let $f(x) = e^{cx}$, for some number c, and let $f(2) = 3$.

37. Use the law of exponents to find $f(8)$.

38. Use the law of exponents to find $f(-8)$.

39. If a thermometer is moved from inside a house where it reads 70°F, to outside the house where it is 10°F, the temperature of the thermometer will read

$$T(t) = 10 + 60e^{-0.46t}$$

where t is in minutes and $T(t)$ is in degrees Fahrenheit.
(a) What temperature does the thermometer read after 5 minutes?
(b) What temperature will the thermometer eventually reach?

40. If $1000 is deposited into an account that pays 5% interest compounded annually, the future amount is given by

$$A(t) = 1000(1.05)^t$$

where t is in years and $A(t)$ is in dollars. What is the future amount after 10 years?

41. It is estimated that the population of a certain city will be given by the equation

$$P(t) = \dfrac{5}{2 + e^{-0.1t}}$$

where t is in years after 1990 and $P(t)$ is in millions of inhabitants.
(a) Sketch the graph of $P(t)$.
(b) Is the population increasing or decreasing?
(c) What will happen to the population in the distant future?

42. Suppose a disease spreads according to the equation

$$E(t) = \dfrac{8}{2 + 5e^{-0.1t}}$$

where t is in days after the outbreak of the disease and $E(t)$ is in thousands of people.
(a) Sketch the graph of $E(t)$.
(b) How many people contracted the disease after 10 days?
(c) In the long run, how many people will contract the disease?

••• 43. Let $f(x) = (-1)^x$. Plot the points on the graph of $f(x)$ for x an integer from 0 to 4. Can the entire graph be drawn by connecting these points with a smooth curve?

44. (a) Use a calculator to approximate this infinite sum to five decimal places.

$$1 + \dfrac{1}{1!} + \dfrac{1}{2!} + \dfrac{1}{3!} + \dfrac{1}{4!} + \cdots$$

(b) What number do you think this sum is equal to?

45. Let $f(x) = e^x$. Evaluate each limit by using a calculator to form a table in which one row

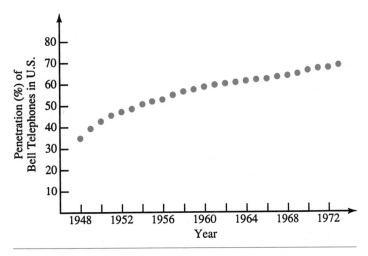

FIGURE 7

contains values of h and the other row contains the corresponding difference quotient.

(a) $\lim\limits_{h \to 0} \dfrac{f(3 + h) - f(3)}{h}$

(b) $\lim\limits_{h \to 0} \dfrac{f(4 + h) - f(4)}{h}$

46. A study of a game called a *lottery* derived a way to distinguish between different types of risk takers.* Consider a given lottery L whose utility is denoted by $U(L) = 0.817$. (*Utility* is a technical term whose meaning is not necessary to define here). A decision maker is said to be *risk-averse* with respect to another lottery G if $U(G) < U(L)$, is *risk-neutral* if $U(G) = U(L)$, and is *risk-prone* if $U(G) > U(L)$. Let G be a lottery in which a player wins \$2 90% of the time and loses \$2 otherwise. The expected value is $E = (2)(.9) + (-2)(.1) = 1.6$. The utility of G is defined by

$$U(G) = (.9)f(2) + (.1)f(-2)$$

where

$$f(x) = \frac{1 - e^{-E(.1x + .5)}}{1 - e^{-E}}$$

Calculate $U(G)$, and determine whether the decision maker is risk-averse, risk-prone, or risk-neutral with respect to the lottery L.

47. An article on forecasting market development showed that two growth curves could be applied in a wide variety of settings.† A *logistic curve* is of the form

$$l(x) = \frac{a}{1 + ce^{-bx}}$$

and a *Gompertz curve* is of the form

$$g(x) = ae^{-c(e^{-bx})}$$

where a, b, and c are positive numbers. Figure 7 shows the percentage of households in the United States that had Bell telephones in the period 1948–1972, while Figure 8 shows the percentage of households in the United Kingdom that had color television sets in the period 1968–1980. Which figure is modeled by which

*Joao L. Becker and Rakesh K. Sarin, "Lottery Dependent Utility," *Management Science*, November 1987, pp. 1367–1382.

†Nigel Meade, "Forecasting Using Growth Curves—An Adaptive Approach," *Journal of the Operational Research Society*, December 1985, pp. 1103–1115.

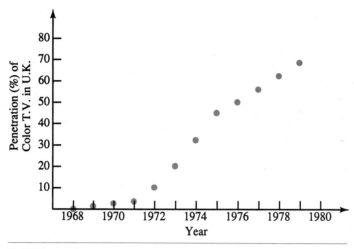

FIGURE 8

curve? (*Hint:* set $a = b = c = 1$, and sketch the graphs of the resulting logistic curve and Gompertz curve.)

48. A study of the spread of automatic teller machines (ATMs) in the banking industry found that the proportion of banks that had adopted the innovation could be expressed as the logistic function

$$P(t) = \frac{1}{1 + e^{5.293 - 0.698t}}$$

where t denotes the number of years after the ATMs were first introduced in 1972.‡ Determine the proportion of banks which used ATMs in each listed year by calculating $P(t)$ for the appropriate value of t.
(a) 1975   (b) 1980   (c) 1988

‡Timothy H. Hannan and John M. McDowell, "Market Concentration and the Diffusion of New Technology in the Banking Industry," *The Review of Economics and Statistics,* November 1984, pp. 686–691.

CASE STUDY   The Frustration of Dieting*

Who among us has not had to curtail eating habits in order to keep the body as trim as possible? It is true that some people are able to eat as much as they like, and as often as they desire, yet still maintain a slim figure. But others have to work hard to keep their weight down, whether the reason is serious (due to health) or cosmetic (due to vanity).

This Case Study examines dieting from a mathematical perspective. It draws one inescapable conclusion: there is a mathematical reason behind the fact that dieting is a long and arduous—in a word, frustrating—ordeal.

Equilibrium weight   A person's weight depends primarily on two factors: energy intake and energy consumption. There are several other important factors, such

*Adapted from Arthur C. Segal, "A Linear Diet Model," *College Mathematics Journal,* January 1987, pp. 44–45.

as age, sex, and metabolic rate, but we restrict our consideration to the two energy factors. We also assume that each of the two main factors remains constant each day, meaning that a diet consists of the same number of calories per day and that the energy consumption is the same number of calories per pound per day.

Most people's energy consumption varies between 15 and 20 calories per pound per day. *We will assume that the energy consumption is 17.5 calories per pound per day.*

Let I denote the daily energy intake (in calories) and C denote the daily energy consumption (in calories per pound). A person who follows such a regimen will maintain a *weight equilibrium* of I/C pounds (lb).

$$w_{eq} = \frac{I}{C}$$

For instance, a person who eats 2625 calories a day will maintain a weight equilibrium of $2625/17.5 = 150$ lb. Since the energy consumption C remains constant, if I increases above 2625 calories, then the person will gain weight, while if I decreases below 2625 calories, then the person will lose weight.

The weight equation   One of the basic assumptions in physiology is that the rate of change of a person's weight w is directly proportional to the difference between the energy intake I and the energy consumption C. Since the rate of change is the derivative $\dfrac{dw}{dt}$, this assumption can be written as an equation.

$$\frac{dw}{dt} = k(I - C)$$

The most commonly used dietetic conversion factor is $k = 1/3500$ lb per calorie, so

$$\frac{dw}{dt} = \frac{1}{3500}(I - C)$$

In Section 15–4 it will be possible to verify that the solution to this equation is

$$w(t) = w_{eq} + (w_0 - w_{eq})e^{-0.005t}$$

This equation is called the *weight equation*, where $w(t)$ represents the weight after t days and w_0 represents the initial weight. It is an important equation in many people's lives. Let us illustrate the weight equation for a hypothetical person named Rufus.

Rufus presently weighs 200 lb, so $w_0 = 200$. This means that his usual daily energy intake is $I = 200 \cdot 17.5 = 3500$ calories a day. He goes on a 3150 calories-per-day diet in order to lose weight. Now since $I = 3150$, the weight equilibrium will be

$$w_{eq} = \frac{3150}{17.5} \approx 180$$

Therefore, by adopting this diet, Rufus will eventually reach a weight of 180 lb and will maintain it.

How long will it take him to achieve this goal? To answer this question, set up the weight equation.

$$w(t) = w_{eq} + (w_0 - w_{eq})e^{-0.005t}$$
$$= 180 + (200 - 180)e^{-0.005t}$$
$$= 180 + 20e^{-0.005t}$$

Rufus' weight after $t = 7$ days is

$$w(7) = 180 + 20e^{-0.005\cdot7}$$

Use a calculator to evaluate this expression: $w(7) = 199.3$ (rounded to one decimal place). Thus Rufus will lose less than 1 lb during the first week of the diet. After 4 weeks Rufus will weigh about 197 lb because

$$w(28) = 180 + 20e^{-0.005\cdot28} \approx 197.4$$

Therefore after almost a month Rufus will have lost less than 3 lb. After a year, however, his weight will be $w(365) = 183.2$ lb, which is about 3 lb from his goal.

Does this mathematical model convince you of the difficulty involved in losing weight strictly by dieting? Poor Rufus stuck to his diet evey day for a year, yet he was still 3 lb shy of his goal. Figure 1 shows the asymptotic nature of the weight equation. No wonder that losing weight is a long and arduous struggle!

Example 1 considers another person's valiant fight to remove extra pounds.

FIGURE 1

EXAMPLE 1

Tina weighs 130 lb. She wants to lose 10 lb.

Problem

(a) What should Tina's daily caloric intake be?
(b) If she maintains the diet for 100 days, how much weight will she lose?

SOLUTION (a) Set $w_0 = 130$. Since Tina wants to lose 10 lb, her equilibrium weight is $w_{eq} = 120$. It is assumed that $C = 17.5$. Since $w_{eq} = I/C$, the daily caloric intake is $I = C \cdot w_{eq} = 17.5 \cdot 120 = 2100$. Therefore Tina must go on a 2100 calories-per-day diet.
(b) Tina's weight equation is

$$w(t) = 120 + 10e^{-0.005t}$$

If she adheres to this diet for $t = 100$ days, her weight will be

$$w(100) = 120 + 10e^{-0.5}$$
$$= 120 + 10 \cdot 0.6$$
$$= 126$$

Since Tina will drop from 130 to 126 lb, she will end up losing 4 lb in 100 days.

Notice that Tina's weight can never reach exactly 120 lb on this diet because the weight equation $w(t) = 120 + 10e^{-0.005t}$ is always greater than 120. (Recall that e^x is positive for all values of x.) However, it is possible to determine when her weight will reach $w(t) = 121$ lb by solving the weight equation for t.

$$121 = 120 + 10e^{-0.005t}$$

The solution requires the logarithmic function, which will be discussed in the next section. By direct substitution into the weight equation, Tina will drop to 121 lb after 460 days. Now that's an ordeal!

Considering the cases of Rufus and Tina, is it any wonder that so many dieters give up in frustration?

Case Study Exercises

1. Rufus' initial weight is 200 pounds (lb). If he goes on a 2500 calories-per-day diet, how much weight will he lose in 100 days?

2. Rufus' initial weight is 200 lb. If he goes on a 2000 calories-per-day diet, how much weight will he lose in 100 days?

3. Tina weighs 120 lb and wants to lose 10 lb.

   (a) What should her daily caloric intake be?

   (b) If she maintains the diet for 100 days, how much weight will she lose?

4. Joan weighs 125 lb and wants to lose 15 lb.
   (a) What should her daily caloric intake be?
   (b) If she maintains the diet for 100 days, how much weight will she lose?

5. For Chris, $w_0 = 150$ and $I = 2275$. How much weight will Chris lose in 100 days?

6. Let your present weight be w_{eq}. Assume an average daily consumption of 17.5 calories per pound. (a) What is your daily energy intake I? (b) If you reduce I by 500 calories a day, what will your new equilibrium weight be?

7. (a) What is your weight equation for losing 10 lb? (b) What will you weigh in 100 days if you stick to the diet?

5–2 LOGARITHMIC FUNCTIONS

This section is organized like Section 5–1. First a new function, the natural logarithmic function, is introduced in terms of more familiar functions. Then its graph is drawn between the graphs of two similar functions and several of its properties are derived.

This approach is appropriate since the definition of the natural logarithmic function is based on the exponential function $f(x) = e^x$ and the natural logarithmic function is shown to be the inverse of $f(x)$. This inverse relationship is exploited in a discussion of the doubling time of an investment.

Graphs

Logarithmic functions are defined in terms of exponential functions: $y = \log_2 x$ if and only if y is the number such that $2^y = x$. Therefore $3 = \log_2 8$ because $2^3 = 8$. This leads to the definition of a *logarithmic function*.

$$f(x) = \log_2 x \qquad \text{if and only if} \qquad 2^{f(x)} = x$$

(Usually $\log_2 x$ is read "the logarithm of x to the base 2.") Notice that the logarithm of a number is an exponent. The definition yields $f(8) = 3$, so the point $(8, 3)$ lies on the graph of $f(x) = \log_2 x$. Additional points are listed in Table 1. The graph is drawn in Figure 1.

TABLE 1

x	1	2	4	8	16	32
$\log_2 x$	0	1	2	3	4	5

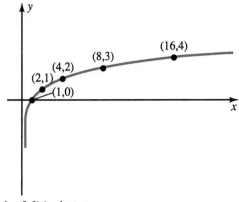

The graph of $f(x) = \log_2 x$

FIGURE 1

--- **Definition** ---

Let b be a positive number with $b \neq 1$. The **logarithmic function with base b** is defined by

$$f(x) = \log_b x \quad \text{if and only if} \quad b^{f(x)} = x$$

The domain is the set of positive numbers and the range consists of all real numbers.

EXAMPLE 1

Problem

Sketch the graph of $f(x) = \log_{10} x$.

SOLUTION By definition $y = \log_{10} x$ if $10^y = x$. Substitute some numbers for y and solve for x. For instance, if $y = 2$, then $x = 10^2 = 100$. Thus $(100, 2)$ is a point on the graph. Additional points are listed in the table. The graph is sketched in Figure 2.

x	0.01	0.1	1	10	100	1000
$\log_{10} x$	-2	-1	0	1	2	3

Logarithms to the base 10 are called **common logarithms.** They are usually written $\log x$ instead of $\log_{10} x$. Common logarithms were invented about 1600 as an aid for astronomers in performing involved calculations. Scientific calculators, however, have rendered this use entirely dispensable.

The graphs of $f(x) = \log_2 x$ and $f(x) = \log_{10} x$ are typical of the graphs of logarithmic functions $f(x) = \log_b x$ for $b > 1$. They pass through the point $(1, 0)$, they always increase, they are negative for $0 < x < 1$, and they are positive for $x > 1$.

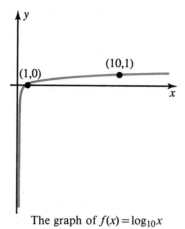

The graph of $f(x) = \log_{10} x$

FIGURE 2

The Natural Logarithm

Figure 3 shows the graphs of the functions $f(x) = 2^x$ and $f(x) = \log_2 x$ on the same coordinate system. The graphs are symmetric about the dashed line $y = x$, meaning that if one of them is flipped about this line, then it would land precisely on the other. As an example, the point $(3, 8)$ lies on the graph of $f(x) = 2^x$, the point $(8, 3)$ lies on the graph of $f(x) = \log_2 x$, and the distance from $(3, 8)$ to the dashed line is equal to the distance from $(8, 3)$ to the dashed line. When such a relation holds between two functions, each one is called the **inverse function** of the other. Thus $f(x) = \log_2 x$ is the inverse of $f(x) = 2^x$.

FIGURE 3

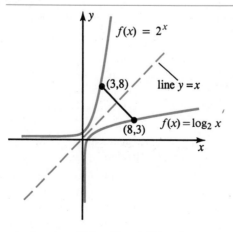

The functions $f(x) = 2^x$ and $f(x) = \log_2 x$
are inverses of each other

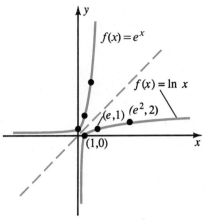

ln x is the inverse of e^x

FIGURE 4

Figure 4 shows the graph of the exponential function $f(x) = e^x$ and its inverse, which is called the **natural logarithmic function.** The natural logarithm, which is denoted by ln x instead of $\log_e x$, is defined by

$$y = \ln x \qquad \text{if and only if} \qquad e^y = x$$

Notice that ln $e = 1$ (since $e^1 = e$) and ln $1 = 0$ (since $e^0 = 1$). However, ln 0 is not defined. The graph of $f(x) = \ln x$ is similar to the graph of $f(x) = \log_2 x$. It can be obtained from the graph of $f(x) = e^x$ as in Figure 4, or from Table 2. (In Table 2 the values of ln x are rounded to two decimal places. Each value of ln x in the table is an irrational number except for $x = 1$.) A more complete table can be found in the Appendix. Many calculators have $\boxed{\ln}$ and $\boxed{\log}$ keys. The key sequence $\boxed{.5}$ $\boxed{\ln}$ produces the value -0.69 in Table 2.

TABLE 2

x	0.05	0.1	0.25	0.5	1	2	3	4
ln x	-3.00	-2.30	-1.39	-0.69	0	0.69	1.10	1.39

Example 2 sketches the graph of a function that is defined in terms of the natural logarithmic function. Its domain is the set of all real numbers and its graph is unlike Figures 1 and 2.

EXAMPLE 2

Problem

Sketch the graph of the function

$$f(x) = \ln(x^2 + 1)$$

SOLUTION Construct a table using a calculator or the table of logarithmic functions in the Appendix. Plot the points and draw a smooth curve through them. This is carried out in Figure 5. Notice that $f(x)$ is defined for all real numbers.

x	-3	-2	-1	0	1	2	3
$\ln(x^2 + 1)$	2.30	1.61	0.69	0	0.69	1.61	2.30

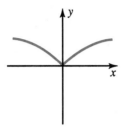

Graph of $f(x) = \ln(x^2 + 1)$

FIGURE 5

Properties

We derive several properties of logarithmic functions from the properties of exponential functions. Although these properties are stated in terms of the natural logarithmic function $f(x) = \ln x$, they hold for all logarithmic functions of the form $f(x) = \log_b x$ as well (for $b > 0$ and $b \neq 1$). In properties 4 to 7 assume that $r > 0$, $s > 0$, and k is any number.

1. $e^{\ln x} = x$ for $x > 0$
2. $\ln e^x = x$ for all x
3. $\ln 1 = 0$
4. If $\ln r = \ln s$, then $r = s$
5. $\ln rs = \ln r + \ln s$
6. $\ln \dfrac{r}{s} = \ln r - \ln s$
7. $\ln r^k = k \cdot \ln r$

The natural logarithm is defined by

$$y = \ln x \qquad \text{if and only if} \qquad e^y = x$$

Property 1 is obtained by substituting $\ln x$ for y in the right-hand equality. Similarly, substituting e^y for x in $y = \ln x$ yields $y = \ln e^y$. Writing this equality in the variable x gives property 2. Property 3 follows by substituting $x = 0$ in property

2. Property 4 can be inferred from the graph of $f(x) = \ln x$. Alternately, if $\ln r = \ln s$, then $e^{\ln r} = e^{\ln s}$. By property 1, $r = e^{\ln r}$ and $s = e^{\ln s}$, so $r = s$.

To prove property 5, use property 1 to write $r = e^{\ln r}$ and $s = e^{\ln s}$. Then $rs = e^{\ln r} \cdot e^{\ln s} = e^{\ln r + \ln s}$. By property 1, $rs = e^{\ln rs}$, so $e^{\ln rs} = e^{\ln r + \ln s}$. Property 5 then follows from property 1 of exponential functions. Properties 6 and 7 can be derived similarly.

Examples 3 and 4 illustrate these properties.

EXAMPLE 3 | **Problem**

Write these expressions in reduced form: (a) $\ln(4x^3)$ and (b) $\ln\left(\dfrac{6x^2y}{x + y}\right)$.

SOLUTION (a) Using properties 4 and 6, we obtain $\ln(4x^3) = \ln(2^2x^3) = \ln(2^2) + \ln(x^3) = (2 \ln 2) + (3 \ln x)$.
(b) Again by using properties 4 and 6, we have $\ln(6x^2y) = \ln 6 + 2 \ln x + \ln y$. Using property 5, we get

$$\ln\left(\frac{6x^2y}{x + y}\right) = \ln(6x^2y) - \ln(x + y)$$
$$= \ln 6 + 2 \ln x + \ln y - \ln(x + y)$$

WARNING: It is *not* true that $\ln(x + y) = \ln x + \ln y$. Do not try to use such an equality to simplify Example 3(b) further than it is.

EXAMPLE 4 | **Problem**

Simplify (a) $\ln e^{(3x+4)}$ and (b) $e^{\ln(3x+4)}$.

SOLUTION (a) Rewrite property 2 in the variable v, so that $\ln e^v = v$. Now set $v = 3x + 4$. Then $\ln e^{(3x+4)} = 3x + 4$.
(b) Similarly rewrite property 1 in the variable v, so that $e^{\ln v} = v$. Once again set $v = 3x + 4$. Then $e^{\ln(3x+4)} = 3x + 4$.

The same reasoning used in Example 4 leads to a generalization of properties 1 and 2:

1′. $e^{\ln f(x)} = f(x)$ for $f(x) > 0$
2′. $\ln e^{f(x)} = f(x)$ for all $f(x)$

Doubling Time

One way to measure the effectiveness of an investment is by means of **doubling time,** which is the time required for the initial investment to double. By the Compound Interest Formula, if $1 is invested at an interest rate of $100r\%$ compounded annually (where r is written as a decimal), the formula for the amount after t years is given by

$$A(t) = (1 + r)^t$$

This is an exponential function. The doubling time is computed by solving the equation $A(t) = 2$ for t.

$$2 = (1 + r)^t$$

Since the logarithmic function is the inverse of the exponential function, take the natural logarithm of both sides.

$$\ln 2 = \ln(1 + r)^t$$

$$\ln 2 = t \cdot \ln(1 + r)$$

$$t = \frac{\ln 2}{\ln(1 + r)}$$

For instance, the doubling time of an 8% investment is 9 years because

$$t = \frac{\ln 2}{\ln 1.08} = \frac{0.69315}{0.07696} \approx 9$$

EXAMPLE 5 | **Problem**

What is the doubling time for an investment that earns 16% compounded annually?

SOLUTION By the Compound Interest Formula

$$A(t) = (1 + r)^t$$

Here $A(t) = 2$ and $r = 0.16$, so

$$2 = (1.16)^t$$

$$\ln 2 = t \cdot \ln(1.16)$$

$$t = \frac{\ln 2}{\ln 1.16} = \frac{0.69315}{0.14842} \approx 4.67017$$

Therefore the doubling time is about $4\frac{2}{3}$ years.

EXERCISE SET 5–2

● In Problems 1 to 4 sketch the graph of the stated function by the plotting method, using the definition of the logarithmic function.

1. $f(x) = \log_{10} 2x$

2. $f(x) = \log_{10} \dfrac{x}{2}$

3. $f(x) = \log_3 x$

4. $f(x) = \log_{20} x$

In Problems 5 to 8 sketch the graph of the stated function by the plotting method, using the table in the Appendix or a calculator.

5. $f(x) = \ln(3x)$

6. $f(x) = \ln(x/2)$

7. $f(x) = \ln(x^2 + 3)$

8. $f(x) = \ln(x^2 + 4)$

In Problems 9 to 12 write the given expression in reduced form.

9. $\ln(3x^4)$

10. $\ln(2x^5)$

11. $\ln\left(\dfrac{3xy^2}{x-y}\right)$ 12. $\ln\left(\dfrac{x^3y^2}{x+y}\right)$

In Problems 13 to 16 evaluate the expression.

13. (a) $\ln e^{3x}$ (b) $e^{\ln(5-x)}$

14. (a) $\ln e^{-2}$ (b) $e^{\ln 4}$

15. (a) $\ln e^5$ (b) $e^{\ln 0.5}$

16. (a) $\ln e^{3-8x}$ (b) $e^{\ln(2x-6)}$

In Problems 17 to 20 determine the doubling time for an investment that earns the stated interest compounded annually?

17. 6%   18. 10%   19. 20%   20. 5%

•• In Problems 21 to 26 sketch the graph of the stated function by the plotting method, using the table in the Appendix or a calculator.

21. $f(x) = \ln\dfrac{1}{x^2}$

22. $f(x) = \ln\left(\dfrac{1}{x^2+2}\right)$

23. $f(x) = \ln(2x+3)$ 24. $f(x) = \ln(x-4)$

25. $f(x) = \ln|x|$ 26. $f(x) = \ln\sqrt{x}$

In Problems 27 to 30 state the domain of the function in the problem stated above.

27. Problem 21   28. Problem 23

29. Problem 25   30. Problem 26

In Problems 31 to 34 write the given expression in reduced form.

31. $\ln\sqrt{xy}$ 32. $\ln\sqrt{x+y}$

33. $\ln\dfrac{1}{\sqrt{x^3y}}$ 34. $\ln\dfrac{1}{xy^2}$

In Problems 35 and 36 let $f(x) = \ln x$ and $g(x) = e^x$.

35. Compute $(f \circ g)(x)$.

36. Compute $(g \circ f)(x)$.

In Problems 37 to 44 solve for x.

37. $e^{2x} = 3$ 38. $e^{5x} = 1$

39. $18 = e^{-x} - 2$ 40. $5 = \dfrac{1}{1-2e^{-x}}$

41. $5 \ln x = 1$ 42. $4 \ln(1/x) = 1$

43. $3^x = e^2$ 44. $2^x = e^5$

In Problems 45 to 48 evaluate the expression without using a calculator.

45. (a) $\ln e$ (b) $e^{\ln 4}$

46. (a) $\ln 1$ (b) $e^{3 \ln 2}$

47. (a) $\ln \sqrt[3]{e}$ (b) $e^{4 \ln 3 - 3 \ln 4}$

48. (a) $\ln(1/\sqrt{e})$ (b) $e^{-\ln 2}$

49. Use a calculator to try to evaluate
(a) $\ln 0$ (b) $\ln -2$

50. Use a calculator to show that $\ln(2 + 3) \neq \ln 2 + \ln 3$.

51. If a thermometer is moved from inside a house where it reads 70°F, to outside the house where it is 10°F, the temperature of the thermometer will read

$$T(t) = 10 + 60e^{-0.46t}$$

where t is in minutes and $T(t)$ is in degrees Fahrenheit. If the thermometer is digital (meaning that it only displays whole degrees), when will the reading first reach 10°F?

52. Answer true or false.
(a) $\ln(x^3 + 5x^2) = 3 \ln x + 10 \ln x$
(b) $\ln(x^5) = (\ln x)^5$

••• 53. In 1987 the world's population reached 5 billion.* If it continues to grow 2% each year, in what year will it reach 10 billion?

54. Determine how long it will take for an investment that earns 8% interest compounded annually to triple?

55. (a) Prove property 6 of logarithmic functions.
(b) Prove property 7 of logarithmic functions.

56. The weight equation for a 140-pound (lb) person who wants to lose 10 lb is

$$f(t) = 130 + 10e^{-0.005t}$$

*Dan Shannon, "World Population at 5 Billion," *Los Angeles Times*, July 12, 1987.

where $f(t)$ is the weight (in lb) after being on the diet for t days. How long will it take the person to lose 5 lb? (This weight equation is derived in the Case Study in this chapter titled "The Frustration of Dieting.")

Consider a job in which a number of workers produce a certain amount of products, say Boeing 707 jets. When the job is performed for the first time it cannot be expected to be completed in an acceptable amount of time. A certain period of time must elapse before the workers become familiar with the tasks involved and learn how to do them in a shorter amount of time. A *learning curve* is the graph of a function which measures the mastery of the job. If one hour is needed to produce the first unit, then the number of hours needed to produce the xth unit is given by

$$Y(x) = x^{\log L/\log 2}$$

where L is the learning rate of the particular job.† Data from the airframe manufacturing industry reveal that $L = 0.8$ is a typical learning rate. Assume such a value in Problems 57 and 58.

57. Compute the time (as a decimal part of an hour) needed to produce (a) the second unit, (b) the third unit, (c) the tenth unit, (d) the one-hundredth unit.

58. How many units have to be produced before a unit can be produced in each of these time periods?
   (a) 30 minutes (0.5 hour)
   (b) 6 minutes (0.1 hour)

Simple products and machine-paced processes were shown to exhibit learning rates between 0.9 and 1.0.‡

59. Repeat Problem 57 for a learning rate of $L = 0.95$.

60. Repeat Problem 58 for a learning rate of $L = 0.95$.

In the "classical secretary problem" each applicant for a job is ranked by a manager according to some quality. As each applicant appears, her rank relative to those preceding her is recorded and the manager decides either to select or reject her. The object is to hire the best applicant, and the manager is rewarded accordingly. A Japanese specialist in operations research has solved this problem for various kinds of selections that the manager can make.§ Problems 61 and 62 are taken from this study.

61. The reward for the model "without promotion" is $R(x) = x(1 - x)$, where x is a solution to the equation

$$2(1 - x) + \ln x = 0 \qquad 0 < x < 1$$

Solve the equation to one decimal place (by trial-and-error) and determine the reward.

62. The reward for the model "with promotion" is $R(x) = x\left(\dfrac{2}{\sqrt{e}} - x\right)$, where x is a solution to the equation

$$\sqrt{e}(1 + x) - \ln x = 3.5 \qquad 0 < x < 1$$

Solve the equation to one decimal place (by trial-and-error) and determine the reward.

---

†Louis E. Yelle, "The Learning Curve: Historical Review and Comprehensive Survey," *Decision Sciences*, April 1979, pp. 302–328.

‡Timothy L. Smunt, "A Comparison of Learning Curve Analysis and Moving Average Ratio Analysis for Detailed Operational Planning," *Decision Sciences*, Fall 1986, pp. 475–495.

§Mitsushi Tamaki, "A Generalized Problem of Optimal Selection and Assignment," *Operations Research*, May–June 1986, pp. 486–493.

CASE STUDY   Sears, Clothespins, and Unnatural Logarithms*

Consider the clothespin. No, not the one shown in Figure 1. That one is actually a Claes Oldenberg sculpture that stands in downtown Philadelphia. Instead,

*Adapted from Elliott W. Montroll, "On the Dynamics and Evolution of Some Sociotechnical Systems," *Bulletin of the American Mathematical Society*, January 1987, pp. 1–46.

FIGURE 1 A photo of a Claes Oldenberg sculpture of a clothespin that stands in downtown Philadelphia.

regard the usual clothespin, the one that has graced porches and backyards for over a century. It will serve as a central prop in our study of the American economy.

Sears catalogs also play an important role. Sears Roebuck and Co. began its mail-order operation at the turn of the 20th century based on these catalogs. The first ones were distributed in rural areas as a marketing strategy to sell items that were otherwise unavailable. Their overwhelming success spawned a vast nationwide network of over 800 stores, which are woven into the lives of millions of people.

In 1987 Sears produced its first videocassette catalog to market some of its products. We will not attempt to predict whether this venture will be as successful in the 21st century as the standard catalogs have been in the 20th century. Instead, we will document several ways in which the standard catalogs have been used to study the American way of life over the past 100 years. We will see that they provide an economic index that measures the cost of living and indicates changes in the American life-style.

TABLE 1

P	$\log_2 P$	$(\log_2 P - M)^2$
128.00	7	$2.94^2 = 8.6436$
39.40	5.3	$1.24^2 = 1.5376$
16.00	4	$-0.06^2 = 0.0036$
8.00	3	$-1.06^2 = 1.1236$
2.00	1	$-3.06^2 = 9.3636$
Sum	20.3	20.672

$$M = \frac{20.3}{5} = 4.06 \qquad S = \sqrt{\frac{20.672}{5}} \approx 2.03$$

Cost of living   On a given page in a Sears catalog you might see bicycles, VCRs, and—yes—clothespins. Since the clothespin is an invariant item, the 1988 model being indistinguishable from the 1900 model, its change in price reflects one kind of measure of the cost of living. Another measure results from examining bicycles. Since the models of the 1990s are considerably different from the models from 1900, the variation in price reflects an evolving technology and a varying public taste. VCRs, on the other hand, certainly were not featured in the 1900 catalog. Conversely, you are not likely to find buggy whips in the 1988 catalog. Yet in spite of such variations, one constant emerges. We describe it now.

Consider a page in a Sears catalog containing five items whose sales prices are $128, $39.40, $16, $8, and $2. It is possible to compute the mean and standard deviation for the items on this page. However, instead of regarding each price P as a data entry, we use its logarithm to the base 2. For $P = \$128$ the data entry becomes $\log_2 P = \log_2 128 = 7$ because $2^7 = 128$. The remaining entries are listed in the second column of Table 1.

The mean of the $\log_2 P$ entries is $20.3/5 = 4.06$. We denote this mean by M, to distinguish it from the mean of the prices P, and refer to it as the *log-mean*. Similarly the standard deviation of the $\log_2 P$ entries is denoted by S and called the *log-standard deviation*. The third column of Table 1 reveals that $S = \sqrt{20.672/5} \approx 2.03$.

The values that we have obtained for this hypothetical page actually correspond to the log-mean M and the log-standard deviation S for all items that were listed in the 1975–1976 Sears catalog. Table 2 contains the values of M and S for selected annual Sears catalogs since 1900.

One fact stands out in Table 2. Although M varies considerably, S remains almost the same. For this reason S has been described as an "economic constant of motion" for marketing. For all Sears catalogs, which have ranged over various economic climates and two world wars, the standard deviation of the log-standard deviations is only 0.17. This reflects a remarkable uniformity and a very consistent marketing strategy.

Example 1 provides more practice in computing the log-standard deviation S.

TABLE 2

Year	M	S
1900	0.150	2.43
1908	−0.023	2.29
1916	−0.068	2.38
1924–1925	0.422	2.32
1932–1933	0.691	1.91
1939–1940	0.627	2.62
1951–1952	1.785	2.34
1962	2.403	2.24
1975–1976	4.060	2.03

EXAMPLE 1

A page of a Sears catalog lists five items at these prices: $78.25, $3.36, $1.68, $1.06, and $1.04.

Problem

Which annual Sears catalog has a log-mean and a log-standard deviation that are similar to the log-mean and the log-standard deviation of these five items?

SOLUTION The prices P of the items are listed in the left-hand column of Table 3. The associated $\log_2$ prices are listed in column 2. Each number in the second column can be obtained by using the natural logarithm. For instance,

$$\log_2 78.25 = \frac{\ln 78.25}{\ln 2} = \frac{4.3599}{0.6931} = 6.29$$

The remaining calculations, which are shown at the bottom of Table 3, reveal that $M = 1.786$ and $S \approx 2.33$. These numbers approximate the values of M and S in the 1951–1952 Sears catalog.

TABLE 3

P	$\log_2 P$	$(\log_2 P - M)^2$
78.25	6.29	$4.50^2 \approx 20.2860$
3.36	1.75	$-0.04^2 \approx 0.0013$
1.68	0.75	$-1.04^2 \approx 1.0733$
1.06	0.08	$-1.71^2 \approx 2.9104$
1.04	0.06	$-1.73^2 \approx 2.9791$
Sum	8.93	27.2501

$$M = \frac{8.93}{5} = 1.786 \qquad S = \sqrt{\frac{27.25}{5}} \approx 2.33$$

Unnatural logarithms   The analysis of Sears catalogs has shown one use of a logarithm other than the natural logarithm. Another "unnatural" use involves the graphing of data. We show two ways in which such graphs arise.

In most graphs if the x-axis and the y-axis have the same units, then these units are marked uniformly along both axes. Sometimes it is more beneficial to scale one of the axes differently. Consider Figure 2, for instance. The y-axis is scaled uniformly, with each mark indicating 2% of the items in the 1916 Sears catalog. However, the units along the x-axis represent $\log_2 P$, where P is the price of an item. Therefore the interval from $x = 0$ to $x = 2$ represents prices in the range from $P = \$1$ to $P = \$4$ (since $\log_2 1 = 0$ and $\log_2 4 = 2$), while the interval from $x = 2$ to 4 represents prices in the range from $P = \$4$ to $P = \$16$.

Each histogram in Figure 2 is very close to a normal distribution. The same type of histogram results for every Sears catalog when $\log_2 P$ is used for the x-axis. In their marketing wisdom the Sears Roebuck Company has created catalogs whose price distribution year after year has maximized profit by maximizing a function associated with $\log_2 P$ (the so-called entropy function).

EXAMPLE 2

Problem

Use Figure 2 to determine which price range was listed most frequently in the 1916 catalog.

SOLUTION The maximum interval of the histogram occurs from $x = -1$ to $x = 0$. The associated prices are \$0.50 (since $\log_2 0.5 = -1$) and \$1. Therefore items costing between \$0.50 and \$1 were listed most frequently. (They comprise about 17% of all listings.)

Figure 3 shows the graph of the log-mean M of the Sears catalogs. The years are marked uniformly along the x-axis and the values of $\log_2 M$ are marked along the y-axis. Therefore the point (1975, 4) lies on the graph because the average item in the 1975–1976 Sears catalog listed for about $M = \$16$ and $\log_2 16 = 4$.

The Sears catalogs list many diverse products, so the variation in the annual log-mean price represents an average over numerous technologies. The graph, called *the SR index,* has become a standard for comparison. For example, Figure 3 includes the average factory wholesale price to automobile dealers in the United States. The graph is similar to the SR index after the initial drop in prices until about 1935. (The computer industry has also experienced such a period of decrease in prices since its inception.)

EXAMPLE 3

Problem

Use Figure 3 to approximate those years when the average factory wholesale price to automobile dealers was about \$1000.

SOLUTION The units along the y-axis of Figure 3 refer to $\log_2 M$, where M is the price to automobile dealers. Since $\log_2 1000 \approx 10$, we can locate the approximate years by drawing a horizontal line at $y = 10$. (It is the dashed line in Figure 3.) The approximate years are 1912, 1920, and 1945.

FIGURE 2 Histogram of distribution of prices in Sears Roebuck catalogs for years 1916, 1924–1925, and 1974–1975. *Source:* M. F. M. Osborne, *Operations Research,* 7(145), 1959.

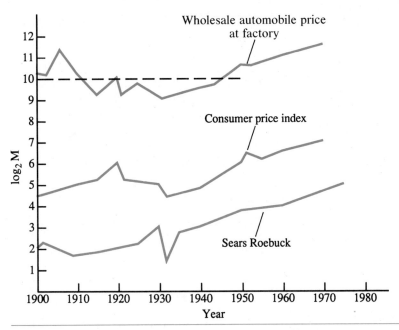

FIGURE 3   *Source:* From annual volumes, "Automobile Facts and Figures," American Automobile Association.

Case Study Exercises

1. Compute the log-mean and log-standard deviation for the five items whose sales prices are $128, $32, $8, $2, and $1.

2. Which annual Sears catalog (from Table 2) has a log-mean and a log-standard deviation that are similar to the log-mean and the log-standard deviation of these five items: $39.40, $32, $1.62, $1.41, and $1.41?

3. Use Figure 2 to determine which price range was listed most frequently in the 1924–1925 catalog.

4. Use Figure 2 to determine which price range was listed most frequently in the 1974–1975 catalog.

Problems 5 to 8 refer to Figure 3.

5. Approximate the average sales price of an item in the 1970 Sears catalog.

6. Approximate the average factory wholesale price to automobile dealers in 1970.

7. What was the Consumer Price Index in 1970?

8. Approximate the years in which the Consumer Price Index was about $64.

5–3 DIFFERENTIATION OF LOGARITHMIC FUNCTIONS

The definition of the derivative has been used thus far for two purposes. First, in order to demonstrate its meaning, we applied it to compute the derivatives of several simple functions. Then it was used to derive the standard techniques that were applied to a wide range of functions. The derivative of the logarithmic function, however, cannot be derived by these techniques, so in this section we return to the definition of the derivative. We begin by appealing to the definition to suggest a rule for the derivative of $f(x) = \ln x$. This rule will then be generalized, illustrated, and proved more rigorously.

The domain of the function $\ln x$ is $(0, \infty)$. In order to include all nonzero values of x in the domain, we often use the function $\ln |x|$. Since including absolute value signs complicates the formulas and problems, in the first part of the section we assume that the domain of each function is the largest possible set of real numbers. At the end of the section we handle functions like $\ln |x|$.

The Derivative of ln x

Let us compute the derivative of $f(x) = \ln x$ at $x = a$ using the definition of the derivative. By definition

$$f'(a) = \lim_{h \to 0} \frac{f(a + h) - f(a)}{h}$$

$$= \lim_{h \to 0} \frac{\ln(a + h) - \ln(a)}{h}$$

In previous similar problems the next step is to simplify the difference quotient. However, this quotient does not simplify easily. At the end of the section a rigorous method for calculating this limit is given, but for now we evaluate the limit for two particular values of a and various choices of h that approach 0. Using a calculator, we obtain the following tables of values for $a = 5$ and $a = 8$, respectively, which give an indication of what the rule is for the derivative of $f(x) = \ln x$.

TABLE 1

	$a = 5$			$a = 8$	
h	$\ln(5 + h)$	$\dfrac{\ln(5 + h) - \ln 5}{h}$	h	$\ln(8 + h)$	$\dfrac{\ln(8 + h) - \ln 8}{h}$
0.1	1.629241	0.198026	0.1	2.091864	0.124225
0.01	1.611436	0.199800	0.01	2.080691	0.124922
0.001	1.609638	0.199980	0.001	2.079567	0.124992
0.0001	1.609458	0.199998	0.0001	2.079454	0.124999

For $a = 5$, as h approaches 0 the values of the quotient approach 0.2. This suggests that $f'(5) = 0.2$. It is instructive to write

$$f'(5) = \frac{1}{5}$$

In the same way, for $a = 8$, the table suggests that $f'(8) = 0.125$. Once again, write

$$f'(8) = \frac{1}{8}$$

These two results suggest that $f'(a) = 1/a$. This is the rule for the derivative of $f(x) = \ln x$. The formal proof at the end of the section is based in part on this discussion.

The Derivative of the Logarithmic Function

$$D_x \ln x = \frac{1}{x}$$

EXAMPLE 1

Problem

Find the derivative of $f(x) = x^3 - 5 \ln x$.

SOLUTION Use the rule for the derivative of a sum and the rule for a constant times a function.

$$D_x(x^3 - 5 \ln x) = D_x(x^3) - 5D_x(\ln x)$$
$$= 3x^2 - 5\left(\frac{1}{x}\right) = 3x^2 - \frac{5}{x}$$

The Logarithmic Function Version of the Chain Rule

Most logarithmic functions found in applications are of the form $\ln u$ where $u = g(x)$ is a function of x. If $g'(x)$ exists, the chain rule can be used to find the derivative of $\ln u$. Recall that the chain rule states that if $y = f(g(x))$, then

$$y' = f'(g(x))g'(x)$$

To compute the derivative of $y = \ln u$, let $u = g(x)$ and $f(x) = \ln x$ in the formula. Then $y = f(g(x)) = f(u) = \ln u = \ln g(x)$. From the rule for the derivative of the logarithmic function, $f'(u) = f'(g(x)) = 1/u = 1/g(x)$. From the chain rule we get

$$y' = f'(g(x))g'(x) = \left(\frac{1}{g(x)}\right)g'(x) = \frac{g'(x)}{g(x)}$$

The form of the chain rule applied to the logarithmic function $f(x) = \ln g(x)$ is the formula

---

Derivative of Logarithmic Functions

$$D_x \ln g(x) = \frac{g'(x)}{g(x)}$$

EXAMPLE 2

Problem

Find the derivative of (a) $f(x) = \ln(3x^2 + 5)$ and (b) $f(x) = \ln(\ln x)$.

SOLUTION (a) Let $g(x) = 3x^2 + 5$. Then $g'(x) = 6x$ and we get

$$D_x \ln(3x^2 + 5) = \frac{g'(x)}{g(x)}$$

$$= \frac{6x}{3x^2 + 5}$$

(b) Let $g(x) = \ln x$. Then $g'(x) = 1/x$ and we get

$$D_x \ln(\ln x) = \frac{g'(x)}{g(x)}$$

$$= \frac{(1/x)}{\ln x} = \frac{1}{x \ln x}$$

The next example demonstrates how one derivative can require two versions of the chain rule, the generalized power rule and the logarithmic function version.

EXAMPLE 3

Problem

Find the derivative of

$$f(x) = [2 + \ln(1 + 3x)]^3$$

SOLUTION First apply the generalized power rule. We get

$$f'(x) = 3[2 + \ln(1 + 3x)]^2[D_x(2 + \ln(1 + 3x))]$$

The derivative of $2 + \ln(1 + 3x)$ requires the derivative of a sum with $g(x) = 1 + 3x$. Then $g'(x) = 3$ and $D_x \ln(1 + 3x) = 3/(1 + 3x)$. Then

$$f'(x) = 3[2 + \ln(1 + 3x)]^2[D_x(2 + \ln(1 + 3x))]$$

$$= 3[2 + \ln(1 + 3x)]^2 \left[\frac{3}{1 + 3x}\right]$$

$$= \frac{9[2 + \ln(1 + 3x)]^2}{1 + 3x}$$

Graphing Logarithmic Functions

The curve-sketching techniques of the previous chapter can be used to graph logarithmic functions. The methods are illustrated in the next example.

EXAMPLE 4 | **Problem**

Find the relative extrema and the points of inflection of the function $f(x) = x - \ln x$. Then draw the graph.

SOLUTION Find the extrema via the second derivative test. Compute $f'(x)$ and $f''(x)$.

$$f'(x) = 1 - \frac{1}{x} \qquad f''(x) = \frac{1}{x^2}$$

Solve $f'(x) = 0$; $f'(x) = 1 - 1/x = 0$. Then $x = 1$, so the only critical value is $x = 1$. Although $f'(0)$ does not exist, $x = 0$ is not a critical value because 0 is not in the domain of $f(x)$. This is because $\ln x$ is not defined at $x = 0$.

Since $f''(1) = 1$, $f(x)$ has a relative minimum at $x = 1$. Since $f''(x) > 0$ for all x, $f(x)$ has no point of inflection. As x approaches 0, $\ln x$ approaches $-\infty$, and so $-\ln x$ approaches ∞, meaning that $f(x)$ has a vertical asymptote at $x = 0$. Note that $f(1) = 1 - \ln 1 = 1 - 0 = 1$. These facts can be used to draw the graph of $f(x)$ shown in Figure 1.

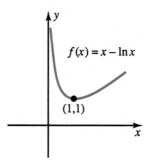

FIGURE 1

Consider the function $y = \ln |x|$ whose domain is the set of all real numbers except $x = 0$. The definition of this function is

$$\ln |x| = \begin{cases} \ln x & \text{if } x > 0 \\ \ln(-x) & \text{if } x < 0 \end{cases}$$

If $x > 0$, then $y = \ln x$ and $y' = 1/x$. Suppose $x < 0$. Then $y = \ln(-x)$ and to find y' we use the rule for the derivative of logarithmic functions with $g(x) = -x$. The derivative of $\ln(-x)$ is $-1/-x = 1/x$. Thus the derivative of $\ln(-x)$ is equal to the derivative of $\ln x$. A similar argument shows that the derivative of $\ln g(x)$ is the same as $\ln(-g(x))$.

$$D_x \ln |x| = \frac{1}{x}$$

$$D_x \ln |g(x)| = \frac{g'(x)}{g(x)}$$

EXAMPLE 5

Problem

Find the derivative of (a) $f(x) = \ln |3x - 2|$ and (b) $f(x) = x \ln |1 + x^{-3}|$.

SOLUTION (a) Let $g(x) = 3x - 2$. From the formula we get

$$f'(x) = \frac{3}{3x - 2}$$

(b) Let $g(x) = 1 + x^{-3}$ and apply the product rule.

$$f'(x) = xD_x \ln |1 + x^{-3}| + \ln |1 + x^{-3}|D_x x$$

$$= x \left[\frac{-3x^{-4}}{1 + x^{-3}} \right] + \ln |1 + x^{-3}(1)|$$

$$= \frac{-3x^{-3}}{(1 + x^{-3})} + \ln |1 + x^{-3}|$$

Proof of the Rule for the Derivative of the Logarithmic Function

Proof. If $f(x) = \ln x$, then

$$f'(x) = \lim_{h \to 0} \left[\frac{f(x + h) - f(x)}{h} \right] = \lim_{h \to 0} \left[\frac{\ln(x + h) - \ln x}{h} \right]$$

$$= \lim_{h \to 0} \left(\frac{1}{h} \right) \ln \left[\frac{x + h}{x} \right]$$

$$= \lim_{h \to 0} \ln \left[\frac{x + h}{x} \right]^{1/h}$$

We first deal with the case where $h > 0$. The case where $h < 0$ is handled in a similar way. To simplify this expression, let $m = x/h$. In this expression x is being held constant while h varies. As $h \to 0$, note that $m \to \infty$ and vice versa. In the last limit above write $(x + h)/x = 1 + h/x = 1 + 1/m$. The limit then becomes

$$f'(x) = \lim_{m \to \infty} \ln \left[1 + \frac{1}{m} \right]^{m/x}$$

$$= \lim_{m \to \infty} \ln \left[\left(1 + \frac{1}{m} \right)^m \right]^{1/x}$$

$$= \lim_{m \to \infty} \frac{1}{x} \cdot \ln \left[1 + \frac{1}{m} \right]^m$$

$$= \frac{1}{x} \cdot \lim_{m \to \infty} \ln \left[1 + \frac{1}{m} \right]^m$$

$$= \frac{1}{x} \cdot \ln \left[\lim_{m \to \infty} \left(1 + \frac{1}{m} \right)^m \right]$$

From the definition of e we have

$$\lim_{m \to \infty} \left(1 + \frac{1}{m} \right)^m = e$$

Therefore

$$f'(x) = \frac{1}{x} \cdot \ln \left[\lim_{m \to \infty} \left(1 + \frac{1}{m} \right)^m \right] = \frac{1}{x} \ln e = \frac{1}{x} \cdot 1 = \frac{1}{x}$$

Thus $D_x \ln x = 1/x$, as desired.

Elasticity

Economists define the **rate of growth** $Rf(x)$ of the function $f(x)$ as the ratio

$$Rf(x) = f'(x)/f(x)$$

It measures the relative rate of change of $f(x)$. For example, suppose an automobile manufacturer sells its car for $20,000 and another sells its car for $8000. Suppose both manufacturers increase the price, the first to $21,000 and the second to $8,500. So the first manufacturer's increase is $1000 and the second manufacturer's increase is $500. Which is the larger relative increase, meaning which is the larger percentage increase? Let the average price per car be represented by $f(x)$ for the first manufacturer and $g(x)$ for the second manufacturer. Let $x = 0$ represent the time when the original prices of $20,000 and $8000 were in effect. Then $f(0) = 20,000$ and $g(0) = 8000$. The rate of increase for each function is given by $f'(0) = 1000$ and $g'(0) = 500$. The rate-of-growth functions are

$$Rf(0) = f'(0)/f(0) = 1000/20000 = 1/20 = .05$$

$$Rg(0) = g'(0)/g(0) = 500/8000 = 1/16 = .0625$$

This means that the relative rate of increase for the lower priced car is greater than that of the higher priced car, even though the actual size of the increase is only half that of the higher priced car.

Another way to view the rate-of-growth function is to express it in terms of the derivative of the logarithm of $f(x)$:

$$Rf(x) = f'(x)/f(x) = D_x \ln f(x)$$

It is called the *logarithmic derivative of* $f(x)$. Economists use the rate-of-growth

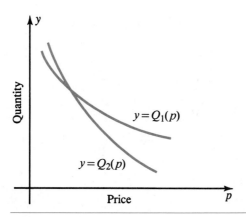

FIGURE 2

function to measure the relative change of various quantities, such as revenue and demand. In this way the relative changes of a product made by different companies can be compared. For instance, suppose two companies have demand curves $y = Q_1(p)$ and $y = Q_2(p)$, whose graphs are given in Figure 2. Because the graph of $Q_2(p)$ is steeper than the graph of $Q_1(p)$, a small change in the price p will result in a larger change in the quantity $Q_2(p)$ demanded versus the quantity $Q_1(p)$. From an intuitive point of view, economists refer to a steep demand curve such as the graph of $y = Q_2(p)$ as elastic while the graph of $y = Q_1(p)$ is called inelastic.

We now give a concrete definition of the **elasticity** $E(p)$ of a demand curve $y = Q(p)$. It is defined as the negative of the ratio of the rate of growth of $Q(p)$ to the rate of growth of p. That is:

$$E(p) = \frac{-(\text{rate of growth of } Q(p))}{\text{rate of growth of } p}$$

$$= \frac{-D_p \ln Q(p)}{D_p \ln p}$$

$$= \frac{-Q'(p)/Q(p)}{1/p} = \frac{-pQ'(p)}{Q(p)}$$

Elasticity is defined with a negative sign so that $E(p)$ is always positive. Notice that p and $Q(p)$ are always positive and $Q'(p)$ is always negative because $y = Q(p)$ is a decreasing function. If $E(p) > 1$ the demand curve is called **elastic** at p and if $E(p) < 1$ it is called **inelastic** at p.

EXAMPLE 6

The demand for oil is given by the formula

$$Q(p) = 60 - 2p$$

where p is the price per barrel and $Q(p)$ is the number of barrels demanded at price p in millions of barrels, for p between \$5 and \$30.

Problem

Find the values of p where $Q(p)$ is (a) elastic and (b) inelastic.

SOLUTION First compute $Q'(p) = -2$. Then

$$E(p) = \frac{-p(-2)}{60 - 2p} = \frac{2p}{60 - 2p}$$

Determine the value of p where $E(p) = 1$ by solving

$$E(p) = \frac{2p}{60 - 2p} = 1$$

This simpifies to $2p = 60 - 2p$, and then $4p = 60$, which implies that $p = 15$. $E(p)$ is a continuous function that crosses the p-axis at $p = 15$. This means that $E(p)$ is either greater than 1 at each point in the interval to the left of 15, or it is less than 1 at each point in the interval. The same is true for the interval greater than 15. This means that we must determine whether $E(p)$ is greater than or less than 1 on the intervals (5, 15) and (15, 30). Choose a test point in each interval, say 10 and 20, respectively: $E(10) = 20/40 = 1/2$ and $E(20) = 40/20 = 2$.
(a) Since 10 is in the interval (5, 15) and $E(10) < 1$, $E(p) < 1$ in (5, 15) and so $Q(p)$ is inelastic there.
(b) Since 20 is in the interval (15, 30) and $E(20) > 1$, $E(p) > 1$ in (15, 30) and so $Q(p)$ is elastic there.

If the demand curve is inelastic, then an increase in p results in an increase in revenue while a decrease in p results in a decrease in revenue. This means that revenue is an increasing function in an interval where the demand is inelastic. If the demand curve is elastic, the opposite is true: an increase in p results in a decrease in revenue, meaning that revenue is decreasing in such an interval. Let us see why this is so using the demand curve in Example 1. If $Q(p) = 60 - 2p$, then the revenue $R(p)$, defined by (price $\times$ quantity), is

$$R(p) = pQ(p) = p(60 - 2p) = 60p - 2p^2$$

The question is to determine when $R(p)$ is increasing, so we compute $R'(p)$ and solve $R'(p) > 0$:

$$R'(p) = 60 - 4p > 0$$
$$4(15 - p) > 0$$

This is true when $p < 15$, meaning that $R(p)$ is increasing in (5, 15). Therefore, from Example 1, in the interval (5, 15), $E(p) < 1$, and thus $Q(p)$ is inelastic in this interval. The previous computation shows that $R(p)$ is increasing in this interval. In a similar way it can be shown that in the interval (15, 30) the demand curve is elastic and $R(p)$ is decreasing.

EXERCISE SET 5–3

• In Problems 1 to 24 find $f'(x)$.

1. $f(x) = \ln 6x$
2. $f(x) = \ln 9x$
3. $f(x) = \ln 3x + 5$
4. $f(x) = \ln 5x - 6$
5. $f(x) = \ln(x + 5)$
6. $f(x) = \ln(x - 7)$
7. $f(x) = \ln(x^2 + 3)$
8. $f(x) = \ln(3x^2 - 5)$
9. $f(x) = \ln(1 + 4x^2)$
10. $f(x) = \ln(3 - 5x^2)$
11. $f(x) = \ln(2x^3 + x^{-1})$
12. $f(x) = \ln(4 - x^{1/2} + x)$
13. $f(x) = x \ln 4x$
14. $f(x) = 3x \ln 7x$
15. $f(x) = x^{-3} \ln(x^2 - 1)$
16. $f(x) = 5x^{2/3} \ln\left(2x - \dfrac{3}{2}\right)$
17. $f(x) = (\ln 3x)/(2x - 5)$
18. $f(x) = (x^5 - 11)/\ln 3x$
19. $f(x) = \ln|3 - 5x|$
20. $f(x) = \ln|2x^3 + 1|$
21. $f(x) = \ln|4 - x^{5/2}|$
22. $f(x) = \ln|2x^3 + x|$
23. $f(x) = x \ln|7x|$
24. $f(x) = \dfrac{x^3}{\ln|x^2 - 1|}$

In Problems 25 to 28 find the relative extrema and sketch the graph.

25. $f(x) = x \ln x$
26. $f(x) = x \ln|x|$
27. $f(x) = \dfrac{\ln x}{x}$
28. $f(x) = x - \ln|x|$

In Problems 29 and 30 the demand function $Q(p)$ is given defined in the given interval. Find where $Q(p)$ is (a) elastic and (b) inelastic.

29. $Q(p) = 80 - 5p$, $[5, 20]$
30. $Q(p) = 100 - 5p$, $[5, 30]$

•• In Problems 31 to 36 find $f'(x)$.

31. $f(x) = (x + \ln x)^2$

32. $f(x) = (4x + \ln 3x)^{-3}$
33. $f(x) = (\ln x + x^2)/\ln x$
34. $f(x) = x^2(x - \ln x)^2$
35. $f(x) = \ln(x^2 + \ln x)$
36. $f(x) = \ln(x + x \ln x)$

37. Sound intensity is measured in *decibels*. If $D(x)$ is the number of decibels when x is the power of the sound, measured in watts/cm^2, then the relationship between D and x is given by

$$D(x) = 10 \log_{10}(10^{16}x)$$

Find $D'(x)$.

38. The relationship between cost C and labor x is often given by a logarithmic equation. It is a form of a Cobb–Douglas function. Suppose a firm determines that

$$\ln C(x) = 8.3x^{-0.17}$$

Find $C'(x)$.

••• 39. Economists define the *rate of growth* of a function $f(x)$ as the ratio $F(x)$ given by

$$F(x) = \frac{f'(x)}{f(x)}$$

Show that $F(x) = D_x(\ln f(x))$.

40. A manufacturer has a cost function

$$C(x) = 5000 + 600x - 0.3x^2$$

Find the rate of growth of $C(x)$ when $x = 100$.

41. Let $f(x) = \ln x$. Find a formula for $f^{(n)}(x)$ for n a positive integer by computing the first few derivatives of $f(x)$ and finding a pattern.

42. Find $f'(x)$ for $f(x) = \ln(\ln(\ln x))$.

43. Show that $f(x) = \ln x^2$ is not the same function as $g(x) = (\ln x)^2$ by first showing that their functional values are different for a particular value of x and then showing that their derivatives are different.

44. The relationship between home range and body

size of mammals is believed to be as important to carnivores as it is to herbivores. In an article studying this relationship, Lindstedt et al. use a formula that relates home range A, measured in hectares, and body mass m, measured in kilograms.* The relationship is

$$\ln A = 1 + \ln m$$

Express A in terms of m directly, and find $A'(1)$.

45. Accurate leaf-area measurements are critical for estimating fluxes of carbon, solar energy, and water in forests. In an article comparing various methods of estimating leaf areas in fir trees, Marshall and Waring show that the leaf-area index L, which is the projected surface area of foliage per unit ground area, is a function of the irradiance q below the canopy, which is the light level below the tops of the trees.† The relationship for a Douglas-fir forest is estimated as

$$L(q) = 2.5(2q - \ln q)$$

Find the value of q where $L(q)$ is a minimum.

*Stan L. Lindstedt et al., "Home Range and Body Size in Mammals," *Ecology,* Vol. 67, 1986, pp. 413–418.

†J. D. Marshall and R. H. Waring, "Comparison of Methods of Estimating Leaf-area Index in Old-growth Douglas Fir," *Ecology,* Vol. 67, 1986, pp. 413–418.

5–4 DIFFERENTIATION OF EXPONENTIAL FUNCTIONS

Recall that the natural logarithm function and the exponential function are inverses of each other. This means that $\ln e^x = e^{\ln x} = x$. This property enables us to derive the derivatives of the functions $y = e^x$ and $y = e^{g(x)}$.

The Derivative of $y = e^x$

Because e^x and $\ln x$ are inverses of each other, the following property holds:

$$\ln e^x = x$$

Take the derivative of each side.

$$D_x(\ln e^x) = D_x(x)$$

On the left-hand side use the rule for the derivative of logarithmic functions, with $g(x) = e^x$.

$$\frac{D_x e^x}{e^x} = 1$$

Multiply by e^x to get

$$D_x e^x = e^x$$

The Derivative of the Exponential Function

$$D_x e^x = e^x$$

EXAMPLE 1

Problem

Find $f'(x)$ for

(a) $f(x) = x^4 e^x - 5e^x$ and (b) $f(x) = \left(\dfrac{e^x}{x^2 + 5x}\right)$.

SOLUTION (a) Use the rule for the derivative of a sum, the rule for the derivative of a product, and the rule for a constant times a function.

$$
\begin{aligned}
D_x(x^4 e^x - 5e^x) &= D_x(x^4 e^x) - 5D_x(e^x) \\
&= x^4 D_x(e^x) + e^x D_x(x^4) - 5D_x(e^x) \\
&= x^4 e^x + e^x 4x^3 - 5e^x \\
&= e^x(x^4 + 4x^3 - 5)
\end{aligned}
$$

(b) Use the rule for the derivative of a quotient.

$$
\begin{aligned}
D_x\left(\frac{e^x}{x^2 + 5x}\right) &= \frac{(x^2 + 5x)D_x(e^x) - e^x D_x(x^2 + 5x)}{(x^2 + 5x)^2} \\
&= \frac{(x^2 + 5x)e^x - e^x(2x + 5)}{(x^2 + 5x)^2} \\
&= \frac{e^x(x^2 + 5x - 2x - 5)}{(x^2 + 5x)^2} \\
&= \frac{e^x(x^2 + 3x - 5)}{(x^2 + 5x)^2}
\end{aligned}
$$

The Exponential Function Version of the Chain Rule

The general exponential function is expressed in the form e^u where $u = g(x)$ is a function of x. If $g'(x)$ exists, the chain rule can be used to find the derivative of e^u. Recall that the chain rule states that if $y = f(g(x))$, then

$$y' = f'(g(x))g'(x)$$

To compute the derivative of $y = e^u$, let $u = g(x)$ and $f(x) = e^x$ in the formula. Then $y = f(g(x)) = f(u) = e^u = e^{g(x)}$. From the rule for the derivative of the exponential function $f'(u) = e^u$ or $f'(g(x)) = e^{g(x)}$. From the chain rule we get

$$y' = f'(g(x))g'(x) = e^{g(x)} \cdot g'(x) = g'(x)e^{g(x)}$$

The form of the chain rule applied to the exponential function

$$f(x) = e^{g(x)}$$

is the formula

Derivative of Exponential Functions

$$D_x e^{g(x)} = g'(x)e^{g(x)}$$

Example 2 illustrates how to apply the exponential function version of the chain rule.

EXAMPLE 2 | **Problem**

Find the derivative of (a) $f(x) = e^{2x}$ and (b) $f(x) = e^{3x^2 + 7}$.

SOLUTION (a) Let $g(x) = 2x$. Then $g'(x) = 2$ and

$$D_x e^{2x} = g'(x)e^{g(x)} = 2e^{2x}$$

(b) Let $g(x) = 3x^2 + 7$. Then $g'(x) = 6x$ and

$$D_x e^{(3x^2 + 7)} = g'(x)e^{g(x)}$$
$$= 6xe^{3x^2 + 7}$$

The next example demonstrates how one derivative can require two versions of the chain rule, the generalized power rule, and the exponential function version.

EXAMPLE 3 | **Problem**

Find the derivative of

$$f(x) = (x^{-1} + e^{5x + 2})^3$$

SOLUTION First apply the generalized power rule. We get

$$f'(x) = 3(x^{-1} + e^{5x + 2})^2 \cdot D_x(x^{-1} + e^{5x + 2})$$

The derivative of $x^{-1} + e^{5x + 2}$ requires the derivative of a sum. Let $g(x) = 5x + 2$. Then $g'(x) = 5$ and $D_x e^{5x + 2} = 5e^{5x + 2}$. Thus

$$f'(x) = 3(x^{-1} + e^{5x + 2})^2(-x^{-2} + 5e^{5x + 2})$$

Graphing Exponential Functions

The curve-sketching techniques of the previous chapter can be used to graph exponential functions. The methods are illustrated in the next two examples.

EXAMPLE 4 | **Problem**

Graph the function $f(x) = \dfrac{1}{1 + e^{-x}}$.

SOLUTION To find $f'(x)$, write $f(x) = (1 + e^{-x})^{-1}$. Then

$$f'(x) = -(1 + e^{-x})^{-2}(-e^{-x})$$
$$= e^{-x}(1 + e^{-x})^{-2}$$

$$= \frac{1}{e^x(1 + e^{-x})^2}$$

$$= \frac{1}{e^x(1 + 2e^{-x} + e^{-2x})}$$

$$= \frac{1}{2 + e^x + e^{-x}}$$

The numerator is never 0, so there is no value of x such that $f'(x) = 0$. Each term in the denominator is positive so there is no value of x such that $f'(x)$ does not exist. Thus there is no relative extremum. To test for a point of inflection, we calculate $f''(x)$ using the quotient rule.

$$f''(x) = \frac{-(e^x - e^{-x})}{(2 + e^x + e^{-x})^2}$$

$$= \frac{e^{-x} - e^x}{(2 + e^x + e^{-x})^2}$$

Solving $f''(x) = 0$ yields $e^{-x} - e^x = 0 = e^{-x}(e^{2x} - 1)$. This is satisfied when $e^{2x} = 1$, which is true when $x = 0$. Therefore $f(x)$ has a point of inflection at $x = 0$. When $x < 0$, $f''(x) > 0$ since $e^{-x} > e^x$. When $x > 0$, $f''(x) < 0$. This means that $f(x)$ is concave up in $(-\infty, 0)$ and concave down in $(0, \infty)$. Since $f'(x) > 0$ for all values of x, the graph is always increasing. As x gets very large e^{-x} approaches 0, so $f(x)$ approaches $1/(1 + 0) = 1$. Similarly as x approaches $-\infty$, $f(x)$ approaches 0. Putting this information together yields the graph in Figure 1.

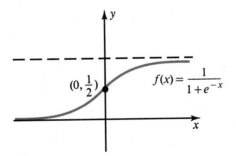

FIGURE 1

EXAMPLE 5 | **Problem**

Graph the function $f(x) = \dfrac{1}{\sqrt{2\pi}} e^{-x^2/2}$.

SOLUTION First find $f'(x)$.

$$f'(x) = \frac{1}{\sqrt{2\pi}} \left(\frac{-2x}{2}\right) e^{-x^2/2}$$

$$= \left(\frac{-x}{\sqrt{2\pi}}\right) e^{-x^2/2}$$

The only value of x such that $f'(x) = 0$ is $x = 0$ since $e^{-x^2/2}$ is always positive. Thus the only critical value is $x = 0$. Applying the first derivative test shows that $f(x)$ is increasing in $(-\infty, 0)$ and decreasing in $(0, \infty)$ because $e^{-x^2/2}$ is always positive, and so $f'(x)$ is positive and negative when x is positive and negative. Therefore $f(x)$ has a relative maximum of $f(0) = 1/\sqrt{2\pi}$. The points of inflection are found by solving $f''(x) = 0$.

$$f''(x) = \left(\frac{x^2}{\sqrt{2\pi}}\right) e^{-x^2/2} - \frac{1}{\sqrt{2\pi}} e^{-x^2/2}$$

$$= \left(\frac{1}{\sqrt{2\pi}}\right)(x^2 - 1)e^{-x^2/2}$$

The function has points of inflection at $x = -1$ and $x = 1$ and $f(x)$ is concave up in the intervals $(-\infty, -1)$ and $(1, \infty)$ and concave down in the interval $(-1, 1)$. The graph of $f(x)$ is given in Figure 2.

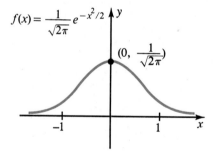

FIGURE 2

The function in Example 5 is the well-known "bell-shaped" curve that is used extensively in statistics. This function is known as the **standard probability density function.**

The Logistic Curve

Psychologists use various techniques to model the learning process. One type of model that uses exponential functions of the form $y(x) = a - be^{-ct}$, was introduced in Section 5–1. A more complex model of learning uses an exponential function of the form

$$y(x) = \frac{A}{1 + Be^{-kx}}$$

It is called the general form of the **logistic equation.** An example of a specific logistic equation is the function in Example 4, where $A = B = k = 1$. It is the function

$$f(x) = \frac{1}{1 + e^{-x}}$$

Its graph is given in Figure 1.

The first known application using the logistic equation was a study of world population made by the Belgian mathematician P. Verhulst. In 1840 Verhulst created a very accurate model of world population. Its prediction was off by only 1% a century later.

One of the first applications of the logistic curve in the social sciences was made by Clark L. Hull.* He created a model of the learning process that has had a wide variety of uses in many disciplines.

The graph of a logistic equation is an S-shaped curve, as shown in Figure 3. When $x = 0$ the denominator is $1 + B$, so

$$y(0) = \frac{A}{1 + B}$$

As x increases without bound, the denominator approaches 1 since e^{-kx} approaches 0. This means that $y(x)$ approaches the value $A/1$ or simply A. It never reaches A, but as x increases, $y(x)$ gets arbitrarily close to A. When the logistic equation is applied to population growth, the constant A is the maximum level of the population that the environment can sustain.

In Hull's model of learning A represents the maximum amount of learning. For example, suppose a puzzle is given to a large group of individuals. Let $f(x)$ be the average amount of learning, measured in percent, per time x spent solving the puzzle. Then A is 1, or 100%. It represents total mastery of the puzzle. One

*Clark L. Hull, *A Behavior System,* New Haven, CT, Yale University Press, 1952.

FIGURE 3

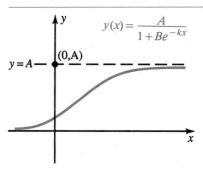

such puzzle consists of arranging several different cubes with variously colored sides in a specific order. An individual's learning consists in discovering the various characteristics of the cubes. The puzzle is usually solved the first time merely by chance. Little actual learning has taken place. When the cubes are mixed up, it takes the person almost the same amount of time to solve it again. Once some information is acquired about the cubes, say their individual color schemes or which one must come first in the sequence, then learning takes place. Hull found that when solving such tasks, learning takes place very slowly at first. The rate of learning then rises rapidly. When mastery of the task is approached the rate of learning slows again. These are the characteristics of the logistic equation.

The maximum learning rate is the maximum of the derivative of the logistic equation, $y = f(x)$. It occurs at the point of inflection of $f(x)$. The next example looks at a logistic curve similar to one used by Hull.

EXAMPLE 6

A psychologist gives a puzzle to a large group of individuals. In the directions of the puzzle each subject was told 20% of the solution. On the average, after 10 minutes a subject had learned 50% of the solution.

Problem

Find the specific form of the logistic equation that measures the percent of learning $f(x)$ after x minutes of solving the puzzle.

SOLUTION The maximum amount of learning is 100%, so $A = 1$. At $x = 0$ the time when the subject starts to solve the puzzle, 20% of the solution is known, so $f(0) = 0.2$. Therefore

$$f(0) = \frac{1}{1 + Be^0} = 0.2$$

Thus $1 = 0.2 + 0.2B$ and $0.8 = 0.2B$, so $B = 4$. Now solve the equation for k using the fact that $f(10) = 0.5$.

$$f(10) = \frac{1}{1 + 4e^{-10k}} = 0.5$$
$$1 = 0.5 + 2e^{-10k}$$
$$0.5 = 2e^{-10k}$$
$$0.25 = e^{-10k}$$
$$\ln(0.25) = \ln(e^{-10k}) = -10k$$
$$k = \frac{1.39}{10} = 0.139$$

Therefore the equation is

$$f(x) = \frac{1}{1 + 4e^{-0.139x}}$$

The logistic equation is also used in business to model a wide variety of situations. For example, Hannan and McDowell† studied how information spreads in an industry, with the banking industry as a specific example. They found that information spreads slowly at first, meaning that the proportion of firms using a new technological process remains small for a period of time. Then as firms using the process enjoy success with it the proportion of firms adopting it increases rapidly. To model the growth of automatic teller machines in the banking industry, they studied 89 local banks and developed the logistic equation

$$f(x) = \frac{1}{1 + 1.97e^{-0.33x}}$$

†Timothy H. Hannan and John M. McDowell, "Market Concentration and the Diffusion of New Technology in the Banking Industry," *Review of Economics and Statistics*, Vol. 66, 1984, pp. 686–691.

EXERCISE SET 5–4

• In Problems 1 to 24 find $f'(x)$.

1. $f(x) = e^{6x}$

2. $f(x) = e^{9x}$

3. $f(x) = e^{3x+5}$

4. $f(x) = e^{5x-6}$

5. $f(x) = 4e^{7x+5}$

6. $f(x) = 6e^{4-7x}$

7. $f(x) = e^{x^2+3}$

8. $f(x) = e^{3x^2-5}$

9. $f(x) = 2e^{1+4x^2}$

10. $f(x) = -3e^{3-5x^2}$

11. $f(x) = 4 - e^{2x^3+x^{-1}}$

12. $f(x) = 3 + e^{4-x^{1/2}+x}$

13. $f(x) = xe^{4x}$

14. $f(x) = 3xe^{7x}$

15. $f(x) = x^{-3}e^{x^2-1}$

16. $f(x) = 5x^{2/3}e^{2-3/2}$

17. $f(x) = (e^{3x})/(2x - 5)$

18. $f(x) = x^5 - 11/e^{3x}$

19. $f(x) = e^{5x} + e^{-5x}$

20. $f(x) = e^{3x} - 3e^{-3x}$

21. $f(x) = \dfrac{1}{2 + e^{-x}}$

22. $f(x) = \dfrac{1}{1 + e^{-2x}}$

23. $f(x) = \dfrac{1}{1 + e^x + e^{-2x}}$

24. $f(x) = \dfrac{2}{e^{2x} + e^{-2x}}$

In Problems 25 to 28 sketch the graph.

25. $f(x) = xe^x$

26. $f(x) = x + e^x$

27. $f(x) = 1/(2 + e^{-2x})$

28. $f(x) = 4/(1 + e^{-3x})$

•• In Problems 29 to 36 find $f'(x)$.

29. $f(x) = \ln(x + e^{4x})$

30. $f(x) = \ln(3 + xe^{2x})$

31. $f(x) = (\ln x + e^{5x})^2$

32. $f(x) = (2 \ln x + e^{3x})^{-3}$

33. $f(x) = (e^{x^2} + x^2)/e^x$

34. $f(x) = (\ln x)/(x - e^{3x})$

35. $f(x) = e^{x^3} \ln x$

36. $f(x) = e^{3x} \ln(x + e^x)$

••• 37. Let $f(x) = e^{2x}$. Find a formula for $f^{(n)}(x)$ for n a positive integer by computing the first few derivatives of $f(x)$ and finding a pattern.

38. Find $f'(x)$ for $f(x) = xe^x \ln x$.

39. In an article* studying the social cost of pop-

*Norman R. Glass et al., "Human Ecology and Educational Crises," in *Is There an Optimum Level of Population?*, Fred Singer (ed.), New York, McGraw-Hill, 1971, pp. 205–218.

ulation growth, researchers used educational tax revenue as a measure of the cost of an expanding population. They focused on California's 1965 tax profile as a baseline for a community's ability to fund fully its educational system. The proportion of the tax paid by various age groups increased from age 16 to 65. Because many people retired at age 65, the proportion of tax paid per age decreased linearly from age 65 to age 75, after which it was always 0. The data are given by the formula $y = f(x)$ where $f(x)$ is the proportion of the total educational tax paid by those of age x. In the interval $[16, 65]$ the data are given as an exponential function, and as a linear function in the interval $(65, 75]$.

$$f(x) = \begin{cases} 0.03e^{-0.0008(x-65)^2} & x \in [16, 65] \\ -0.003x + 0.225 & x \in (65, 75] \end{cases}$$

Graph $f(x)$ in $[16, 75]$.

40. Thermal recovery of the skin after cooling is not only an important medical treatment but it is also a key indicator of normal versus pathological circulation. In a paper studying thermal recovery, Steketee and Van Der Hoek cooled the skin of the forehead of patients and measured the time it took to reach a steady-state temperature.[†] The temperature T of the forehead, measured in degrees centigrade, was found to be a function of time t, measured in minutes. For one group of patients, the functional relationship was found to be

$$T(t) = 35 + 2.1e^{-.017t}$$

The steady-state temperature was defined to be that value of t when $T'(t) = -.028$. Find the steady-state temperature for this group of patients.

41. The problem of finding an optimal advertising policy over a period of time is one of the most important questions in the field of marketing. In a paper studying oligopoly models, that is, business problems involving more than one firm, Teng and Thompson studied how firms market new products.[‡] The price of the new product p, measured in dollars, was a function of time t, measured in months. In one market the new product sales price was found to obey the formula

$$p(t) = 100e^{.01t}$$

The object of the firm is to allow the price to increase to the point at which the increase is so great that it influences customers to purchase competitors' products. This point was reached when $p'(t) = 1.02$. Find this value of t.

[†]J. Steketee, and M. J. Van Der Hoek, "Thermal Recovery of the Skin after Cooling," *Journal of Physics and Medical Biology,* Vol. 24, 1979, pp. 583–592.

[‡]Jinn-tsair Teng and Gerald L. Thompson, "Oligopoly Models for Optimal Advertising when Production Costs Obey a Learning Curve," *Management Science,* Vol. 29, 1983, pp. 1087–1104.

CASE STUDY   Denver Fire Department Deployment Policy Analysis*

In 1972 the mayor and the chief of the Fire Department of Denver, Colorado, agreed that a study of the fire department was in order in the face of mounting costs. They decided to apply modern operations analysis by contracting for a research study with faculty from the Department of Operations Research at the University of Colorado. The study was funded by HUD's Office of Policy Development and Research at a level of $140,000. The challenge given the

*Adapted from D. E. Monarchi et al., "Simulation for Fire Department Deployment Policy Analysis," *Decision Sciences,* Vol. 8, 1977, pp. 211–227.

research team was: Can the fire department provide approximately the same level of fire suppression service at a lower cost?

The most critical cost was maintenance. Denver spent in excess of $250,000 per year to staff each fire engine or ladder truck around the clock. In contrast, the cost of the vehicle itself was about $60,000, and a fire house had a one-time cost of $400,000. Thus the number of vehicles and their location accounted for the bulk of the city's expenditure for fire protection.

In the first phase of the analysis the research team used linear programming to study fire company locations. The objective function minimized cost and the constraints were a set of response time requirements. The latter included the assumptions that every vehicle was always available to respond to a fire incident, and that *alarm rates,* the number of incidents per time period, were less than or equal to a specific threshhold. The conclusion of the study was that the level of service would be unchanged with a reduction of five companies—from 44 to 39 fire stations. This reduction would take place by closing obsolete stations and building new houses at new locations.

The recommendations were to take place over a period of seven years. But strong opposition, especially from the fire fighters' union, prevented immediate implementation. The union contended that if alarm rates increased dramatically, suppression service would significantly decline because of the fewer number of stations.

The second challenge for the researchers was to provide evidence supporting their conclusion. They needed to produce realistic data depicting possible future trends of alarm rates. Their approach was to use a dynamic analysis that compared the service provided by the existing fire company locations versus the proposed configuration. The term "dynamic" means that every possible fire incident situation in the city would be considered. For each situation it would be determined whether a vehicle could respond. They first had to study suppression service with the current alarm rate and then determine the effect on service if the alarm rate increased. The important question was whether the new configuration of fire stations recommended by the research team would provide the same service as the current one, both for the current alarm rate and for possible increased rates in the future.

The specific method used for this dynamic analysis was a computer simulation model developed by the New York City-Rand Institute. By using simulation they could explore the effects of increasing alarm rates on fire department performance for each configuration.

The simulation model had three parts. It created an "incident generator," which developed 1000 separate types of incidents, reflecting what, when, and where fires occur and the equipment required to service them. A random number generator was used to simulate the occurrence of the incidents, depending on the probability of occurrence assigned to each incident. The simulator then monitored the equipment—it dispatched the necessary equipment, assigned a time that the equipment would expect to be in service, and then "returned" the vehicle to await the next call.

The third part of the model compared the performance of two fire station

FIGURE 1

configurations, the current one versus the proposed one. The simulation was run for various alarm rates.

One of the key measures of performance is the occurrence of long response times. If a vehicle takes longer than 5 minutes to respond, its performance is poor. For a given average alarm rate the number of responses greater than 5 minutes (for the first vehicle to respond) was measured. The data for the new configuration are shown in Figure 1. As the average alarm rate increases, the number of long responses increases. For example, for the lowest alarm rate $t = 2.5$ (or 2.5 alarms per hour), $N = 3$, meaning there were 3 long responses out of 1000. For $t = 5$, $N = 14$, so when the alarm rate doubles from $t = 2.5$ to $t = 5$, N increases more than fourfold from 3 to 14.

A key point is that N increases at a fast rate from 2.5 to 5, but then the rate of increase of N slows dramatically. That is, while N increases as t increases, the rate of increase, measured by the derivative, decreases when t approaches 5. The data suggest that the best type of function to model the behavior of $N(t)$ is a *logistic curve,* whose general equation is

$$N(t) = \frac{a}{b + e^{-kt}}$$

Analysis of the data from the simulation suggests that the logistic equation of best fit for the proposed configuration is

$$N(t) = \frac{0.375}{0.013 + e^{-0.87t}}$$

This function is used to determine when $N(t)$ starts to level off. At that point, with the new configuration, the fire department could maintain service even as the average alarm rate increases. That is, even though the alarm rate might increase, the number of long responses remains almost constant. The value of t

where $N(t)$ starts to level off is the point where the derivative is a maximum, that is, at the point of inflection. It is found by solving $N''(t) = 0$. To compute this number, we use the quotient rule and the chain rule.

$$N'(t) = \frac{0.326e^{-0.87t}}{(0.013 + e^{-0.87t})^2}$$

$$N''(t) = \frac{-0.284e^{-0.87t}(0.013 - e^{-0.87t})}{(0.013 + e^{-0.87t})^3}$$

When we set $N''(t) = 0$ we get $0.013 = e^{-0.87t}$, and so $t = (\ln 0.013)/(-0.87) = 5.1$.

The simulation showed that $N(t)$ leveled off from $t = 5$ to 6, as in the logistic equation model, but when the average alarm rate was increased beyond 6, $N(t)$ began to rise appreciably. Hence the value $t = 5$ is critical. If Denver's average alarm rate ever reaches it, service can be maintained until t reaches 6, but by then the city would have to implement some kind of strategy, such as building new stations, to keep the same level of service. These relatively high alarm rates occur infrequently, however, as Denver's average fire company is available to respond from its quarters about 95 percent of the time.

When the simulation was run for the current configuration the results were almost identical. The only significant discrepency occurred when t increased beyond 6, for then N increased at a far greater rate for the new configuration than for the current one. The research team recommended that if the new configuration was implemented, then the alarm rate would have to be closely monitored. If and when it reached $t = 5$, the city still had time until it reached $t = 6$ to take further steps to maintain service, like building new stations.

In summary, the research team used linear programming to determine that the fire company's cost would be minimized, subject to the present level of service being maintained, by closing five fire stations. The team used dynamic simulation at first to test the system to see if there were any hidden elements that would cause problems of service, such as poor response time for some combination of incidents. None was discovered. Then, when their conclusions were challenged the fact that the city had been modeled in a dynamic environment gave the team a "more believable" presentation. The predictions of the study had more credibility because the simulation effectively replicated the historical fire suppression data of the city.

CHAPTER REVIEW

Key Terms

5–1 **Exponential Functions**
Exponential Function
Base

e
Learning Curve

Summary of Important Concepts

Formulas

$$e = \lim_{x \to \infty} \left(1 + \frac{1}{x} \right)^x$$

$$D_x \ln x = \frac{1}{x}$$

$$D_x \ln g(x) = \frac{g'(x)}{g(x)}$$

$$D_x e^x = e^x$$

$$D_x e^{g(x)} = g'(x) e^{g(x)}$$

REVIEW PROBLEMS

In Problems 1 to 5 sketch the graph of the function.

1. $f(x) = 2^{x+3}$

2. $f(x) = 5 + e^{-0.5x}$

3. $f(x) = \dfrac{1}{\sqrt{6}} e^{-x^2/2}$

4. $f(x) = \log_{1/e}(x - 3)$

5. $f(x) = \ln(4 - x^2)$

6. Evaluate (a) $\ln e^{-6.2}$ and (b) $e^{\ln 0.5}$

7. Solve for x.

$$10 = \frac{1}{e^{-x} - 2}$$

8. Evaluate

$$\lim_{h \to 0} \frac{\ln(10 + h) - \ln 10}{h}$$

9. Evaluate

$$\lim_{h \to 0} \frac{e^{2+h} - e^2}{h}$$

In Problems 10 to 12 find the derivative of the function.

10. $f(x) = 3 \ln x - \dfrac{1}{\ln x}$

11. $f(x) = [\ln(1 + 3x)^4]^5$

12. $f(x) = x \ln |x^2 - 1|$

13. Evaluate the derivative of the function $f(x) = \ln \sqrt{2x}$ in two ways: using the formula and using the properties of logarithmic functions.

14. Find the relative extrema and the points of inflection of the function $f(x) = \dfrac{\ln x}{x}$. Sketch the graph.

In Problems 15 and 16 find the derivative of the function.

15. $f(x) = x^2 e^x - e^{-x}$

16. $f(x) = (x^2 - e^{2x-1})^4$

17. Find the relative extrema and the points of inflection of the function $f(x) = e^x + e^{-x}$. Sketch the graph.

6 The Integral

CHAPTER OVERVIEW   Calculus is divided into two branches, differential calculus, which was discussed in the previous four chapters, and integral calculus, which we study in this chapter. Just as the derivative of a function $f(x)$ is a new function derived from $f(x)$, so too the indefinite integral of $f(x)$ is another function computed from $f(x)$. The process of "finding the integral" of $f(x)$ is the reverse operation of differentiation. From a geometric point of view the derivative measures the slope of the tangent line, while the definite integral measures the area bounded by $f(x)$.

In the first section we study the reverse process of differentiation called antidifferentiation. A function whose derivative is equal to $f(x)$ is called an antiderivative of $f(x)$. Section 6–2 uses the limit process to calculate certain types of areas in the plane that are bounded by curves. Section 6–3 shows the close connection between antiderivatives and the areas studied in Section 6–2. The fourth section extends this concept of computing areas to more intricate regions.

CASE STUDY PREVIEW

What happens to a stable industry when a new, innovative production technique is introduced? How can the effect on the industry be measured? One method is called consumers' surplus, which is a measure of the gain to consumers as a result of the new mode of business. The Case Study centers on the effect on the Louisiana crawfish industry when fish farms using private ponds compete with standard fisheries using public watersheds, such as lakes and rivers. We see that the consumer gains via significant price decreases at the expense of the standard fisheries.

6–1 ANTIDERIVATIVES

In this section several analogs of differentiation rules are presented. Each differentiation rule shows how to start with a function and then find its derivative. This section deals with the reverse operation: start with the derivative and find a function that has it as its derivative. These new rules will be applied to the geometry of curves, the velocity of an object, and carbon-14 dating.

The differentiation rules considered in this section are the power rule, the exponential and logarithmic function rules, the sum of functions rule, and the constant multiple rule. Later sections will look at the analogs of various other differentiation rules.

Antiderivatives in Applications

There are many applications where the rate of change of a function is computed rather than computing the function directly. For instance, it is usually easier to compute the rate of increase of a population than the actual size of the population. In archaeology it is easier to estimate the rate of change of the aging process of items found at an ancient site, rather than computing the age of the site directly.

Suppose a firm computes that the marginal cost $MC(x)$ of a product is given by

$$MC(x) = x + 1$$

The marginal cost is the derivative of the cost function $C(x)$. Can $C(x)$ be computed from $MC(x)$? This is the type of question we answer in this section.

Antiderivatives

We know that the derivative of $f(x) = x^2$ is $2x$. Here we start with x^2 and compute $2x$ via the power rule. The reverse operation starts with the derivative, $2x$, and computes a function, x^2, whose derivative is $2x$. The process is called **antidifferentiation** and the function $f(x) = x^2$ is called an *antiderivative* of $2x$. In general antidifferentiation answers the question: Given a function $y = f(x)$, what is a function whose derivative is $f(x)$?

_____ **Definition** _____

Given a function $f(x)$, if $F'(x) = f(x)$, then $F(x)$ is an **antiderivative** of $f(x)$.

_____

Example 1 illustrates not only the definition, but also why we said "an" instead of "the" antiderivative.

EXAMPLE 1

Problem

Verify that three antiderivatives of $f(x) = 3x^2$ are x^3, $x^3 + 1$, and $x^3 - 2$.

SOLUTION The function x^3 is an antiderivative of $3x^2$ because $D_x x^3 = 3x^2$. The two other functions are also antiderivatives because $D_x(x^3 + 1) = D_x(x^3 - 2) = 3x^2$.

Example 1 uses the fact that the derivative of a constant function is 0. The converse of this is true and is a useful result. The proofs of this theorem and the next one are outlined in the exercises. Theorem 1 states that if a function has derivative 0, then it must be a constant function.

_____ **Theorem 1** _____

If $F'(x) = 0$ for all x, then $F(x) = C$ for some real number C.

In Example 1 three functions are given whose derivatives are equal. Since the derivative measures the slope of the tangent line, the graphs of the three functions have the same slope at every point. See Figure 1.

The three antiderivatives of $3x^2$ mentioned in Example 1 are similar because each is of the form x^3 plus a constant. In fact any function of the form $x^3 + C$, where C is a constant, is an antiderivative of $3x^2$.

In general if $F(x)$ is an antiderivative of $f(x)$, then $F(x) + C$ is also an antiderivative of $f(x)$. Are there other functions besides these that are antiderivatives? The next theorem shows that all antiderivatives of $f(x)$ are of the form $F(x) + C$. The constant C is called the **constant of integration.**

FIGURE 1

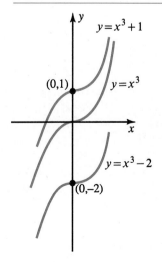

_____ Theorem 2 _____

If $F(x)$ and $G(x)$ are both antiderivatives of $f(x)$, then there is a constant C such that

$$F(x) = G(x) + C$$

_____

Another way to interpret this theorem is to say that any two antiderivatives of the same function differ by a constant.

The Indefinite Integral

Theorem 2 states that once one antiderivative $F(x)$ of $f(x)$ is known, then all antiderivatives are of the form $F(x) + C$. The **integral sign** $\int$ is used to express this.

$$\int f(x)\, dx = F(x) + C$$

This is called the general form of the antiderivative of $f(x)$. It states that the antiderivatives of $f(x)$ are the functions $F(x) + C$. The integral is similar to the derivative, $\dfrac{dy}{dx}$, in that it indicates an operation on the function. The meaning of the notation $\dfrac{dy}{dx}$ is to find the derivative, while the meaning of the notation $\int f(x)\, dx$ is to find all antiderivatives of $f(x)$. The term "dx" serves the same purpose in each notation. It indicates what the variable is. Thus $\int 2x\, dx$ is different from $\int 2x\, dt$. The expression $\int f(x)\, dx$ is called the **indefinite integral,** and $f(x)$ is called the **integrand.** To find the indefinite integral means to find all antiderivatives of the integrand.

Example 2 shows how to use this notation when finding the indefinite integral of another power function.

EXAMPLE 2 | **Problem**

Find the indefinite integral $\int x^3\, dx$.

SOLUTION Work backward as before. The function whose derivative is almost x^3 is x^4. But the derivative of x^4 is $4x^3$; that is, $D_x(x^4) = 4x^3$. Recall that the derivative of a constant times a function is the constant times the derivative. This means that the differentiation statement above can be divided by 4 to get

$$\frac{1}{4} D_x(x^4) = D_x\left(\frac{x^4}{4}\right) = x^3$$

Therefore

$$\int x^3\, dx = \frac{x^4}{4} + C$$

The Power Rule for Antiderivatives

Example 2 illustrates how to compute the indefinite integral of all power functions x^n, except $n = -1$. To find the function whose derivative is x^n, start with x^{n+1}. The derivative of x^{n+1} has a factor of $n + 1$, so divide the function by that factor. Thus an antiderivative of x^n is $\dfrac{x^{n+1}}{n+1}$. This is the power rule for antiderivatives. It holds for all real numbers n except -1, because $n = -1$ produces division by 0.

The Power Rule

$$\int x^n \, dx = \frac{x^{n+1}}{n+1} + C \qquad n \neq -1$$

EXAMPLE 3

Problem

Find the indefinite integrals (a) $\displaystyle\int x^{-4} \, dx$, (b) $\displaystyle\int \sqrt{x} \, dx$, and (c) $\displaystyle\int dx$.

SOLUTION (a) Use the power rule with $n = -4$.

$$\int x^{-4} \, dx = \frac{x^{-4+1}}{-4+1} + C = \frac{x^{-3}}{-3} + C = -\frac{1}{3}x^{-3} + C$$

(b) Use the power rule with $n = 1/2$.

$$\int x^{1/2} \, dx = \frac{x^{(1/2)+1}}{1/2 + 1} + C = \frac{x^{3/2}}{3/2} + C = \frac{2x^{3/2}}{3} + C$$

(c) The integrand is understood to be $1 = x^0$. From the power rule with $n = 0$ we have

$$\int dx = \int x^0 \, dx = \frac{x^{0+1}}{0+1} + C = x + C$$

Exponential and Logarithmic Functions

The derivative rules for exponential and logarithmic functions can also be expressed as indefinite integrals. Recall the rules

$$D_x(\ln |x|) = \frac{1}{x} \qquad D_x e^{rx} = r e^{rx}$$

The power rule for antiderivatives excludes $r = -1$, so it does not give the antiderivatives of x^{-1}. However, since $D_x(\ln x) = 1/x = x^{-1}$, the general form of the antiderivative of x^{-1} is expressed in the logarithmic function rule.

Logarithmic Function Rule

Antiderivative of $f(x) = \dfrac{1}{x} = x^{-1}$:

$$\int x^{-1}\,dx = \ln|x| + C$$

From the formula $D_x(e^{rx}) = re^{rx}$ divide by r to get the exponential function rule.

Exponential Function Rule

Antiderivative of $f(x) = e^{rx}$:

$$\int e^{rx}\,dx = \frac{e^{rx}}{r} + C$$

EXAMPLE 4 | **Problem**

Find the indefinite integral of $f(x) = e^{5x}$.

SOLUTION From the antiderivative rule of $f(x) = e^{rx}$ with $r = 5$ we have

$$\int e^{5x}\,dx = \frac{e^{5x}}{5} + C$$

Addition and Constant Rules

The first three antidifferentiation rules presented so far can be derived by working backward from their corresponding differentiation rules. In a similar way the formulas that state (a) "the derivative of a sum is the sum of the derivatives" and (b) "the derivative of a constant times a function is the constant times the derivative of the function" can be expressed as indefinite integrals.

The Sum Rule

$$\int [f(x) + g(x)]\,dx = \int f(x)\,dx + \int g(x)\,dx$$

> **The Constant Times a Function Rule**
>
> $$\int kf(x)\,dx = k\int f(x)\,dx \qquad \text{for any constant } k$$

A function with more than one term can be antidifferentiated term by term. It can be extended to a finite number of functions, so that

$$\int [f_1(x) + \cdots + f_n(x)]\,dx = \int f_1(x)\,dx + \cdots + \int f_n(x)\,dx$$

The constant times a function rule states that a constant can be moved through the integral sign in the same way that it can be moved through a derivative sign. The next two examples illustrate these rules.

EXAMPLE 5

Problem

Find the indefinite integrals (a) $\displaystyle\int (x^4 + e^{-x})\,dx$ and (b) $\displaystyle\int \left(3x^{-5} - \frac{4}{x}\right) dx$.

SOLUTION

(a) $\displaystyle\int (x^4 + e^{-x})\,dx = \int x^4\,dx + \int e^{-x}\,dx$

$$= \frac{x^{4+1}}{4+1} + \frac{e^{-x}}{-1} + C$$

$$= \frac{x^5}{5} - e^{-x} + C$$

(b) $\displaystyle\int \left(3x^{-5} - \frac{4}{x}\right) dx = \int 3x^{-5}\,dx + \int \left(\frac{-4}{x}\right) dx$

$$= 3\int x^{-5}\,dx + (-4)\int \left(\frac{1}{x}\right) dx$$

$$= \frac{3x^{(-5)+1}}{-5+1} - 4\int \frac{1}{x}\,dx + C$$

$$= \frac{-3x^{-4}}{4} - 4\ln|x| + C$$

Sometimes algebraic manipulations can be used to express the integrand in a form that matches the antidifferentiation rules. The next example demonstrates one such manipulation.

EXAMPLE 6 | **Problem**

Find the indefinite integral $\int \sqrt{x}(x^2 - 1) \, dx$.

SOLUTION The integrand does not match any of those in the antidifferentiation rules, so an algebraic simplification is called for. Multiply through the parentheses to get

$$\int \sqrt{x}(x^2 - 1) \, dx = \int (\sqrt{x}\, x^2 - \sqrt{x}) \, dx = \int (x^{5/2} - x^{1/2}) \, dx$$

$$= \int (x^{5/2}) \, dx - \int (x^{1/2}) \, dx$$

$$= \frac{x^{7/2}}{7/2} - \frac{x^{3/2}}{3/2} + C$$

$$= \frac{2x^{7/2}}{7} - \frac{2x^{3/2}}{3} + C$$

EXAMPLE 7 | **Problem**

Find the function whose slope of the tangent line at x is $f'(x) = 3x^2 - 1$ and whose graph passes through the point $(0, 1)$.

SOLUTION The first step is to find the indefinite integral of $f'(x)$. Use the fact that $f(x)$ is the antiderivative of $f'(x)$.

$$f(x) = \int f'(x) \, dx = \int (3x^2 - 1) \, dx = x^3 - x + C$$

This gives all functions whose slope is the given formula. To find which of these functions also passes through $(0, 1)$, substitute $f(0) = 1$ into the formula.

$$1 = f(0) = 0^3 - 0 + C = C$$

The function whose constant of integration is $C = 1$ is the solution, which is $f(x) = x^3 - x + 1$.

Recall that a production function $P(x)$ measures the production level P, or the output, of a firm using x amount of labor. The derivative of $P(x)$ is called marginal product, $MP(x)$. Antiderivatives can be used to compute the production function from the marginal product.

EXAMPLE 8 | **Problem**

A firm determines that its marginal product $MP(x)$ is given by

$$MP(x) = x^2 - 10x + 20$$

where $MP(x)$ is measured in thousands of dollars and x in hundreds of workers. Compute the production function.

SOLUTION Since $P'(x) = MP(x)$, we have

$$P(x) = \int MP(x)\, dx$$
$$= \int (x^2 - 10x + 20)\, dx = \frac{x^3}{3} - 5x^2 + 20x + C$$

Similar statements hold for marginal cost, marginal revenue, and marginal profit. If $R(x)$ is the revenue function of a firm, then $R'(x)$ is the marginal revenue. $R(x)$ can be computed from marginal revenue by the formula

$$R(x) = \int R'(x)\, dx$$

The Growth Equation

Many interesting applications of calculus lead to equations involving derivatives. They are called **differential equations.** In fact whenever an antiderivative is computed we are solving a differential equation of the form $f'(x) = g(x)$, where $g(x)$ is given and $f(x)$ is unknown. A common and powerful type of differential equation is

$$f'(x) = kf(x)$$

for some constant k. It is called the **growth equation.** If k is positive, then k is called the *growth constant*. If k is negative, then k is called the *decay constant*. The next two examples illustrate an application from archaeology, where the growth equation plays a key role.

The rate at which radioactive material changes to lead is proportional to the amount of the material present; that is, if $f(t)$ is the amount present, then $f'(t) = kf(t)$ for some constant k. Hence $f(t)$ is a function whose derivative is a constant times itself. The only type of function with that property is $f(t) = Ce^{kt}$. To check that $f(t)$ satisfies the differential equation, we compute $f'(t) = Cke^{kt}$, which is equal to $kf(t)$ as needed.

EXAMPLE 9

Living plant and animal tissues contain the radioactive substance carbon-14, which has a half-life of 5568 years (at the end of that time one-half of that amount remains). When the organism dies, no new carbon is received.

Problem

Find a formula for the amount of carbon-14 remaining after t years.

SOLUTION From above, $f(t) = Ce^{kt}$. Since $f(0) = Ce^0 = C$, the amount present at time $t = 0$ is $C = f(0)$. When $t = 5568$, $f(0)/2$ is present, and thus

$$f(5568) = f(0)e^{5568k} = \frac{f(0)}{2}$$

Therefore $e^{5568k} = 1/2$, and so $\ln e^{5568k} = \ln 1/2$, which gives $5568k = \ln 1/2$ and $k = (1/5568)\ln 1/2$. Therefore

$$f(t) = Ce^{(1/5568)(\ln 1/2)t}$$

Archaeologists use the function in Example 9 to date uncovered ruins.

EXAMPLE 10   Suppose an analysis of ashes found in an unearthed settlement reveals that $1/5$ of the carbon-14 present in the original ashes has decomposed, so that $4/5$ of $f(0)$ remains.

Problem

How old is the settlement?

SOLUTION   The task is to find t such that $f(t) = (4/5)f(0)$. Then

$$(4/5)f(0) = f(0)e^{kt} \qquad \text{implies} \qquad 4/5 = e^{kt}$$

Hence $\ln 4/5 = \ln e^{kt} = kt$. Thus

$$t = \left(\frac{1}{k}\right)\ln 4/5 = \frac{5568(\ln 4/5)}{\ln 1/2}$$

$$= \frac{5568(\ln 4 - \ln 5)}{\ln 1 - \ln 2}$$

$$\approx \frac{5568(1.386 - 1.609)}{(0 - 0.693)}$$

$$\approx 1792$$

Thus the settlement is approximately 1792 years old.

EXERCISE SET 6–1

• In Problems 1 to 24 find the antiderivative.

1. $\int (3x + 5)\, dx$

2. $\int (4x - 2)\, dx$

3. $\int (2x^4 - 7)\, dx$

4. $\int (3x^5 + 1)\, dx$

5. $\int (6x^{-2} + 5x)\, dx$

6. $\int (5x^{-3} - x^3 + 1)\, dx$

7. $\int (2x^{1/2} + 3x^{1/3} - 5x)\, dx$

8. $\int (5x^{1/5} - x^{2/3} + 7x)\, dx$

9. $\int (3x^{-1/3} + 7x^{-1/4} - 5)\, dx$

10. $\int (6x^{-1/6} - 3x^{-3/7} - 6x)\, dx$

11. $\int (x^{0.2} + 3x^{1.4})\, dx$

12. $\int (1.2x^{-1.6} - 4.1x^{-2.2})\, dx$

13. $\int (1/x^2 + 5/x^4)\, dx$

14. $\int (2/x^{0.5} - 4/x^{2.2})\, dx$

15. $\int (1/x^7 + 4/x^{1/4} + 4x^{1/4})\, dx$

16. $\int (3/x^{2.2} - 3/x^{2/5} - 2x^{2/5})\, dx$

17. $\int (3e^{2x} + 5/x)\, dx$

18. $\int (5e^{3x} - 2/x)\, dx$

19. $\int (x^3 + 2e^{-x} + x^{1/2} - 5/x)\, dx$

20. $\int (3e^{-4x} - 4x^{-3} - 15/x + x^{-1/2})\, dx$

21. $\int (5\sqrt{x} - 2e^{-4x} + 1/x)\, dx$

22. $\int (8/x + 2/\sqrt{7x} + 3e^{-2x})\, dx$

23. $\int (18/x + 2/\sqrt[3]{4x} - e^{7x})\, dx$

24. $\int (8/\sqrt[4]{3x} + 2/x - e^{-10x})\, dx$

In Problems 25 to 28 find the function whose slope of the tangent line to the curve $y = f(x)$ is given and whose graph passes through the given point.

25. $f'(x) = x^2 - 2$ at $(1, -1)$

26. $f'(x) = 2x^2 - 1$ at $(1, 2)$

27. $f'(x) = 4x^{-1} - 2x$ at $(2, -2)$

28. $f'(x) = 2x^{-2} - x$ at $(1, 1)$

•• In Problems 29 to 34 find $\int f(x)\, dx$ by using algebra

to express the integrand in a form that fits the integration rules.

29. $\int (2x^4 - 7x)/x^2\, dx$

30. $\int (3x^5 + 10x)/x^3\, dx$

31. $\int x^2(6x^{-2} + 5x)\, dx$

32. $\int x^3(5x^{-3} - x^3 + 1)\, dx$

33. $\int (2x^4 - 7x)^2\, dx$ 34. $\int (3x^5 + 10x)^2\, dx$

In Problems 35 and 36 find the position function $s(t)$ given the velocity function $v(t)$ and the initial position $s(0)$.

35. $v(t) = t^2 + 2t$, $s(0) = 0$

36. $v(t) = t^3 + t$, $s(1) = 1$

37. Find the age of a settlement where ashes found at the site are assumed to be 2/3 of the original amount; that is, 1/3 of the original amount has decomposed.

38. Do Problem 37 with the fraction 2/3 replaced by 3/4.

••• 39. Give a reason for each step in the following proof of Theorem 1:
Proof of Theorem 1: Suppose $F'(x) = 0$ for all x. Then the graph of $F(x)$ has slope 0 at every point. Thus the tangent line is horizontal at every point. This means that the graph of $F(x)$ is a horizontal line, since a horizontal line is the only graph with that property.

40. Give a reason for each step in the following proof of Theorem 2:
Proof of Theorem 2: Suppose $F(x)$ and $G(x)$ are two antiderivatives of $f(x)$. We want to show that the function $F(x) - G(x)$ is a constant function, since then $F(x) - G(x) = C$ for some C and so $F(x) = G(x) + C$. Consider the function $H(x) = F(x) - G(x)$. Its derivative is

$$H'(x) = F'(x) - G'(x) = f(x) - f(x) = 0$$

Therefore $H(x)$ must be a constant function by Theorem 1. Hence $F(x) - G(x) = C$ and so $F(x) = G(x) + C$.

41. In order to illustrate the application of dating methods in archaeology covered in his book, Joseph Michels discusses one of the archaeological expeditions he organized with students from the Pennsylvania State University.* The site was Sheep Rock Shelter in Huntingdon County, PA. For thousands of years, the shelter, located on the Juniata River, provided small platforms for various living mammals. Artifacts

*Joseph W. Michels, *Dating Methods in Archaeology*, New York, Seminar Press, 1973.

were found in the rock strata as deep as 14 feet below the present surface. In one level it was determined that of the original amount of carbon-14 contained in certain artifacts, 42% remained. Find the age of the artifacts.

In Problems 42 and 43 an article is quoted that discusses growth rates. Find the growth rate $f'(x) = kf(x)$ that is satisfied by the growth equation $f(x)$ given in the article.

42. In an article studying the causes and effects of wind damage to trees such as sugar maples, King investigates the factor influencing wind damage due to the rapid decline in wind speed beneath the canopy, or top, of the forest.† The

relationship between the wind speed $V(h)$ and the height h of the wind is given by

$$V(h) = 22e^{1.4h}$$

43. Because the bald eagle is an endangered species, the study of the growth rates of sibling eagles is very important to conservationists. In an article investigating the growth rates of nestlings, or broods, with one versus more than one young eagle, Bortolotti calculates the following growth equation that measures the difference in the size of two young eagles in the same nestling, born one day apart.‡ Let m be the difference in the mass, measured in percent, at time t, measured in days. The growth equation is

$$m(t) = 38.1e^{-.068t}$$

---

†David A. King, "Tree Form, Height Growth, and Susceptibility to Wind Damage in *Acer saccharum*," *Ecology*, Vol. 67, 1986, pp. 980–990.

‡Gary R. Bortolotti, "Evolution of Growth Rates in Eagles: Sibling Competition vs. Energy Considerations," *Ecology*, Vol. 67, 1986, pp. 182–194.

6–2 AREA AND THE DEFINITE INTEGRAL

One application of the definite integral is the measure of the area under a curve, which appears in many real-world problems. Areas of certain types of regions can be found using only geometry, such as squares, rectangles, and triangles. The classical Greeks even knew how to find the area of some regions bounded by specific curves. But the general problem of finding the area under an arbitrary continuous curve was not solved until about 1700 with the advent of calculus.

An Application from Economics

The area under a curve has a wide variety of uses in many disciplines, one of which is economics. One application analyzes the distribution of wealth and income in a society. For centuries economists, politicians, and social scientists have debated the question, "What constitutes a fair and equitable distribution of income and wealth?" Wealth is the value of the goods and property people own, while income is the gain derived from the use of human and material resources. For measuring inequalities in the distribution of income and wealth the most commonly used device is a graph called a **Lorentz diagram.** A Lorentz diagram illustrates what percentage of families received what percentage of a nation's total income. It is used to compare the income distribution of different countries as well as to evaluate an individual country's system.

Consider the Lorentz diagram for families in the United States in Figure 1,

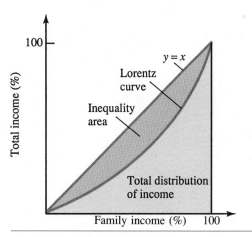

FIGURE 1 *Source:* Milton H. Spenser, *Contemporary Economics,* New York, Worth Publishers, Inc., 1986, pp. 53–55.

which is adapted from *Contemporary Economics* by Milton H. Spenser.* There are two curves in a Lorentz diagram. The lower one, called a *Lorentz curve,* measures the percentage of a nation's total income that is shared by those families with a certain income level. As an example of how to interpret the curve, consider $x = 40$ on the horizontal axis. It denotes those families with the 40% lowest family incomes. The vertical axis represents percentages of the total income of the United States. Corresponding to $x = 40$ on the Lorentz curve is $y = 16$, which means that the families with the lowest 40% of family incomes have 16% of the total income of the United States. The second curve in the diagram is $y = x$. It represents complete equality of distribution of income, because if income were distributed equally, families with the lowest x% of income would have $y = x$% of the total income. Economists call the area between these two curves the *inequality area* because it represents the extent of departure between an equal distribution of income and the actual distribution of income. The area of the region bounded above by the Lorentz curve and below by the x-axis is called the total distribution of income. The area of the region bounded above by $y = x$ and below by the x-axis is the area of complete equality of income distribution.

In this section we study how to compute the latter two areas. Each is the area of a region bounded above by a curve and below by the x-axis. Section 16–4 will show how to evaluate the area between two curves. Then we will compute the inequality income area for the United States.

Area Under a Curve

To find the area under a curve, we approximate the area by using rectangles whose area can be easily computed. The process entails fitting rectangles into the region. The first approximation is often not very close to the actual area. Better approxi-

---

*Milton H. Spenser, *Contemporary Economics,* New York, Worth Publishers, Inc., 1986, pp. 53–55.

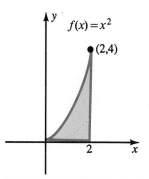

FIGURE 2

mations are usually found by using more rectangles. The actual area is found by using a limit process.

We start by finding an initial approximation to the area under the curve $f(x) = x^2$ bounded by $x = 0$ to $x = 2$ and above the x-axis. Figure 2 shows this region. A rough approximation uses two rectangles as in Figure 3. Divide the interval $[0, 2]$ into two subintervals of equal width and choose the right-hand end point to calculate the height of each rectangle. Each width is 1 and the two heights are $f(1) = 1^2 = 1$ and $f(2) = 2^2 = 4$. Thus the sum of the areas of the two rectangles is

$$1 \cdot f(1) + 1 \cdot f(2) = 1 + 4 = 5$$

Figure 3 shows that this approximation is greater than the actual area, but it illustrates how to compute further approximations that are closer. To approximate the area of the region bounded by $y = f(x)$ (for now we must assume that $f(x) \geq 0$), the x-axis, $x = a$, and $x = b$, first divide the interval $[a, b]$ into n equal subintervals, each with width $(b - a)/n$. Evaluate $f(x)$ at the right-hand end point to find the height of each rectangle. Then the approximation is the sum of the areas of these rectangles. Example 1 illustrates this procedure.

FIGURE 3

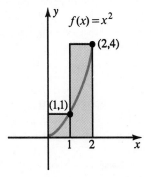

EXAMPLE 1

Problem

Find the approximation of the area of the region bounded by $f(x) = x^2$, $x = 0$, $x = 2$, and the x-axis, using four subintervals.

SOLUTION The width of each subinterval is $(2 - 0)/4 = 1/2$. The right-hand end points are 1/2, 1, 3/2, and 2. The approximation is the sum of the areas of the rectangles. (See Figure 4.)

$$\frac{1}{2} \cdot f(1/2) + \frac{1}{2} \cdot f(1) + \frac{1}{2} \cdot f(3/2) + \frac{1}{2} \cdot f(2)$$

$$= \frac{1}{2} \cdot \frac{1}{4} + \frac{1}{2} \cdot 1 + \frac{1}{2} \cdot \frac{9}{4} + \frac{1}{2} \cdot 4$$

$$= \frac{1}{2}\left(\frac{1}{4} + 1 + \frac{9}{4} + 4\right) = \left(\frac{1}{2}\right)\left(\frac{30}{4}\right) = \frac{15}{4}$$

Area and the Definite Integral

From Figure 4 it is clear that the approximation of 15/4 is still greater than the area, but it is closer than the previous approximation of 5. To get even closer approximations, use more subintervals with smaller widths. In general the process of approximating the area entails dividing [0, 2] into n subintervals and computing

FIGURE 4

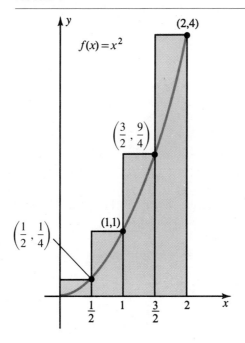

the sum of the areas of the rectangles. Then we let n get large to see if the process has a limiting value. Each subinterval has width $(2 - 0)/n = 2/n$. The right-hand end points are $2/n$, $4/n$, . . . , 2. For example, if $n = 8$, the end points are 1/4, 1/2, 3/4, 1, 5/4, 3/2, 7/4, 2. A computer was used to find the approximations to the area for the following values of n:

n	2	4	8	100	1000	10000	100000
area	5	3.75	3.188	2.707	2.671	2.667	2.6667

The areas of the approximations approach the number 2 2/3 = 8/3. In fact the larger the number of rectangles that are used, the closer the approximation is to 8/3. This means that as n increases without bound, the limit is equal to 8/3. This is expressed as

$$\lim_{n \to \infty} (\text{sum of the areas of the } n \text{ rectangles}) = 8/3$$

Therefore the area of the region bounded by $f(x) = x^2$, the x-axis, $x = 0$, and $x = 2$ is equal to 8/3. This is what is meant by the **definite integral** of $f(x) = x^2$ from $x = 0$ to $x = 2$. It is written as

$$\int_0^2 x^2 \, dx = 8/3$$

The function $f(x) = x^2$ is called the *integrand*. The number 2 is called the **upper limit of integration** and the number 0 is called the **lower limit of integration.**

    For any continuous function $f(x)$, with $f(x) \geq 0$ in $[a, b]$, as the number n of subintervals of $[a, b]$ gets larger, the corresponding approximations get closer to the area under the curve. This can be seen from Figure 5. The limit of these approximations is the definite integral of $f(x)$ from a to b.

―――――――――――――――――――― **Definition** ――――――――――――――――――――

If $f(x)$ is continuous in the interval $[a, b]$ and $[a, b]$ is divided into n subintervals whose right-hand end points are $x_1, x_2, \ldots, x_n$, then the *definite integral of $f(x)$ from a to b* is

$$\int_a^b f(x) \, dx = \lim_{n \to \infty} \frac{(b - a)}{n} [f(x_1) + \cdots + f(x_n)]$$

    The next example shows how this limit behaves with a linear function as the integrand. We use the fact that

$$1 + 2 + \cdots + n = \frac{n(n + 1)}{2}$$

EXAMPLE 2 | **Problem**

Find $\int_0^4 2x \, dx$.

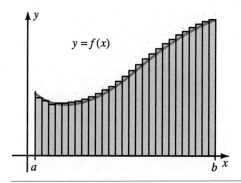

$y = f(x)$

FIGURE 5

SOLUTION Divide $[0, 4]$ into n subintervals, each of width $4/n$. The right-hand end points are $4/n, 8/n, \ldots, 4$. From the definition with $a = 0$ and $b = 4$ we have

$$\int_0^4 2x\,dx = \lim_{n \to \infty} \frac{(b - a)}{n} [f(x_1) + \cdots + f(x_n)]$$

$$= \lim_{n \to \infty} \frac{(4 - 0)}{n} \left[f\left(\frac{4}{n}\right) + f\left(\frac{8}{n}\right) + \cdots + f(4) \right]$$

$$= \lim_{n \to \infty} \frac{4}{n} \left[\frac{8}{n} + \frac{16}{n} + \cdots + 8 \right]$$

$$= \lim_{n \to \infty} \frac{4}{n} \left(\frac{8}{n}\right) [1 + 2 + \cdots + n]$$

$$= \lim_{n \to \infty} \frac{32}{n^2} \left[\frac{n(n + 1)}{2} \right] = \lim_{n \to \infty} \frac{16(n + 1)}{n} = \lim_{n \to \infty} \left(16 + \frac{16}{n} \right)$$

$$= 16$$

Thus $\int_0^4 2x\,dx = 16$.

The definite integral is defined in terms of a limit process. The definition does not stipulate that the function $f(x)$ be nonnegative in the interval, but when the definite integral is linked with the area under a curve this condition is important. When $f(x) \geq 0$ in $[a, b]$, then the area of the region bounded by $f(x)$, the x-axis, $x = a$, and $x = b$ is equal to the definite integral, $\int_a^b f(x)\,dx$. But if $f(x)$ is less than 0 in the interval, so that its graph is below the x-axis somewhere in the interval, then the two are not equal. The next section examines this situation.

_____ **Area Under a Curve** _____

If $f(x) \geq 0$ in $[a, b]$, then the area bounded by $f(x)$, the x-axis, $x = a$, and $x = b$ is given by

$$\int_a^b f(x) \, dx$$

_____

A more general definition of the definite integral can be used that does not restrict the subintervals to have equal length. Also, the height of each rectangle can be computed by using any point in the subinterval, rather than the right-hand end point. The definition given here is equivalent to the less restrictive one and is easier to use.

EXAMPLE 3 | **Problem**

Compute $\int_0^1 (3x + 1) \, dx$.

SOLUTION Divide $[0, 1]$ into n subintervals, each of width $1/n$. The right-hand end points are $1/n, 2/n, \ldots, 1$. From the definition with $a = 0$ and $b = 1$ we have

$$\int_0^1 (3x + 1) \, dx = \lim_{n \to \infty} \frac{(b - a)}{n} [f(x_1) + \cdots + f(x_n)]$$

$$= \lim_{n \to \infty} \frac{(1 - 0)}{n} \left[f\left(\frac{1}{n}\right) + f\left(\frac{2}{n}\right) + \cdots + f(1) \right]$$

$$= \lim_{n \to \infty} \frac{1}{n} \left[3\left(\frac{1}{n}\right) + 1 + 3\left(\frac{2}{n}\right) + 1 + \cdots + 3(1) + 1 \right]$$

The term "1" appears n times and we can factor out 3 and $1/n$ from each of the remaining terms to get

$$= \lim_{n \to \infty} \frac{1}{n} \left[\frac{3}{n} (1 + 2 + \cdots + n) + n \right]$$

Now use the identity for $1 + 2 + \cdots + n$.

$$= \lim_{n \to \infty} \frac{1}{n} \left[\frac{3}{n} \frac{n(n + 1)}{2} + n \right]$$

$$= \lim_{n \to \infty} \left[\frac{3(n + 1)}{2n} + \frac{n}{n} \right]$$

$$= \lim_{n \to \infty} \left[\frac{3}{2} \frac{(n + 1)}{n} + 1 \right]$$

$$= \frac{3}{2} + 1 = \frac{5}{2}$$

Thus

$$\int_0^1 (3x + 1)\, dx = \frac{5}{2}$$

Lorentz Diagrams Revisited

The area of complete equality of income distribution in a Lorentz diagram is the area of the region bounded by $y = x$ and the x-axis. It is the definite integral

$$\int_0^1 x\, dx$$

We use the definition to compute this area. Divide [0, 1] into n subintervals, each of width $1/n$. The right-hand end points are $1/n$, $2/n$, $\ldots$, 1. With $a = 0$, $b = 1$, and $f(x) = x$, we have

$$\int_0^1 x\, dx = \lim_{n\to\infty} \frac{(b - a)}{n} [f(x_1) + \cdots + f(x_n)]$$

$$= \lim_{n\to\infty} \frac{(1 - 0)}{n} \left[f\left(\frac{1}{n}\right) + f\left(\frac{2}{n}\right) + \cdots + f(1) \right]$$

$$= \lim_{n\to\infty} \left(\frac{1}{n}\right) \left[\frac{1}{n} + \frac{2}{n} + \cdots + 1 \right]$$

Factoring $1/n$ from each term, we get

$$= \lim_{n\to\infty} \left(\frac{1}{n^2}\right) [1 + 2 + \cdots + n]$$

Now use the identity for $1 + 2 + \cdots + n$.

$$= \lim_{n\to\infty} \left(\frac{1}{n^2}\right) \left[\frac{n(n + 1)}{2} \right]$$

$$= \lim_{n\to\infty} \frac{n + 1}{2n}$$

$$= \frac{1}{2}$$

Thus

$$\int_0^1 x\, dx = \frac{1}{2}$$

This is the area of complete equality of income distribution. A country's actual income distribution is compared with it. For income in the United States the Lorentz curve given in Spenser can be approximated by the curve $y = x^2$. (See Figure 1.)

The area of the region bounded by this curve and the x-axis is called the area of total income distribution. For this Lorentz curve it is

$$\int_0^1 x^2 \, dx$$

In the next section we will learn how to compute this area from the fundamental theorem of calculus.

A Few Concluding Remarks

The introduction of this section states that the ancient Greeks knew how to find the area under a curve for several specific curves. One such type of curve is a straight line, the graph of a linear function. This is because the area under a linear function can be divided into triangles and rectangles whose areas are easily computed. For instance, Example 3 shows that the definite integral of the linear function $f(x) = 3x + 1$ from $x = 0$ to $x = 1$ is 5/2. In the interval [0, 1] the function $f(x) = 3x + 1 > 0$, so the value of the definite integral should be equal to the area under the curve. From Figure 6 the area under the curve can be divided into a square with area 1 and a triangle with area $(1/2)(1)(3) = 3/2$. So the area under the curve is $1 + 3/2 = 5/2$, which is the same as the definite integral computed in the example.

In a similar way consider the Lorentz diagram in Figure 1: the area of total income distribution is the area of the region bounded by the linear function $y = x$. This region is a triangle and its area is $(1/2)(1)(1) = 1/2$, which is the same number computed from the definition. The definition of area is much more

FIGURE 6

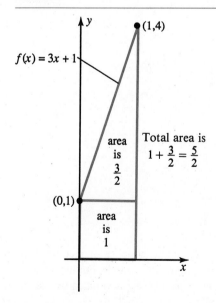

difficult to apply for nonlinear functions. The next section will give us an easier way to compute definite integrals.

The notation for the indefinite integral and the definite integral are similar and it is easy to confuse them. In the next section we will discover the close connection between the two concepts—basically the indefinite integral can be used to evaluate the definite integral. Remember that the definite integral $\int_a^b f(x)\,dx$ is a real number, whereas the indefinite integral $\int f(x)\,dx$, whose notation does not include limits of integration, is the set of functions consisting of all antiderivatives, $F(x) + C$ for $F(x)$ a particular antiderivative of $f(x)$.

EXERCISE SET 6–2

• In Problems 1 to 6 let $f(x) = x^2$. For the given choice of n, use n rectangles to approximate the area under the curve $y = f(x)$ from $x = a$ to $x = b$.

1. $n = 2$, $a = 1$, $b = 2$

2. $n = 2$, $a = 1$, $b = 4$

3. $n = 4$, $a = 1$, $b = 2$

4. $n = 4$, $a = 1$, $b = 4$

5. $n = 6$, $a = 0$, $b = 3$

6. $n = 8$, $a = 0$, $b = 4$

In Problems 7 to 12 fill in the following table for the given choice of $f(x)$ by dividing $[0, 2]$ into n subintervals and computing the sum of the areas of the rectangles.

n	2	4	8	10
area				

7. $f(x) = 2x + 1$

8. $f(x) = 3x - 1$

9. $f(x) = x^2 + 1$

10. $f(x) = x^2 - 2$

11. $f(x) = 2x^2 - 2$

12. $f(x) = 2 + x^2$

In Problems 13 to 16 use the definition to find the definite integral.

13. $\int_0^4 3x\,dx$

14. $\int_0^8 2x\,dx$

15. $\int_0^4 (2x + 1)\,dx$

16. $\int_0^8 (3x + 1)\,dx$

In Problems 17 to 20 find the area under $f(x)$ from $x = a$ to $x = b$.

17. $f(x) = 2x + 1$, $a = 0$, $b = 2$

18. $f(x) = 3x - 1$, $a = 0$, $b = 1$

19. $f(x) = 3 - 2x$, $a = 0$, $b = 1$

20. $f(x) = 7 - 3x$, $a = 0$, $b = 2$

•• In Problems 21 to 24 use the definition to find the definite integral.

21. $\int_3^4 3x\,dx$

22. $\int_6^8 2x\,dx$

23. $\int_{-2}^0 (1 - 2x)\,dx$

24. $\int_7^8 (3x - 1)\,dx$

In Problems 25 to 30 use the definition to find the definite integral by using the identity

$$1^2 + 2^2 + \cdots + n^2 = \frac{n(n + 1)(n + 2)}{6}$$

25. $\int_0^4 x^2\,dx$

26. $\int_{-2}^2 2x^2\,dx$

27. $\int_0^4 (2x^2 + 1)\,dx$

28. $\int_1^5 (3x^2 - 1)\,dx$

29. $\int_0^1 (x^2 + x)\,dx$

30. $\int_0^1 (3x^2 + x)\,dx$

31. $\int_1^3 \frac{1}{x}\,dx$

32. $\int_1^4 \frac{2}{x}\,dx$

••• 33. Use the definition to show that $\int_a^b c\,dx = c(b - a)$ for the constant function $f(x) = c$.

34. Verify the result of Example 3 geometrically.

35. In Problem 23 verify that the definite integral is the area under the curve by finding the area geometrically.

36. In an article studying the "well-being" of farm families, Kinsey* quotes one of the original articles† in economics that applied Lorentz curves to income inequality. The latter paper by Carlin and Reinsel constructed a Lorentz curve for the family income of farmers. It can be approximated by

$$y = \frac{x^2}{3} + \frac{2x}{3}.$$

Graph this function using the facts that it passes

through $(0, 0)$ and $(1, 1)$ and is similar in shape to $y = x^2$. Shade the inequality area.

37. In an article studying how farmland prices are determined, Brown and Brown discuss the possible options for a prospective seller of a tract of farmland.‡ They define $f(x)$ to be the likelihood that the seller will receive a bid for the land if the price is x thousand dollars. For a particular tract, $f(x) = 1/500$. They define the likelihood that the seller will receive a bid between \$1,900 and \$2,000 to be

$$\int_{1900}^{2000} \frac{1}{500}\, dx$$

Calculate this integral.

*Jean Kinsey, "Measuring the Well-being of Farm Households: Farm, Off-farm and In-kind Sources of Income: Discussion," *American Journal of Agricultural Economics,* Vol. 67, 1985, pp. 1105–1107.

†Thomas A. Carlin and Edward I. Reinsel, "Combining Income and Wealth—An Analysis of Farm Family 'Well-being'," *American Journal of Agricultural Economics,* Vol. 55, 1973, pp. 38–46.

‡Keith C. Brown and Deborah J. Brown, "Heterogenous Expectations and Farmland Prices," *American Journal of Agricultural Economics,* Vol. 66, 1984, pp. 164–169.

6–3 THE FUNDAMENTAL THEOREM OF CALCULUS

The definition of the definite integral involves a somewhat complicated limit process. It would be cumbersome if the definition were necessary to compute each definite integral. However, the **fundamental theorem of calculus** eases the burden by tying together the two branches of calculus, integral and differential calculus. This is done by using antiderivatives, or indefinite integrals, to calculate definite integrals.

The Fundamental Theorem of Calculus

We first present the theorem with an example. Its proof will follow.

—————————— The Fundamental Theorem of Calculus ——————————
Let $f(x)$ be continuous on the interval $[a, b]$ and let $F(x)$ be any antiderivative of $f(x)$. Then

$$\int_a^b f(x)\, dx = F(b) - F(a)$$

To represent the expression $F(b) - F(a)$ in a more compact form, we use the symbol $F(x) \Big|_a^b$. Thus the fundamental theorem states

$$\int_a^b f(x)\,dx = F(x)\,\Big|_a^b = F(b) - F(a)$$

Example 1 illustrates how to use the theorem.

EXAMPLE 1

Problem

Use the fundamental theorem of calculus to find the definite integrals

(a) $\int_2^4 2x\,dx$ and (b) $\int_{-1}^2 (x^2 + 4x^3)\,dx$.

SOLUTION (a) An antiderivative of $2x$ is x^2, so

$$\int_2^4 2x\,dx = x^2\,\Big|_2^4 = 4^2 - 2^2 = 16 - 4 = 12$$

(b) An antiderivative of $x^2 + 4x^3$ is $x^3/3 + x^4$, so

$$\int_{-1}^2 (x^2 + 4x^3)\,dx = (x^3/3 + x^4)\,\Big|_{-1}^2$$

$$= (8/3 + 16) - (-1/3 + 1) = 18$$

The fundamental theorem states that any antiderivative can be used. To demonstrate what this means consider Example 1(a). The general antiderivative is $x^2 + C$. Using this antiderivative the computation is

$$\int_2^4 2x\,dx = x^2 + C\,\Big|_2^4 = 4^2 + C - (2^2 + C)$$

$$= (16 + C) - (4 + C) = 12$$

The constant drops out in the calculation, showing that any antiderivative can be used. Also, note that the theorem does not stipulate that $f(x) \geq 0$ in the interval as was required in calculating area in the previous section. Only when we use the definite integral to compute area will this requirement be important.

Verification of the Fundamental Theorem

The verification of the fundamental theorem of calculus provides insight into the connection between area and the definite integral. It starts by looking at areas under the graph of $f(x)$ and quickly shifts to the antiderivative.

Proof of the Fundamental Theorem of Calculus

Proof: Let $f(x) \geq 0$. The proof for $f(x) < 0$ is similar. Define the function $A(x)$ to be the area bounded by $f(x)$ and the x-axis between a and x, where $a \leq x \leq b$. We want to show that $A(x)$ is an antiderivative of $f(x)$, or that $A'(x) = f(x)$. Let h be a small positive number. Then $A(x + h)$ is the area from a to $x + h$. Since $A(x)$ is the area from a to x (Figure 1), $A(x + h) - A(x)$ is

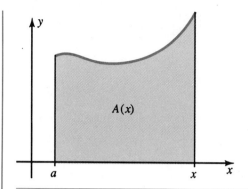

FIGURE 1

the area shaded in Figure 2. This area can be approximated by a rectangle of width h and height $f(x)$, whose area is $h \cdot f(x)$. Thus

$$A(x + h) - A(x) \simeq h \cdot f(x)$$

The approximation improves as h gets close to 0. Dividing by h yields

$$\frac{A(x + h) - A(x)}{h} \simeq f(x)$$

Take the limit of both sides as h approaches 0. Since x is not affected by h, the limit of the right-hand side is $f(x)$. Because the approximation improves as h gets close to 0, taking the limit means the approximation sign can be replaced by an equal sign. Thus the limit of the right-hand side is equal to $f(x)$. But the limit of the left-hand side is the definition of the derivative of $A(x)$. Therefore $A'(x) = f(x)$.

FIGURE 2

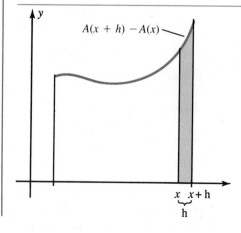

There is one more step in the proof. Since x represents any number between a and b, set $x = a$. Then $A(a) = 0$. Also, setting $x = b$ gives $A(b)$, which is the desired area. Since $A(x)$ is an antiderivative of $f(x)$, we have

$$\int_a^b f(x)\,dx = A(x)\,\Big|_a^b = A(b) - A(a) = A(b)$$

Therefore the desired area is the definite integral.

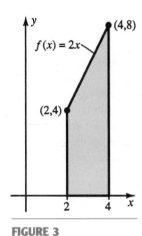

FIGURE 3

Area

Let us look at an example of the relationship between the definite integral and the area under a curve. Consider the function $f(x) = 2x$ in Example 1(a). Its graph is in Figure 3. The area under the curve bounded by the two lines $x = 2$ and $x = 4$ consists of two areas, the rectangle bounded above by the line $y = 4$ and the triangle above it. The area of the rectangle is $4\cdot2 = 8$. The area of the triangle is $1/2\ \text{(base)(height)} = (1/2)\cdot2\cdot4 = 4$. So the total area under $y = 2x$ from $x = 2$ to $x = 4$ is 12. Notice that this is equal to the definite integral $\int_2^4 2x\,dx$, computed in Example 1(a).

What is the area of the region between the x-axis and $f(x) = 2x$ from $x = -1$ to $x = 2$? It might seem that the answer is $\int_{-1}^2 2x\,dx$. However

$$\int_{-1}^2 2x\,dx = x^2\,\Big|_{-1}^2 = 4 - 1 = 3$$

From Figure 4 the total area is the sum of the area from $x = -1$ to $x = 0$ and the area from $x = 0$ to $x = 2$; that is, $1 + 4 = 5$.

What went wrong? Why did the definite integral yield an incorrect answer? Because part of the graph is below the x-axis. An area problem must be divided into subproblems, those where the graph is above the x-axis and those where it is below. The definite integral from $x = -1$ to $x = 0$ gives the negative of the area.

$$\int_{-1}^0 2x\,dx = x^2\,\Big|_{-1}^0 = 0 - (-1)^2 = -1$$

FIGURE 4

This leads to the next important property of definite integrals.

_____ Areas Under the x-axis _____

If $f(x) \le 0$ in $[a, b]$, then the area bounded by $f(x)$, the x-axis, $x = a$, and $x = b$ is given by

$$-\int_a^b f(x)\,dx$$

Therefore if $f(x) \ge 0$, the area and the definite integral are equal, while if $f(x) \le 0$, the area is the negative of the definite integral. For example, to find the area bounded by $f(x) = 2x$, the x-axis, $x = -1$, and $x = 2$, we need to divide the region into two separate regions: the region from $x = -1$ to $x = 0$ where $f(x) \leq 0$ and the region from $x = 0$ to $x = 2$ where $f(x) \geq 0$. The area of the first region is then the negative of the definite integral from -1 to 0. The area of the second region is the definite integral from 0 to 2. The total area is then

$$-\int_{-1}^{0} 2x\, dx + \int_{0}^{2} 2x\, dx = -(-1) + 4 = 5$$

Procedure for Finding the Area Under a Curve

To compute the area of the region bounded by $f(x)$, the x-axis, $x = a$, and $x = b$:

1. From the graph of $f(x)$ find the intervals where $f(x)$ is above the x-axis and where $f(x)$ is below the x-axis by locating the x-intercepts in $[a, b]$.
2. Compute the definite integral of $f(x)$ between each pair of x-intercepts.
3. If $f(x)$ is greater than 0 in an interval, the area is equal to the definite integral. If $f(x)$ is less than 0 in an interval, the area is the negative of the definite integral.
4. The total area is the sum of the areas in step 3.

The next example illustrates the procedure.

EXAMPLE 2

Problem

Find the area bounded by $f(x) = 2 - 2x$, the x-axis, $x = 0$, and $x = 4$.

SOLUTION The graph is in Figure 5. The region consists of two triangles, one above the x-axis and one below it. From $x = 0$ to $x = 1$, $f(x)$ is above the x-axis, so the corresponding area is equal to the definite integral

$$\int_{0}^{1} (2 - 2x)\, dx = (2x - x^2) \Big|_{0}^{1} = (2 - 1) - 0 = 1$$

From $x = 1$ to $x = 4$ the graph is below the x-axis, so the area under the curve is equal to the negative of the definite integral.

$$-\int_{1}^{4} (2 - 2x)\, dx = -(2x - x^2) \Big|_{1}^{4} = -[(8 - 16) - (2 - 1)]$$

$$= -(-8 - 1) = -(-9) = 9$$

Therefore the total area is $1 + 9 = 10$.

Properties of Definite Integrals

Some properties of antiderivatives or indefinite integrals are given in Section 6–1. Similar properties hold for definite integrals. We list them on the next page.

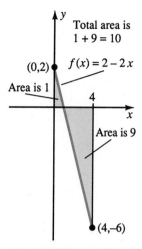

FIGURE 5

Let $f(x)$ and $g(x)$ be functions and a, b, c, and k be real numbers. Then

1. $\displaystyle\int_a^b kf(x)\, dx = k \int_a^b f(x)\, dx$

2. $\displaystyle\int_a^b f(x)\, dx + \int_a^b g(x)\, dx = \int_a^b [f(x) + g(x)]\, dx$

3. $\displaystyle\int_a^a f(x)\, dx = 0$

4. $\displaystyle\int_a^c f(x)\, dx + \int_c^b f(x)\, dx = \int_a^b f(x)\, dx \qquad \text{for} \qquad a \leq c \leq b$

The first two properties hold for indefinite integrals so they are also valid for definite integrals. Since the distance from a to a is 0, the third property states that the area under the curve between a and a is 0. The fourth property can be illustrated geometrically. The area bounded by the x-axis, $y = f(x)$ from $x = a$ to $x = b$ can be viewed as the sum of the area from $x = a$ to $x = c$ and the area from $x = c$ to $x = b$. This geometric argument works only for $f(x) > 0$, but a similar argument holds for the case when the graph of $f(x)$ is below the x-axis.

The next example shows how to use the properties to find more complicated definite integrals.

EXAMPLE 3 | **Problem**

Find

$$\int_0^1 (4x^3 + 4x - 2e^x)\, dx$$

SOLUTION

$$\int_0^1 (4x^3 + 4x - 2e^x)\, dx = \int_0^1 4x^3\, dx + \int_0^1 4x\, dx - \int_0^1 2e^x\, dx$$

$$= x^4 \Big|_0^1 + 2x^2 \Big|_0^1 - 2e^x \Big|_0^1$$

$$= 1^4 - 0^4 + 2(1^2) - 2(0^2) - 2e^1 + 2e^0$$

$$= 1 + 2 - 2e + 2 = 5 - 2e$$

Wages and Rent

A fundamental principle in economics is that "wages equal marginal product." This means that the employer will pay all employees doing similar work for similar pay, a wage equal to the marginal product of the last worker hired. This is because the last worker's value to the firm is that worker's marginal product, or the amount of increase in production the worker generates.

Along with this idea there are two additional principles. Consider the graph in Figure 6. Its curve is the marginal product function, which is equal to wages. That is, $f(x)$ is the wage of each employee when x employees are hired. This curve is decreasing, meaning that as more workers are hired, the marginal product of the last worker decreases, and therefore the wage of the workers also decreases. If the number of employees is $x = a$, then the wage for each employee is $f(a)$. The total wages TW paid to all employees is the area bounded by the constant function $y = f(a)$, which is

$$TW = \int_0^a f(a)\, dx = f(a)x \Big|_0^a = f(a)(a - 0) = af(a)$$

In other words, the region defined by this integral is the rectangle whose length is a and whose height is $f(a)$, so its area is $af(a)$. The **total product** is the total area

FIGURE 6

of the region bounded by $y = f(x)$ and the x-axis from $x = 0$ to $x = a$. Therefore total product TP is defined to be

$$TP = \int_0^a f(x)\, dx$$

There is another region in Figure 6. It is the region bounded by $y = f(x)$ and the horizontal line $y = f(a)$ from $x = 0$ to $x = a$. It represents a savings for the employer because it measures the amount of total product that is not paid out in wages. Economists call the area of this region the employer's *rent*.* From Figure 6 we see that rent is the total product minus total wages. That is

rent = $TP - TW$

When economists study the effects of a new innovation or product on an industry, the concept of rent provides a measure of how much the industry has gained from the innovation. See Exercise 39 and the Case Study for examples of this type of use of rent.

EXAMPLE 4

A firm determines that the marginal product $MP(x)$ for a level of labor x is

$$MP(x) = 0.03x^2 - 0.6x + 23$$

$MP(x)$ is defined in the interval $[0, 10]$ and is measured in thousands of dollars, while x is measured in hundreds of employees. The number of employees hired is $x = 10$ hundred.

Problem

Find (a) total wages TW, (b) total product TP, and (c) the employer's rent.

SOLUTION The wage paid to each employee is the marginal product of the last employee hired, which is $MP(10) = 3 - 6 + 23 = 20$.
(a) Total wages is defined by

$$TW = \int_0^{10} MP(10)\, dx = \int_0^{10} 20\, dx = 20x \Big|_0^{10} = 200$$

(b) The total product TP is defined by

$$TP = \int_0^{10} (0.03x^2 - 0.6x + 23)\, dx$$

$$= 0.01x^3 - 0.3x^2 + 23x) \Big|_0^{10} = 10 - 30 + 230 = 210$$

(c) The employer's rent is the area of the region in Figure 7. It is the area of the region under $y = MP(x)$ and above $y = 20$ from $x = 0$ to $x = 10$. This area is TP minus TW.

rent = $TP - TW = 210 - 200 = 10$

*Paul A. Samuelson, *Economics,* 11th ed., New York, McGraw-Hill Book Co., 1980, pp. 465–467.

290   6 The Integral

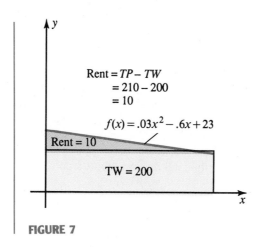

Rent = TP – TW
= 210 – 200
= 10

$f(x) = .03x^2 - .6x + 23$

Rent = 10

TW = 200

FIGURE 7

EXERCISE SET 6–3

• In Problems 1 to 10 find the definite integral by using the fundamental theorem of calculus.

1. $\int_0^1 (3x + 5)\, dx$ 2. $\int_1^3 (4x - 2)\, dx$

3. $\int_0^1 (x^2 + 1)\, dx$ 4. $\int_{-2}^2 (3x^2 + 1)\, dx$

5. $\int_{-3}^2 (x^2 + 5x)\, dx$ 6. $\int_1^2 (x^3 + x)\, dx$

7. $\int_0^8 (-x^2 + \sqrt[3]{x})\, dx$

8. $\int_1^4 (-x^2 + 1/\sqrt{x} + 7e^x)\, dx$

9. $\int_1^9 (x^{-3} + 7x^{-4} + 5\sqrt{x})\, dx$

10. $\int_1^4 (x^{-3} - 3x^{-2} - 3\sqrt{x})\, dx$

In Problems 11 and 12 find the definite integral by using geometric properties of the integrand. Verify by using the fundamental theorem of calculus.

11. $\int_{-2}^4 (x - 1)\, dx$ 12. $\int_{-1}^4 (2 - x)\, dx$

In Problems 13 to 20 use the definite integral to find the area bounded by $f(x)$, the x-axis, $x = a$, and $x = b$.

13. $f(x) = x - 1,\quad a = -3,\quad b = 4$
14. $f(x) = 1 - x,\quad a = -3,\quad b = 3$
15. $f(x) = x^2 - 4,\quad a = -2,\quad b = 4$
16. $f(x) = 4 - x^2,\quad a = -3,\quad b = 2$
17. $f(x) = x^2 - x,\quad a = 0,\quad b = 2$
18. $f(x) = 4x - x^2,\quad a = -1,\quad b = 2$
19. $f(x) = x^3 - x,\quad a = -1,\quad b = 1$
20. $f(x) = x^3 - x^2,\quad a = 0,\quad b = 2$

•• In Problems 21 to 24 assume $\int_0^1 f(x)\, dx = 10$, $\int_1^2 f(x)\, dx = 15$, and $\int_0^2 g(x)\, dx = -5$. Use the properties of definite integrals to evaluate the integral.

21. $\int_0^2 f(x)\, dx$

22. $\int_0^2 3f(x)\, dx$

23. $\int_0^2 [f(x) + g(x)]\, dx$

24. $\int_0^2 [2f(x) + 3g(x)]\, dx$

In Problems 25 to 30 find the definite integral.

25. $\int_0^1 (x^{0.2} + 3x^{1.4})\, dx$

26. $\int_0^1 (1.2x^{-1.6} - 4.1x^{-2.2})\, dx$

27. $\int_1^2 (3e^{2x} + 1/x)\, dx$

28. $\int_1^2 (5e^{3x} - 2/x)\, dx$

29. $\int_1^2 (x^{1.3} - 3e^{1.2x} + 1/2x)\, dx$

30. $\int_1^2 (-e^{1-3x} - e/x)\, dx$

In Problems 31 to 36 use the definite integral to find the area bounded by $f(x)$, the x-axis, $x = a$, and $x = b$.

31. $f(x) = x^3 - x$, $a = -2$, $b = 2$

32. $f(x) = x^3 - x^2$, $a = -1$, $b = 2$

33. $f(x) = x^3 - 2x^2$, $a = 0$, $b = 3$

34. $f(x) = x^3 - 4x$, $a = -3$, $b = 5$

35. $f(x) = 1 - 2/x$, $a = 1$, $b = 3$

36. $f(x) = 2 - 3/x$, $a = 1$, $b = 3$

••• 37. Prove the fundamental theorem for $f(x) < 0$.

38. What procedure would you use to find the area of the region bounded by $f(x) = x^2 - 4x + 3$ and the x-axis?

39. Television is an example of an industry whose firms reap benefits from regulation. In an article studying the television industry, Fournier used rents to measure the effects of regulation on the industry.* The marginal cost function $MP(x)$ can be approximated by

$MP(x) = 3x^2 - 60x + 800$

where x is the average price of a show. Find the rent when $x = 10$. Here total product is income and total wages is total price.

40. In an article investigating the size of prey versus an optimal diet for carnivorous fish, Bence and Murdoch studied the relationship between the success rate $S(s)$ of the predator catching the prey versus the size s of the prey.† For one type of fish the relationship was calculated to be

$S(s) = 1 + .861s - 91.57s^2 + 322.7s^3$

The total success of attacks of prey between a mm (millimeters) and b mm is given by

$$\int_a^b S(s)\,ds$$

Find the total success for prey between 1 mm and 2 mm.

41. It is often difficult for economists to measure the benefits and costs of the uses of natural resources, such as federally owned lakes. In an article investigating ways to estimate the value of these resources, Seller et al. studied lakes in East Texas.‡ They calculated the demand curve for Lake Livingston to be

$V(x) = 10.04 - .12x$

where $V(x)$ is the value of the land when the cost of a family visit is x dollars. One measure of the benefit of the resource involved computing the consumers' surplus, an economic measure defined in Section 6–4. It entails computing the integral of $V(x)$ from $x = 0$ to $x = 2$. Compute this integral.

*Gary M. Fournier, "Nonprice Competition and the Dissipation of Rents from Television Regulation," *Southern Economics Journal*, Vol. 51, 1985, pp. 754–765.

†James R. Bence and W. W. Murdoch, "Prey Size Aselection by the Mosquitofish: Relation to Optimal Diet Theory," *Ecology*, Vol. 67, 1986, pp. 324–336.

‡Christine Seller et al., "Validation of Empirical Measures of Welfare Change: A Comparison of Nonmarket Techniques," *Land Economics*, Vol. 61, 1985, pp. 156–175.

6-4 AREA BOUNDED BY CURVES

Thus far we have considered regions bounded by a single curve and the x-axis. Often in applications a region is bounded both above and below by graphs of functions. In this section the concept of the area between a curve and the x-axis is generalized to include the area between two curves.

Areas Between Two Curves

Consider the area of the region bounded above by $f(x) = x + 2$ and below by $g(x) = x^2$. This region is graphed in Figure 1. View the region in two parts: the first part, region I, is the area below $f(x)$ from $x = -1$ to $x = 2$; the second part, region II, is the area below $g(x)$ from $x = -1$ to $x = 2$. Then the desired region is (region I–region II). (See Figure 2.) Its area is the difference between the two definite integrals.

$$
\begin{aligned}
\text{(the desired area)} &= \text{region I} \qquad - \text{region II} \\
&= (\text{area below } f(x)) - (\text{area below } g(x)) \\
&= \int_{-1}^{2} (x + 2)\, dx - \int_{-1}^{2} x^2\, dx \\
&= \left(\frac{x^2}{2} + 2x\right)\Bigg|_{-1}^{2} - \frac{x^3}{3}\Bigg|_{-1}^{2} = \frac{15}{2} - 3 = \frac{9}{2}
\end{aligned}
$$

To compute this area, we can also use the third property from the previous section and get

$$
\begin{aligned}
&= \int_{-1}^{2} (x + 2)\, dx - \int_{-1}^{2} x^2\, dx = \int_{-1}^{2} (x + 2 - x^2)\, dx \\
&= \left(\frac{x^2}{2} + 2x - \frac{x^3}{3}\right)\Bigg|_{-1}^{2} = \frac{9}{2}
\end{aligned}
$$

In general the area between two curves is the difference between the two definite integrals. We then use a property of definite integrals to simplify the integration.

FIGURE 1

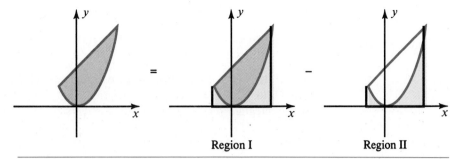

FIGURE 2

—————————————— **The Area Bounded by Two Curves** ——————————

If a region is bounded above by $f(x)$ and below by $g(x)$ between $x = a$ and $x = b$, then the area of the region is

$$\int_a^b [f(x) - g(x)]\, dx$$

Lorentz Diagrams

In Section 6–2 we defined a Lorentz diagram. The important part of the diagram is the area between the two curves $y = x$, the curve of complete equality of income distribution, and $y = x^2$, the Lorentz curve of the distribution of income in the United States. This area is called the area of inequality and measures the extent of departure between an equal distribution of income and the actual distribution of income. The area between these two curves is the difference between the area of the region bounded by $y = x$ and the area of the region bounded by $y = x^2$. We can now compute this area. It is

$$\text{area} = \int_0^1 x\, dx - \int_0^1 x^2\, dx$$

$$= \int_0^1 (x - x^2)\, dx$$

$$= \left(\frac{x^2}{2} - \frac{x^3}{3} \right) \Big|_0^1$$

$$= \frac{1}{2} - \frac{1}{3} = \frac{1}{6}$$

Economists compare this value with the inequality areas of other countries to determine which has a more equitable income distribution.

The formula for the area between two curves does not mention whether the functions are positive or negative. The only stipulation is that $f(x) \geq g(x)$ in $[a, b]$. The next example demonstrates what to do when two curves cross each other in the desired area.

EXAMPLE 1 | **Problem**

Find the area of the region bounded by $f(x) = 3x^2 - 2x$, and $g(x) = x$ from $x = 0$ to $x = 2$.

SOLUTION This region consists of two separate regions as shown in Figure 3. From 0 to 1, $g(x) \geq f(x)$, so we integrate $g(x) - f(x)$. From 1 to 2, $f(x) \geq g(x)$, so we integrate $f(x) - g(x)$. The area is the sum of the two integrals.

$$
\text{area} = \int_0^1 [x - (3x^2 - 2x)]\, dx + \int_1^2 [(3x^2 - 2x) - x]\, dx
$$

$$
= \int_0^1 (3x - 3x^2)\, dx + \int_1^2 (3x^2 - 3x)\, dx
$$

$$
= \left(\frac{3x^2}{2} - x^3 \right) \Big|_0^1 + \left(x^3 - \frac{3x^2}{2} \right) \Big|_1^2
$$

$$
= \left(\frac{3}{2} - 1 \right) + \left[(8 - 6) - \left(1 - \frac{3}{2} \right) \right] = 3
$$

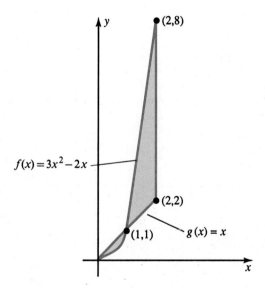

FIGURE 3

End Points

In some problems the end points of the region are given and in others they are not. If the end points are not specified, they are found by computing the first coordinates of the points of intersection of the two curves. So the first step is to graph the region and locate the points of intersection. The next example shows how to do this.

EXAMPLE 2 | **Problem**

Find the area of the region bounded by $y = x^3 - 2x$ and $y = x^2$.

SOLUTION First sketch the curves. In order to refer to each curve, we name them. Let $f(x) = x^3 - 2x$ and $g(x) = x^2$. To find the points of intersection, solve $f(x) = g(x)$, so $x^3 - 2x = x^2$ or $x^3 - x^2 - 2x = 0$. Factoring yields $x(x - 2)(x + 1) = 0$. Thus the solutions are $x = -1, 0, 2$, and the curves intersect at $(-1, 1)$, $(0, 0)$, and $(2, 4)$. See Figure 4. From $x = -1$ to $x = 0$, $f(x) \geq g(x)$, while from $x = 0$ to $x = 2$, $g(x) \geq f(x)$. Therefore the area of the region bounded by the two curves is

$$\int_{-1}^{0} [(x^3 - 2x) - x^2] \, dx + \int_{0}^{2} [x^2 - (x^3 - 2x)] \, dx$$

$$= \int_{-1}^{0} [x^3 - 2x - x^2] \, dx + \int_{0}^{2} [x^2 - x^3 + 2x] \, dx$$

$$= \left[\frac{x^4}{4} - x^2 - \frac{x^3}{3} \right] \Big|_{-1}^{0} + \left[\frac{x^3}{3} - \frac{x^4}{4} + x^2 \right] \Big|_{0}^{2}$$

$$= 0 - \left(\frac{1}{4} - 1 + \frac{1}{3} \right) + \frac{8}{3} - 4 + 4 - 0$$

$$= \frac{5}{12} + \frac{8}{3} = \frac{37}{12}$$

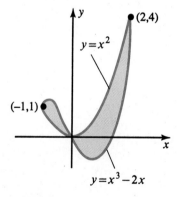

FIGURE 4

Procedure for Finding the Area of a Region Bounded by Two Curves

Find the area of a region bounded by two curves, $y = f(x)$ and $y = g(x)$, with the following steps:

1. Sketch the region. Find the points of intersection.
2. Find the end points, $x = a$ and $x = b$. If they are not given, then they are points of intersection.

3. Find the intervals where $f(x) \geq g(x)$.

  (i)   If $f(x) \geq g(x)$ in $[a, b]$, then the area is

$$\int_a^b [f(x) - g(x)] \, dx$$

  (ii)   If $g(x) \geq f(x)$ in $[a, b]$, then the area is

$$\int_a^b [g(x) - f(x)] \, dx$$

  (iii)   If there is a point c between a and b such that $f(x) \geq g(x)$ in $[a, c]$ and $g(x) \geq f(x)$ in $[c, b]$, then the area is

$$\int_a^c [f(x) - g(x)] \, dx + \int_c^b [g(x) - f(x)] \, dx$$

Consumers' Surplus and Producers' Surplus

A pair of economic concepts that are best explained by calculus are *consumers' surplus* and *producers' surplus*. Consider the demand function $y = D(x)$ and the supply function $y = S(x)$, where x is the quantity of a product and y is the price. At price $D(x)$ consumers will demand x amount of the product. As x increases, $D(x)$ decreases, meaning that consumers will purchase more of the product at lower prices. This implies that $D(x)$ is a decreasing function. At price $S(x)$ producers are willing to supply x amount of the product. As x increases, $S(x)$ also increases because producers will be willing to supply more of the product at higher prices. Therefore $S(x)$ is an increasing function. In other words, $D(x)$ is the price that consumers are willing to pay for x amount of the product, and producers will supply x amount at price $S(x)$. In a free competitive market the actual selling price for the product will tend to be that price where supply equals demand, that is, where $S(x) = D(x)$. It is called the **equilibrium price,** and it is that quantity $x = a$ where $S(a) = D(a)$.

    Figure 5 shows that the demand $D(x)$ is larger than $D(a)$ for values of x less than a. This means that some consumers would be willing to pay a higher price than the equilibrium price. In a certain sense these consumers have gained because the price they pay, the equilibrium price, is lower than what they would have been willing to pay. Economists refer to this gain as **consumers' surplus.** It is defined to be the area of the region bounded above by $y = D(x)$ and below by the constant function $y = D(a)$, from $x = 0$ to $x = a$. This area is

$$\text{consumers' surplus} = \int_0^a [D(x) - D(a)] \, dx$$

EXAMPLE 3   Suppose $D(x) = x^2 - 8x + 20$ for x in $[0, 4]$ and the equilibrium price occurs at $a = 2$.

$y = D(x)$ is the demand function

Consumers' surplus

Equilibrium price

$y = S(x)$ is the supply function

Price

a

Quantity

x

FIGURE 5

Problem

Compute the consumers' surplus.

SOLUTION The equilibrium price is $D(2) = 8$. The consumers' surplus is defined to be

$$\int_0^2 [(x^2 - 8x + 20) - 8] \, dx = \frac{x^3}{3} - 4x^2 + 12x \Big|_0^2$$

$$= \frac{8}{3} - 16 + 24 = \frac{32}{3}$$

Similarly the producers who were willing to supply goods at a price smaller than $S(a)$ gain because they would have made less revenue at the lower price. Economists call the sum of their gains the **producers' surplus.** It is defined to be the area of the region bounded above by $y = S(a)$ and below by $y = S(x)$ from $x = 0$ to $x = a$. It is

producers' surplus $= \int_0^a [S(a) - S(x)] \, dx$

EXAMPLE 4

Suppose $S(x) = x^2 + 2x$ for x in $[0, 4]$ and the equilibrium price occurs at $a = 2$.

Problem

Compute the producers' surplus.

SOLUTION The equilibrium price is $S(2) = 8$. The producer's surplus is defined to be

$$\int_0^2 [8 - (x^2 + 2x)]\, dx = \left(8x - \frac{x^3}{3} - x^2 \right) \Big|_0^2$$

$$= 16 - \frac{8}{3} - 4 = \frac{28}{3}$$

Figure 6 shows the supply and demand curves in Examples 3 and 4 as well as the areas defining the consumers' surplus and producers' surplus.

The concept of consumers' surplus measures the benefit consumers gain from the power of buying at low prices and not having to pay the higher price a monopolist may insist on. Another use of consumers' surplus is in measuring "social welfare." Sometimes economic value, revenue, or profit cannot be used to account for the value of a good. For instance, the economic value of air is negligible but its social welfare is great. The next example demonstrates how economists use consumers' surplus to help make a decision where the intangible quantity of utility is the only "profit."

EXAMPLE 5 A new community is debating whether to self-impose a tax to install street lights. The community has 3000 homes that will be taxed the same amount for the improvement. Contractors are willing to supply the lights according to the schedule

$$S(x) = x^2 + 9x$$

The community is willing to pay according to the schedule

$$D(x) = x^2 - 18x + 81$$

FIGURE 6

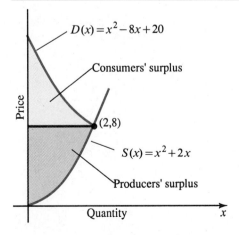

$D(x) = x^2 - 8x + 20$

Consumers' surplus

$(2,8)$

$S(x) = x^2 + 2x$

Producers' surplus

Price

Quantity x

Price is measured in $100,000 units and x is measured in units of 100 street lights. To find the equilibrium price, solve $S(x) = D(x)$. This yields $a = 3$, so the equilibrium price is $S(3) = D(3) = 36$. Thus each homeowner must pay a tax of $3,600,000/3000 = $1,200$ for the improvement. Is it worth it to the individual? Since there are no revenues from this project, the only profit is the increased utility achieved. Hence the consumers' surplus gives an indication of the worth of the project.

Problem

Compute the consumers' surplus.

SOLUTION From the definition the consumers' surplus is

$$\int_0^3 [(x^2 - 18x + 81) - 36]\, dx = \frac{x^3}{3} - 9x^2 + 45x \Big|_0^3$$

$$= 9 - 81 + 135 = 63$$

Thus the consumers' surplus is $6,300,000. Each homeowner gets a utility of $6,300,000/3000 = $2,100$. Therefore the consumers' surplus is far greater than the amount spent on the improvement. The project is worthwhile from this point of view.

EXERCISE SET 6–4

• In Problems 1 to 30 find the area between the curves.

1. $x = 0, \quad x = 2,$
   $f(x) = x^2 + 1, \quad g(x) = 0$

2. $x = 0, \quad x = 3,$
   $f(x) = x^2 + 4, \quad g(x) = 0$

3. $x = 0, \quad x = 2,$
   $f(x) = x^2 + 2x, \quad g(x) = 0$

4. $x = 0, \quad x = 3,$
   $f(x) = x^2 + x, \quad g(x) = 0$

5. $x = 0, \quad x = 2,$
   $f(x) = x^3 + 3x, \quad g(x) = 0$

6. $x = 0, \quad x = 4,$
   $f(x) = x^3 + 6x, \quad g(x) = 0$

7. $x = 0, \quad x = 1,$
   $f(x) = x^2, \quad g(x) = x$

8. $x = 0, \quad x = 1,$
   $f(x) = x^2, \quad g(x) = 2x$

9. $x = 0, \quad x = 1,$
   $f(x) = x^2 + x, \quad g(x) = 2x$

10. $x = 0, \quad x = 1,$
    $f(x) = x^2 + 4x, \quad g(x) = 5x$

11. $x = 0, \quad x = 1,$
    $f(x) = 3x - x^2, \quad g(x) = 2x$

12. $x = 0, \quad x = 1,$
    $f(x) = 5x - x^2, \quad g(x) = 6x$

13. $f(x) = x^2 + x, \quad g(x) = 2x$

14. $f(x) = x^2 + 4x, \quad g(x) = 5x$

15. $f(x) = 3x - x^2, \quad g(x) = 2x$

16. $f(x) = 5x - x^2, \quad g(x) = 6x$

17. $f(x) = x^2 - 2x, \quad g(x) = 2x$

18. $f(x) = x^2 - 4x, \quad g(x) = 5x$

19. $f(x) = 2 + x - x^2,$
    $g(x) = 2 - 3x$

20. $f(x) = 3 + 2x - x^2$,
    $g(x) = 3 - 4x$

21. $f(x) = x^3$, $g(x) = x^2$

22. $f(x) = x^3 - 4x$,
    $g(x) = x^2 - 4x$

•• 23. $x = 0$, $x = 4$, $f(x) = x + 2$, $g(x) = x^2$

24. $x = 0$, $x = 6$, $f(x) = 3x + 2$,
    $g(x) = 2x^2$

25. $x = -1$, $x = 4$, $f(x) = 2x + 3$,
    $g(x) = x^2$

26. $x = -1$, $x = 5$, $f(x) = 3x + 4$,
    $g(x) = x^2 + x + 1$

27. $x = -2$, $x = 3$, $f(x) = 1 - 2x$,
    $g(x) = x^2 - 2x - 3$

28. $x = -3$, $x = 3$, $f(x) = 3x + 4$,
    $g(x) = x^2 + 3x$

29. $x = -2$, $x = 2$, $f(x) = 3x - x^2$,
    $g(x) = x^2 - x$

30. $x = 0$, $x = 4$, $f(x) = 4x - x^2$,
    $g(x) = 2x^2 - 5x$

In Problems 31 to 34 compute the consumers' surplus and the producers' surplus.

31. $D(x) = x^2 - 8x + 24$, $S(x) = x^2 + 4x$

32. $D(x) = x^2 - 12x + 48$, $S(x) = x^2 + 4x$

33. $D(x) = 48 - 10x - x^2$, $S(x) = x^2 + 10x$

34. $D(x) = 39 - 12x - x^2$, $S(x) = x^2 + 8x$

••• In Problems 35 to 40 find the area between the curves.

35. $x = 1$, $x = 2$, $f(x) = \dfrac{1}{x}$, $g(x) = 1$

36. $x = 2$, $x = 3$, $f(x) = \dfrac{1}{x}$, $g(x) = x$

37. $x = 1$, $x = 5$, $f(x) = \dfrac{4}{x}$, $g(x) = x$

38. $x = 0$, $x = 3$, $f(x) = \dfrac{2}{x + 1}$,
    $g(x) = x - 1$

39. $x = 0$, $x = 3$, $f(x) = \dfrac{5}{x + 2}$,
    $g(x) = x - 2$

40. $x = 1$, $x = 5$, $f(x) = \dfrac{8}{x}$, $g(x) = x^2$

41. A new trash-to-steam plant is proposed for a community to reduce refuse. The community is willing to pay according to the schedule

$$D(x) = 1000 - 50x - x^2$$

Contractors are willing to build the plant according to the schedule

$$S(x) = x^2 + 30x$$

Compute the consumers' surplus and determine if the project is "profitable" in the sense that the consumers' surplus is greater than the cost.

42. In an article studying the adverse effects of ozone and other air pollutants on crop yields, Adams et al. assessed the benefits to agriculture from reductions in these pollutants.* The measurements used were consumers' surplus and producers' surplus. In one model, where a 10% reduction is assumed, the supply and demand curves are approximated by

$$S(x) = 0.03x^2 + 1.2x$$
$$D(x) = 0.03x^2 - 2.7x + 28.2$$

where $S(x)$ and $D(x)$ are measured in billions of dollars. The total benefit was the sum of the consumers' surplus and the producers' surplus. Compute the total benefit.

43. In many areas where soil is good but rainfall is poor, water irrigation projects are developed to support farm products. Often taxpayers are required to subsidize these projects. In 1978, California rice producers were given a subsidy in the form of reduced prices for water to irrigate the rice crop. In an article studying the effect of this subsidy on producers and consumers, Foster et al. used producers' surplus and consumers' surplus to measure the benefit of the

*R. M. Adams et al., "The Benefits of Pollution Control: The Case of Ozone and U.S. Agriculture," *American Journal of Agricultural Economics*, Vol. 68, 1986, pp. 886–893.

subsidy to each group.[†] The demand curve $D(x)$ and the supply curve $S(x)$ at price x, measured in thousands of dollars, were approximated by

$D(x) = 34 - 28x$
$S(x) = 30 + 2x + 16x^2$

Compute the producers' surplus and the consumers' surplus.

44. In his book *Applied Mathematical Demography*, Nathan Keyfitz argues that there is an urgent need for haste in lowering the birth rate in less developed countries by showing the lag between

the birth rates and death rates of such countries.[‡] If the birth rate is represented by $b(t)$ and the death rate by $d(t)$, then the total lag $l(t)$ between the two rates over T years is defined by

$$l(t) = \int_0^T b(t)dt - \int_0^T d(t)dt$$

The birth rate $b(t)$ and the death rate $d(t)$ of one country are approximated by

$b(t) = .01t^4 - .18t^2 + 1.3$
$d(t) = 1.2 - .06t$

Compute $l(t)$ for $T = 10$.

[†]William E. Foster et al., "Distribution Welfare Implications of an Irrigation Water Subsidy," *American Journal of Agricultural Economics*, Vol. 68, 1986, pp. 778–786.
[‡]Nathan Keyfitz, *Applied Mathematical Demography*, New York, Springer-Verlag, 1985, pp. 1–26.

CASE STUDY   Consumers' Surplus and Louisiana Fish Farming

Jim Larkin runs a fleet of small fishing boats in the Atchafalaya River basin in southern Louisiana. Jim's brother Larry owns a small farm nearby. Two brothers with two different careers. But Larry's innovative use of his ponds had a devastating effect on Jim's livelihood in the late 1970s.

The bulk of Jim's catch consists of shellfish, such as shrimp, crabs, and crawfish. Each type of shellfish has a distinct season, depending on its own natural characteristics. For instance, crawfish produce only one brood per year, and young crawfish grow rapidly, doubling in weight each month. Sexual maturity is reached from March to July, and the cycle is repeated. The size of the crawfish population is determined by water levels and temperature. When surface water of their habitat disappears in the fall, crawfish burrow into the ground. In addition, natural predators are reduced in this period. The lower the water level is in the fall, the higher will be the population the next year. This means there will be an abundance of crawfish as the water level rises from January to March. Production of good catches requires that the commercial fishing fleet be in synchronization with water changes and crawfish life cycles.

Rearing crawfish in a small pond is an entirely different story. As early as 1950 a few farmers were harvesting crawfish in ponds, but it was a haphazard effort. The pond-reared crawfish catch amounted to less than 1% of the total crawfish market. By 1980 the total crawfish market had increased from about 1 million pounds in 1950 to about 30 million pounds, a 3000% increase. But here's the catch—the pond production of crawfish accounted for a whopping 85% of the total harvest. Fish farming had become big business at the expense of the commercial fleet. The pond producers could circumvent the environmental barriers facing the commercial fleet. For instance, they could maintain the water level in the ponds year-round as well as supply a constant source of food for the fish.

   The innovative aquaculture techniques of the pond producers have had a remarkable impact on the crawfish market, especially on the commercial fleet. But what effect does this have on the public? To what extent has pond-reared crawfish benefited consumers?

   Economist Frederick W. Bell describes this phenomenon in an article studying the effects of new and innovative production techniques in the marketplace.* Economists use consumers' surplus to measure the benefit of such techniques to the public. One obvious benefit is that there are more crawfish to consume. As demand increases, the price of crawfish falls. Bell uses 1978 as a representative year in his study. He estimates the demand function in 1978 to be

$$D(x) = -0.0055x + 0.39$$

where x is the quantity of crawfish harvested in millions of pounds and $D(x)$ is dollars per pound. The demand function is graphed in Figure 1. Bell estimates that without pond-reared crawfish, only 11.9 pounds would be produced at a price of $0.32. Thus $D(11.9) = 0.32$. With the addition of pond-reared crawfish, 35.2 lb would be harvested at a price of $0.20. Thus $D(35.2) = 0.20$. The total effect of pond-reared crawfish is measured by the difference in the two values of consumers' surplus, with and without pond-reared crawfish.

$$\begin{array}{l} \text{total effect} \\ \text{of pond fish} \end{array} = \left(\begin{array}{c} \text{consumers' surplus} \\ \text{with pond fish} \end{array}\right) - \left(\begin{array}{c} \text{consumers' surplus} \\ \text{without pond fish} \end{array}\right)$$

$$= \int_0^{35.2} (-0.0055x + 0.39)\, dx - 35.2(0.20)$$

$$- \left[\int_0^{11.9} (-0.0055x + 0.39)\, dx - 11.9(0.32) \right]$$

$$= (-0.00275x^2 + 0.39x) \Big|_0^{35.2} - 7.04$$

$$- \left[(-0.00275x^2 + 0.39x) \Big|_0^{11.9} - 3.81 \right]$$

$$= -0.00275(35.2)^2 + 0.39(35.2) - 7.04$$

$$- [-0.00275(11.9)^2 + 0.39(11.9) - 3.81]$$

$$= -3.41 + 13.72 - 7.04 + 0.39 - 4.64 + 3.81$$

$$= 2.83$$

This shows that pond-reared fishing greatly increases the benefit to consumers. The fish farmers have gained income as well as expanding the market for

*Frederick W. Bell, "Competition from Fish Farming in Influencing Rent Dissipation: The Crawfish Industry," *American Journal of Agricultural Economics*, Vol. 68, 1986, pp. 95–101.

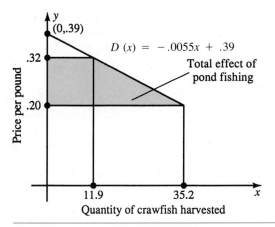

FIGURE 1

crawfish and increasing demand. On the other hand, commercial fishermen like Jim Larkin have experienced a decline in revenue. The brothers seldom share crawfish stories.

Case Study Exercises

In Problems 1 to 4 compute the total effect of pond fishing for the given demand function and the quantities of crawfish harvested without pond fishing and those harvested with pond fishing.

1. $D(x) = -0.0055x + 0.30$,
   $x = 12$ and $x = 40$

2. $D(x) = -0.0055x + 0.30$,
   $x = 15$ and $x = 30$

3. $D(x) = -0.011x + 0.40$,
   $x = 10$ and $x = 50$

4. $D(x) = -0.011x + 0.40$,
   $x = 20$ and $x = 40$

In this article Bell computes the marginal cost function to be

$$MC(x) = \frac{7.26}{(13.7 - x)^2}$$

He defines "social welfare loss" by the formula

$$\int_a^b MC(x)\, dx - P(b - a)$$

where a and b are the quantities of crawfish harvested and P is the price per pound. In Problems 5 and 6 compute the social welfare loss for the given values taken from Bell.

5. $a = 7.71$, $b = 10.2$,
   $P = 0.20$

6. $a = 9.2$, $b = 11.9$, $P = 0.32$

CHAPTER REVIEW

Key Terms

6–1 Antiderivatives

Antidifferentiation	Indefinite Integral
Antiderivative	Integrand
Constant of Integration	Differential Equation
Integral Sign	Growth Equation

6–2 Area and the Definite Integral

Lorentz Diagram	Limits of Integration
Definite Integral	Area Under a Curve

6–3 The Fundamental Theorem of Calculus

Fundamental Theorem of Calculus

6–4 Area Bounded by Curves

Area Bounded by Two Curves	Consumers' Surplus
Equilibrium Price	Producers' Surplus

Summary of Important Concepts

Formulas

Growth equation $f(x) = kf(x)$

The fundamental theorem of calculus $\displaystyle\int_a^b f(x)\, dx = F(b) - F(a)$

REVIEW PROBLEMS

In Problems 1 and 2 find the antiderivative.

1. $\displaystyle\int (5x^{-1/5} - 2x^{-3/5} - 3x)\, dx$

2. $\displaystyle\int \frac{x^7 + 3x}{x^3}\, dx$

3. Find the function whose slope of the tangent line is $y' = x^{-2} - x$ and whose graph passes through $(1, 0)$.

4. Use the definition to find the definite integral.

$\displaystyle\int_0^4 (2x + 1)\, dx$

5. Find the area under $f(x)$ from $x = 0$ to $x = 2$ where $f(x) = 7 - 3x$.

6. Use the definition to find the definite integral

$\displaystyle\int_0^2 x^2\, dx$

by using the identity

$1^2 + 2^2 + \cdots + n^2 = \dfrac{n(n + 1)(n + 2)}{6}$

7. Find the definite integral by using the fundamental theorem of calculus.

$\displaystyle\int_1^4 \left(-x^2 + \frac{1}{\sqrt{x}} + 7e^x\right) dx$

8. Use the definite integral to find the area bounded by $f(x)$, the x-axis, $x = -4$, and $x = 2$ where $f(x) = 4 - x^2$.

9. Find the definite integral.

$$\int_1^2 \left(5e^{3x} - \frac{2}{x} \right) dx$$

In Problems 10 to 12 find the area between the curves.

10. $x = 0$, $x = 1$, $f(x) = x^2 + 4x$, $g(x) = 5x$

11. $f(x) = 5x - x^2$, $g(x) = 6x$

12. $x = 0$, $x = 6$, $f(x) = 3x + 2$, $g(x) = 2x^2$

7 Techniques of Integration

CHAPTER OVERVIEW

The previous chapter showed how to compute the integral of various simple functions. This chapter shows how to integrate more complicated functions. There is no one method of integration that works for all functions. Rather, there are various techniques that are applied on a trial-and-error basis.

The first section deals with the simplest technique. It uses a substitution of a variable to reduce the integrand to a familiar form whose integral can be found by a method that was demonstrated earlier. Section 7–2 considers more difficult functions. It relies on the chain rule. The third section explains how to apply a table of integrals. The table gives several formulas involving a diverse number of different integrands. Sometimes an exact integrand cannot be found. In this case it is often preferable to approximate the definite integral by means of one of several techniques. This is called numerical integration, which is covered in Section 7–4.

CASE STUDY PREVIEW

Many people mistakenly believe that the taxes they pay to the social security system each year are put into a savings account to be held until their retirement. In reality, the money paid by the work force goes directly to fund those who are retired. Those in the age range of 20 to 64 support those retired over 64. The plan works well when there is a large work force to support a much smaller retired segment of the population.

The Case Study demonstrates that the birthrate is starting to lag well behind the death rate. And the gap is growing. The demographic consequences show that either the system will prove to be a great burden on the work force or the benefits paid to retirees will be greatly reduced.

307

7–1 SUBSTITUTION

The fundamental theorem of calculus states that antiderivatives can be used to compute integrals. Thus far we have integrated only relatively simple functions. Just as there are various rules for finding derivatives that depend on the type of function, so too there are several methods for finding indefinite integrals of different types of functions. In this section we discuss the method of integration called **substitution.** To apply this method the integrand must be in a particular form. Determining whether a function fits the description is part of the technique.

Integration by Substitution

Let us start with an example. By the chain rule for derivatives

$$D_x(x^2 + 1)^4 = 4(x^2 + 1)^3 2x = 8x(x^2 + 1)^3$$

The corresponding indefinite integral is

$$\int 8x(x^2 + 1)^3 \, dx = (x^2 + 1)^4 + C$$

Finding this integral was straightforward because we knew $D_x(x^2 + 1)^4$ beforehand. The method of substitution allows us to perform the integration of such functions without knowing the corresponding derivative beforehand. **Integration by substitution** entails expressing the integrand such as the one above as the product of two functions, one of which is the derivative of the other, or at least almost its derivative. In the integrand above, the function $f(x) = x^2 + 1$ is part of the integrand, and its derivative, $f'(x) = 2x$, is also present. The integrand consists of $f(x)$ raised to the third power times $f'(x)$, with the factor 4 left over. Viewed this way, it is in the proper form for substitution. We write

$$\int 8x(x^2 + 1)^3 \, dx = \int 4 \cdot (x^2 + 1)^3 \cdot 2x \, dx$$

Make the substitution $u = x^2 + 1$. Then $\dfrac{du}{dx} = 2x$. Now formally multiply through by dx to get $du = 2x \, dx$. Then the integrand is written

$$\int 8x(x^2 + 1)^3 \, dx = \int 4 \cdot u^3 \cdot du$$

The new integrand is expressed in terms of the new variable u and is easily integrated.

$$\int 4u^3 \, du = u^4 + C$$

Making the reverse substitution for u in terms of x yields the the same integral computed earlier.

$$\int 8x(x^2 + 1)^3 \, dx = (x^2 + 1)^4 + C$$

There are two key steps in applying substitution in this type of problem. First, it must be possible to express the integrand as the product of one function raised to a power and another function that is the derivative of the first function. The entire integrand, with the sole exception of a constant left over, must be described this way. The next step is to let the new variable u equal this function and then compute du, the *differential of u*, which is defined as follows:

If $u = g(x)$, then the **differential of u** is $du = g'(x)\, dx$.

The differential actually has a deeper meaning and a wider application than is used in this text. Since we use differentials only in this chapter to explain techniques of integration, we will not expand on this definition.

Once the substitution is made, the integrand is expressed in terms of u and constants. Next compute the integral in terms of u if possible. Finally, substitute back for x to find the integral. The result can always be checked by differentiating the result to see if it equals the integrand. Let us look at an example to illustrate the procedure.

EXAMPLE 1

Problem

Compute $\int x(1 - x^2)^5\, dx$.

SOLUTION Let $u = 1 - x^2$. Then $du = -2x\, dx$. The differential is not present exactly, but it varies only by a constant, -2. In this case the differential can be divided by the constant to get $(-1/2)\, du = x\, dx$. Now the substitution can be made that expresses the integrand entirely in terms of u.

$$\int x(1 - x^2)^5\, dx = \int (1 - x^2)^5\, x\, dx$$

$$= \int u^5 \left(-\frac{1}{2}\right) du = -\frac{1}{2} \int u^5\, du$$

$$= -\frac{1}{2}\frac{u^6}{6} + C = -\frac{1}{12}(1 - x^2)^6 + C$$

This result can be verified by computing the derivative of $(-1/12)(1 - x^2)^6 + C$ and checking that it equals the original integrand, $x(1 - x^2)^5$.

Notice how the constant $-1/2$ can be taken out of the integrand because of the rule that states that the integral of a constant times a function is the constant times the integral. This is why the expression for du can differ by a constant when deciding what substitution to make for u.

The next example shows how to use substitution to find definite integrals.

EXAMPLE 2 | **Problem**

Find $\displaystyle\int_0^1 (x + 1)(x^2 + 2x + 1)^{-1/2}\, dx$

SOLUTION First find the indefinite integral of $(x + 1)(x^2 + 2x + 1)^{-1/2}$. Let $u = x^2 + 2x + 1$. Then $du = (2x + 2)\, dx = 2(x + 1)\, dx$. The integrand contains only $(x + 1)\, dx$, so we must divide by the constant 2. Dividing by 2 yields $(1/2)du = (x + 1)\, dx$. By the method of substitution

$$\int (x + 1)(x^2 + 2x + 1)^{-1/2}\, dx$$

$$= \int u^{-1/2}\, \frac{1}{2}\, du$$

$$= \frac{1}{2}\int u^{-1/2}\, du = \frac{1}{2}\frac{u^{1/2}}{1/2} + C = u^{1/2} + C$$

$$= (x^2 + 2x + 1)^{1/2} + C$$

Now insert the limits of integration.

$$\int_0^1 (x + 1)(x^2 + 2x + 1)^{-1/2}\, dx = (x^2 + 2x + 1)^{1/2}\, \Big|_0^1$$

$$= 4^{1/2} - 1^{1/2} = 1$$

Sometimes the choice of substitution for u does not work. It might not express the entire integral in terms of u or the resulting integrand may not be easily integrated. Another substitution may be tried. Thus the method often involves trial and error. But not all integrals can be solved by substitution. For example, consider

$$\int (x^2 + 1)^{1/2}\, dx$$

The natural choice would be $u = x^2 + 1$. Then $du = 2x\, dx$. The factor of x is the problem. If the differential differed from the integrand by only a constant, then we could divide through by the constant. But here the differential and the integrand differ by the factor x. If we try to divide through by x, then the left-hand side of the equation for the differential would have two variables, u and x. So the substitution would not entirely replace the variable x by the variable u. Whenever du varies from the integrand by a constant the substitution can be made.

In the examples presented so far the function in the integrand chosen for u has been raised to a power. If we let u be that function, then u' must also be present, at least up to a constant. The general formula governing this type of substitution problem can be expressed as follows, with $u = f(x)$ and $n \neq -1$:

$$\int [f(x)]^n f'(x)\, dx = \int u^n\, du = \frac{u^{n+1}}{n+1} + C$$

$$= \frac{f(x)^{n+1}}{n+1} + C, \qquad n \neq 1$$

There are other types of functions that can be integrated by substitution. The next type that we consider are certain exponential functions. If e is raised to the power $f(x)$ and $f'(x)$ is also present in the integrand, set $u = f(x)$ and find du to complete the substitution.

The general formula, with $u = f(x)$, is

$$\int e^{f(x)} f'(x)\, dx = \int e^u\, du = e^u + C = e^{f(x)} + C$$

The formula can be verified using the definition of the antiderivative.

$$D_x(e^{f(x)} + C) = e^{f(x)} f'(x)$$

EXAMPLE 3

Problem

Find $\int x^2 e^{x^3}\, dx$.

SOLUTION Let $u = x^3$. Then $du = 3x^2\, dx$ and $(1/3)\, du = x^2\, dx$. Thus

$$\int x^2 e^{x^3}\, dx = \int e^{x^3}(x^2\, dx)$$

$$= \int e^u \frac{1}{3}\, du = \frac{1}{3} e^u + C$$

$$= \frac{1}{3} e^{x^3} + C$$

The third form of the substitution method is derived from the derivative formula $D_x \ln |x| = 1/x = x^{-1}$. Thus $\int x^{-1}\, dx = \ln |x| + C$. If a function is raised to the power -1 and its derivative is present in the integrand, then the substitution can be made and the integral involves a logarithmic function. The same is true if the integrand is a rational function whose numerator is a constant multiple of the derivative of the denominator. This means that the denominator can be expressed in the numerator as a function raised to the power -1.

EXAMPLE 4 | **Problem**

Find $\int x(x^2 - 4)^{-1} \, dx$.

SOLUTION Let $u = x^2 - 4$. Then $du = 2x \, dx$ and $(1/2) \, du = x \, dx$. Thus

$$\int x(x^2 - 4)^{-1} \, dx = \int (x^2 - 4)^{-1}(x \, dx)$$

$$= \int u^{-1} \frac{1}{2} \, du$$

$$= \frac{1}{2} \int u^{-1} \, du = \frac{1}{2} \ln |u| + C$$

$$= \frac{1}{2} \ln |x^2 - 4| + C$$

The general formula, with $u = f(x)$, is

$$\int [f(x)]^{-1} f'(x) \, dx = \int \frac{f'(x)}{f(x)} \, dx$$

$$= \int u^{-1} \, du = \int \frac{du}{u}$$

$$= \ln |u| + C = \ln |f(x)| + C$$

The next example illustrates how to use the exponential and logarithmic forms of the substitution method of integration in which the choice for the substitution is not obvious.

EXAMPLE 5 | **Problem**

Find (a) $\int \frac{e^{x^{-1}}}{x^2} \, dx$ and (b) $\int \frac{x^{1/2}}{x^{3/2} - 1} \, dx$

SOLUTION (a) Let $u = x^{-1}$. Then $du = -x^{-2} \, dx$ and $-du = x^{-2} \, dx$. Thus

$$\int \frac{e^{x^{-1}}}{x^2} \, dx = \int x^{-2} e^{x^{-1}} \, dx$$

$$= -\int e^u \, du = -e^u + C = -e^{x^{-1}} + C$$

(b) Let $u = x^{3/2} - 1$. Then $du = (3/2)x^{1/2}\, dx$ and $(2/3)\, du = x^{1/2}\, dx$. Thus

$$\int \frac{x^{1/2}}{x^{3/2} - 1}\, dx = \int \frac{1}{u} \left(\frac{2}{3}\right) du = \frac{2}{3} \int \frac{du}{u}$$

$$= \frac{2}{3} \ln |u| + C = \frac{2}{3} \ln |x^{3/2} - 1| + C$$

The method of substitution is versatile. The next example shows how to apply it in a case where the substitution is subtle.

EXAMPLE 6   **Problem**

Find $\displaystyle \int_0^1 \frac{x\, dx}{x + 1}$.

SOLUTION First find the antiderivative and then insert the limits of integration. Let $u = x + 1$. Then $x = u - 1$ and $du = dx$, so

$$\int \frac{x\, dx}{x + 1} = \int \frac{u - 1}{u}\, du = \int \left(1 - \frac{1}{u}\right) du$$

$$= u - \ln |u| + C$$

$$= x + 1 - \ln |x + 1| + C$$

Insert the limits of integration.

$$\int_0^1 \frac{x\, dx}{x + 1} = x + 1 - \ln |x + 1| \Big|_0^1 = 2 - \ln 2 - (1 - \ln 1)$$

$$= 1 - \ln 2$$

An often troublesome aspect of substitution is the role of the differential, du. The differential is a vital ingredient in putting the integrand in proper form, but when the antiderivative is found it "disappears," in the sense that it merely indicates the variable of integration. That is, once the substitution is made the sole role of du is to label the variable, so it does not appear in the final solution.

EXERCISE SET 7–1

• In Problems 1 to 26 find the antiderivative.

5. $\displaystyle \int x(3x^2 + 5)^3\, dx$ 6. $\displaystyle \int x(2x^2 - 3)^5\, dx$

1. $\displaystyle \int 2x(x^2 + 5)^3\, dx$ 2. $\displaystyle \int 2x(x^2 - 2)^4\, dx$ 7. $\displaystyle \int x^3(3x^4 - 5)^3\, dx$ 8. $\displaystyle \int x^4(4x^5 + 1)^4\, dx$

3. $\displaystyle \int 4x^3(x^4 - 7)^2\, dx$ 4. $\displaystyle \int 5x^4(x^5 + 1)^2\, dx$ 9. $\displaystyle \int x^5(3x^6 - 5)^{3/2}\, dx$

10. $\int x^3(5x^4 + 1)^{3/4} \, dx$

11. $\int 2x^3(x^4 - 7)^{-3} \, dx$

12. $\int 6x^4(x^5 + 10)^{-4} \, dx$

13. $\int x^{-3}(6x^{-2} + 5)^{-2} \, dx$

14. $\int x^{-4}(5x^{-3} - 1)^{-5} \, dx$

15. $\int (2x + 5)(x^2 + 5x)^{-3} \, dx$

16. $\int (2x - 2)(x^2 - 2x)^{1/4} \, dx$

17. $\int (x^3 - 1)(x^4 - 4x)^{1/2} \, dx$

18. $\int (x^4 + 2x)(x^5 + 5x^2)^{2/5} \, dx$

19. $\int (e^{2x} + 5e^{4x}) \, dx$

20. $\int (5e^{3x} - e^{-3x}) \, dx$

21. $\int (x^3 e^{x^4} + 2e^{-x}) \, dx$

22. $\int (x^{-3} e^{x^{-2}}) \, dx$

23. $\int \dfrac{2}{3x + 1} \, dx$

24. $\int \dfrac{x}{x^2 - 1} \, dx$

25. $\int x^2(x^3 - 1)^{-1} \, dx$

26. $\int (x^2 + 1)(x^3 + 3x)^{-1} \, dx$

In Problems 27 to 30 find the definite integral.

27. $\int_0^1 x(x^2 + 1)^2 \, dx$ 28. $\int_0^1 x(x^2 - 2)^3 \, dx$

29. $\int_{-1}^1 x^3(x^4 + 1)^2 \, dx$

30. $\int_0^2 x^4(x^5 + 1)^4 \, dx$

•• In Problems 31 to 34 find the definite integral.

31. $\int_0^3 x(x^2 + 1)^{1/2} \, dx$

32. $\int_0^1 x(1 - x^2)^{1/3} \, dx$

33. $\int_1^2 (x^3 + 1)(x^4 + 4x)^{-2} \, dx$

34. $\int_0^1 (x^4 + 5)(x^5 + 5x)^{-2} \, dx$

In Problems 35 to 42 find the antiderivative.

35. $\int (e^x + 1)e^x \, dx$

36. $\int (e^{2x+1} + 10) \, e^{2x} \, dx$

37. $\int \dfrac{\ln x}{x} \, dx$ 38. $\int \dfrac{\ln x^2}{x} \, dx$

39. $\int x(x - 1)^2 \, dx$ 40. $\int x(3x + 4)^4 \, dx$

41. $\int x(1 - 5x)^{1/2} \, dx$ 42. $\int 5x(x + 4)^{-4} \, dx$

••• In Problems 43 to 46 find the antiderivative.

43. $\int x^3(x^2 - 1)^{2/3} \, dx$

44. $\int x^3(2 - 3x^2)^{-3} \, dx$

45. $\int \dfrac{x \ln(1 + x^2)}{1 + x^2} \, dx$

46. $\int \dfrac{e^x \ln(1 + e^x)}{1 + e^x} \, dx$

47. In an article studying the possible environmental impact of uranium mining in northern Australia, Vardavas investigated how contaminants were transported by surface water and ground water.* This study is meant to predict human radiological exposure from contaminants from mine sites.

*Ilias Mihail Vardavas, "Modelling the Seasonal Variation of Net All-wave Radiation Flux and Evaporation in a Tropical Wet-dry Region," *Ecological Modelling*, Vol. 39, 1987, pp. 247–268.

One part of the study requires the calculation of the total water evaporation per day at Manton Dam. The evaporation rate $E(x)$ for a given wind speed x can be approximated by

$$E(x) = \frac{.5x + .8}{x + 1}$$

The total evaporation is then computed by finding the antiderivative of $E(x)$. Find this antiderivative.

48. Talent searches identifying mathematically gifted high school youth have shown remarkably consistent sex differences. In an article studying the effect of a sex-linked gene that might facilitate high mathematical test score performance, Thomas considered data from the Iowa Test of Basic Skills and the SAT-mathematics test.[†] One part of the study investigated the properties of the function

$$f(x) = \frac{x}{x^2 + 1}$$

where $f(x)$ is a psychological measure, called the Mills ratio, and x is a percentile. Find the antiderivative of $f(x)$.

[†]Hoben Thomas, "A Theory of High Mathematical Aptitude," *Journal of Mathematical Psychology,* Vol. 29, 1985, pp. 231–242.

7–2 INTEGRATION BY PARTS

In the previous section we mentioned that substitution is a trial-and-error method that does not work in all cases. In this section we consider another method of integration, integration by parts. To apply the technique, think of the integrand as formed by two distinct functions, or "parts."

Integration by Parts

The product rule for derivatives cannot be transformed directly into an integration formula, but it can be revised to give a powerful rule. Suppose u and v are functions of x. The product rule states that

$$D_x uv = u D_x v + v D_x u$$

Expressing this formula in terms of differentials yields

$$d(uv) = u\, dv + v\, du$$

It is now in a form that is more amenable for integration. But first solve for $u\, dv$.

$$u\, dv = d(uv) - v\, du$$

Taking the integral of both sides yields

$$\int u\, dv = \int d(uv) - \int v\, du$$

$$= uv - \int v\, du$$

This is the formula for the technique of integration called **integration by parts.**

> **Integration by Parts**
>
> If u and v are differentiable functions, then
>
> $$\int u \, dv = uv - \int v \, du$$

The term "parts" refers to the way that the integrand is separated into two functions, u and dv. Thus there are two parts to the integrand. The underlying idea of the technique is that the integral $\int u \, dv$ is difficult, if not impossible, to calculate. The goal is to choose u and v so that the integral $\int v \, du$ is easier to calculate. The formula for integration by parts shows how to use $\int v \, du$ to compute $\int u \, dv$. The first example illustrates how to use the formula.

EXAMPLE 1

Problem

Compute $\displaystyle\int xe^x \, dx$

SOLUTION Separate the integrand into u and dv as follows:

$$u = x \qquad dv = e^x \, dx$$

Then

$$du = dx \qquad v = \int dv = \int e^x \, dx = e^x$$

Since the constant of integration is usually only included in the final answer, it is not mentioned in the intermediate steps. From the formula

$$\int xe^x \, dx = \int u \, dv = uv - \int v \, du$$

$$= xe^x - \int e^x \, dx$$

$$= xe^x - e^x + C$$

    It is important to realize that integration by parts is another trial-and-error method. For example, in Example 1 if the choice for u and dv is $u = e^x$ and $dv = x \, dx$, then $du = e^x \, dx$ and $v = \int x \, dx = x^2/2$. Then the formula yields $\int xe^x \, dx = x^2 e^x/2 - \int x^2 e^x/2 \, dx$. The latter integral is more complex than the original one. This choice of u and dv does not work. The object is to choose u and dv so that

1. It is possible to compute $v = \int dv$, and
2. It is possible to compute $\int v \, du$, or $\int v \, du$ is simpler than $\int u \, dv$.

The next example shows how to compute a seemingly simple integral by a judicious choice of u and dv.

EXAMPLE 2 | **Problem**

Compute $\int \ln x \, dx$.

SOLUTION Separate the integrand into u and dv as follows:

$$u = \ln x \qquad dv = dx$$

Then

$$du = \frac{1}{x} \, dx \qquad v = \int dv = \int dx = x$$

Substituting these expressions in the formula, we get

$$\int \ln x \, dx = \int u \, dv = uv - \int v \, du$$

$$= (\ln x)x - \int x\left(\frac{1}{x}\right) dx = x \ln x - \int dx$$

$$= x \ln x - x + C$$

Sometimes it is necessary to apply the formula more than once to compute the integral, as demonstrated in Example 3.

EXAMPLE 3 | **Problem**

Compute $\int x^2 e^{3x} \, dx$.

SOLUTION The idea is to eliminate the x^2 from the integrand. This is not possible in one application of the technique, but we can replace x^2 by x to simplify the integrand. Separate the integrand into u and dv as follows:

$$u = x^2 \qquad dv = e^{3x} \, dx$$

$$du = 2x \, dx \qquad v = \int dv = \int e^{3x} \, dx = \frac{1}{3} e^{3x}$$

$$\int x^2 e^{3x} \, dx = uv - \int v \, du$$

$$= x^2 \frac{1}{3} e^{3x} - \int \frac{1}{3} e^{3x} 2x \, dx$$

$$= \frac{1}{3} x^2 e^{3x} - \frac{2}{3} \int x e^{3x} \, dx \tag{1}$$

The problem has been reduced to finding $\int x e^{3x} \, dx$. To compute the latter integral, use integration by parts with

$$u = x \qquad dv = e^{3x}$$

$$du = dx \qquad v = \frac{1}{3} e^{3x}$$

This is similar to Example 1. The formula yields

$$\int xe^{3x}\,dx = x\frac{1}{3}e^{3x} - \int \frac{1}{3}e^{3x}\,dx$$

$$= \frac{1}{3}xe^{3x} - \frac{1}{9}e^{3x}$$

Now substitute this result into equation 1 to get the final answer.

$$\int x^2 e^{3x}\,dx = \frac{1}{3}x^2 e^{3x} - \frac{2}{3}\int xe^{3x}\,dx$$

$$= \frac{1}{3}x^2 e^{3x} - \frac{2}{3}\left[\frac{1}{3}xe^{3x} - \frac{1}{9}e^{3x}\right]$$

$$= \frac{1}{3}x^2 e^{3x} - \frac{2}{9}xe^{3x} + \frac{2}{27}e^{3x} + C$$

The next example shows how we sometimes need to employ more than one technique of integration in a problem. It uses substitution and integration by parts.

EXAMPLE 4

Problem

Compute $\int x(2x + 1)^{1/2}\,dx$.

SOLUTION Choose u and dv as follows:

$$u = x \qquad dv = (2x + 1)^{1/2}\,dx$$

$$du = dx \qquad v = \int (2x + 1)^{1/2}\,dx$$

To compute v, we must use substitution. Let $w = 2x + 1$. Then $dw = 2\,dx$, $(1/2)dw = dx$, and so

$$v = \int (2x + 1)^{1/2}\,dx = \int w^{1/2}\left(\frac{1}{2}\right)dw = \frac{1}{2}\left(\frac{2}{3}\right)w^{3/2}$$

$$= \frac{1}{3}(2x + 1)^{3/2}$$

Substituting into the formula for integration by parts yields

$$\int x(2x + 1)^{1/2}\,dx = \int u\,dv = uv - \int v\,du$$

$$= x\left(\frac{1}{3}\right)(2x + 1)^{3/2} - \int \frac{1}{3}(2x + 1)^{3/2}\,dx$$

$$= \frac{1}{3}x(2x + 1)^{3/2} - \frac{1}{3}\int (2x + 1)^{3/2}\,dx$$

Again, the substitution $w = 2x + 1$ is needed in the last integral. The final answer is

$$\int x(2x + 1)^{1/2} \, dx = \frac{1}{3} x(2x + 1)^{3/2} - \frac{1}{15} (2x + 1)^{5/2} + C$$

The integral in Example 4 could also be computed by using the substitution $u = 2x + 1$. In fact substitution is the first method to try. We used integration by parts to illustrate the method and to show that the two methods can be used on the same problem.

Definite Integrals and Parts

To find a definite integral using integration by parts, first find the antiderivative and then insert the limits of integration. This yields the following formula:

$$\int_a^b u \, dv = uv \, \Big|_a^b - \int_a^b v \, du$$

The next example shows that first we can calculate $\int v \, du$ and then insert the limits.

EXAMPLE 5

Problem

Compute $\int_0^4 x(2x + 1)^{1/2} \, dx$.

SOLUTION First find the indefinite integral. This was done in Example 4. Then insert the limits of integration.

$$\int_0^4 x(2x + 1)^{1/2} \, dx = \left[\frac{1}{3} x(2x + 1)^{3/2} - \frac{1}{15} (2x + 1)^{5/2} \right] \Big|_0^4$$

$$= \frac{1}{3} 4(9)^{3/2} - \frac{1}{15} (9)^{5/2} - 0 + \frac{1}{15}$$

$$= \frac{108}{3} - \frac{243}{15} + \frac{1}{15}$$

$$= \frac{298}{15}$$

EXERCISE SET 7–2

• In Problems 1 to 20 find the antiderivative.

1. $\int 2xe^x \, dx$

2. $\int 5xe^x \, dx$

3. $\int xe^{3x} \, dx$

4. $\int 5xe^{4x} \, dx$

5. $\int x(3x + 5)^{3/2} \, dx$

6. $\int x(2x - 3)^{5/3} \, dx$

7. $\int 3x(5x - 2)^{-3} \, dx$

8. $\int 2x(4x + 1)^{-4} \, dx$

9. $\int x \ln x \, dx$

10. $\int x \ln 5x \, dx$

11. $\int x^2 \ln x \, dx$

12. $\int x^4 \ln x \, dx$

13. $\int x^2 e^x \, dx$

14. $\int x^2 e^{5x-3} \, dx$

15. $\int (2x + 5)(x + 5)^{-3} \, dx$

16. $\int (x - 2)(4x - 3)^{-1/2} \, dx$

17. $\int x^2(x - 4)^{1/2} \, dx$ 18. $\int x^2(1 - x)^{2/3} \, dx$

19. $\int x^2(1 - 3x)^{-2} \, dx$

20. $\int 4x^2(2 + 3x)^{-3} \, dx$

In Problems 21 to 26 find the definite integral.

21. $\int_0^1 xe^{2x} \, dx$ 22. $\int_0^1 xe^{x+2} \, dx$

23. $\int_0^2 x(2x + 5)^{1/2} \, dx$

24. $\int_0^2 x(4x + 1)^{-1/2} \, dx$

25. $\int_0^2 x^2(x + 1)^{-4} \, dx$

26. $\int_0^2 x^2(2x + 3)^{-3} \, dx$

•• In Problems 27 to 34 find the antiderivative.

27. $\int x^2(x + 1)^{1/2} \, dx$

28. $\int x^2(1 - 3x)^{-1/2} \, dx$

29. $\int x^3(x + 1)^{1/2} \, dx$

30. $\int x^3(1 - 3x)^{-1/2} \, dx$

31. $\int x^3(x^2 + 1)^{1/2} \, dx$

32. $\int x^3(2 - 3x^2)^{-1/2} \, dx$

33. $\int x^3 e^{x^2} \, dx$ 34. $\int x^5 e^{x^3} \, dx$

••• In Problems 35 and 36 find the antiderivative.

35. $\int_0^2 x^5(2x^2 + 1)^{1/2} \, dx$

36. $\int_0^1 x^5(x^2 + 3)^{-4} \, dx$

37. In an article studying organic contaminants in natural environments, especially industrial and municipal water wastes, Chang and Rittmann focus attention on the contaminant, activated carbon.* Let $q(r)$ be the surface concentration of activated carbon on a circular surface r units from the center of the concentration. The computation of the total concentration involves evaluating the definite integral of $r^2 q(r)$ from $r = 0$ to $r = R$, the radius of the circular region. Compute this integral for $q(r) = e^{-.1r}$ and $R = 1$ mm.

38. In business the term "mark-up" refers to the difference between the selling price to the customer and the cost the retailer paid for the item. The mark-up takes into account the retailer's cost to market and sell the item and the retailer's profit. In an article studying how mark-up varies during various phases of the business cycle, Goldstein applied his model to 20 manufacturing industries from 1949 to 1980.† The goal of each industry is to maximize profit subject to cost restrictions. Goldstein assumed that profit $P(t)$ varied with time t according to the relationship $P'(t) = e^{-rt}f(t)$, where r is the prevailing interest rate and $f(t)$ depended on selling price, market share, and competitors' prices. He defined total profit $P(t)$ as the antiderivative of $P'(t)$. Find $P(t)$ for $f(t) = t$.

*Ted Chang and Bruce E. Rittmann, "Mathematical Modelling of Biofilm on Activated Carbon," *Environmental Science and Technology*, Vol. 21, 1987, pp. 273–280.

†Jonathan Goldstein, "Mark-up Pricing over the Business Cycle: The Microfoundations of the Variable Mark-up," *Southern Economics Journal*, Vol. 53, 1987, pp. 233–246.

7–3 TABLES OF INTEGRALS

Most of the integrands encountered thus far have required a direct application of one particular technique of integration. Often, however, an integration problem requires lengthy and cumbersome manipulation. For such problems a table of integrals is usually helpful.

The table in the Appendix provides a list of antiderivatives that is intended to acquaint the student with such tables. More extensive tables can be found, such as in the *Mathematical Handbook of Formulas and Tables,* which contains more than 500 antiderivatives.

Using a table of integrals is usually straightforward, but there are a few pitfalls that need to be pointed out. Tables of integrals are organized by similarity of the integrands. The first few listings are generally the elementary formulas. These are usually followed by integrands involving $ax + b$ for constants a and b, which in turn are followed by integrands involving $ax^2 + b$, $(ax + b)^2$, and so forth.

Because the table is arranged in this way, it is easy to mistake the correct formula for a similar one. For example, consider the integral

$$\int \frac{dx}{x(3x + 2)}$$

Referring to the table, we find that formulas 12, 13, and 15 are all similar. Each could be mistaken for the proper formula, namely, formula 14 with $a = 3$ and $b = 2$. Formula 12 is incorrect because there is no x appearing in the denominator of the formula's integrand. Similarly, formulas 13 and 15 are incorrect because $(ax + b)^2$ appears in the integrand rather than $(ax + b)$.

The first example shows how to find an antiderivative when the integrand fits the formula directly. Later examples show how some integrands need to be expressed in a different form to apply a formula.

EXAMPLE 1

Problem

Find $\displaystyle\int \frac{dx}{x(3x + 2)^2}$.

SOLUTION The integrand fits formula 15 with $a = 3$ and $b = 2$. Substituting these values into formula 15 yields

$$\int \frac{dx}{x(3x + 2)^2} = \frac{1}{2(3x + 2)} + \frac{1}{4} \ln \left| \frac{x}{3x + 2} \right| + C$$

Often some algebraic manipulation is necessary to see which formula is correct, as illustrated in the next example.

EXAMPLE 2 | **Problem**

Find

$$\int \frac{dx}{1 - 4x^2}.$$

SOLUTION The integrand fits formula 6, but it cannot be applied immediately because the coefficient of x^2 in the problem is 4, while the coefficient of x^2 in the formula is 1. The coefficient must be factored out of the denominator and the integral sign because the coefficients of x^2 in the problem and the formula must agree. Use the fact that $1 - 4x^2 = 4(1/4 - x^2)$. Hence

$$\int \frac{dx}{1 - 4x^2} = \int \frac{dx}{4(1/4 - x^2)} = \frac{1}{4} \int \frac{dx}{1/4 - x^2}$$

Now formula 6 can be applied with $a = 1/2$.

$$\int \frac{dx}{1 - 4x^2} = \frac{1}{4} \int \frac{dx}{1/4 - x^2} = \frac{1}{4} \ln \left| \frac{1/2 + x}{1/2 - x} \right| + C$$

This formula holds only for $x^2 < a^2$; otherwise the antiderivative is not defined.

Most tables do not include integrals that can be simplified by a simple substitution. Sometimes a simple substitution of variable is required before applying the appropriate formula.

EXAMPLE 3 | **Problem**

Find $\int x(x^4 + 4)^{1/2} \, dx$.

SOLUTION The integrand is similar to formula 16, but it cannot be applied immediately because the problem contains x^4 instead of x^2 in the formula. The x outside the parentheses permits the substitution $u = x^2$. Then $du = 2x \, dx$ and so $(1/2)du = x \, dx$. Since $u^2 = x^4$, this substitution yields

$$\int x(x^4 + 4)^{1/2} \, dx = \frac{1}{2} \int (u^2 + 4)^{1/2} \, du$$

Now apply formula 16 with $a = 2$.

$$\int x(x^4 + 4)^{1/2} \, dx = \frac{1}{2} \int (u^2 + 4)^{1/2} \, du$$

$$= \frac{1}{2} \left[\frac{1}{2} u(u^2 + 4)^{1/2} + \frac{4}{2} \ln |u + (u^2 + 4)^{1/2}| \right] + C$$

$$= \frac{1}{4} x^2(x^4 + 4)^{1/2} + \ln |x^2 + (x^4 + 4)^{1/2}| + C$$

Formula 18 is a recursive formula, meaning that it might have to be applied several times to solve a problem.

EXAMPLE 4

Problem

Compute $\int x^3 e^{2x} \, dx$.

SOLUTION Apply formula 18 with $n = 3$ and $a = 2$.

$$\int x^3 e^{2x} \, dx = \frac{1}{2} x^3 e^{2x} - \frac{3}{2} \int x^2 e^{2x} \, dx \tag{1}$$

To compute the integral on the right, again apply formula 18, this time with $n = 2$ and $a = 2$.

$$\int x^2 e^{2x} \, dx = \frac{1}{2} x^2 e^{2x} - \frac{2}{2} \int x e^{2x} \, dx \tag{2}$$

Once again apply formula 18, now with $n = 1$ and $a = 2$.

$$\int x e^{2x} \, dx = \frac{1}{2} x e^{2x} - \frac{1}{2} \int e^{2x} \, dx$$

$$= \frac{1}{2} x e^{2x} - \frac{1}{4} e^{2x} \tag{3}$$

Now put all the pieces of the puzzle together. Substitute equation 3 into equation 2, and substitute the result into equation 1. The final answer is

$$\int x^3 e^{2x} \, dx = \frac{1}{2} x^3 e^{2x} - \frac{3}{2} \left[\frac{1}{2} x^2 e^{2x} - \frac{1}{2} x e^{2x} + \frac{1}{4} e^{2x} \right]$$

$$= \frac{1}{2} x^3 e^{2x} - \frac{3}{4} x^2 e^{2x} + \frac{3}{4} x e^{2x} - \frac{3}{8} e^{2x}$$

To evaluate a definite integral via the table of integrals, first find the indefinite integral and then evaluate the antiderivative at the limits of integration.

EXAMPLE 5

Problem

Find $\int_2^3 \dfrac{dx}{x(2x - 3)}$.

SOLUTION The integrand fits formula 14 with $a = 2$ and $b = -3$. Substitute these values into the formula without the limits of integration.

$$\int \frac{dx}{x(2x - 3)} = \frac{1}{-3} \ln \left| \frac{x}{(2x - 3)} \right| + C$$

Inserting the limits of integration yields

$$\int_2^3 \frac{dx}{x(2x - 3)} = -\frac{1}{3} \ln \left| \frac{x}{(2x - 3)} \right| \bigg|_2^3$$

$$= -\frac{1}{3} \ln \left| \frac{3}{3} \right| + \frac{1}{3} \ln \left| \frac{2}{1} \right|$$

$$= \frac{1}{3} \ln 2$$

EXERCISE SET 7–3

• In Problems 1 to 38 find the integral.

1. $\int \dfrac{dx}{x(3x + 2)^2}$

2. $\int \dfrac{dx}{x(5x + 6)^2}$

3. $\int \dfrac{dx}{x(3x - 4)^2}$

4. $\int \dfrac{dx}{x(5x - 7)^2}$

5. $\int \dfrac{dx}{1 - 9x^2}$

6. $\int \dfrac{dx}{4 - 9x^2}$

7. $\int \dfrac{dx}{16 - 9x^2}$

8. $\int \dfrac{dx}{4 - 25x^2}$

9. $\int \dfrac{dx}{36 - 4x^2}$

10. $\int \dfrac{dx}{25 - 4x^2}$

11. $\int x(x^4 + 9)^{1/2}\, dx$

12. $\int x(x^4 + 16)^{1/2}\, dx$

13. $\int x(4x^4 + 9)^{1/2}\, dx$

14. $\int x(9x^4 + 16)^{1/2}\, dx$

15. $\int 2x(4x^4 - 25)^{1/2}\, dx$

16. $\int 3x(9x^4 - 49)^{1/2}\, dx$

17. $\int_2^3 \dfrac{dx}{x(3x - 5)}$

18. $\int_2^3 \dfrac{dx}{x(6 - 5x)}$

19. $\int_0^1 \dfrac{dx}{x(3x + 2)^2}$

20. $\int_1^3 \dfrac{dx}{x(5x + 6)^2}$

21. $\int_0^1 x(x^4 + 1)^{1/2}\, dx$

22. $\int_0^1 x(x^4 + 4)^{1/2}\, dx$

23. $\int x^3 e^{3x}\, dx$

24. $\int x^4 e^{5x}\, dx$

•• 25. $\int \dfrac{x\, dx}{(5x + 2)^2}$

26. $\int \dfrac{x\, dx}{(7x + 6)^2}$

27. $\int \dfrac{dx}{x(x - 4)}$

28. $\int \dfrac{x\, dx}{5x - 7}$

29. $\int \dfrac{dx}{x\sqrt{1 - 9x^2}}$

30. $\int \dfrac{dx}{\sqrt{4 - 9x^2}}$

31. $\int \dfrac{x\, dx}{\sqrt{16 + 9x^4}}$

32. $\int \dfrac{dx}{x\sqrt{4 - x^4}}$

33. $\int \dfrac{dx}{x\sqrt{9 + 4x^2}}$

34. $\int \dfrac{x\, dx}{49 - 4x^4}$

••• 35. $\int \dfrac{x^2\, dx}{9 - 4x^6}$

36. $\int \dfrac{x^3\, dx}{9 - 4x^4}$

37. $\int x^5 \ln x^2\, dx$

38. $\int x^5 e^{x^2}\, dx$

39. The study of waiting situations, the so-called "queueing problems," usually assumes that people dislike having to wait. In an article studying the resulting psychological cost of waiting, Osuna defines the total stress of waiting $TH(t)$

as the integral of $H(t)$, the stress at time t.* Find the indefinite integral of $H(t)$ for

$$H(t) = \frac{t^2}{10 - t}$$

40. Purchasing land always involves risk, but the purchase of farmland usually involves a greater amount of risk than other types of land because there are so many additional factors. These include complex tax laws and government price supports that affect not only profit but land value. In an article studying farmland pricing strategies, Brown and Brown define the expected earnings per acre $E(a, b)$ of selling a piece of land for a price between $x = a$ and $x = b$ hundred dollars per acre to be

$$E(a, b) = \int_a^b f(x)\, dx$$

where $f(x)$ is the likelihood that the seller will get a bid of price x.† Find $E(1, 1.5)$ for

$$f(x) = \frac{x}{x + 1}$$

*Edgar Elias Osuna, "The Psychological Cost of Waiting," *Journal of Mathematical Psychology,* Vol. 29, 1985, pp. 82–106.

†Keith C. Brown and Deborah J. Brown, "Heterogenous Expectations and Farmland Prices," *American Journal of Agricultural Economics,* Vol. 66, 1984, pp. 164–169.

7–4 NUMERICAL INTEGRATION

The definite integral is defined by using approximations to the area under the curve that consisted of sums of areas of certain rectangles. A limit process was applied to define the definite integral. The fundamental theorem of calculus permits us to calculate integrals by using antiderivatives, thus avoiding the need to use approximations. But sometimes it is impractical or even impossible to find the correct antiderivative. In this case we need to approximate the area under the curve in order to approximate the definite integral.

There are many ways to approximate the definite integral. They are called **numerical integration** methods because they require several numerical calculations. With the advent of calculators and computers the tedium of these computations has been significantly reduced.

One method of approximation is to use rectangles, as is done when defining the definite integral. In this section we will present another method, called the **trapezoidal rule.** It usually produces a closer approximation than the method using merely rectangles. Another method, **Simpson's rule,** is described in the exercises.

An Example

Rectangles can yield good approximations to the area under a curve, but their upper leg is always horizontal. A trapezoid's upper leg can be chosen so that it usually comes closer to the curve. This is why trapezoids usually produce closer approximations to the area under a curve. In Figure 1 we see that the sum of the areas of the trapezoids is a better approximation to the area under the curve than the sum of the areas of the rectangles. Each trapezoid comes closer to the area under the curve.

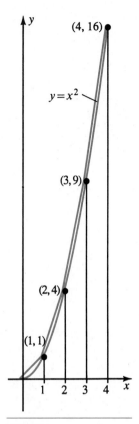

FIGURE 1

Let us look at an example of how to approximate the area under a curve with trapezoids. Then we will develop the formula for approximating $\int_a^b f(x)\,dx$. Consider the area under $f(x) = x^2$ from $x = 0$ to $x = 4$. The first step is to divide the interval $[0, 4]$ into subintervals. We choose a convenient number, say, 4, and select the subintervals so that they have equal length. Then the end points of the subintervals are 0, 1, 2, 3, and 4. Each trapezoid has width 1. The heights of the trapezoids are $f(0) = 0$, $f(1) = 1$, $f(2) = 4$, $f(3) = 9$, and $f(4) = 16$. (See Figure 1.) The formula for the area of a trapezoid with heights h and k and width w is

$$\frac{w}{2}(h + k)$$

Let the area of the four trapezoids be A_1, A_2, A_3, and A_4. Then

$$A_1 = \frac{1}{2}[f(0) + f(1)] = \frac{1}{2}[0 + 1] = \frac{1}{2}$$

$$A_2 = \frac{1}{2}[f(1) + f(2)] = \frac{1}{2}[1 + 4] = \frac{5}{2}$$

$$A_3 = \frac{1}{2}[f(2) + f(3)] = \frac{1}{2}[4 + 9] = \frac{13}{2}$$

$$A_4 = \frac{1}{2}[f(3) + f(4)] = \frac{1}{2}[9 + 16] = \frac{25}{2}$$

The approximation is the sum of these areas. It is

$$A_1 + A_2 + A_3 + A_4 = \frac{1}{2} + \frac{5}{2} + \frac{13}{2} + \frac{25}{2} = \frac{44}{2} = 22$$

The actual area can be found using the fundamental theorem. It is

$$\int_0^4 x^2\,dx = \frac{x^3}{3}\bigg|_0^4 = \frac{64}{3}$$

As a decimal rounded to two places the actual integral is 21.33. This compares favorably with the approximation via the trapezoidal rule with four subintervals, which is 22. A better approximation could be obtained by using more subintervals.

Trapezoidal Rule

We now present the formula for the trapezoidal rule. Assume that $f(x) \geq 0$ in the interval $[a, b]$ so that the area under the curve is equal to the definite integral. To approximate the definite integral $\int_a^b f(x)\,dx$, first divide the interval $[a, b]$ into n equal subintervals, each having width $(b - a)/n$. The elements of the subdivision are

$$a = x_0, x_1, \ldots, x_n = b$$

Observe that $x_k = a + k(b - a)/n$, for $k = 0, 1, \ldots, n$. The area A_k of a trapezoid whose base is $[x_{k-1}, x_k]$ and whose heights are $f(x_{k-1})$ and $f(x_k)$ is given by

$$A_k = \frac{b - a}{2n} [f(x_{k-1}) + f(x_k)]$$

To approximate the area under the curve, we sum all the areas of the trapezoids. Thus

$$\int_a^b f(x)\, dx \simeq A_1 + A_2 + \cdots + A_n$$

$$= \frac{b - a}{2n} [f(x_0) + f(x_1)] + \frac{b - a}{2n} [f(x_1) + f(x_2)]$$

$$+ \cdots + \frac{b - a}{2n} [f(x_{n-1}) + f(x_n)]$$

The expression $(b - a)/2n$ can be factored from each term. Also, each term $f(x_k)$ occurs twice, except the first and last terms, $f(b)$ and $f(a)$. These simplifications result in the following formula, known as the *trapezoidal rule*.

Trapezoidal Rule

Suppose $f(x)$ is a continuous function on the interval $[a, b]$ and $f(x) \geq 0$ on $[a, b]$. To approximate the definite integral of $f(x)$ from a to b by the trapezoidal rule, divide $[a, b]$ into n equal subintervals whose end points are $a = x_0, x_1, \ldots, x_n = b$, then

$$\int_a^b f(x)\, dx \simeq \frac{(b - a)}{2n} [f(a) + 2f(x_1) + 2f(x_2)$$

$$+ \cdots + 2f(x_{n-1}) + f(b)]$$

The first example shows that the formula yields the same result as the earlier computations.

EXAMPLE 1

Problem

Use the trapezoidal rule with $n = 4$ to approximate the integral $\int_0^4 x^2\, dx$.

SOLUTION Each subinterval has width 1, and the heights of the trapezoids are $f(0) = 0$, $f(1) = 1$, $f(2) = 4$, $f(3) = 9$, and $f(4) = 16$. The formula yields

$$\int_0^4 x^2 \, dx \simeq \frac{4}{8} [f(0) + 2f(1) + 2f(2) + 2f(3) + f(4)]$$

$$= \frac{1}{2} [0 + 2 + 8 + 18 + 16]$$

$$= \frac{44}{2} = 22$$

This agrees with earlier computations.

The next example shows that better approximations are usually obtained by increasing the number of subintervals.

EXAMPLE 2

Problem

Use the trapezoidal rule with $n = 8$ to approximate the integral $\int_0^4 x^2 \, dx$.

SOLUTION Each subinterval has width $1/2$ since $b = 4$, $a = 0$, and $n = 8$. The heights of the trapezoids are $f(0) = 0$, $f(1/2) = 1/4$, $f(1) = 1$, $f(3/2) = 9/4$, $f(2) = 4$, $f(5/2) = 25/4$, $f(3) = 9$, $f(7/2) = 49/4$, and $f(4) = 16$. The formula yields

$$\int_0^4 x^2 dx \simeq \frac{1}{4} \left[f(0) + 2f\left(\frac{1}{2}\right) + 2f(1) + \cdots + f(4) \right]$$

$$= \frac{1}{4} \left[0 + \frac{1}{2} + 2 + \frac{9}{2} + 8 + \frac{25}{2} + 18 + \frac{49}{2} + 16 \right]$$

$$= \frac{1}{4} \left(\frac{172}{2} \right)$$

$$= \frac{43}{2} = 21.5$$

This is closer to the actual area, $21.33 \ldots$, than the earlier approximation using four subintervals.

The next two examples show how to use the trapezoidal rule when no anti-derivative is attainable in a simple form.

EXAMPLE 3

Use the trapezoidal rule with $n = 4$ to approximate $\int_0^2 (x^3 + 1)^{1/2} \, dx$.

SOLUTION Let $n = 4$, $a = 0$, and $b = 2$. Then

$$f(x_0) = f(0) = 1$$

$$f(x_1) = f\left(\frac{1}{2}\right) = \sqrt{\frac{9}{8}} \simeq 1.06$$

$$f(x_2) = f(1) = \sqrt{2} \simeq 1.41$$

$$f(x_3) = f\left(\frac{3}{2}\right) = \sqrt{\frac{35}{8}} \simeq 2.09$$

$$f(x_4) = f(2) = \sqrt{9} \simeq 3$$

Substituting these values into the formula yields

$$\int_0^2 (x^3 + 1)^{1/2} \, dx \simeq \frac{1}{4}\left[f(0) + 2f\left(\frac{1}{2}\right) + 2f(1) + 2f\left(\frac{3}{2}\right) + f(2) \right]$$

$$= \frac{1}{4}(1 + 2.12 + 2.82 + 4.18 + 3)$$

$$= \frac{1}{4}(13.12)$$

$$= 3.28$$

The function e^{-x^2} has no simple antiderivative that can be used to evaluate a definite integral. Thus to evaluate $\int_0^1 e^{-x^2} \, dx$, we must use a numerical method of integration as in the next example.

EXAMPLE 4

Problem

Use the trapezoidal rule with $n = 5$ to approximate $\int_0^1 e^{-x^2} \, dx$.

SOLUTION Let $n = 5$, $a = 0$, and $b = 1$. Then

$$f(x_0) = f(0) = 1$$

$$f(x_1) = f\left(\frac{1}{5}\right) = e^{-1/25} \simeq 0.961$$

$$f(x_2) = f\left(\frac{2}{5}\right) = e^{-4/25} \simeq 0.852$$

$$f(x_3) = f\left(\frac{3}{5}\right) = e^{-9/25} \simeq 0.698$$

$$f(x_4) = f\left(\frac{4}{5}\right) = e^{-16/25} \simeq 0.527$$

$$f(x_5) = f(1) = e^{-1} \simeq 0.368$$

Substituting these values into the formula yields

$$\int_0^1 e^{-x^2}\, dx$$

$$\approx \frac{1}{10}\left[f(0) + 2f\left(\frac{1}{5}\right) + 2f\left(\frac{2}{5}\right) + 2f\left(\frac{3}{5}\right) + 2f\left(\frac{4}{5}\right) + f(1)\right]$$

$$= \frac{1}{10}(1 + 1.922 + 1.704 + 1.396 + 1.054 + 0.368)$$

$$= \frac{1}{10}(7.444) = 0.7444$$

EXERCISE SET 7–4

• In Problems 1 to 4 approximate the integral $\int_a^b x^2\, dx$ for the given choice of a and b by using the trapezoidal rule with n subintervals.

1. $a = 0, b = 3, n = 3$
2. $a = 0, b = 3, n = 6$
3. $a = 0, b = 6, n = 6$
4. $a = 0, b = 6, n = 12$

In Problems 5 to 8 approximate the integral $\int_a^b x^3\, dx$ for the given choice of a and b using the trapezoidal rule with n subintervals.

5. $a = 0, b = 3, n = 3$
6. $a = 0, b = 3, n = 6$
7. $a = 0, b = 6, n = 6$
8. $a = 0, b = 6, n = 12$

In Problems 9 and 10 approximate $\int_a^b (x^3 + 1)^{1/2}\, dx$ for the given choice of a and b using the trapezoidal rule with n subintervals.

9. $a = 0, b = 5, n = 5$
10. $a = 0, b = 5, n = 10$

In Problems 11 and 12 approximate $\int_a^b e^{-x^2}\, dx$ for the given choice of a and b using the trapezoidal rule with n subintervals.

11. $a = 0, b = 5, n = 5$
12. $a = 0, b = 5, n = 10$

In Problems 13 to 16 approximate $\int_a^b \frac{1}{x+1}\, dx$ for the given choice of a and b using the trapezoidal rule with n subintervals.

13. $a = 0, b = 4, n = 4$
14. $a = 0, b = 4, n = 8$
15. $a = 0, b = 4, n = 6$
16. $a = 0, b = 4, n = 10$

In Problems 17 to 20 approximate $\int_a^b \frac{1}{x^2}\, dx$ for the given choice of a and b using the trapezoidal rule with n subintervals.

17. $a = 1, b = 5, n = 4$
18. $a = 1, b = 5, n = 8$
19. $a = 1, b = 3, n = 8$
20. $a = 1, b = 5, n = 10$

•• In Problems 21 to 24 approximate the definite integral using the trapezoidal rule with four subintervals.

21. $\int_0^2 (x^2 + x)^{1/2}\, dx$ 22. $\int_0^2 (x^3 + x)^{1/3}\, dx$

23. $\int_0^4 (e^x + x^{1/3})\, dx$ 24. $\int_1^5 (\ln x + x)^2\, dx$

In Problems 25 to 28 approximate the definite integral using the trapezoidal rule with eight subintervals.

25. $\int_0^8 (x^2 + 2x)^{1/2} \, dx$ 26. $\int_0^4 (x^3 + x)^{1/4} \, dx$

27. $\int_0^4 (e^{-x} + 2x^{1/2}) \, dx$ 28. $\int_1^5 (x \ln x + x)^2 \, dx$

●●● Simpson's rule is another method of numerical integration that approximates sections of the curve by using segments of parabolas as opposed to line segments in the trapezoidal rule. Usually an approximation by Simpson's rule yields a closer approximation than the trapezoidal rule for the same number of subintervals. It is necessary to have an even number of subintervals for Simpson's rule. The formula for Simpson's rule follows:

Simpson's Rule

Suppose $f(x)$ is a continuous function on the interval $[a, b]$ and suppose $f(x) \geq 0$ on $[a, b]$. To approximate the definite integral of $f(x)$ from a to b by Simpson's rule, divide $[a, b]$ into n (an even number) equal subintervals whose end points are $x_0 = a, x_1, \ldots, x_n = b$, then

$$\int_a^b f(x) \, dx \approx \frac{(b - a)}{3n} [f(a) + 4f(x_1)$$

$$+ 2f(x_2) + 4f(x_3)$$

$$+ \cdots + 2f(x_{n-2})$$

$$+ 4f(x_{n-1}) + f(b)]$$

In Problems 29 to 32 approximate the definite integral by using Simpson's rule with four subintervals.

29. $\int_0^2 (x^2 + x)^{1/2} \, dx$ 30. $\int_0^2 (x^3 + x)^{1/3} \, dx$

31. $\int_0^4 (e^x + x^{1/3}) \, dx$ 32. $\int_1^5 (\ln x + x)^2 \, dx$

In Problems 33 and 34 approximate $\int_0^4 \frac{1}{x + 1} \, dx$ using Simpson's rule with n subintervals. Compare this answer with the approximations in Problems 13 and 14. Then compare both approximations with the actual indefinite integral computed by using the fundamental theorem of calculus.

33. $n = 4$ 34. $n = 8$

35. To ensure that air quality satisfies set standards, air pollution control authorities must determine allowable emission rates for various types of pollutants. An important variable is the cost to treat the pollutants. The total treatment cost $C(x)$ is a function of the emission rate x. In an article investigating how pollution control authorities can establish an optimum scheme for calculating pollution cost functions, Hashimoto and Kimura studied the emission of nitrogen oxide in Tokyo City, where the primary pollutants were utility and industrial boilers and industrial internal combustion engines.* They were concerned with the case when the total cost could not be directly assessed, but the marginal cost $g(x)$ could be computed for various zones of the city. Then $C(x)$ is defined to be the integral of $g(x)$, which was computed by numerical integration. In one part of the study seven zones were used. The measurements are given in the table.

zone	1	2	3	4	5	6	7
$g(x)$	101	105	119	163	143	180	105

Take $g(0) = 0$ and approximate the total cost using the trapezoidal rule.

36. In 1962 Pierre Wenger discovered two persons who were contaminated with radium-226 and strontium-90. They agreed to have their bodily functions closely monitored for 10 years in order to study the long-range effects of contamination. In an article comparing the observed data with the predictions of various models, Wenger and Cosandey centered on the effects of contami-

---

*Akihiro Hashimoto and Yuri Kimura, "Determining the Optimal Scheme of Zoned Effluent Charges for the Control of Air Pollution," *Socio-Economic Planning Science*, Vol. 14, 1980, pp. 197–208.

nation on excretion rates, where the units are in levels of radium-226 excreted per day.[†] They took periodic measurements, and they used numerical integration to compute the sum of the bodily excretion. Their measurements are summarized in the following table.

year	1962	1964	1966	1968	1970	1972
excretion rate	.91	.65	.28	.26	.20	.18

Approximate the sum of the bodily excretion using the trapezoidal rule.

37. In an article investigating the costs and benefits of sexual versus asexual reproduction in various plants in different environments and mating systems, Michaels and Bazzaz studied a perennial herb growing in Kickapoo State Park, Danville, Illinois.[‡] In one part of the study the survival rate was given by $f(t) = 118e^{-.05t^2}$ for $t = 0$ to $t = 5$ months. Approximate the total number of survivors by integrating $f(t)$ from $t = 0$ to $t = 5$ using the trapezoidal rule with 5 intervals.

[†]Pierre Wenger and Maurice Cosandey, "Retention and Excretion of Radium-226 and Strontium-90 in Two Doubly Contaminated Persons," *Health Physics,* Vol. 31, 1976, pp. 225–229.

[‡]H. J. Michaels and F. A. Bazzaz, "Resource Allocation and Demography of Sexual and Apomictic *Antennaria parlinii*," *Ecology,* Vol. 67, 1986, pp. 27–36.

CASE STUDY   Demography and the Social Security System

Most people believe that the taxes they pay to the social security system each year are put into a savings account to be held until their retirement. Not so! The U.S. social security system is a pay-as-you-go plan, meaning that the money paid by the work force goes directly to fund those who are retired. In essence, the plan calls for those in the age range of 20 to 64 to support those over 64. The plan has worked well since its inception in 1935 because there has always been a large work force to support a much smaller retired segment of the population.

But times are changing. The present system faces grave consequences in the near future, primarily because both the birthrate and death rate are declining. This means that in a short while fewer people will be entering the work force while a larger number will be retiring. The latter is because of advances in health and medicine.

Demography   Demography answers many age-related questions about various populations, such as what is the effect of a lowered death rate on the retired segment of society, and what is the effect of abortions on the birthrate? The insurance industry was the first industry to use demography extensively, but now many diverse fields, including government, apply demography to a multitude of problems.

In his book on applied demography Nathan Keyfitz[*] uses calculus to give concise explanations of the theory of demography. As a specific application, Keyfitz describes the ramifications of demographic projections on the U.S. social security system.

[*]Nathan Keyfitz, *Applied Mathematical Demography,* 2nd ed., New York, Springer-Verlag, 1985.

Definitions of life table functions A life table for a given population is a table that contains the percentage of people who are expected to survive to a certain age. These percentages are computed from past records and mathematical projections. The probability of surviving from birth to age x is defined to be $l(x)$. Even though births and deaths occur in single increments, demographers assume $l(x)$ is a continuous function in order to use calculus for their projections. If the population is large, this is a valid assumption.

Consider those who have attained the age of t. The fraction of years survived by these people in the next n years is the "sum" of the probabilities of their surviving at each of these ages from t to $t + n$. Since $l(x)$ is a continuous function, this sum is the integral

$$\int_t^{t+n} l(x)\, dx$$

The definite integral of $l(x)$ from 0 to infinity, or at least to the largest age that anyone lives, say, w, is the sum of all the probabilities, so it is 1.

Example 1 shows how numerical integration is used to compute the number of people in a segment of a population. The example studies the female population of Mexico in 1970.

The following partial life table gives the values of $l(x)$ of the female population of Mexico in 1970.

Female Population, Mexico, 1970

Age	Fraction of Those Surviving to a Specific Age, $l(x)$
60	0.008939
62	0.008036
64	0.007071

EXAMPLE 1

Problem

Use the trapezoidal rule and the accompanying table to compute the fraction of females in Mexico between the ages of 60 and 64 in 1970.

SOLUTION Find the integral of $l(x)$ from 60 to 64. The trapezoidal rule, with $n = 2$, states

$$\int_{60}^{64} l(x)\, dx \simeq \frac{64\text{-}60}{4} [l(60) + 2l(62) + l(64)]$$

$$= l(60) + 2 \cdot l(62) + l(64)$$
$$= 0.008939 + 0.016072 + 0.007071$$
$$= 0.032082$$

Therefore the percentage of the female population in Mexico in 1970 between the ages of 60 and 64 was about 3.2%.

Pension cost   Consider a pay-as-you-go pension plan in which each person over age 64 is paid the same amount, say, one unit. Also assume that the pension is funded by those aged 20 to 64 and that their salaries are the same amount, one unit. Suppose the total number of individuals in the population is P. Then the amount disbursed to the pensioners is P times the fraction of individuals over 64:

$$P \int_{64}^{w} l(x) \, dx$$

The amount of money paid into the fund will be a fraction of the workers' salaries. Let this fraction be g. Thus g is the premium paid by each worker. The amount of money paid into the fund by the workers is

$$gP \int_{20}^{64} l(x) \, dx$$

Since the fund is a pay-as-you-go plan the amount paid into the fund must equal the amount paid out to the pensioners. This means the two expressions must be equal. Setting them equal and solving for g yields

$$g = \frac{\displaystyle\int_{64}^{w} l(x) \, dx}{\displaystyle\int_{20}^{64} l(x) \, dx}$$

This means that the premium paid by the workers is a percentage of their salary equal to the ratio of the number of individuals over age 64 divided by the number of individuals aged 20 to 64.

The assumptions seem restrictive, but adapting the model to salaries and pension payments that vary with age and even within given ages, as well as other practical considerations—including accounting for some younger persons receiving pensions—complicates the formula for g without altering the principles of the argument. More realistic models have several additional terms that give minor adjustments to the computation of the premium. But the primary conclusion remains: the premium depends on the ratio of those over 64 to those between 20 and 64.

The formula for the premium g illustrates how demographic projections affect pension payments. The premium increases in two ways: if the numerator increases or if the denominator decreases. This means that if the number of pensioners increases, or if the number of individuals in the work force decreases, the premium paid by the workers increases.

Projections for the social security system   The U.S. social security system is a pay-as-you-go plan. The premium paid by the work force is dependent on the ratio of the number of pensioners to the number of individuals in the work force. This type of plan has demographic problems if this ratio has a significant increase. The premium paid by each worker will increase as well. The accompanying table shows the demographic projections of the ratio of pensioners to workers. The values in the table from 1990 on were computed by using numerical integration.

Ratio of Individuals of Pensionable Age Versus Working Age

Year	Age 21–64	Age Over 64	Ratio Over 64/21–64
1950	85,944	12,397	0.1442
1960	92,181	16,675	0.1809
1970	103,939	20,085	0.1932
1980	122,115	24,523	0.2008
1990	137,500	28,933	0.2104
2000	148,589	30,600	0.2059
2025	146,645	45,715	0.3137
2050	147,635	45,805	0.3103

Significant pressure on the social security plan will come after the year 2000. The ratio will increase by more than 50% in the years from 2000 to 2025, from 0.20 to 0.31. The only thing that could avert the swing is a large number of births before the year 2000 that would increase the work force in the early 21st century. This seems unlikely.

The social security system can be understood as a method of borrowing from future generations. When initiated, it was reasonable because there were no start-up funds to pay the retirees during its first few years. In the ensuing years the plan also was viable because the ratio of pensioners to the work force was a relatively small fraction, so that the premium paid by each worker was manageable. In addition, the premium was thought of as insurance for one's own retirement. But the ratio will increase dramatically soon so that either the system will prove to be a great burden on the work force or the benefits paid to retirees will be greatly reduced.

Case Study Exercises

In Problems 1 to 4 use the following table and the trapezoidal rule to find the integral of $l(x)$ from the given ages.

Female Population, Mexico, 1970

Age	Fraction of Those Surviving to a Specific Age, $l(x)$
60	0.008939
61	0.008570
62	0.008036
63	0.007512
64	0.007071

1. From age 60 to 62

2. From age 61 to 63

3. From age 60 to 63

4. From age 60 to 64

5. Compare the answer in Exercise 4 to Example 1. Which most likely gives the better approximation?

6. Give a reason why the percentage in the ratio of individuals of pensionable age versus working age increases so dramatically from the year 2000 to 2025.

CHAPTER REVIEW

Key Terms

7–1   Substitution
Substitution of Variable   Differential of u
Integration by Substitution

7–2   Integration by Parts
Integration by Parts

7–4   Numerical Integration
Numerical Integration   Simpson's Rule
Trapezoidal Rule

Summary of Important Concepts

Formulas

If $u = g(x)$ then $du = g'(x)\, dx$

$$\int [f(x)]^n\, f'(x)\, dx = \frac{f(x)^{n+1}}{n+1} + C, \quad n \neq 1$$

$$\int e^{f(x)}\, f'(x)\, dx = e^{f(x)} + C$$

$$\int [f(x)]^{-1}\, f'(x)\, dx = \ln |f(x)| + C$$

$$\int u\, dv = uv - \int v\, du$$

REVIEW PROBLEMS

1. Find the antiderivative of

$$\int (2x + 1)(x^2 + x)^{-3}\, dx$$

2. Find $\displaystyle\int (e^{2x} + 3)^{-2} e^{2x}\, dx$

3. Find $\displaystyle\int_0^1 x(x^2 - 2)^3\, dx.$

4. Find $\displaystyle\int 3xe^{4x}\, dx$

5. Find $\displaystyle\int x^2(1 - x)^{1/3}\, dx$

6. Find the definite integral $\displaystyle\int_0^2 x(4x + 1)^{-1/2}\, dx.$

In Problems 7 to 9 use the table of integrals to find the definite integral.

7. $\displaystyle\int \frac{dx}{x(5x - 7)^2}$ **8.** $\displaystyle\int x(9x^4 + 16)^{1/2}\, dx$

9. $\displaystyle\int \frac{dx}{14 - 9x^2}$

10. Use the trapezoidal rule with four subintervals to approximate $\displaystyle\int_0^4 x^3\, dx.$

11. Use the trapezoidal rule with six subintervals to approximate $\displaystyle\int_0^3 e^{-x^2}\, dx.$

8 Functions of Several Variables

CHAPTER OVERVIEW

Many applications of calculus use variable quantities that are functions of more than one variable. For example, profit is a function of cost, price, and sales volume. The total population of a species in a given environment depends on its rate of reproduction, its death rate, the availability of food, and the population size of its predators.

In this chapter we study functions of more than one variable. The bulk of the material is on functions of two variables. Such functions are graphed in three dimensions so that it is possible to view them geometrically. In the first section we cover some examples of functions of several variables. The later sections show that many of the concepts that are developed for functions of one variable have analogous statements for functions of several variables.

In Section 8–2 we study the concept of rate of change of a function of several variables. It is an extension of the idea of the derivative of a function of one variable. Likewise in Section 8–3 the idea of relative extrema is covered. In Section 8–4 a constraint is added to the problem of finding relative extrema. These constraints, called Lagrange multipliers, have many applications to business and other disciplines.

The multivariable analog of the integral of a function of one variable is covered in the next section. We consider only functions of two variables, so the section is called "Double Integrals."

CASE STUDY PREVIEW

Jojoba oil is used for producing amides which, in turn, are an essential ingredient in many chemical processes. The procedure of extracting amides from jojoba oil is time-consuming and expensive. First the oil is refined and purified. Then other chemicals are added and the mixture is heated for several hours. Once the material becomes cakelike the amides can be extracted.

Researchers have tried various methods to maximize the amide yield while conserving cost to produce the cakelike material. The cost is a function of the two variables—temperature and time. The Case Study describes how the chemical industry uses this function of two variables to minimize the cost of producing amides.

8–1 EXAMPLES OF FUNCTIONS OF SEVERAL VARIABLES

Thus far we have considered only functions of one variable. Such functions are of the form $y = f(x)$ where x is the single independent variable. One value of the independent variable is substituted into the formula and a corresponding number is produced. For a function of two variables two numbers, given as an ordered pair, are substituted into a formula and another number is produced. In this section we consider examples of functions of two and more variables.

The Definition of a Function of Two Variables

An example of a function of two variables is

$$z = f(x, y) = 2x + 3y$$

The name of the function is f and z is called the **dependent variable.** The two variables x and y are **independent variables,** meaning that x and y can be chosen to be any numbers, while the corresponding value of z depends on which choice is made for x and y. For instance, let $x = 1$ and $y = -4$, then

$$z = f(1, -4) = 2(1) + 3(-4) = 2 - 12 = -10$$

Thus $z = -10$ when $x = 1$ and $y = -4$. This is succinctly stated by $f(1, -4) = -10$.

This leads to the definition of a function of two variables. The definition of functions of more than two variables is similar.

_____ **Definition of a Function of Two Variables** _____

The formula $z = f(x, y)$ defines f as a function of the two independent variables x and y if, for each ordered pair of real numbers (x, y) for which the formula is defined, there exists a unique corresponding value for the variable z.

The set of all ordered pairs of numbers for which $f(x, y)$ exists is called the *domain of f*. The set of all values of z that correspond to an ordered pair (x, y) is called the *range of f*. Usually the domain of a given function is not stated, and it is assumed that the domain is the largest set of ordered pairs for which the formula is defined. Occasionally, however, a restricted domain is given.

Sometimes the name of the function is omitted when only one function is being discussed. For example, the function above could have been given as simply $z = 2x + 3y$. If the dependent variable is not specifically mentioned, we assume its name is either z or the name of the function. That is, if the function above would have been given by the formula $f(x, y) = 2x + 3y$, where the dependent variable is not mentioned, it could be named either f or z.

EXAMPLE 1

Problem

If possible, evaluate the given functions at the points $(0, 0)$, $(-1, 2)$, and $(2, -3)$: (a) $f(x, y) = 4x - 5y + 2$ and (b) $g(x, y) = 2x^2 + xy - \dfrac{1}{y + 3}$.

SOLUTION (a) $f(0, 0) = 4(0) - 5(0) + 2 = 2$; $f(-1, 2) = 4(-1) - 5(2) + 2 = -12$; $f(2, -3) = 4(2) - 5(-3) + 2 = 8 + 15 + 2 = 25$.
(b) $g(0, 0) = 2(0)^2 + (0)(0) - 1/(0 + 3) = -1/3$; $g(-1, 2) = 2(-1)^2 + (-1)(2) - 1/(2 + 3) = 2 - 2 - 1/5 = -1/5$; $g(2, -3) = 2(2)^2 + (2)(-3) - 1/(-3 + 3) = 8 - 6 - 1/0$, which is not defined because $1/0$ is not defined, so $(2, -3)$ is not in the domain of g.

A function of three variables is defined similarly to a function of two variables. It differs only in that an ordered triple is substituted into the formula that has three variables. The next example gives an illustration of a function of three variables and one of four variables.

EXAMPLE 2

Define the functions

$$f(x, y, z) = x^2 - 4yz \qquad g(x, y, z, w) = xy + \frac{4z^2}{w}$$

Problem

(a) Find $f(2, 1, -1)$ and (b) $g(1, 2, -3, 2)$.

SOLUTION (a) $f(2, 1, -1) = (2)^2 - 4(1)(-1) = 4 + 4 = 8$.
(b) $g(1, 2, -3, 2) = (1)(2) + 4(-3)^2/2 = 2 + 18 = 20$.

Three-Dimensional Coordinate Systems

The graph of a function of two variables requires a three-dimensional coordinate system. This is because a point in the domain (a, b) is plotted in the xy-plane and

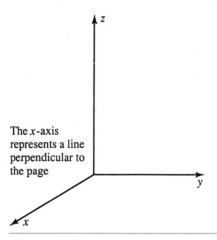

The x-axis represents a line perpendicular to the page

FIGURE 1

then its functional value c is plotted on a third axis. Thus the point (a, b, c) is plotted in a three-dimensional system. Figure 1 demonstrates the usual convention for labeling the three axes, called the x-, y-, and z-axes. The z-axis is vertical, the y-axis is horizontal, and the x-axis represents a line that is perpendicular to the plane of the paper. Since the figure is a two-dimensional depiction of a three-dimensional object, a perspective drawing is needed. That is why the x-axis is drawn at an angle. Only the positive half of each axis is drawn for clarity.

The point where the three axes intersect is called the *origin*. A point (a, b) in the xy-plane is given the three-dimensional coordinates $(a, b, 0)$. A point (a, b, c) is $|c|$ units above or below the point $(a, b, 0)$, depending on whether c is positive or negative. The coordinate system is divided into eight regions, called *octants*, four above the xy-plane and four below. Only one octant is named. It is the region where all three coordinates are positive, called the *first octant*. Several points in various octants are plotted in Figure 2.

Graphs of Functions of Two Variables

Since the domain of the function $z = f(x, y)$ is a set of ordered pairs (x, y), the graph of the function entails locating the points in the domain in the xy-plane. Then their corresponding functional values are plotted on the z-axis. Thus a point on the graph is an ordered triple (x, y, z). For example, consider the function $z = f(x, y) = 2x + 3y$. To locate the point on the graph corresponding to $(1, 2)$, first plot $(1, 2)$ in the xy-plane and locate $z = f(1, 2) = 8$ on the z-axis. Thus the point on the graph is $(1, 2, 8)$. This point is one unit out on the x-axis, two units to the right on the y-axis and eight units up on the z-axis. This point along with several others is plotted in Figure 3.

There is no easy method for graphing functions of two variables. The problem entails working algebraically with the formula to determine what type of two-dimensional figures arise when the graph is intersected with the graph of vertical

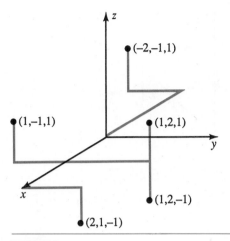

FIGURE 2

and horizontal planes. These intersections are then pieced together to get a perspective figure. The planes used in the perspective drawings are the coordinate planes and those that are parallel to the coordinate planes. For instance, the set of points with z-coordinate 1 is the plane one unit above the xy-plane. Its equation is $z = 1$. Likewise

> The graph of $x = a$ is a plane parallel to the yz-plane.
> The graph of $y = b$ is a plane parallel to the xz-plane.
> The graph of $z = c$ is a plane parallel to the xy-plane.

Figure 4 illustrates these planes.

The intersection of one of the coordinate planes with the graph of the function

FIGURE 3

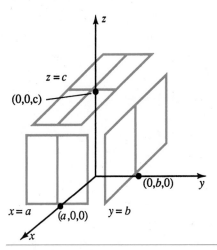

FIGURE 4

is called a **trace** of the graph. A **section** of a surface $z = f(x, y)$ is the intersection of the surface with a plane that is parallel to a coordinate plane. A section of the graph is found by solving simultaneously the two equations, the equation of the function and the equation of the plane. Thus a section is a two-dimensional figure in one of the planes $x = a$, $y = b$, or $z = c$.

The next example illustrates how to use sections to graph functions of two variables.

EXAMPLE 3

Problem

Graph the functions.
(a) $z = f(x, y) = 2x + 3y$, (b) $z = h(x, y) = x^2 + 4y^2$, and
(c) $z = g(x, y) = y^2 - x^2$.

SOLUTION (a) The sections of the graph corresponding to $x = a$ are the straight lines $z = 2a + 3y$, each in the plane $x = a$, which is parallel to the yz-plane. The sections of the graph corresponding to $y = b$ are the straight lines $z = 2x + 3b$, each in the plane $y = b$, which is parallel to the xz-plane. The sections of the graph corresponding to $z = c$ are the straight lines $c = 2x + 3y$, each in the plane $z = c$, which is parallel to the xy-plane. The graph is the plane sketched in Figure 5.
(b) The trace in the yz-plane is found by letting $x = 0$ in the equation. It is the parabola $z = 4y^2$. The trace in the xz-plane is the parabola $z = x^2$. The sections of the graph corresponding to $x = a$ are the parabolas $z = a^2 + 4y^2$. The sections of the graph corresponding to $y = b$ are the parabolas $z = x^2 + 4b^2$. The sections of the graph corresponding to $z = c$ are the ellipses $c = x^2 + 4y^2$. The graph is an infinite bowl-shaped surface, sketched in Figure 6. It is called an elliptic paraboloid.
(c) The trace in the yz-plane is the parabola $z = y^2$ (see Figure 7a) and the trace in the xz-plane is the parabola $z = -x^2$ (see Figure 7b). The sections of the graph corresponding to $x = a$ are the parabolas $z = y^2 - a^2$. The sections of

the graph corresponding to $y = b$ are the parabolas $z = b^2 - x^2$. The sections of the graph corresponding to $z = c$ are the hyperbolas $c = y^2 - x^2$. The graph is a saddle-shaped surface, sketched in Figure 7c. It is called a hyperbolic paraboloid.

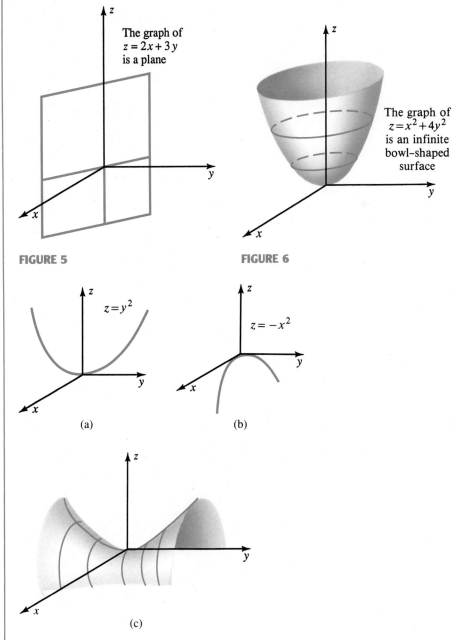

FIGURE 5

The graph of
$z = 2x + 3y$
is a plane

FIGURE 6

The graph of
$z = x^2 + 4y^2$
is an infinite
bowl-shaped
surface

$z = y^2$

(a)

$z = -x^2$

(b)

(c)

FIGURE 7 (a) The trace in the xy-plane is $z = y^2$. (b) The trace in the xz-plane is $z = -x^2$. (c) The graph of $z = y^2 - x^2$ is a saddle-shaped surface.

Applications of Sections

To see how sections are used to give a visual structure to a two-dimensional graph of a three-dimensional surface, consider the two graphs in Figure 8. The first graph appears in a book that investigates the future of the world supply of oil.* The

*Peter R. Odell and Kenneth E. Rosing, *The Future of Oil,* New York, Nichols Publishing, 1980, p. 182.

FIGURE 8

(a)

(b)

(a)   (b)

FIGURE 9

second is taken from a book describing the structure of sound in music.* The surface in Figure 8a describes the relationship between the likelihood that the world supply of oil will exceed demand. This likelihood depends on the amount of resources available (already taken from the earth) and the growth rate of demand. The surface in Figure 8b describes the relationship among amplitude, frequency, and time of a musical note.

Level Curves

Another method used to describe the graph of a function of two variables is to sketch the sections $z = c$ in the xy-plane. Each graph is called a **level curve.** Level curves are commonly used on television weather reports and in contour maps, also called topographical maps. For example, Figure 9a illustrates a picture of a mountain range and Figure 9b depicts its corresponding topographical map. The level curves are the intersections of horizontal planes ($z = c$) and the earth, and thus describe the heights of the mountains.

For the function $z = x^2 + 4y^2$, whose graph is sketched in Figure 6, the level curves are the sections corresponding to $z = c$. They are the ellipses graphed in Figure 10.

*Robert Erickson, *Sound Structure in Music*, Berkeley, CA, University of California Press, 1975, p. 62.

FIGURE 10

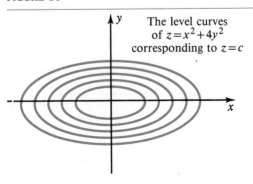

The level curves
of $z = x^2 + 4y^2$
corresponding to $z = c$

EXERCISE SET 8–1

• In Problems 1 to 4 find the functional value of $f(x, y) = 2x^2 - y^2$ at the given point.

1. $(1, 1)$ 2. $(-1, 2)$

3. $(0, -3)$ 4. $(-4, 0)$

In Problems 5 to 8 find the functional value of $f(x, y) = y(x^3 + 1)^{1/2} + xy^3$ at the given point.

5. $(0, 2)$ 6. $(0, -3)$

7. $(2, 3)$ 8. $(2, -3)$

In Problems 9 to 12 find the domain of the function.

9. $f(x, y) = \dfrac{y}{x - 2}$

10. $f(x, y) = \dfrac{2}{x(y^2 - 1)}$

11. $f(x, y) = \dfrac{y}{(x^3 + 1)^{1/2}}$

12. $f(x, y) = \dfrac{2}{x(y - 1)^{1/2}}$

In Problems 13 to 20 graph the functions, giving the traces and at least one section that is parallel to each coordinate plane.

13. $f(x, y) = 3x + 5y$ 14. $f(x, y) = y - 4x$

15. $f(x, y) = 2 + 2x - 3y$

16. $f(x, y) = y - 2x - 3$

17. $f(x, y) = 4 + x^2 + 5y^2$

18. $f(x, y) = 1 + 2x^2 + y^2$

19. $f(x, y) = y^2 - 5x^2$ 20. $f(x, y) = x^2 - 2y^2$

•• In Problems 21 to 26 graph the functions.

21. $f(x, y) = 1 - x^2 - y^2$

22. $f(x, y) = 2 - 2x^2 - y^2$

23. $f(x, y) = x^2 + 2x + y^2 + 4y$

24. $f(x, y) = 1 + x^2 + 2x + y^2$

25. $f(x, y) = 2x + x^2$ 26. $f(x, y) = y^2 - 4y$

27. The equation of the sphere having a radius of one unit and its center at the origin is $x^2 + y^2 + z^2 = 1$. Is it possible to represent this sphere as a function?

28. Solve the equation $x^2 + y^2 + z^2 = 4$ for z and then keep only the positive square root. Graph the resulting function using the fact about spheres stated in Problem 27.

29. Give a geometrical test using lines perpendicular to the xy-plane to determine whether a given surface is a function $z = f(x, y)$. Recall the geometric test to determine whether a two-dimensional curve is a function.

30. A salesperson hires two employees. One earns \$10 per hour and the second earns \$12 per hour. Let the number of hours worked per week by the first employee be represented by x and the number of hours worked by the second employee be represented by y. If the salesperson's fixed cost is \$300 per week, find a formula for the cost per week.

31. A manufacturing company makes three types of desks, small, medium, and large. If the profit is \$40 for each small desk sold, \$50 for each medium desk sold, and \$60 for each large desk sold, find a formula for the profit by letting x represent the number of small desks sold, y the number of medium desks sold, and z the number of large desks sold.

••• 32. Accurate forecasting of ambulance demand directly affects staffing, inventories, and assignment of units within the region served by the ambulance corps. In an article studying ambulance service to four counties in South Carolina, Baker and Fitzpatrick developed a model that accurately forecasted ambulance demand while minimizing cost.* The cost function to be minimized was

---

*J. R. Baker and K. E. Fitzpatrick, "Determination of an Optimal Forecast Model for Ambulance Demand Using Goal Programming," *Journal of the Operations Research Society*, Vol. 37, 1986, pp. 1047–1059.

$$C(x, y, z, w) = 103x + 109y + 116z + 117w$$

where the cost per call for an ambulance was $103 in Spartanburg County, $109 in Clarendon County, $116 in Horry County, and $117 in Florence County; and x, y, z and w were the number of calls per day in each county, respectively. For a given week in 1983 in the study, the number of calls in each county was as follows: (a) on Monday, Spartanburg had 670 calls, Clarendon had 370 calls, Horry had 37 calls, Florence had 320 calls; (b) on Tuesday, Spartanburg had 690 calls, Clarendon had 300 calls, Horry had 38 calls, Florence had 290 calls; (c) on Wednesday, Spartanburg had 680 calls, Clarendon had 320 calls, Horry had 40 calls, Florence had 300 calls. Which day produced the least total cost?

33. Most large industries utilize research and development to improve their products and their marketing techniques. In an article evaluating the cost-effectiveness of research in the Florida citrus industry, Stranahan and Shonkwiler centered on the Florida frozen concentrate orange juice market.[†] The cost C was assumed to be a function of the output Y, the quantity of research R, and the price P. In one part of the study the cost was computed to be

$$C(Y, R, P) = 1.1Y^{.5}R^{.43}P^{-.009}$$

Compute $C(Y, R, P)$ when $Y = 10$ million gallons per day, $R = .2$ million dollars per day, and $P = \$1$ per gallon.

34. Many studies have been done on the harmful effects of large doses of radiation, such as the tragic accident at Chernobyl, U.S.S.R., in 1987. Johnson studied harmful effects of exposure to radiation at very low level dose rates which result from various environmental sources of radiation.[‡] These dosages are called high natural radiation background. They occur from diverse sources such as sun rays and wristwatches. The article investigated data from the World Health Organization based on radiation reports from 45 countries. One part of the article studied cancer rates. The number of years required for detection y was found to be a function of the background radiation rate r, measured in rads per year, and the average cancer rate p, measured in cases per million persons per year. Thus

$$y(r, p) = \frac{kp}{r^2}$$

where k is a constant depending upon the country or region being studied. Compute $y(.88, .02)$ for $k = .04$.

[†]H. A. Stranahan and J. S. Shonkwiler, "Evaluating the Returns to Postharvest Research in the Florida Citrus Processing Subsector," *American Journal of Agricultural Economics,* Vol. 68, 1986, pp. 88–94.

[‡]Richard E. Johnson, "Problems Involved in Detecting Increased Malignancy Rates in Areas of High Natural Radiation Background," *Health Physics,* Vol. 31, 1976, pp. 148–160.

8–2 PARTIAL DERIVATIVES

One of the most important concepts of calculus is rate of change. For a function of one variable, rate of change is measured by the derivative. For a function of two variables, rate of change can be measured in many ways so that the situation is more complicated. In this section we study the three-dimensional analog of the derivative, the partial derivative.

An Application from Business

Rate of change is used extensively in business. The demand function for a particular company or industry often is expressed as a function of more than one variable.

For instance, in an article studying the effect of the ''health scare'' on the cigarette industry, Bishop and Yoo consider the quantity demanded Q as a function of disposable income D and cigarette advertising expenditures A.* To model the industry's data, the authors use a Cobb–Douglas function of the form

$$Q(D, A) = 231.7D^{0.66}A^{0.07}$$

The article studies the effect on Q when D and A experience increases and decreases. Bishop and Yoo use several types of measures, each of which depends on the partial derivatives of Q. At the end of the section we compute these partial derivatives.

Partial Derivatives: A Geometric Description

The derivative of $y = f(x)$ measures how a change in x affects y. For a function of two variables $z = f(x, y)$ the partial derivatives measure how a change in x or in y affects z. Geometrically the derivative $f'(a)$ is the slope of the tangent line to the curve $y = f(x)$ at $x = a$. But a surface in three dimensions can have a tangent line in any direction, so there are an infinite number of tangent lines. (See Figure 1.) It suffices to study tangent lines in only two directions, the directions parallel to the x-axis and parallel to the y-axis. This is because the tangent lines in every other direction can be expressed in terms of these two.

Consider the direction parallel to the x-axis. How can we interpret the rate of change of $z = f(x, y)$ at the point (a, b) in this direction? The section of the graph in this direction is a two-dimensional curve obtained by taking the intersection of $z = f(x, y)$ with the plane $y = b$. The equation of this curve is $z = f(x, b)$. Since this curve is two-dimensional, the slope of the tangent line at (a, b) is the derivative of the function of one variable, $z = f(x, b)$. This is called the partial derivative of $z = f(x, y)$ with respect to x. It is found by holding y constant and finding the derivative with respect to x.

*John A. Bishop and Jang H. Yoo, '' 'Health Scare,' Excise Taxes and Advertising in the Cigarette Demand and Supply,'' *Southern Economics Journal*, Vol. 52, 1985, pp. 402–411.

FIGURE 1

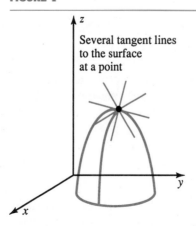

z

Several tangent lines
to the surface
at a point

y

x

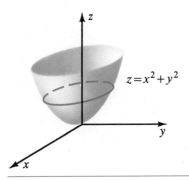

FIGURE 2

For example, consider the function

$$z = f(x, y) = x^2 + y^2$$

Its graph, given in Figure 2, is an infinite bowl-shaped surface. To find the rate of change at $(1, 0)$ in the direction that is parallel to the x-axis, first find the intersection of $y = 0$ and the curve. It is $z = f(x, 0) = x^2 + 0 = x^2$. The derivative of this two-dimensional function is $2x$. This is expressed as

$$f_x(x, y) = 2x$$

It is called the *partial derivative of f with respect to x*. To get the slope of the tangent line at $(1, 0)$, substitute $x = 1$ into the formula for f_x and get $f_x(1, 0) = 2$. From Figure 3 we see that the function $z = f(x, 0)$ is a parabola in the xz plane whose tangent line at $(1, 0)$ is 2.

Partial Derivatives: Formal Definition

The definition of the partial derivative uses the fact that if one variable is held constant the resulting expression is a function of one variable, and so its derivative

FIGURE 3

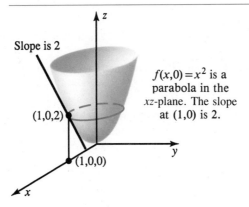

$f(x,0) = x^2$ is a parabola in the xz-plane. The slope at $(1,0)$ is 2.

can be computed. Thus a partial derivative of $f(x, y)$ is the derivative of a function of one variable that is obtained by holding constant one of the two variables of $f(x, y)$.

Partial Derivative

The **partial derivative** of $f(x, y)$ with respect to x is defined as

$$f_x(x, y) = \lim_{h \to 0} \frac{f(x + h, y) - f(x, y)}{h}$$

The partial derivative of $f(x, y)$ with respect to y is defined as

$$f_y(x, y) = \lim_{h \to 0} \frac{f(x, y + h) - f(x, y)}{h}$$

There are several alternate notations for the partial derivatives. If it is clear that f is a function of the two variables x and y, then sometimes $f_x(x, y)$ is expressed as simply f_x. Similarly f_y represents the partial derivative of f with respect to y. The notation for the partial derivative that is similar to the $\frac{dy}{dx}$ notation for the derivative is

$$\frac{\partial f}{\partial x} \quad \text{and} \quad \frac{\partial f}{\partial y}$$

which represent the partial derivative with respect to x and y, respectively. Example 1 illustrates how to compute the partial derivatives of functions of two variables.

EXAMPLE 1

Problem

Compute $f_x(x, y)$ and $f_y(x, y)$ for the functions (a) $f(x, y) = x^3 + xy - y^{-1}$ and (b) $f(x, y) = x^2 y^{-3} - e^{2xy} + 3$.

SOLUTION (a) To find f_x treat y as a constant and differentiate with respect to x.

$$f_x(x, y) = 3x^2 + y$$

Since y is held constant, the term y^{-1} is treated as a constant so that the partial derivative of it is 0.

To find f_y, treat x as a constant and differentiate with respect to y.

$$f_y(x, y) = x - (-1)y^{-2} = x + y^{-2}$$

(b) To find f_x, treat y as a constant and differentiate with respect to x.

$$f_x(x, y) = 2xy^{-3} - 2ye^{2xy}$$

To find f_y, treat x as a constant and differentiate with respect to y.

$$f_y(x, y) = -3x^2 y^{-4} - 2xe^{2xy}$$

The next example demonstrates how to incorporate the chain rule with partial derivatives.

EXAMPLE 2

Problem

Consider the function $f(x, y) = (x^3 + xy^{-1})^4$. Compute (a) $f_x(x, y)$ and (b) $f_y(x, y)$.

SOLUTION (a) Use the extended power rule. Let $u = x^3 + xy^{-1}$ so that the partial derivative of u with respect to x is $3x^2 + y^{-1}$. Treat y as a constant and differentiate with respect to x.

$$f_x(x, y) = 4(x^3 + xy^{-1})^3(3x^2 + y^{-1})$$

(b) Again let $u = x^3 + xy^{-1}$. Then the partial derivative of u with respect to y is $-xy^{-2}$. Treat x as a constant and differentiate with respect to y.

$$f_y(x, y) = 4(x^3 + xy^{-1})^3(-xy^{-2})$$

If the partial derivative with respect to x is to be computed at a particular point (a, b), then the notation used is

$$f_x(a, b)$$

Likewise $f_y(a, b)$ represents the partial derivative with respect to y evaluated at the point (a, b). The next example illustrates this notation.

EXAMPLE 3

Problem

Consider the function $f(x, y) = 2x^3 + xy^2 - y$. Compute (a) $f_x(1, 3)$ and (b) $f_y(-1, 2)$.

SOLUTION (a) First find $f_x(x, y)$.

$$f_x(x, y) = 6x^2 + y^2$$

Next substitute the values $x = 1$ and $y = 3$.

$$f_x(1, 3) = 6(1)^2 + (3)^2 = 6 + 9 = 15$$

(b) Find $f_y(x, y)$.

$$f_y(x, y) = 2xy - 1$$

Next substitute the values $x = -1$ and $y = 2$.

$$f_y(-1, 2) = 2(-1)(2) - 1 = -4 - 1 = -5$$

Second-Order Partial Derivatives

Just as we can define higher-order derivatives of functions of one variable, so too we can define higher-order partial derivatives. The partial derivative of a partial derivative is called a second-order partial derivative, or simply a **second partial derivative.** The four second partial derivatives of $f(x, y)$ are

$f_{xx}(x, y)$
$f_{xy}(x, y)$
$f_{yx}(x, y)$
$f_{yy}(x, y)$

Just as with first-order partial derivatives, the shortened form of $f_{xx}(x, y)$ is f_{xx}. Also f_{xy}, f_{yx}, and f_{yy} are defined in a similar way. The functions f_{xy} and f_{yx} are called **mixed partial derivatives.** The notation for f_{xy} states that the partial derivative of $f(x, y)$ with respect to x is to be computed first, followed by computing the partial derivative of $f_x(x, y)$ with respect to y; f_{yx} calls for the reverse order of differentiation. For most functions that we will study these two functions are equal. The next example illustrates how to compute second partial derivatives.

EXAMPLE 4 | **Problem**

Consider the function $f(x, y) = x^3 y + xy^{-2} - ye^{3x}$. Compute (a) $f_{xx}(x, y)$, (b) $f_{xy}(x, y)$, (c) $f_{yx}(x, y)$, (d) $f_{yy}(x, y)$, (e) $f_{xx}(1, 3)$, and (f) $f_{xy}(0, 1)$.

SOLUTION First find $f_x(x, y)$ and $f_y(x, y)$.

$f_x(x, y) = 3x^2 y + y^{-2} - 3ye^{3x}$
$f_y(x, y) = x^3 - 2xy^{-3} - e^{3x}$

(a) To find f_{xx}, compute the partial derivative of f_x with respect to x.

$$f_{xx}(x, y) = \frac{\partial}{\partial x} f_x(x, y) = \frac{\partial}{\partial x} (3x^2 y + y^{-2} - 3ye^{3x})$$

$$= 6xy - 9ye^{3x}$$

(b) To find f_{xy} compute the partial derivative of f_x with respect to y.

$$f_{xy}(x, y) = \frac{\partial}{\partial y} f_x(x, y) = \frac{\partial}{\partial y} (3x^2 y + y^{-2} - 3ye^{3x})$$

$$= 3x^2 - 2y^{-3} - 3e^{3x}$$

(c) To find f_{yx}, compute the partial derivative of f_y with respect to x.

$$f_{yx}(x, y) = \frac{\partial}{\partial x} f_y(x, y) = \frac{\partial}{\partial x} (x^3 - 2xy^{-3} - e^{3x})$$

$$= 3x^2 - 2y^{-3} - 3e^{3x}$$

Notice that $f_{xy} = f_{yx}$.
(d) To find f_{yy}, compute the partial derivative of f_y with respect to y.

$$f_{yy}(x, y) = \frac{\partial}{\partial y} f_y(x, y) = \frac{\partial}{\partial y} (x^3 - 2xy^{-3} - e^{3x})$$

$$= 6xy^{-4}$$

(e) To find $f_{xx}(1, 3)$ substitute the values $x = 1$ and $y = 3$ into $f_{xx}(x, y)$.

$$f_{xx}(1, 3) = 6(1)(3) - 9(3)e^{3(1)}$$
$$= 18 - 27e^3$$

(f) To find $f_{xy}(0, 1)$, substitute the values $x = 0$ and $y = 1$ into $f_{xy}(x, y)$.

$$f_{xy}(0, 1) = 3(0)^2 - 2(1)^{-3} - 3e^{3(0)}$$
$$= -2 - 3 = -5$$

The Cigarette Industry Revisited

The demand function computed by Bishop and Yoo for the cigarette industry was given in the beginning of the section as

$$Q(D, A) = 231.7D^{0.66}A^{0.07}$$

The analysis of the industry in the article uses several measures of the effect on Q by changes in D and A. These measures are based on the partial derivatives of Q. The investigators reached various conclusions concerning when it would be most advantageous for the industry to increase advertising in order to increase demand. The next example computes these partial derivatives.

EXAMPLE 5 | **Problem**

Find (a) Q_D and (b) Q_A for the above demand function.

SOLUTION (a) Treat A as a constant.

$$Q_D = 231.7(0.66)D^{0.66 - 1}A^{0.07}$$
$$= 152.9D^{-0.34}A^{0.07}$$

(b) Treat D as a constant.

$$Q_A = 231.7(0.07)D^{0.66}A^{0.07 - 1}$$
$$= 16.22D^{0.66}A^{-0.93}$$

EXERCISE SET 8–2

● In Problems 1 to 14 find $f_x(x, y)$ and $f_y(x, y)$.

1. $f(x, y) = 4 + 3x^2 + 5y^{1/2}$

2. $f(x, y) = 1 + 2x^{2/3} + y^{-2}$

3. $f(x, y) = xy^2 - 5x^2$

4. $f(x, y) = x^2 - 2x^3y + y^3$

5. $f(x, y) = xy^2 - 3x^2y$

6. $f(x, y) = x^2y - 2x^2y^3$

7. $f(x, y) = x^{-2} + xy^{-2} - x^2y^3 + 4y$

8. $f(x, y) = x^{1/2} - x^{1/3}y + 3xy^{-3} + y$

9. $f(x, y) = xe^{2y} - \ln(2x + 3y)$

10. $f(x, y) = ye^{3x} + 4e^{(x-y)}$

11. $f(x, y) = (2x + 3y)^4$

12. $f(x, y) = (x^2 + 5y)^3$

13. $f(x, y) = (2x + 4xy)^{-2}$

14. $f(x, y) = (x^2y - 5x)^{1/3}$

In Problems 15 to 18 find $f_x(1, 0)$ and $f_y(2, 1)$.

15. $f(x, y) = 3xy^3 + 5x^{-1}y$

16. $f(x, y) = 4xy^4 - 6x^{-2}y + 3$

17. $f(x, y) = 2x^{1/2} - 3xy^3 + e^y$

18. $f(x, y) = x^3y + 4x^2y^2 + 6e^{2y}$

In Problems 19 to 22 find all second-order partial derivatives.

19. $f(x, y) = 3x^2 + 5y^{-2}$

20. $f(x, y) = 2x^{2/3} + y^{-1/2}$

21. $f(x, y) = 4 + 3x^2y + 5xe^{2y}$

22. $f(x, y) = 1 + 2x^{2/3}y + x \ln(1 + y^{-2})$

•• In Problems 23 and 24 find all second-order partial derivatives.

23. $f(x, y) = xy^2e^x + e^{2x}$

24. $f(x, y) = 2x^3ye^{3y}$

In Problems 25 and 26 find the values of x and y such that $f_x(x, y) = 0$ and $f_y(x, y) = 0$.

25. $f(x, y) = 1 - x^2 + y^2 + 2xy - 4y$

26. $f(x, y) = 2 - 2x^2 - y^2 + xy + 7x$

In Problems 27 to 30 find f_{xxx}, f_{xxy}, and f_{xyy}.

27. $f(x, y) = x^4 - x^3 + 2x + 2y^5 + 4y^2$

28. $f(x, y) = 1 + x^5 + x^{-1} + y^2$

29. $f(x, y) = 2xy + x^2y - xe^{2y}$

30. $f(x, y) = x^{-2} + xy^{-2} - \ln(y - 1)$

••• 31. Give a definition of the partial derivative with respect to x of a function of three variables $w = f(x, y, z)$.

In Problems 32 and 33 find f_x and f_y by holding constant all variables except the one with which the partial derivative is being taken.

32. $f(x, y, z) = x^2 + x^3z - 2yz^3 + xyz$

33. $f(x, y, z) = 3x^{-2} + x^{1/3}z - x^2z^3 + xyz$

34. Find $f_{xyz}(1, 2, 3)$ for $f(x, y, z) = x^2 - yze^x - 4xyz^2$

35. Show that the function $f(x, y) = x^3 + 5x^2 + e^y$ satisfies the partial differential equation $f_{xy} = 0$.

36. Show that the function $f(x, y) = x^3 - 3xy^2$ satisfies the partial differential equation $f_{xx} + f_{yy} = 0$.

37. In an article studying the economics of urban bus transportation Obeng* gathered data from 62 bus systems. The cost C was treated as a function of demand Q, fuel price F, the price of labor L, and the price of capital P. The data can be described by a Cobb–Douglas function of the form $C = 1.7Q^{1.1}F^{1.5}L^{0.36}P^{0.5}$. Find the partial derivatives of C.

38. In an article studying the optimal use of cobalt beams in treating cancer, Gantchew defines the dose distribution d along a straight line perpendicular to the central ray as a function of the width x of the ray and the length of penetration y as follows:†

$$d(x, y) = 0.93e^{[70x/(60+y)]^{2.7}}$$

Compute d_x.

39. In their article on advertising in the cigarette industry, Bishop and Yoo also expressed price P as a function of disposable income D and cigarette advertising expenditures A.‡ In one part of the article the following function was defined:

$$P(D, A) = 0.125D^{0.56}A^{0.06}$$

Compute (a) P_D and (b) P_A.

---

*Kofi Obeng, "The Economics of Bus Transit Operation," *The Logistics and Transportation Review*, Vol. 20, 1986, pp. 45–63.

†M. G. Gantchew, "On the Application of a Simple Model of Cobalt-60 Beams," *Physics in Medicine and Biology*, Vol. 24, 1979, pp. 443–446.

‡John A. Bishop and Jang H. Yoo, " 'Health Scare,' Excise Taxes and Advertising Ban in the Cigarette Demand and Supply," *Southern Economics Journal*, Vol. 52, 1985, pp. 402–411.

8–3 RELATIVE EXTREMA

Many applications of functions of more than one variable involve finding a value of a function that is a relative maximum or minimum. We concentrate on functions of two variables because they have a firm geometrical base. The concepts can be extended to functions of more than two variables. The ideas in this section are similar to corresponding results for functions of one variable. For instance, to compute relative extrema we use a theorem with first and second partial derivatives that is similar to the second derivative test for functions of one variable.

Definitions

The definition of a relative maximum and a relative minimum for a function of two variables is similar to the definitions for a function of one variable. The difference is that there are two independent variables instead of one. This means that $f(a, b)$ is a relative maximum if it is larger than all functional values of points (x, y) close to (a, b), whereby "close to" is meant within a circular region about (a, b). Geometrically, a relative maximum resembles a mountain peak or the top of a dome. A relative minimum resembles the bottom of a valley.

_____ Relative Maximum and Relative Minimum _____

A function $f(x, y)$ has a **relative maximum** at (a, b) if

$$f(a, b) \geq f(x, y)$$

for all (x, y) in some circular region around (a, b); $f(x, y)$ has a **relative minimum** at (a, b) if

$$f(a, b) \leq f(x, y)$$

for all (x, y) in some circular region around (a, b). The term **relative extremum** refers to either a relative maximum or a relative minimum.

Example 1 illustrates the definition.

EXAMPLE 1 | **Problem**

Determine from the graphs given in Figure 1 if the function has a relative extremum at $(0, 0)$: (a) $f(x, y) = x^2 + 4y^2$ and (b) $f(x, y) = y^2 - x^2$.

SOLUTION (a) The graph in Figure 1a shows that all functional values $f(x, y) \geq 0$ for all (x, y), so that any circular region can be chosen to show that $f(x, y)$ has a relative maximum at $(0, 0)$.
(b) Figure 1b shows that for points $(x, 0)$ on the x-axis, $f(x, 0) = -x^2 \leq 0$, and for points $(0, y)$ on the y-axis, $f(0, y) = y^2 \geq 0$. Any circular region containing $(0, 0)$ must contain points on each axis, so that some functional values will be greater than $f(0, 0) = 0$ and some will be smaller. Therefore $f(x, y)$ has neither a relative maximum nor a relative minimum at $(0, 0)$.

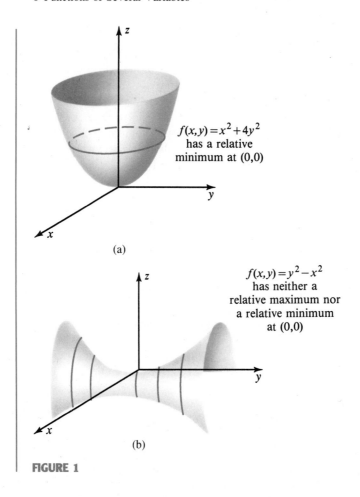

$f(x,y)=x^2+4y^2$
has a relative
minimum at (0,0)

(a)

$f(x,y)=y^2-x^2$
has neither a
relative maximum nor
a relative minimum
at (0,0)

(b)

FIGURE 1

Critical Points

Recall that a function of one variable has a relative extremum at a point only if its derivative is 0 at the point, or if the derivative fails to exist. Figure 2 demonstrates that if a function of two variables has a relative maximum at (a, b), then the tangent lines to the graph at (a, b) must be horizontal, meaning that their slopes are 0. Since the partial derivative measures the slope of the tangent line, $f_x(a, b)$ and $f_y(a, b)$ must be 0. The same holds for relative minima. This is recorded in the following result.

If $f(x, y)$ has a relative extremum at (a, b) and if both partial derivatives exist, then

$$f_x(a, b) = 0 = f_y(a, b)$$

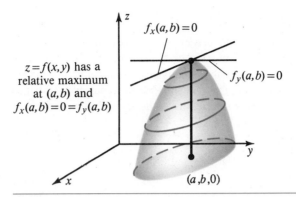

$z=f(x,y)$ has a
relative maximum
at (a,b) and
$f_x(a,b)=0=f_y(a,b)$

$f_x(a,b)=0$

$f_y(a,b)=0$

$(a,b,0)$

FIGURE 2

Extrema can also occur when a partial derivative fails to exist, but we will not consider this case.

A point (a, b) such that $f_x(a, b) = 0 = f_y(a, b)$ is called a **critical point.** The first step in locating relative extrema is to find the critical points. However, just as with functions of one variable, not all critical points correspond to relative extrema. For example, consider the function in Example 1(b), $f(x, y) = y^2 - x^2$. The partial derivatives are $f_x(x, y) = -2x$ and $f_y(x, y) = 2y$, so that each is 0 at $(0, 0)$, but the graph in Figure 1b shows that f has neither a relative maximum nor a relative minimum. This is an example of a **saddle point,** that is, a point where the partial derivatives are 0 but the function does not have a relative extremum at the point.

To find the critical points of a function, we need to solve $f_x = 0$ and $f_y = 0$ simultaneously, meaning that the same point (a, b) must satisfy both equations. This often requires that a system of two equations in two unknowns be solved, as illustrated in the next example.

EXAMPLE 2

Problem

Find the critical points of the function

$$f(x, y) = x^2 - 4x + y^2 - 6y$$

SOLUTION First compute the partial derivatives.

$$f_x(x, y) = 2x - 4$$
$$f_y(x, y) = 2y - 6$$

Set each partial derivative equal to 0.

$$f_x(x, y) = 2x - 4 = 0$$
$$f_y(x, y) = 2y - 6 = 0$$

This is a system of two equations in two unknowns. The first equation implies that $x = 2$ and the second implies $y = 3$. Thus $(2, 3)$ is the solution of the system of equations. Therefore $f_x(2, 3) = 0$ and $f_y(2, 3) = 0$, which means that $(2, 3)$ is the only critical point.

The Second Partial Derivative Test

According to the earlier result, if the function $f(x, y)$ in Example 2 has a relative extremum, it must occur at the critical value $(2, 3)$. But this does not mean that there is necessarily a relative extremum at $(2, 3)$; the point could be a saddle point. The following theorem shows how to determine whether critical points are relative extrema. It is referred to as the **second partial derivative test.**

Second Partial Derivative Test

Suppose $z = f(x, y)$ has first and second partial derivatives in a circular region around (a, b) where (a, b) is a critical point; that is

$$f_x(a, b) = 0 = f_y(a, b)$$

To simplify notation, let

$$A = f_{xx}(a, b), \quad B = f_{yy}(a, b) \quad \text{and} \quad C = f_{xy}(a, b)$$

Then
1. If $A > 0$ and $AB - C^2 > 0$, then $f(a, b)$ is a relative minimum.
2. If $A < 0$ and $AB - C^2 > 0$, then $f(a, b)$ is a relative maximum.
3. If $AB - C^2 < 0$, then $f(a, b)$ is not a relative extremum; it is a saddle point.
4. If $AB - C^2 = 0$, then the test yields no conclusion.

The next three examples show how to apply the test.

EXAMPLE 3

Problem

Find the relative extrema of the function $f(x, y) = x^2 - 4x + y^2 - 6y$.

SOLUTION In Example 2 it was shown that the only critical point of $f(x, y)$ is $(2, 3)$. To apply the second partial derivative test, compute $f_{xx}(x, y) = 2$, $f_{yy}(x, y) = 2$, and $f_{xy}(x, y) = 0$. Then evaluate each at the critical point, which is immediate because each partial derivative is a constant.

$$A = f_{xx}(2, 3) = 2$$
$$B = f_{yy}(2, 3) = 2$$
$$C = f_{xy}(2, 3) = 0$$

Since $A > 0$ and $AB - C^2 = 2 \cdot 2 - 0 = 4 > 0$, part 1 of the test applies so that $f(2, 3) = -13$ is a relative minimum.

EXAMPLE 4

Problem

Find the relative extrema of the function

$$f(x, y) = x^2 + 4x - 2y^3 + 3y^2$$

SOLUTION First find the critical points by setting the partial derivatives equal to 0.

$$f_x(x, y) = 2x + 4$$
$$f_y(x, y) = -6y^2 + 6y$$
$$f_x = 0 = 2x + 4 \tag{1}$$
$$f_y = 0 = -6y^2 + 6y \tag{2}$$

Equation 1 implies that $x = -2$ and Equation 2 can be factored into

$$-6y(y - 1) = 0$$

This implies that $y = 0, 1$. Thus there are two critical points, $(-2, 0)$ and $(-2, 1)$. We handle each critical point separately after the second partial derivatives are computed.

$$f_{xx}(x, y) = 2$$
$$f_{yy}(x, y) = -12y + 6$$
$$f_{xy}(x, y) = 0$$

(a) Consider the critical point $(-2, 0)$. Then

$$A = f_{xx}(-2, 0) = 2$$
$$B = f_{yy}(-2, 0) = 6$$
$$C = f_{xy}(-2, 0) = 0$$

Since $A > 0$ and $AB - C^2 = 2 \cdot 6 - 0 = 12 > 0$, part 1 of the test applies so that $f(-2, 0) = -4$ is a relative minimum.

(b) Consider the critical point $(-2, 1)$. Then

$$A = f_{xx}(-2, 1) = 2$$
$$B = f_{yy}(-2, 1) = -6$$
$$C = f_{xy}(-2, 1) = 0$$

Then $AB - C^2 = 2(-6) - 0 = -12 < 0$, and so part 3 of the test applies. Thus $f(x, y)$ has a saddle point at $(-2, 1)$.

Sometimes the expressions for f_x and f_y contain both x and y, which makes solving for the critical points a bit more complicated, as illustrated in the next example.

EXAMPLE 5

Problem

Find the relative extrema of the function $f(x, y) = x^2 - 5x - y^2 + xy$.

SOLUTION First find the critical points by setting the partial derivatives equal to 0.

$$f_x(x, y) = 2x - 5 + y = 0$$
$$f_y(x, y) = x - 2y = 0$$

The second equation gives $x = 2y$. Substituting this equation into the equation $f_x = 0$ yields

$$2(2y) - 5 + y = 0$$
$$5y = 5$$
$$y = 1$$

Substituting $y = 1$ into $x = 2y$ yields $x = 2$. Therefore the critical point is $(2, 1)$. Next compute

$$A = f_{xx}(2, 1) = 2$$
$$B = f_{yy}(2, 1) = -2$$
$$C = f_{xy}(2, 1) = 1$$

Then $AB - C^2 = 2(-2) - 1 = -5 < 0$, and so part 3 of the test applies. Thus $f(x, y)$ has a saddle point at $(2, 1)$.

Part 4 of the test needs clarification. If $AB - C^2 = 0$, the second partial derivative test cannot determine whether the critical point is a relative extremum. It is necessary to use another test. For instance, all of the points close to the critical point can be tested. If they are all greater than $f(a, b)$, then $f(a, b)$ is a relative minimum; if they are all less than $f(a, b)$, then $f(a, b)$ is a relative maximum; if there are always some that are greater and some that are less than $f(a, b)$, then $f(x, y)$ has a saddle point at (a, b). This is usually a difficult task, however, and we will not pursue it.

EXERCISE SET 8–3

In Problems 1 to 6 determine whether the graph of the function has a relative extremum at $(0, 0)$.

1.

2.

3.

6.

4.

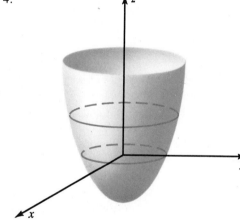

In Problems 7 to 12 find the critical points of the function.

7. $f(x, y) = x^2 + 6x + y^2 + 8y$

8. $f(x, y) = x^2 - 2x + y^2 + 6y$

9. $f(x, y) = 2x^2 + 4xy + y^2 - 6y + 1$

10. $f(x, y) = 3x^2 - 6xy + 2y^2 + 10y - 2$

11. $f(x, y) = x^2 - 2y^3 + 6y$

12. $f(x, y) = x^3 - 3x + 3y^2 + 12y$

In Problems 13 to 24 find the relative extrema and any saddle points of the function.

13. $f(x, y) = x^2 + 4x - 2y^3$

14. $f(x, y) = x^3 + 3y^2 + 12y$

15. $f(x, y) = x^2 - 8x + 2y^3 + 6y^2$

16. $f(x, y) = x^3 - 3x^2 + 3y^2 + 12y$

17. $f(x, y) = x^2 - 2y^3 + 6y$

18. $f(x, y) = x^3 - 3x + 3y^2 + 12y$

19. $f(x, y) = x^2 - 2x - 2xy + 2y^2 - 4y$

20. $f(x, y) = x^2 - 2xy + 3y^2 + 12y$

21. $f(x, y) = x - xy + y^3 - 3y^2$

22. $f(x, y) = x^2 + 4x - 2xy + 12y$

23. $f(x, y) = x^3 - 2x^2 - 2xy + y^2$

24. $f(x, y) = x^2 - 2xy - y^3 + 14y$

5.

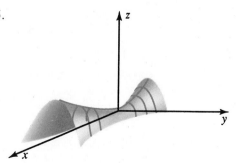

•• In Problems 25 to 30 find the relative extrema of the function.

25. $f(x, y) = x^3 - 2y^3 + 6y$

26. $f(x, y) = x^3 + 3y^3 - 9y$

27. $f(x, y) = x^3 - 6x^2 - 2y^3 + 24y$

28. $f(x, y) = x^3 + 3y^3 - 9y^2$

29. $f(x, y) = x^2 - 8x + 2y^3 + 6y^2 - 3y$

30. $f(x, y) = x^3 - 6x^2 - 8x + 3y^2 + 12y$

31. A firm utilizes x million dollars per year in labor cost and y million dollars per year in production cost. The cost function is

$$C(x) = 20 - 24x - 12y + x^2 + 3y^2$$

Find the amount that the firm should spend on labor and production each year to minimize cost.

32. A firm produces two types of radios, AM and AM–FM. If x thousand of AM radios and y thousand AM–FM radios are made per month, the revenue $R(x)$ and cost $C(x)$ functions are given as follows, where $R(x)$ and $C(x)$ are measured in millions of dollars:

$$R(x) = 5x + 6y$$
$$C(x) = 10 - 5x - 12y + x^2 + 2y^2$$

Find how many radios of each type should be produced to maximize profit.

33. A manufacturing firm wants to package its product in a rectangular box with no top. If the volume of the box is to be 64 cubic inches, find the dimensions that the box should be to use the least amount of material to construct it.

34. A firm wants to package its product in a rectangular box with no top and two intersecting partitions. If the volume of the box is to be 64 cubic inches, find the dimensions that the box should be to use the least amount of material to construct it.

35. Extractive industries are those that extract minerals from the ground, such as metals or petroleum. In an article analyzing environmental controls on extractive industries, Stollery defines a welfare function W in terms of the quantity of minerals extracted x and the quantity of pollutants released y.[*] The welfare function measures the benefit to society based on the positive benefit due to extraction versus the negative benefit due to pollution. Stollery defines the function

$$W(x, y) = ax - bx^2 - cy + d$$

where a, b, c and d are constants depending on the industry and the method of extraction. Compute the values of x and y that maximize $W(x, y)$.

36. The design of an effective ramp is one of the most important aspects of highway design. When a car merges traffic on a highway from a ramp, the car enters traffic in a gap between two vehicles. The distance between these two vehicles is called the gap size. Pignataro shows that the average gap size g is a function of the angle of approach x of the ramp, measured in degrees, and the length of the ramp y, measured in 100-foot sections.[†] The relationship is

$$g(x, y) = 5.5 + 0.83x - 0.04x^2 - 1.04y + 0.05y^2$$

Find the relative extrema of $g(x, y)$.

[*]Kenneth R. Stollery, ''Environmental Controls in Extractive Industries,'' *Land Economics,* Vol. 61, 1985, pp. 136–144.

[†]Louis J. Pignataro, *Traffic Engineering: Theory and Practice,* Englewood Cliffs, N.J., Prentice-Hall, 1973, p. 169.

8–4 LAGRANGE MULTIPLIERS

The previous section dealt with finding the relative extrema of a function. This section also deals with optimizing functions, but the types of problems will have a little twist. Often in applications the optimum value of a function is needed subject to a **constraint.** For example, it might be necessary to find the maximum revenue subject to cost considerations. Another example might be to find the box with minimum surface area where the volume of the box is given.

An Application: Chemical Optimization at Monsanto

"Optimization" is a hallowed term in business. Sometimes a company's primary goal is to maximize profit, but at other times the objective is to minimize cost. As an example of the latter, consider the situation at Monsanto described by a consultant, Raymond Boykin, who was hired to evaluate Monsanto's chemical production methods.* In 1983 Monsanto opened a new chemical plant that caused capacity to exceed demand for many products. This prompted the company to hire a consultant to determine how to optimize production.

Boykin concentrated on one important product, maleic anhydride, which is used to make polyester resins. These resins are used in the manufacture of boat hulls, shower stalls, autobody parts, and counter tops. The chemical is produced by feeding raw material (butane) into a reactor under pressure. Production and cost are dependent on the raw material feed rate, the reactor velocity, and reactor pressure. Hence both production and cost are functions of three variables.

Because Monsanto is such a large producer of maleic anhydride, even small reductions in cost per pound would save large amounts of money. So Boykin recommended that Monsanto concentrate on minimizing cost once the amount of production was determined. This resulted in a constrained optimization problem: minimize cost subject to a certain level of production.

In this section we study one method for solving such constrained optimization problems of functions of several variables. At the end of the section we will show how Monsanto solves the problem.

Lagrange Multipliers

Constrained optimization problems are often difficult to solve. An 18th-century mathematician, Joseph Louis Lagrange (1736–1813), discovered an ingenious method for solving these types of problems. Let us start with the general problem. Suppose $f(x, y)$ and $g(x, y)$ are functions of two variables. The problem is to find the values of x and y that either maximize or minimize the objective function $f(x, y)$ subject to the constraint $g(x, y) = 0$.

Lagrange's idea is to replace $f(x, y)$ with another function of three variables by introducing a new variable t called a Lagrange multiplier. The new function is

$$F(x, y, t) = f(x, y) + tg(x, y)$$

The variable t is always multiplied by the constraining function $g(x, y)$. The following theorem shows how to use Lagrange's method, called the technique of **Lagrange multipliers.** The proof of the theorem is beyond the scope of this text.

*Raymond F. Boykin, "Optimizing Chemical Production at Monsanto," *Interfaces,* Vol. 15, 1985, pp. 88–95.

Theorem

If the function $f(x, y)$ has an extremum at (a, b), then there is a value of t, say, $t = c$, such that the partial derivatives of $F(x, y, t)$ each equal 0 at (a, b, c).

The theorem states that to find the extrema, first find the critical values of $F(x, y, t)$ by setting the partial derivatives equal to 0 and solving for x, y, and t. An extremum for $f(x, y)$ subject to $g(x, y) = 0$, if it exists, will be among these critical values. Example 1 illustrates how to apply the theorem.

EXAMPLE 1 | **Problem**

Find the minimum value of $f(x, y) = x^2 + y^2$ subject to $g(x, y) = x + 2y - 10 = 0$.

SOLUTION Form the function

$$F(x, y, t) = x^2 + y^2 + t(x + 2y - 10)$$

Find the partial derivatives of $F(x, y, t)$ and set each equal to 0.

$$F_x(x, y, t) = 2x + t = 0 \tag{1}$$

$$F_y(x, y, t) = 2y + 2t = 0 \tag{2}$$

$$F_t(x, y, t) = x + 2y - 10 = 0 \tag{3}$$

Solving equations 1 and 2 for t yields $t = -2x$ and $t = -y$. Thus $y = 2x$. Substituting $y = 2x$ into equation 3 yields

$$x + 2(2x) - 10 = 0$$
$$5x = 10$$
$$x = 2$$

Substitute $x = 2$ into equation 3 to get $y = 4$. Hence the minimum value of the function $f(x, y) = x^2 + y^2$, subject to $g(x, y) = x + 2y - 10 = 0$, occurs at the point $(2, 4)$. The minimum value is $f(2, 4) = 20$.

We have no easy means to verify that $f(2, 4)$ is actually a minimum rather than a maximum or a saddle point. Often it suffices to substitute into $f(x, y)$ several points close to $(2, 4)$ that also satisfy the constraint. Such points will produce functional values that are larger than 20. For instance, to select a point on $g(x, y) = 0$ close to $(2, 4)$, first choose a value of y that is close to 4, say, $y = 4.1$. Then the corresponding value of x is found by substituting $y = 4.1$ into $g(x, y) = 0$, which yields $x = 1.8$. Then $f(1.8, 4.1) = (1.8)^2 + (4.1)^2 = 20.05$.

The Geometric Interpretation

For functions of two variables the method of Lagrange multipliers has a geometric interpretation. To illustrate, graph both $f(x, y) = x^2 + y^2$ and $g(x, y) =$

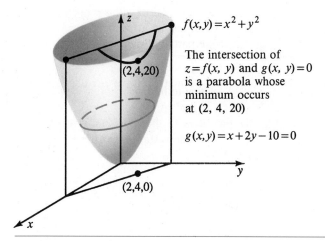

$f(x,y)=x^2+y^2$

The intersection of
$z=f(x, y)$ and $g(x, y)=0$
is a parabola whose
minimum occurs
at (2, 4, 20)

$g(x,y)=x+2y-10=0$

(2,4,20)

(2,4,0)

FIGURE 1

$x + 2y - 10 = 0$ in the same coordinate system. Figure 1 shows that $f(x, y)$ is an infinite bowl and $g(x, y) = 0$ is a plane. Their intersection is a two-dimensional curve, a parabola, whose minimum is at $(2, 4)$. Problem 34 will ask you to substitute algebraically $g(x, y) = x + 2y - 10 = 0$ into $f(x, y) = x^2 + y^2$. This produces the equation of the parabola in Figure 1; it is an equation in one variable whose derivative can be computed and whose minimum will occur at $(2, 4)$.

The Method of Lagrange Multipliers

Before applying the method of Lagrange multipliers to various problems, let us summarize the steps in using the technique.

The Method of Lagrange Multipliers

In order to find the relative extrema of $f(x, y)$ subject to the constraint $g(x, y) = 0$, proceed as follows:

1. Construct the function

   $F(x, y, t) = f(x, y) + tg(x, y)$

2. Compute the partial derivatives of $F(x, y, t)$ and set each equal to 0.
3. Solve the system of equations formed in part 2.
4. The relative extrema of $f(x, y)$ subject to $g(x, y) = 0$ are among the solutions to the system of equations in step 3.

Example 2 illustrates how to apply the method of Lagrange multipliers to a problem where the system is a bit more complicated than the system in Example 1. This example also shows how to deal with a problem in which the constraint is expressed as an equation rather than as a function.

EXAMPLE 2 | **Problem**

Find the relative maximum value of $f(x, y) = 8x - x^2 + 4y - y^2$ subject to $x + y = 10$.

SOLUTION First express the constraint in the proper form, an expression in x and y, $g(x, y)$, set equal to 0. Subtract 10 from each side of the constraining equation to get

$$g(x, y) = x + y - 10 = 0$$

1. Construct the function

$$F(x, y, t) = 8x - x^2 + 4y - y^2 + t(x + y - 10)$$

2. Compute the partial derivatives of $F(x, y, t)$.

$$F_x(x, y, t) = 8 - 2x + t = 0 \qquad (1)$$

$$F_y(x, y, t) = 4 - 2y + t = 0 \qquad (2)$$

$$F_t(x, y, t) = x + y - 10 = 0 \qquad (3)$$

3. Solving equations 1 and 2 for t yields

$$t = 2x - 8 \quad \text{and} \quad t = 2y - 4$$

Thus $2x - 8 = 2y - 4$. This simplifies to the equation

$$x - y = 2 \qquad (4)$$

Form the system of two equations consisting of equations 3 and 4.

$$x + y = 10$$
$$x - y = 2$$

Adding these equations yields $2x = 12$, or $x = 6$. Substituting this value into equation 4 gives $y = 4$.

4. The relative maximum of the value of $f(x, y) = 8x - x^2 + 4y - y^2$ subject to $x + y = 10$ must occur at $(6, 4)$. Testing a few points close to $(6, 4)$ verifies that $f(6, 4) = 12$ is a maximum.

The method of Lagrange multipliers can be applied to functions of more than two variables. Example 3 illustrates how to solve a particular type of word problem by applying Lagrange multipliers to a function of three variables.

EXAMPLE 3 | **Problem**

Find three numbers whose sum is 60 and whose product is a maximum.

SOLUTION First express the problem in a form where Lagrange multipliers can be applied. Let the three numbers be represented by x, y, and z. The problem calls for us to maximize

$$f(x, y, z) = xyz$$

subject to $x + y + z = 60$. Let $g(x, y, z) = x + y + z - 60 = 0$. Apply the steps in the method.

1. Construct the function

$$F(x, y, z, t) = xyz + t(x + y + z - 60)$$

2. Compute the partial derivatives of $F(x, y, z, t)$ and form the system of equations $F_x = 0$, $F_y = 0$, $F_z = 0$, $F_t = 0$.

$$F_x(x, y, z, t) = yz + t = 0 \tag{1}$$

$$F_y(x, y, z, t) = xz + t = 0 \tag{2}$$

$$F_z(x, y, z, t) = xy + t = 0 \tag{3}$$

$$F_t(x, y, z, t) = x + y + z - 60 = 0 \tag{4}$$

3. Solving equations 1, 2, and 3 for t and combining them yields

$$t = -yz = -xz = -xy \tag{5}$$

Assume that none of the numbers is 0, because then the product would be 0, which is not a maximum. Then equation 5 simplifies to $x = y = z$. Substituting these relationships into equation 4 yields $3x = 60$, so that $x = 20$. This means that $x = y = z = 20$ is the solution.

4. The relative maximum value of $f(x, y, z) = xyz$ subject to $x + y + z = 60$ must occur at $(20, 20, 20)$. Testing a few points close to $(20, 20, 20)$ verifies that the maximum value is $f(20, 20, 20) = 8000$.

It is not always easy to test the points close to the critical value. In this case suppose we considered points of the form $(20, 20 + h, 20 - h)$ for a small number h. These points are close to $(20, 20, 20)$ and satisfy the constraint because they sum to 60. Their product is $f(20, 20 + h, 20 - h) = 20(20 + h)(20 - h) = 20(400 - h^2) = 8000 - 20h^2$. This number is always less than 8000. This does not test all the points close to $(20, 20, 20)$, but it gives a good idea of what is required to prove that the maximum occurs at that point.

The Case Study deals with a common type of function in business called a Cobb–Douglas production function, which is of the form

$$f(x, y) = x^a y^b$$

where f is the production or output of a firm while x and y are the inputs, which could be raw goods or labor. The exponents a and b are constants. Example 4 shows how Lagrange multipliers can be used to solve an optimization problem containing a Cobb–Douglas production function.

EXAMPLE 4 | The production capacity of a firm is given by

$$f(x, y) = x^{1/2}y^{1/2}$$

where x is the number of labor-hours available daily and y is the amount of capital invested in machinery. Labor costs are \$10 per labor-hour. Assuming that there are 8 working hours in a day and 260 working days in a year, and that the firm allocates \$5,000,000 in capital for wages and machinery, the constraining equation is

$$10(8)(260)x + y = 5,000,000$$

or

$$20,800x + y = 5,000,000$$

Problem

Find the values of x and y that maximize production capacity.

SOLUTION Express the constraint in the form

$$g(x, y) = 20,800x + y - 5,000,000.$$

Apply the steps in the method.

1. Construct the function

$$F(x, y, t) = x^{1/2}y^{1/2} + t(20,800x + y - 5,000,000)$$

2. Compute the partial derivatives of $F(x, y, t)$ and form the system of equations $F_x = 0$, $F_y = 0$, $F_t = 0$.

$$F_x(x, y, t) = \frac{1}{2}x^{-1/2}y^{1/2} + 20,800t = 0 \tag{1}$$

$$F_y(x, y, t) = \frac{1}{2}x^{1/2}y^{-1/2} + t = 0 \tag{2}$$

$$F_t(x, y, t) - 20,800x + y - 5,000,000 = 0 \tag{3}$$

3. Solving equations 1 and 2 for t and combining them yields

$$-20,800t = \frac{1}{2}x^{-1/2}y^{1/2} = \frac{20,800}{2}x^{1/2}y^{-1/2}$$

Multiply the equation by $2x^{1/2}y^{1/2}$. This yields

$$y = 20,800x$$

Substituting this equation into equation 3 gives $2y = 5,000,000$, or $y = 2,500,000$. Then $x = 2,500,000/20,800 \approx 120.2$.

4. The maximum production occurs at $x \approx 120.2$, or about 120 labor-hours, and $y = \$2,500,000$.

Let us return to the optimization problem facing Monsanto, which was mentioned at the start of the section. The consultant determined that production P and cost C were functions of three variables. Let x represent the raw material feed rate, y the reactor velocity, and z the reactor pressure. The production function has the form $P = kxy/z$ and the cost function $C = ax + by + c/z$. The constants k, a, b, and c depend on which reactor is used. The object is to minimize cost subject to a given level of production. Boykin developed a computer program for Monsanto, which is run five times a day at each plant. The level of production is set and then the program determines the values of x, y, and z that minimize C. The next example illustrates how the program solves the problem.

EXAMPLE 5

Problem

Let $k = 1$, $a = 0.64$, $b = 4$, and $c = 1.6$ in the preceding production and cost functions. Set the production level at $P = 27$. Find the values of x, y, and z that minimize cost.

SOLUTION The problem is to minimize

$$C = 0.64x + 4y + \frac{1.6}{z}$$

Subject to: $\quad \dfrac{xy}{z} - 27 = 0$

Define

$$C(x, y, z, t) = 0.64x + 4y + \frac{1.6}{z} + t\left(\frac{xy}{z} - 27\right)$$

Find the partial derivatives and form the system of equations.

$$C_x(x, y, z, t) = 0.64 + \frac{ty}{z} = 0$$

$$C_y(x, y, z, t) = 4 + \frac{tx}{z} = 0$$

$$C_z(x, y, z, t) = -1.6z^{-2} - txyz^{-2} = 0$$

$$C_t(x, y, z, t) = \frac{xy}{z} - 27 = 0$$

Solve for $-t$ in the first three equations.

$$-t = \frac{0.64z}{y} = \frac{4z}{x} = \frac{1.6}{xy}$$

$$y = 0.16x, \ z = \frac{1}{0.4x}$$

Substituting these values into $C_t = 0$ gives

$$x(0.16x)(0.4x) - 27 = 0$$

$$x^3 = \frac{27}{0.064}$$

$$x = \frac{3}{0.4} = 7.5$$

Therefore the solution is $x = 7.5$, $y = 0.16(7.5) = 1.2$, $z = 1/3$.

EXERCISE SET 8–4

● In Problems 1 to 18 use the method of Lagrange multipliers to find the extrema of the functions with the given constraints.

1. $f(x, y) = x^2 + y^2$
   subject to $g(x, y) = x + y - 1 = 0$

2. $f(x, y) = x^2 + 2y^2$
   subject to $g(x, y) = x + 2y - 3 = 0$

3. $f(x, y) = x^2 + 2x + y^2 - y$
   subject to $g(x, y) = x + 2y = 0$

4. $f(x, y) = x^2 + 4x + y^2 - 6y$
   subject to $g(x, y) = 3x + y - 1 = 0$

5. $f(x, y) = x^3 + 3x^2 + y^2 - 2y$
   subject to $g(x, y) = 3x + 2y + 3 = 0$

6. $f(x, y) = x^3 + 6x^2 + y^2 - 4y$
   subject to $g(x, y) = x + 3y + 1 = 0$

7. $f(x, y) = 2x^3 - 6x^2 + 2y^2 - 8y$
   subject to $x + 2y - 2 = 0$

8. $f(x, y) = 4x^3 + 6x^2 - y^2 - 2y$
   subject to $2x - 3y + 1 = 0$

9. $f(x, y) = 2x - x^2 + 2y^3 - 6y^2$
   subject to $x - 2y = 10$

10. $f(x, y) = 4x^3 + 6x^2 - y^2 - 2y$
    subject to $2x + y = 2$

11. $f(x, y) = xy$
    subject to $g(x, y) = 2x + y - 1 = 0$

12. $f(x, y) = 2xy$
    subject to $g(x, y) = x + 2y - 3 = 0$

13. $f(x, y) = xy + x^2$
    subject to $2x + 3y = 1$

14. $f(x, y) = xy - y^2$
    subject to $x - 2y = 0$

15. $f(x, y) = xyz$
    subject to $g(x, y, z) = 2x + y + z - 2 = 0$

16. $f(x, y) = 2xyz$
    subject to $g(x, y, z) = x + 2y - z - 4 = 0$

17. $f(x, y) = xyz$
    subject to $x + y = z$

18. $f(x, y) = xyz$
    subject to $x = 2y - 3z$

19. Find three numbers that sum to 30 such that their product is a maximum.

20. Find three numbers that sum to 120 such that their product is a maximum.

21. Maximize $f(x, y) = x^{1/2}y^{1/2}$
    subject to $100x + y = 5000$.

22. Maximize $f(x, y) = x^{1/2}y^{1/3}$
    subject to $200x + y = 8000$.

●● 23. Find three numbers such that the sum of two of them is equal to the third and their product is a maximum.

24. Find three numbers such that the sum of two of them is equal to twice the third and their product is a maximum.

25. Find the minimum distance from the line $3x + y = 5$ to the origin $(0, 0)$.

26. Find the minimum distance from the line $x - 2y = 4$ to the point $(1, 0)$.

In Problems 27 to 33 use the method of Lagrange multipliers to find the extrema of the functions with the given constraints.

27. $f(x) = xy^2$ subject to $2x - y = 1$

28. $f(x) = x^2y$ subject to $x + y = 3$

29. $f(x) = xyz^2$ subject to $2x + 3y + z = 2$

30. $f(x) = x^2yz$ subject to $x - 2y - 3z = 4$

31. $f(x) = xy^2 + yz^2$
    subject to $x + y + z = 2$

32. $f(x) = x^2yz + xz$
    subject to $x - y - z = 4$

33. $f(x) = xy^2 + xz^2$
    subject to $xy + yz = 2$

34. Substitute $g(x, y) = x + 2y - 10 = 0$ into $f(x, y) = x^2 + y^2$, and find the extremum of the resulting function.

35. Find the dimensions of the box with maximum volume if the box is to be made from 216 cubic inches of material.

36. Find the minimum distance from the plane $6x - y - 2z = -3$ to the point $(1, 0, -1)$.

37. In an article studying the economics of agricultural production, Chavas et al.* consider the swine industry in depth. The study uses constrained optimization to determine the maximum weight that an animal can attain subject to various limitations, such as cost of grain.

Solve the following problem via Lagrange multipliers that is similar to the problems presented by Chavas: find the values of x and y that maximize $z = 3x + 4y$ subject to $a = 2x^2 + 3y^2 = 0$, where a is the weight of the animal, x is the amount of feed, and y is the amount of veterinary care for the animal.

38. The Swedish roundwood market is a good example of a market in which both buyers and sellers are highly monopolized. This means that various types of simple bargaining models can be tested against the actual results in the marketplace. In one such article, Johannson and Lofgren assume that the owners wish to maximize profit P subject to a utility function u which measures the benefit to the owners when the price of roundwood is greater than the price determined by free competition.† The latter is sometimes possible because of the monopolistic power enjoyed by the sellers. Profit and utility are both functions of the price x and the output y. The two functions are

$$P(x, y) = f(x) - xy$$
$$u(x, y) = x[g(y) - a] - k$$

where $f(x)$ and $g(y)$ are measures of the monopolistic powers of the sellers and the buyers, respectively. Johannson and Lofgren then found conditions when $P(x, y)$ would be maximized subject to the constraint $u(x, y) = 0$ by applying the method of Lagrange multipliers to the function

$$L(x, y, t) = P(x, y) - tu(x, y)$$

Find the maximum of $L(x, y, t)$ for $f(x) = x^2$, $g(y) = y$, $a = 2$ and $k = 1$.

*Jean-Paul Chavas et al., "Modeling Dynamic Agricultural Production Response: The Case of Swine Production," *American Journal of Agricultural Economics,* Vol. 67, 1985, pp. 636–646.

†Per-Olov Johannson and Karl-Gustaf Lofgren, "A Bargaining Approach to the Modeling of the Swedish Roundwood Market," *Land Economics,* Vol. 61, 1985, pp. 65–75.

8–5 DOUBLE INTEGRALS

In the previous sections the concept of the derivative of a function of one variable was generalized to the partial derivative of functions of two or more variables. In this section we investigate how to generalize the integral of a function of one variable to functions of two variables.

The concept of the integral relies on the antiderivative, which is the reverse operation of finding the derivative. Thus the generalization of the integral relies on the reverse operation of computing the partial derivative.

The definite integral of a function of one variable can be used to compute the area under a curve. In a similar way its generalization can be used to compute volumes in three-space. At first we compute volumes bounded by rectangular regions in the xy-plane, and then we consider other types of regions.

Partial Differentiation Reversed

We start with an example of reversing the process of partial differentiation. To find the partial derivative with respect to x of $f(x) = 6xy^2 - 5x^4y$, we treat y as a constant and differentiate with respect to x. To reverse the process, hold y constant and find the antiderivative with respect to x. This yields

$$\int (6xy^2 - 5x^4y)\, dx = 6\left(\frac{x^2}{2}\right)y^2 - 5\left(\frac{x^5}{5}\right)y + C(y)$$

$$= 3x^2y^2 - x^5y + C(y)$$

The constant of integration $C(y)$ represents not just a constant, but an arbitrary function of y. The reason for this can be seen when checking the answer by computing the partial derivative with respect to x of $3x^2y^2 - x^5y + C(y)$ which is $6xy^2 - 5x^4y$. Notice that the partial derivative with respect to x of $C(y)$ is 0.

We write $\int f(x, y)\, dx$ to indicate the operation of computing the partial antiderivative with respect to x of $f(x, y)$ holding y constant. Similarly $\int f(x, y)\, dy$ means find the partial antiderivative with respect to y of $f(x, y)$, holding x constant. The first example further illustrates how to do this.

EXAMPLE 1

Problem

Compute (a) $\displaystyle\int (6xy^5 + 12x^{-2}y^3 - 7)\, dx$ and

(b) $\displaystyle\int (6xy^5 + 12x^{-2}y^3 - 7)\, dy.$

SOLUTION (a) Hold y constant and find the antiderivative with respect to x.

$$\int (6xy^5 + 12x^{-2}y^3 - 7)\, dx = 6\left(\frac{x^2}{2}\right)y^5 + 12\left(\frac{x^{-1}}{-1}\right)y^3 - 7x + C(y)$$

$$= 3x^2y^5 - 12x^{-1}y^3 - 7x + C(y)$$

(b) Hold x constant and find the antiderivative with respect to y.

$$\int (6xy^5 + 12x^{-2}y^3 - 7)\, dy = 6x\left(\frac{y^6}{6}\right) + 12x^{-2}\left(\frac{y^4}{4}\right) - 7y + C(x)$$

$$= xy^6 + 3x^{-2}y^4 - 7y + C(x)$$

The generalization of antidifferentiation to functions of several variables can be used to compute definite integrals of the form

$$\int_a^b f(x, y)\, dx \quad \text{and} \quad \int_c^d f(x, y)\, dy$$

The limits of integration are actually functions. In the first integral the limits must be thought of as $x = a$ and $x = b$. Similarly in the second integral the limits are $y = c$ and $y = d$. The next example demonstrates this computation.

EXAMPLE 2 | **Problem**

Compute (a) $\int_1^2 (6xy^5 + 12x^{-2}y^3 - 7)\, dx$ and

(b) $\int_0^2 (6xy^5 + 12x^{-2}y^3 - 7)\, dy$.

SOLUTION (a) The first step is to find the antiderivative with respect to x. This was done in Example 1. Just as with functions of one variable, any antiderivative can be used when computing definite integrals, so we take $C(y) = 0$. Then insert the limits of integration, remembering that they are $x = 1$ and $x = 2$.

$$\int_1^2 (6xy^5 + 12x^{-2}y^3 - 7)\, dx$$

$$= (3x^2y^5 - 12x^{-1}y^3 - 7x)\, \Big|_{x=1}^{x=2}$$

$$= 3(4)y^5 - 12\left(\frac{1}{2}\right)y^3 - 7(2) - [3(1)y^5 - 12(1)y^3 - 7(1)]$$

$$= 12y^5 - 6y^3 - 14 - 3y^5 + 12y^3 + 7 = 9y^5 + 6y^3 - 7$$

(b) The antiderivative with respect to y was found in Example 1. Insert the limits.

$$\int_0^2 (6xy^5 + 12x^{-2}y^3 - 7)\, dy$$

$$= (xy^6 + 3x^{-2}y^4 - 7y)\, \Big|_{y=0}^{y=2}$$

$$= x(64) + 3x^{-2}(16) - 7(2) - [x(0) + 3x^{-2}(0) - 7(0)]$$

$$= 64x + 48x^{-2} - 14$$

In part (a) of Example 2 the definite integral with dx produces a function of y; x is not mentioned because a number is substituted for each x in the antiderivative. Likewise in part (b) the definite integral with dy produces a function of x. Thus the result of a single definite integral of $f(x, y)$ is a function of one variable. This function can then be integrated, resulting in a double integral. This is illustrated in the next example.

EXAMPLE 3 | **Problem**

Compute (a) $\int_0^2 \left[\int_1^2 (6xy^5 + 12x^{-2}y^3 - 7)\, dx \right] dy$ and

(b) $\int_1^2 \left[\int_0^2 (6xy^5 + 12x^{-2}y^3 - 7)\, dy \right] dx$.

SOLUTION (a) First work within the brackets. The definite integral in the brackets was computed in Example 2a. Substitute its value, $9y^5 + 6y^3 - 7$, for the expression in the brackets to get

$$\int_0^2 \left[\int_1^2 (6xy^5 + 12x^{-2}y^3 - 7)\, dx \right] dy$$

$$= \int_0^2 (9y^5 + 6y^3 - 7)\, dy = \left(\frac{9y^6}{6} + \frac{6y^4}{4} - 7y \right) \Bigg|_{y=0}^{y=2}$$

$$= \frac{9(64)}{6} + \frac{6(16)}{4} - 7(2) = 96 + 24 - 14$$

$$= 106$$

(b) The definite integral within the brackets was computed in Example 2b.

$$\int_1^2 \left[\int_0^2 (6xy^5 + 12x^{-2}y^3 - 7)\, dy \right] dx$$

$$= \int_1^2 (64x + 48x^{-2} - 14)\, dx$$

$$= (32x^2 - 48x^{-1} - 14x) \Bigg|_{x=1}^{x=2}$$

$$= 32(4) - 48\left(\frac{1}{2}\right) - 14(2) - (32 - 48 - 14) = 76 - (-30)$$

$$= 106$$

Notice that the answers in parts (a) and (b) of Example 3 are equal. This is not a coincidence. For most functions if the order of integration is reversed as was done in the two parts of Example 3, the double integrals will be the same. This means that for most functions $f(x, y)$

$$\int_a^b \left[\int_c^d f(x, y)\, dx \right] dy = \int_c^d \left[\int_a^b f(x, y)\, dy \right] dx$$

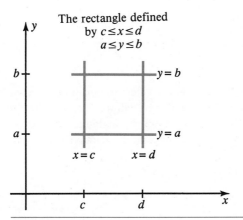

FIGURE 1

This means that the brackets are not needed, which in turn allows us to define the double integral of $f(x, y)$. This is because the limits of integration define a rectangular region in the xy-plane. The region R is the set of points (x, y) satisfying $c \leq x \leq d$ and $a \leq y \leq b$, which is the rectangle graphed in Figure 1. Then the previous integrals are defined over the **region of integration** R. The preceding equality means that either order of integration can be chosen. It is referred to as an **iterated integral** because it is computed by integrating twice. The formal definition is given in the following display.

—————————————— **Double Integral** ——————————————

The **double integral** of $f(x, y)$, called the **integrand,** over the rectangular region of integration R, is expressed as

$$\int \int_R f(x, y) \, dx \, dy \qquad \text{or} \qquad \int \int_R f(x, y) \, dy \, dx$$

and is equal to

$$\int_a^b \left[\int_c^d f(x, y) \, dx \right] dy = \int_c^d \left[\int_a^b f(x, y) \, dy \right] dx$$

As an illustration, consider the region R in the display and let the integrand be the constant function $f(x, y) = C$, for C a positive number. The double integral of $f(x, y)$ over R is the volume of a box with the lengths of its sides $d - c$ and $b - a$, respectively, and with height C. Then

$$\int \int_R C \, dx \, dy = C[\text{area of } R] = C(d - c)(b - a)$$

The next example shows how to evaluate a double integral when the region is given.

EXAMPLE 4

Problem

Evaluate $\displaystyle\int\int_R x^2(1 - y^3)\, dx\, dy$ over the region R defined by $0 \le x \le 3$ and $-1 \le y \le 1$.

SOLUTION Either order of integration can be selected. We choose to integrate first with respect to x and then with respect to y. The reverse order can be used as a check. The boundaries of the region are $x = 0$ to $x = 3$ and $y = -1$ to $y = 1$. These are the limits and the iterated integral is

$$\int_{-1}^{1}\int_{0}^{3} [x^2(1 - y^3)]\, dx\, dy = \int_{-1}^{1}\left[\left[\frac{x^3}{3}(1 - y^3)\right]\Big|_{x=0}^{x=3}\right] dy$$

$$= \int_{-1}^{1} 9(1 - y^3)\, dy = 9\left(y - \frac{y^4}{4}\right)\Big|_{y=-1}^{y=1}$$

$$= 9\left[\left(1 - \frac{1}{4}\right) - \left(-1 - \frac{1}{4}\right)\right] = 9(2) = 18$$

Volume

The definite integral of a function of one variable can be used to compute the area under a curve. The double integral of a function of two variables can be used to find the volume bounded by a surface in three-space. Just as areas in two-space can be approximated by rectangles, so too volumes can be approximated by small boxes. Figure 2 shows the graph of $f(x, y) = x^2 + y^2$ over the region defined by $0 \le x \le 1$ and $0 \le y \le 2$. The volume of the solid bounded above by $f(x, y)$ and below by the region R can be found by computing the double integral of $f(x, y)$ over R. The proof of this result is beyond the scope of this text, but it entails summing the volumes of small boxes whose heights are functional values of $f(x, y)$ and whose length and width are small increments dx and dy. One such box is

FIGURE 2

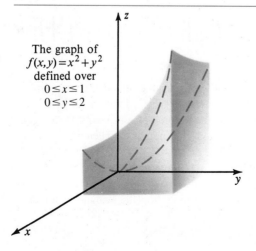

The graph of
$f(x,y)=x^2+y^2$
defined over
$0 \le x \le 1$
$0 \le y \le 2$

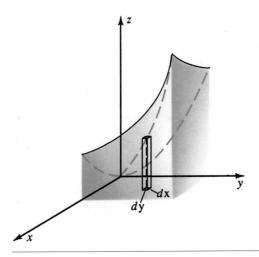

FIGURE 3

graphed in Figure 3. The limit of this sum is then the **volume.** This suggests the following result.

Volume

If $f(x, y) \geq 0$ for all points in the rectangle R defined by $a \leq x \leq b$ and $c \leq y \leq d$, then the volume of the solid bounded above by $f(x, y)$ and below by R (see Figure 4) is equal to

$$\int \int_R f(x, y) \, dx \, dy$$

FIGURE 4

The volume bounded by $z = f(x, y)$ and the region R

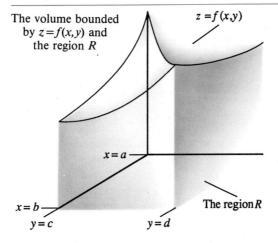

$z = f(x,y)$

$x = a$

$x = b$

$y = c$

$y = d$

The region R

EXAMPLE 5 | **Problem**

Compute the volume of the solid bounded above by $f(x, y) = x^2 + y^2$ and below by the rectangle R defined by $0 \le x \le 1$ and $0 \le y \le 2$. This solid is graphed in Figure 2.

SOLUTION The display states that the volume is the double integral of $f(x, y) = x^2 + y^2$ over R. This is given by

$$\int_0^2 \int_0^1 (x^2 + y^2) \, dx \, dy = \int_0^2 \left[\left(\frac{x^3}{3} + xy^2 \right) \Big|_{x=0}^{x=1} \right] dy$$

$$= \int_0^2 \left(\frac{1}{3} + y^2 \right) dy = \left(\frac{y}{3} + \frac{y^3}{3} \right) \Big|_{y=0}^{y=2}$$

$$= \frac{2}{3} + \frac{8}{3} = \frac{10}{3}$$

Double Integrals Over Other Regions

We have emphasized that the limits of integration of a double integral over a rectangular region are actually functions. In fact they are linear functions of the form $x = a$ and $y = b$. It is also possible to define double integrals whose inner limits of integration are nonlinear functions. However, the outer limits must be of the form $x = a$ or $y = b$, depending on the order of integration. See Figure 4. The next example illustrates how to compute such a double integral.

EXAMPLE 6 | **Problem**

Evaluate $\int_0^1 \int_{x^2}^x 3xy^2 \, dy \, dx$.

SOLUTION Integrate first with respect to y and then with respect to x.

$$\int_0^1 \int_{x^2}^x 3xy^2 \, dy \, dx = \int_0^1 \left[xy^3 \Big|_{y=x^2}^{y=x} \right] dx$$

$$= \int_0^1 (x^4 - x^7) \, dx = \left(\frac{x^5}{5} - \frac{x^8}{8} \right) \Big|_0^1$$

$$= \frac{1}{5} - \frac{1}{8} = \frac{3}{40}$$

The limits of integration in Example 6 define a region R in the xy-plane graphed in Figure 5. It is bounded below by $y = x^2$ and above by $y = x$. Thus the double integral in Example 6 is defined over a region with variable limits. There are two ways to view such a region corresponding to the two orders of integration. In Figure 6 a region is viewed in two different ways in parts (a) and (b). In part

FIGURE 5

(a)

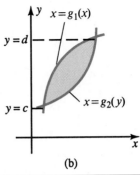

(b)

FIGURE 6

(a) the region is viewed as being bounded above and below by functions of the form $y = f_1(x)$ and $y = f_2(x)$. These functions would then be the inner limits of integration, while the boundaries $x = a$ and $x = b$ would be the outer limits. This corresponds to integrating first with respect to y and then with respect to x. In part (b) the boundaries of the region are viewed from a different perspective. Instead of looking at the region vertically, think of it horizontally. Then the region can be considered as being bounded by $x = g_1(y)$ and $x = g_2(y)$. These functions are the inner limits and the outer limits would be the boundaries $y = c$ and $y = d$. This corresponds to the reverse order of integration, first with respect to x and then y. Example 7 illustrates how to compute such an integral using both orders of integration.

EXAMPLE 7

Problem

Compute the integral $\displaystyle\iint\limits_{R} 2xy \, dx \, dy$ over the region R defined by $x^2 \leq y \leq 4$

and $0 \leq x \leq 2$, which is graphed in Figure 7 using both orders of integration.

SOLUTION (a) Since the variable (nonconstant) boundary of R is given directly as $y = x^2$, the more immediate order of integration is first with respect to y and then x. The inner limits are $y = x^2$ and $y = 4$, while the outer limits are $x = 0$ and $x = 2$. See Figure 7a. The iterated integral is

$$\int_0^2 \int_{x^2}^4 2xy \, dy \, dx = \int_0^2 \left[xy^2 \Big|_{y=x^2}^{y=4} \right] dx = \int_0^2 (16x - x^5) \, dx$$

$$= \left(8x^2 - \frac{x^6}{6} \right) \Big|_{x=0}^{x=2} = \left(32 - \frac{64}{6} \right) - 0 = \frac{64}{3}$$

(b) To switch the order of integration, view the region as being bounded by functions of the form $x = g(y)$. (See Figure 7b.) From this perspective, the function $y = x^2$ has the formula $x = \sqrt{y}$. The boundaries then are the two functions $x = 0$ and $x = \sqrt{y}$. These are the inner limits of integration. The outer limits are $y = 0$ and $y = 4$. The iterated integral is

$$\int_0^4 \int_0^{\sqrt{y}} 2xy \; dx \; dy = \int_0^4 \left[x^2 y \Big|_{x=0}^{x=\sqrt{y}} \right] dx$$

$$= \int_0^4 y^2 \; dy = \frac{y^3}{3} \Big|_0^4 = \frac{64}{3}$$

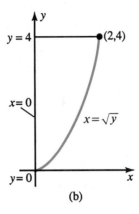

(a)   (b)

FIGURE 7

Notice that the answers in parts (a) and (b) of Example 7 are equal. This is expected because we merely switched the order of integration. Sometimes computing the integral using one order will be harder than using the other order. In fact at times it is impossible to use one order but the computation is easy with the other one.

EXERCISE SET 8–5

- In Problems 1 to 4 compute the antiderivative.

1. $\int (7xy^6 + 6x^3 y^{-2} - 5) \; dx$

2. $\int (5x^4 y + 6x^2 y^{1/2} - 5x + y) \; dx$

3. $\int (8xy^7 - 20x^{-4}y^4 - 3x - y) \; dy$

4. $\int (x^2 y^3 - 12x^{1/4}y^{-2} + x - 2y + 3) \; dy$

In Problems 5 to 8 compute the definite integral.

5. $\int_1^2 (5x^4 y + 2x^{-1}y^3 - 5) \; dx$

6. $\int_1^3 (4x^3 y^2 + x^{-2} - 2y + 1) \; dx$

7. $\int_0^2 (5x^4 y + 2x^{-1}y^3 - 5) \; dy$

8. $\int_1^3 (4x^3 y^2 + x^{-2} - 2y + 1) \; dy$

In Problems 9 to 12 compute the double integral.

9. $\int_1^3 \left[\int_1^2 (3xy^2 - 12x^{-3}y^2 - 1) \; dx \right] dy$

10. $\int_1^2 \left[\int_{-1}^1 (4xy^3 + 6x^2 y^{-1} - y) \; dx \right] dy$

11. $\int_1^2 \left[\int_2^3 (10x^{1/2}y^4 + 12x^{-1/2}y^3 - x) \; dy \right] dx$

12. $\int_0^2 \left[\int_1^3 (3x^{1/3}y^2 + 12x^{-1/3} + x - y)\, dy \right] dx$

In Problems 13 to 16 evaluate the double integral over the rectangular region R.

13. $\iint_R 4x^3(1 - y^2)\, dx\, dy$, where R is defined by

$0 \le x \le 2$ and $0 \le y \le 1$.

14. $\iint_R x^2y(2 - 3y^2)\, dx\, dy$, where R is defined by

$0 \le x \le 2$ and $1 \le y \le 3$.

15. $\iint_R (x^3y + x - 4y^3)\, dx\, dy$, where R is defined

by $1 \le x \le 2$ and $-1 \le y \le 1$.

16. $\iint_R (x^{-2}y + x^{1/2} - 4y)\, dx\, dy$, where R is

defined by $1 \le x \le 2$ and $1 \le y \le 3$.

In Problems 17 and 18 compute the volume of the solid bounded above by $f(x, y)$ and below by the rectangle R.

17. $f(x, y) = x^3y + x - 4y^3$, where R is defined by $1 \le x \le 3$ and $0 \le y \le 1$.

18. $f(x, y) = x^{-2}y + x^{1/2} - 4$, where R is defined by $1 \le x \le 3$ and $1 \le y \le 2$.

In Problems 19 to 24 evaluate the double integral.

19. $\int_0^1 \int_{x^2}^x 4xy^3\, dy\, dx$ 20. $\int_0^2 \int_{x^2}^{2x} 4x^2y^3\, dy\, dx$

21. $\int_0^1 \int_{y^3}^1 xy^2\, dx\, dy$ 22. $\int_0^1 \int_{y^3}^y xy^4\, dx\, dy$

23. $\int_0^2 \int_{-x^2}^{x^2} (x + y^2)\, dy\, dx$

24. $\int_0^2 \int_{1-x^2}^{x^2+1} (x - y)\, dy\, dx$

In Problems 25 to 28 evaluate the double integral over the nonrectangular region R.

25. $\iint_R 4xy\, dx\, dy$, where R is defined by

$x^2 \le y \le 1$ and $0 \le x \le 1$

26. $\iint_R (x + y)\, dx\, dy$, where R is defined by

$x^3 \le y \le 1$ and $0 \le x \le 1$.

27. $\iint_R (2x + 3y)\, dx\, dy$, where R is defined by

$x^3 \le y \le x^2$ and $0 \le x \le 1$.

28. $\iint_R (x - xy)\, dx\, dy$, where R is defined by

$x^4 \le y \le x^2$ and $0 \le x \le 1$.

•• In Problems 29 to 32 compute the double integral.

29. $\int_1^3 \left[\int_1^2 \frac{1}{xy}\, dx \right] dy$

30. $\int_1^3 \left[\int_1^2 \left[\frac{x}{1 + y} \right] dy \right] dx$

31. $\int_0^1 \left[\int_0^2 [x(x^2 + y)^{1/2}]\, dx \right] dy$

32. $\int_0^2 \left[\int_1^2 [xy(x^2 + y^2)^{1/2}]\, dy \right] dx$

In Problems 33 to 36 evaluate the double integral over the region R.

33. $\iint_R \left(\sqrt{x} + y \right) dx\, dy$, where R is defined by

$0 \le y \le 1$ and $0 \le x \le 3y$.

34. $\iint_R \sqrt{1 - x^2}\, dx\, dy$, where R is defined by

$0 \le y \le 1$ and $y \le x \le 1$.

35. $\iint_R \frac{x}{1 + y^2}\, dx\, dy$, where R is defined by

$x^2 \le y \le 4$ and $0 \le x \le 2$.

36. $\iint_R ye^{x^2}\, dx\, dy$, where R is defined by

$0 \le y \le 1$ and $y^2 \le x \le 1$.

37. In an article investigating the effects of urban renewal on rent increases on the land directly affected, Fishelson and Pines define a trans-

portation cost function $C(x, H)$ that measures the dollars per year the average family spends on transportation when the family lives a distance x miles from the center of the urban renewal site and H is the yearly housing expenditure of the family.* The authors compute the marginal cost $MC(x, H)$ to be

$$MC(x, H) = \frac{kx^2}{H}$$

where k is a constant depending on the particular city and the location of the urban renewal site in the city. They then calculate the total transportation cost using double integrals. Find the total cost from $x = 0$ to $x = 5$ miles, from $H = 1$ to $H = 10$.

38. Tax reform has become a priority in recent years.

Many economists have called for tax revisions for farmers that would replace an income tax with a tax on purchases. In an article investigating this proposal, Chambers and Lopez define a cost of labor function $CL(p, t)$ in terms of profit or income p and income tax t.† They compute the marginal cost function $MCL(p, t)$ to be

$$MCL(p, t) = \frac{kp}{1 - t}$$

where k is a constant depending on the particular financial position of the farm. Then the total cost of labor is the double integral of this function. Find the total cost from $t = 0.25$ to $t = 0.30$ when $k = 1$ and p is held constant with $p = 10$ thousand dollars.

*Gideon Fishelson and David Pines, "Market Versus Social Valuation of Redevelopment Projects in an Urban Setting," *Socio-Economic Planning,* Vol. 18, 1984, pp. 419–424.

†Robert G. Chambers and Ramon E. Lopez, "Tax Policies and the Financially Constrained Farm Household," *American Journal of Agricultural Economics,* Vol. 69, 1987, pp. 369–377.

CASE STUDY   **Minimum Cost Production in the Chemical Industry**

Jojoba oil is used for producing amides, which, in turn, are an essential ingredient in many chemical processes. The procedure of extracting amides from jojoba oil is time-consuming and expensive. First the oil is refined and purified. Then other chemicals are added and the mixture is heated for several hours. Once the material becomes cakelike the amides can be extracted.

    Researchers have tried various methods to maximize the amide yield while conserving cost to produce the cakelike material. The two most important variables are temperature and time. In an article* discussing various strategies for improving amide yield, researchers discovered that the relationship between the variables is best described by a Cobb–Douglas production function, which is a power function of the form

$$f(x, y) = ax^b y^c$$

Let $f(x, y)$ represent the amount of amide yield (as a percentage of oil weight) at temperature x (measured in degrees Celsius) and time y (measured in hours). Test results showed that proper production function is

$$f(x, y) = 1.3x^{0.5}y^{0.4}$$

*A. Shani et al., "Synthesis of Jojobamide and Homojojobamide," *Journal of American Oil and Chemical Society,* March 1980, pp. 112–114.

This function is useful in planning production. Various constraints on cost are given and the corresponding production yield is computed. A particular firm would decide what the most reasonable limits on cost would be and then use this function to predict whether the amide yield would be great enough. If not, further study of the cost constraints would be required.

For example, cost considerations require that if x increases, then y must decrease. This means that if the processing time increases, then the temperature would have to decrease to keep the production cost the same. One such constraint is

$$x + 2y = 200$$

This means that when $x = 100$, $y = 50$, and when $x = 110$, $y = 45$; so that the cost of using a temperature of $100°$ for 50 hours is equal to using a temperature of $110°$ for 45 hours. The next example shows how Lagrange multipliers are used to determine the maximum amide yield subject to the given cost constraint.

EXAMPLE | **Problem**

Find the maximum of $f(x, y) = 1.3x^{0.5}y^{0.4}$ subject to the constraint $x + 2y = 200$.

SOLUTION Introduce the Lagrange multiplier t in the function

$$F(x, y, t) = 1.3x^{0.5}y^{0.4} + t(x + 2y - 200)$$

Find the partial derivatives and set them equal to 0.

$$\begin{aligned} F_x(x, y, t) &= 1.3(0.5)x^{-0.5}y^{0.4} + t \\ &= 0.65x^{-0.5}y^{0.4} + t = 0 \end{aligned} \tag{1}$$

$$\begin{aligned} F_y(x, y, t) &= 1.3(0.4)x^{0.5}y^{-0.6} + 2t \\ &= 0.52x^{0.5}y^{-0.6} + 2t = 0 \end{aligned} \tag{2}$$

$$F_t(x, y, t) = x + 2y - 200 = 0 \tag{3}$$

Solve equations 1 and 2 for $-t$.

$$-t = 0.65x^{-0.5}y^{0.4} = 0.26x^{0.5}y^{-0.6}$$

Disregard $-t$ and multiply the resulting equation by $x^{0.5}y^{0.6}$.

$$0.65y = 0.26x$$

Hence $x = 2.5y$. Substitute this relationship into $F_t = 0$.

$$\begin{aligned} 2.5y + 2y - 200 &= 0 \\ 4.5y &= 200 \\ y &\approx 44.4 \\ x &\approx 111 \end{aligned}$$

This means that with the given constraint the maximum amide yield is $f(111, 44.4) \approx 1.3(10.5)(4.56) \approx 62.2\%$ when the material is heated at $111°$ for 44.4 hours.

Case Study Exercise

In Problems 1 to 4 find the maximum amide yield for the given cost function and the production function.

$$f(x, y) = 1.3x^{0.5}y^{0.4}$$

1. $x + 2.4y = 200$
2. $x + 1.8y = 200$
3. $2x + 4.5y = 440$
4. $3x + 5.8y = 600$

In Problems 5 and 6 find the maximum amide yield for the given cost function and the given production function.

5. $x + 2.4y = 200,$
   $f(x, y) = 1.3x^{0.6}y^{0.3}$

6. $x + 1.8y = 200,$
   $f(x, y) = 1.7x^{0.55}y^{0.35}$

CHAPTER REVIEW

Key Terms

8–1 Examples of Functions of Severable Variables

Dependent Variable	Trace
Independent Variable	Section
Functions of Two Variables	Level Curves

8–2 Partial Derivatives

Partial Derivative	Mixed Partial Derivative
Second Partial Derivative	

8–3 Relative Extrema

Relative Extrema	Critical Point
Relative Maximum	Saddle Point
Relative Minimum	Second Partial Derivative Test

8–4 Lagrange Multipliers

Constraint	Lagrange Multipliers

8–5 Double Integrals

Iterated Integral	Region of Integration
Double Integral	Volume
Integrand	

REVIEW PROBLEMS

1. Find the functional value of $f(x, y) = 3x^2 - 4y^2$ at the point $(1, -1)$.

2. Find the domain of the function $f(x, y) = \dfrac{xy}{x^2 - 1}$.

3. Graph the function $f(x, y) = 1 + x^2 + 4y^2$, giving the traces and at least one section parallel to each coordinate plane.

4. Graph the function $f(x, y) = x^2 + 4x + y^2 + 2y$.

5. Find $f_x(x, y)$ and $f_y(x, y)$, where $f(x, y) = xy^3 - (2x + 3y)^4$.

6. Find $f_x(1, 0)$ and $f_y(2, 1)$ for $f(x, y) = 2xy^3 + 3x^{-1}y$.

7. Find all second-order partial derivatives for $f(x, y) = 3x^{2/5} + x^2y^{-2}$.

8. Find the values of x and y such that $f_x(x, y) = 0$ and $f_y(x, y) = 0$ for $f(x, y) = 1 - x^2 + y^2 + 2xy - 4y$.

9. Find the critical points of the function $f(x, y) = 3x^2 - 12xy + y^2 + 10y - 2$.

10. Find the relative extrema and any saddle points of the function $f(x, y) = x - xy + y^3 - 3y^2$.

11. Find the relative extrema of the function $f(x, y) = x^3 - 6x^2 - 2y^3 + 24y$.

12. A firm utilizes x million dollars per year in labor cost and y million dollars per year in production cost. The cost function is

$$C(x) = 100 + 24x + 12y - 2x^2 - 3y^2$$

Find the amount that the firm should spend on labor and production each year to minimize cost.

In Problems 13 and 14 use the method of Lagrange multipliers to find the extrema of the functions with the given constraints.

13. $f(x) = xy$ subject to $g(x, y) = 3x + y - 5 = 0$

14. $f(x) = xyz$ subject to $g(x, y, z) = x + y + 2z - 1 = 0$

15. Find three numbers that sum to 45 such that their product is a maximum.

16. Maximize $f(x, y) = x^{1/2}y^{1/3}$ subject to $200x + y = 8000$.

17. Compute the antiderivative

$$\int (14xy^6 + 6x^3y^{-2} - y + x - 1)\, dx$$

18. Compute the definite integral.

$$\int_1^2 (10x^4y + 2x^{-3}y^3 - 1)\, dx$$

19. Compute the double integral.

$$\int_1^3 \left[\int_1^2 (6xy^2 - 24x^{-3}y^2 - 1)\, dx \right] dy$$

20. Evaluate the double integral over the rectangular region R: $\displaystyle\iint_R 4x^3(1 - y^2)\, dx\, dy$, where R is defined by $0 \le x \le 1$ and $0 \le y \le 1$.

21. Evaluate the double integral over the nonrectangular region R: $\displaystyle\iint_R 4xy\, dx\, dy$, where R is defined by $x^2 \le y \le 1$ and $0 \le x \le 1$.

9 Differential Equations

Section 9–1
Introduction to Differential Equations

Section 9–2
Separation of Variables

Section 9–3
Integrating Factors

Case Study
Models of Growth: World Population

CHAPTER OVERVIEW

Often an application describes a physical phenomenon by an equation that mentions the rate of change of a quantity. This means that the derivative of the quantity is present in the equation. An equation involving properties about an unknown function and its derivatives is called a differential equation. This chapter discusses some of the basic concepts of differential equations.

In Section 9–1 we describe various types of differential equations and their solutions. The next two sections describe two types of techniques of solving differential equations. Section 9–2 studies a type of differential equation whose terms can be arranged in a particular fashion, with the terms containing x on one side of the equation and those containing y' on the other. Section 9–3 considers a technique called solution by integrating factors. This applies to equations whose form is similar to those that can be solved by separation of variables. For this reason solution by integrating factors is often the method attempted if an equation cannot be solved by separation of variables.

CASE STUDY PREVIEW

One of the most fundamental questions of all humankind is, "How do things grow?" This question first occurs to us at the age of 2 or 3. In its most basic form we wonder how many humans can the earth sustain? One of the first comprehensive scientific studies of world population was written by Raymond Pearl in 1925, called *The Biology of Population Growth*. Pearl uses differential equations of growth to model diverse populations such as rats, pumpkins, tadpoles, and yeast. Then he models various countries, such as the United States, Sweden, and Germany before applying the techniques to world population. The Case Study looks at Pearl's techniques and compares them with the way scientists study world population today.

9-1 INTRODUCTION TO DIFFERENTIAL EQUATIONS

Many physical phenomena can best be described by equations involving rates of change, or derivatives. An equation giving information about an unknown function and its derivatives is called a **differential equation.** In this section we describe various types of differential equations and their solutions. The following sections demonstrate several techniques for solving differential equations.

An Application

As an example of how differential equations occur naturally in applications, consider the original work in the field of population growth.

The famous economist Thomas Malthus is credited with the first mathematical model of population growth in 1798. In 1907 A. J. Lotka used Malthus's model to study animal as well as human populations. Even though modern mathematical models of population growth are somewhat more refined and complicated, they are, nonetheless, based on these early versions. Let the population size be represented by $N(t)$ at time t. Mathus's model states that the rate of increase of a population $N'(t)$ is proportional to the size of the population. That is, Malthus assumed that N' is proportional to N. If we let r represent the constant of proportionality, the model is given by the differential equation

$$N'(t) = rN(t)$$

If we let $y = N(t)$, the equation becomes

$$y' = ry$$

With this assumption, Malthus solved the equation for the function $y = N(t)$. He then considered different constants r and reached various conclusions. One of his most fundamental conclusions was that population tends to increase more rapidly than food supplies. While few people today believe in Malthus's theory, his methods provided the springboard for rigorous population study. For instance, Malthus's theory suggested to Charles Darwin the relationship between progress and survival of the fittest.

In this section we show what type of function is a solution of Malthus's model, and in the next section we give the solution technique.

Some Definitions

Examples of some differential equations are

$$y' + 3y = 0$$
$$y' = 6x^2 - 5$$
$$y' + y - 5x = 1$$
$$y'' + yy' = 5$$

The number of the highest derivative that appears in the differential equation is called the **order** of the differential equation. The order of each of the first three equations is one, while the order of the fourth equation is two.

In each equation the unknown is the function y, which is a function of x. The unknown quantity is often referred to merely as y, but if we want to emphasize that it is a function of x, it will be expressed as $y = f(x)$ or simply $f(x)$.

A **solution** of a differential equation is a function $f(x)$ such that when it is substituted into the differential equation for y the result is an identity that is true for all values of x. For example, a solution of the first equation above, $y' + 3y = 0$, is $y = f(x) = 2e^{-3x}$. This is because when the expression for $f(x)$ and $f'(x)$ are substituted into the equation for y and y', respectively, the result is true for all x. Example 1 illustrates this idea in more detail for another solution of this equation.

EXAMPLE 1

Problem

Show that $f(x) = 5e^{-3x}$ is a solution to the differential equation $y' + 3y = 0$.

SOLUTION If $y = 5e^{-3x}$, then $y' = -15e^{-3x}$. Substituting these expressions into the equation yields

$$y' + 3y = 0$$
$$-15e^{-3x} + 3(5e^{-3x}) = 0$$

This expression is true for all values of x. Thus $y = f(x) = 5e^{-3x}$ is a solution of the equation $y' + 3y = 0$.

Example 1 and the statement immediately before it show that a differential equation can have more than one solution. In fact, the differential equation $y' + 3y = 0$ has an infinite number of solutions. They are all of the form $f(x) = Ce^{-3x}$ for any constant C. This is called the **general solution** of the differential equation, and C is called an **arbitrary constant.** A differential equation may have more than one arbitrary constant, as Example 3 will demonstrate. When the arbitrary constants are replaced by specific numbers the resulting function is called a **particular solution.** To illustrate, consider the differential equation $y' + 3y = 0$; the general solution is $y = Ce^{-3x}$, and two particular solutions are $y = 2e^{-3x}$ and $y = 5e^{-3x}$. Since an arbitrary constant can take on any real number, there is an infinite number of particular solutions of a differential equation. The graphs of several particular solutions of $y' + 3y = 0$ are sketched in Figure 1.

There are two fundamental differences between differential equations and algebraic equations like $x^2 - 4x = 0$. First, the unknown in a differential equation represents a function, while the unknown in an algebraic equation represents a number. Also, a differential equation usually has an infinite number of solutions, whereas most algebraic equations have a finite number of solutions.

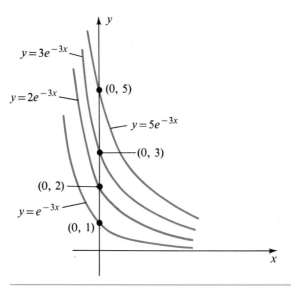

FIGURE 1   Several solutions of $y' + 3y = 0$.

Solving Differential Equations

The theory of differential equations involves studying various types of equations and applying appropriate solution techniques. There is no single method that works in every case. Thus far we have not actually solved a differential equation; we have only stated that a specific function is a solution of a given equation. In the remainder of this section we solve simple differential equations whose solution involves one or two integrations. A key step in many solution techniques uses the fact that

$$\int y' \, dx = y + C$$

This states that the indefinite integral, or the antiderivative, of y' is equal to y plus an arbitrary constant. When a differential equation is integrated, only one arbitrary constant is included, as is demonstrated in the next example. The differential equation in Example 2 shows that antiderivative problems can be expressed as differential equations of the form $y' = g(x)$ for some function g of x.

EXAMPLE 2   **Problem**

Find the general solution and one particular solution of the differential equation $y' = 6x^2 - 5$.

SOLUTION To find the general solution, integrate the differential equation, which means find the antiderivative of each side of the equation.

$$\int y' \, dx = \int (6x^2 - 5) \, dx$$

$$y = 6\left(\frac{x^3}{3}\right) - 5x + C$$

$$= 2x^3 - 5x + C$$

Thus the general solution is $f(x) = 2x^3 - 5x + C$. To find a particular solution, let C equal any specific value. For instance, the particular solution for $C = -1$ is $f(x) = 2x^3 - 5x - 1$.

The next example illustrates how a general solution of a differential equation may have more than one arbitrary constant.

EXAMPLE 3

Problem

Find the general solution and one particular solution of the differential equation $y'' = 12x^2 - 2$.

SOLUTION To find the general solution, we must integrate twice. Observe that the antiderivative of the second derivative is the first derivative. Thus the first integration results in an equation involving y'. Then the solution can be found as in Example 2. The two integrations produce two arbitrary constants. Integrating the original equation yields

$$\int y'' \, dx = \int (12x^2 - 2) \, dx$$

$$y' = 4x^3 - 2x + C$$

Integrating this equation produces

$$\int y' \, dx = \int (4x^3 - 2x + C) \, dx$$

$$= x^4 - x^2 + Cx + K$$

Thus the general solution is $f(x) = x^4 - x^2 + Cx + K$. The two arbitrary constants are C and K. To find a particular solution, let C and K equal any specific values. For instance, the particular solution for $C = -1$ and $K = 8$ is $f(x) = x^4 - x^2 - x + 8$.

Initial Value Problems

One of the most common ways to specify a particular solution of a differential equation is to indicate that the solution has certain properties in the statement of the problem. These properties are called **initial conditions.** Usually an initial condition is a specific functional value of the solution or one of its derivatives. When a problem is to determine the solution of a differential equation with specific initial conditions it is called an **initial value problem.** Example 4 gives two illustrations of initial value problems. The notation $y(a)$ represents $f(a)$ where $y = f(x)$ is the solution of the differential equation.

EXAMPLE 4

Problem

Solve the initial value problems: (a) $y' + 3y = 0$, $y(0) = -10$ and (b) $y'' = 12x^2 - 2$, $y(1) = 0$, $y'(0) = -9$.

SOLUTION (a) The general solution of this differential equation was given immediately after Example 1. It is $y = f(x) = Ce^{-3x}$. The initial condition $y(0) = -10$ means that $f(0) = -10$. Substituting this into the general solution yields

$$-10 = f(0) = Ce^{-3(0)} = Ce^0 = C(1) = C$$

This means that the particular solution corresponding to $y(0) = -10$ is the particular solution with $C = -10$, which is $y = -10e^{-3x}$.
(b) The general solution was found in Example 3. It is $y = f(x) = x^4 - x^2 + Cx + K$. Since there are two arbitrary constants, there must be two initial conditions to specify a particular solution. The second initial condition involves y' so that we need its formula. The formula can be found directly or from Example 3. It is

$$y' = f'(x) = 4x^3 - 2x + C$$

Substituting $y'(0) = -9$ into this equation yields

$$-9 = y'(0) = 4(0)^3 - 2(0) + C = C$$

Thus $C = -9$, which is then substituted into the formula for $f(x)$. This gives $f(x) = x^4 - x^2 - 9x + K$. Next substitute the other initial condition, $y(1) = 0$, into the formula. This yields

$$0 = y(1) = f(1) = (1)^4 - (1)^2 - 9(1) + K$$
$$0 = 1 - 1 - 9 + K$$
$$K = 9$$

Therefore the particular solution corresponding to the given initial conditions is $f(x) = x^4 - x^2 - 9x + 9$.

Some Applications from Biology and Business

The differential equation in Malthus's population growth model, $y' = ry$, is sometimes referred to as the growth equation. The equation in Example 1, $y' + 3y = 0$, is a special case of the growth equation with $r = -3$ because it can be expressed as $y' = -3y$. The general solution of the latter equation is $y = Ce^{-3x}$. This suggests that the general solution of the growth equation is $y = Ce^{rx}$. Malthus used this solution to predict that overpopulation would soon take place because even if the rate of increase r of population were lowered, the rate would still lag behind the rate of increase in food supply, which increases linearly, not exponentially.

A common model in anatomy as well as other disciplines concerns a solution moving along a thin tubular membrane and a solute diffusing across the tube wall. The problem is to find the equation that expresses the concentration $C(x)$ of the solute at a point x of the membrane. Fick's law states that the amount of solute diffusing out across the tube wall is given by the product of the wall surface area, a permeability constant k for the solution, and the average concentration difference over a given short interval. Given the assumptions that the membrane is a right circular cylinder of radius r and length L, and that the average concentration difference can be approximated by $A - Bx^2$, where A and B are constants depending on the velocity of the solution, Fick's law states that

$$C'(x) = 2\pi rLk \, (A - Bx^2)$$

EXAMPLE 5

Problem

Find (a) the general solution of the differential equation in Fick's law for $r = 3$, $L = .1$, $k = 5$, $A = 10$, and $B = 1$, and (b) the particular solution when $C(0) = 0$.

SOLUTION Differentiate the equation

$$C'(x) = 3\pi(10 - x^2)$$

$$\int C'(x) \, dx = \int 3\pi(10 - x^2) \, dx$$

$$C(x) = 3\pi \left(10x - \frac{x^3}{3} \right) + K$$

$$= 30\pi x - \pi x^3 + K$$

(b) When $x = 0$ then $C = 0$, which implies that $K = 0$. Hence the particular solution is

$$C(x) = 30\pi x - \pi x^3$$

In a model of buying behavior Longton* assumes that a customer's decision to buy a brand item depends on the identification the buyer has with the brand. The image that the buyer has of the brand is formed by a set of prescribed elements, including past use of the item, advertising recognition, and recommendations from other users. Let y represent the degree of confidence in the brand, measured by percent, where $y = 100\%$ means the customer would buy the brand immediately. Longton shows that the rate of increase of y over time x obeys the following equation for some products:

$$y' = ay(1 - y) \tag{1}$$

*P. A. Longton, "Mathematics, Models and Marketing," *Human Relations,* Vol. 18, 1965, pp. 289–296.

This type of equation appears often in applications, so we look at one of its solutions in the next example.

EXAMPLE 6 | **Problem**

Show that $y = (1 + e^{-ax})^{-1}$ is a solution to equation 1.

SOLUTION Taking the derivative of the left-hand side of the equation yields

$$
\begin{aligned}
y' &= (-1)(1 + e^{-ax})^{-2}(-ae^{-ax}) \\
&= -(1 + e^{-ax})^{-2}(-ae^{-ax}) \\
&= a(1 + e^{-ax})^{-2}(e^{-ax})
\end{aligned}
\tag{2}
$$

Note that $y^2 = (1 + e^{-ax})^{-2}$ and $y^{-1} - 1 = e^{-ax}$. Make these substitutions in equation 2.

$$
\begin{aligned}
y' &= a(1 + e^{-ax})^{-2}(e^{-ax}) \\
&= ay^2(y^{-1} - 1) \\
&= ay(1 - y)
\end{aligned}
$$

This shows that $y = (1 + e^{-ax})^{-1}$ satisfies equation 1.

EXERCISE SET 9–1

• In Problems 1 to 6 show that the function $f(x)$ is a solution of the differential equation.

1. $y' + 5y = 0$, $f(x) = 2e^{-5x}$
2. $y' - 6y = 0$, $f(x) = -9e^{6x}$
3. $y'' + 2y' - 3y = 0$, $f(x) = e^{-3x}$
4. $(y')^2 - 4y - 4 = 0$, $f(x) = x^2 - 1$
5. $y'' - 2y' + y = 0$, $f(x) = e^x - xe^x$
6. $y'' - 2y' + y = 0$, $f(x) = 2e^x - 3xe^x$

In Problems 7 to 12 find the general solution and a particular solution of the differential equation.

7. $y' = 4x^3 + 2x - 1$
8. $y' = 6x^5 + x^2 - 2$
9. $y' = x^{-3} + e^x$ 10. $y'' = 6x^3 - x^{-3}$
11. $y'' = e^{2x} + x^{1/2}$ 12. $y'' = e^{-3x} + x^{-1/3}$

In Problems 13 to 16 solve the initial value problem.

13. $y' = 8x^3 + 2x^{-3} - 1$, $y(1) = 0$
14. $y' = 6x^5 - x^{1/2} - 2$, $y(0) = 0$
15. $y'' = e^{2x} + x$, $y(0) = -3, y'(0) = 0$
16. $y'' = e^{-3x} + x^{-1/3}$, $y(0) = 2, y'(0) = 1$

In Problems 17 to 22 find the order of the differential equation.

17. $y' = y - x^2 - 2$ 18. $y' + x^{-3} = e^x$
19. $y'' - y' + xy = x^{-3}$ 20. $y''' + e^x + y' = x^{1/2}$
21. $2y''' + y'' - 2y = e^{3x}y$
22. $2y''' + (y')^2 - 2y = e^{3x}y$

•• In Problems 23 to 32 solve the initial value problem.

23. $y' = xe^{x^2}$, $y(0) = 1$
25. $y' = xe^{2x^2}$, $y(0) = 1$
26. $y' = x^2e^x$, $y(0) = 0$
27. $y'' = xe^x$, $y(0) = 0, y'(0) = 1$
28. $y'' = x^2e^{-x}$, $y(0) = 1, y'(0) = 0$
29. $y'' = x \ln x$, $y(1) = 1, y'(1) = -\dfrac{1}{2}$
30. $y'' = x^2 \ln x$, $y(1) = 1, y'(1) = 1$
••• 31. $y'' = ky'$ for a constant k
32. $y''' = ky''$ for a constant k

33. If a new product that uses new technology is introduced to the market before the bugs are

ironed out, that is, before its reliability is established, the usual result is that sales are far less than expected. In an article studying the timing of new products, Kalish and Lilien apply their model to the introduction of a new type of solar battery.* Let the cumulative number of adopters of the product be denoted by X. The authors assume that the rate of change of X with respect to time t is proportional to two quantities: (1) $D_t X$ is proportional to a factor $a + bX$, the so-called "diffusion effect," because the increase in the number of new adopters depends on the present adopters spreading the news about the product's effectiveness, and (2) $D_t X$ is also proportional to the number of potential adopters, which is $N - X$ where N is the total potential adopting population, which is the so-called "saturation effect." Thus they assumed that X satisfied the differential equation

$$D_t X = (a + bX)(N - X)$$
$$= aN + bXN - aX - bX^2$$

Find the order of the differential equation.

34. In radiology the study of the rate of growth of tumors depends on the diffusion or spread of oxygen throughout the tumor. Franko and Freedman studied the diffusion of oxygen in spherical tumor models in order to improve radiation techniques.† They let $v(r)$ represent the concentration of oxygen r mm from the center of the spheroid and still inside the spheroid. They showed that $v(r)$ satisfies the differential equation

$$rv'' + rv' - Q = 0$$

where Q is a constant depending on the consumption of the oxygen in the tissue. Find the order of the differential equation.

35. The study of choice behavior in psychology and economics has a wide range of applications, including marketing and advertising strategies. In one such model, called the "wandering ideal point" model, subjects are asked to compare a pair of stimuli, which are such things as individuals, colors or products, and give a subjective measure of the difference between the two. The difference is then assigned a numerical value. For instance, if the stimuli are the same, the distance is 0. Other distances are assigned on a subjective basis depending on the subjects' responses. An "ideal point" is established, and a distance function $L(x)$ is defined. It measures the distance between each stimuli and the ideal point. The likelihood that stimuli a is preferred over b is $L(a) - L(b)$. If this quantity is negative, then b is preferred over a. In one such study, De Soete, et al., used data from 234 undergraduates who were asked to compare nine celebrities: three actresses, including Elizabeth Taylor; three athletes, including Carl Yastrzemski; and three politicians, including L. B. Johnson.‡ In one scoring method, it was shown that $L(x)$ satisfied the relationship

$$D_x \ln L = x$$

Express this equation as a differential equation in L, L' and x, and find the order of the equation.

*Shlomo Kalish and Gary Lilien, "A Market Entry Timing Model for New Technologies," *Management Science,* Vol. 32, 1986, pp. 194–205.

†A. J. Franko and H. I. Freedman, "Model of Diffusion of Oxygen to Spheroids Grown in Stationary Medium—I. Complete Spherical Symmetry," *Bulletin of Mathematical Biology,* Vol. 46, 1984, pp. 205–217.

‡Geert De Soete, et al., "The Wandering Ideal Point Model: A Probabilistic Multidimensional Unfolding Model for Paired Comparisons Data," *Journal of Mathematical Psychology,* Vol. 30, 1986, pp. 28–41.

9–2 SEPARATION OF VARIABLES

The theory of differential equations involves studying the various types of equations and applying appropriate techniques to solve them. In this section we study the type of differential equation whose terms can be arranged so that all those containing x are on one side of the equation and those containing y are on the other. Also, the terms containing y must have y' as a factor. This means that the equation can be expressed in the form

$$f(y)y' = g(x)$$

Such a differential equation is called **separable** since the variables can be "separated" by the equals sign.

An Application: World Oil Supply

The world's seemingly insatiable thirst for energy makes the continuing search for new oil deposits a necessity. Our need for oil rivals our need for water. One rude fact prevails, however: the earth's supply of oil is finite. When will the supply dry up and how should we use the reserves most economically? In their book *The Future of Oil,* economists Peter Odell and Kenneth Rosing use computer simulations to predict when the earth's oil supplies will diminish.* Their model is based on a differential equation that takes into account future demand as well as various growth rates of the oil supply.

If we let y represent the world's supply of oil, then y is a function of time x. Let r be the growth constant of y, which takes into account demand and discovery potential. Odell and Rosing assume that y' is proportional to two quantities: (1) the more oil on hand, the more we tend to consume, meaning that y' depends on y itself, and (2) as the supply starts to run low, the rate of growth will slow because there is less to discover, meaning that y' is also dependent on the total amount of oil, T, that the earth has to offer. Hence as y approaches T, y' decreases. This means that y' is proportional to $1 - y/T$. (Of course, T is unknown, so the authors make several runs of their model by computer with many different choices of T.) Given these two assumptions, the differential equation in the model is

$$y' = ry\left(1 - \frac{y}{T}\right)$$

This type of differential equation is solved by separation of variables. After we investigate how to apply the method, we will solve the equations for one set of constants r and T that Odell and Rosing used. The solution of this type of differential equation is called the **logistic equation.**

Solving a Separable Differential Equation

Several examples of separable differential equations are

$$y' + 3y = 0$$
$$y^2y' = 2x$$
$$y' + 2xy = 2x$$

*Peter R. Odell and Kenneth E. Rosing, *The Future of Oil,* New York, Nichols Publishing, 1980.

The first equation above was studied in the previous section. We stated that its general solution is $y = Ce^{-3x}$ without explaining how to actually obtain the solution. Now we demonstrate the solution technique. As it stands, the equation is not in the proper form to be solved by the technique of separation of variables. To express it in proper form, subtract the term $3y$ from each side of the equation and then divide by y. This results in

$$y' = -3y$$

$$\frac{1}{y} y' = -3$$

Now the equation is in the proper form, $f(y)y' = g(x)$, where $f(y) = 1/y$ and $g(x) = -3$. The next step is to express y' as $\frac{dy}{dx}$. The equation is then written as

$$\frac{1}{y} \frac{dy}{dx} = -3$$

Integrate both sides of the equation with respect to x. This gives the equation

$$\int \frac{1}{y} \frac{dy}{dx} \, dx = \int -3 \, dx \tag{1}$$

Then think of dy and dx as differentials and formally "cancel the dx's." This results in

$$\int \frac{1}{y} \, dy = \int -3 \, dx \tag{2}$$

Strictly speaking, "canceling the dx's" is equivalent to making the substitution $u = y(x)$, so $du = y' \, dx = \frac{dy}{dx} \, dx$. Thus the left-hand side of equation (1) can be expressed as

$$\int \frac{1}{u} \, du$$

Since any variable can be used during integration, this results in equation 2.

In this form equation (2) can be solved because each side can be integrated. The left-hand side is integrated with respect to y and the right-hand side is integrated with respect to x. This yields

$$\int \frac{1}{y} \, dy = \int -3 \, dx$$

$$\ln y = -3x + C$$

To solve for y, let each side be the exponent of e.

$$e^{\ln y} = e^{-3x+C} \tag{3}$$

Recall that the functions e^x and $\ln x$ are inverses of each other so that $e^{\ln y} = y$. Also, express e^{-3x+C} as $e^{-3x}e^C$ and let $e^C = K$. Then equation (3) becomes

$$y = Ke^{-3x}$$

Let us review the steps in solving a differential equation by this method.

Procedure for Solving a Differential Equation by the Method of Separation of Variables

1. Arrange the equation in the form $f(y)y' = g(x)$.
2. Express y' as $\dfrac{dy}{dx}$ and "multiply the equation by dx." The equation is then in the form $f(y)\, dy = g(x)\, dx$.
3. Integrate each side of the equation, the left-hand side with respect to y and the right-hand side with respect to x.
4. Solve for y.

Example 1 illustrates how to apply the procedure.

EXAMPLE 1

Problem

Solve the differential equation by separation of variables: $y^3y' = 2xy$.

SOLUTION Follow the steps in the procedure.

1. To separate the variables, divide the equation by y to obtain

$$y^2y' = 2x$$

2. Replacing y' by $\dfrac{dy}{dx}$ and multiplying by dx results in

$$y^2\, dy = 2x\, dx$$

3. Integrating yields

$$\int y^2\, dy = \int 2x\, dx$$

$$\frac{y^3}{3} = x^2 + C$$

4. To solve for y, multiply by 3 and then raise each side of the equation to the 1/3 power.

$$y^3 = 3x^2 + K \qquad (\text{where } K = 3C)$$

$$y = \sqrt[3]{3x^2 + K}$$

Let us make two points concerning the constants of integration in Example 1. Only one constant of integration is needed in step 3. In step 4 the arbitrary constant could just as well be expressed as $3C$, but since C represents any value, so too does $3C$, and it is common to drop the number 3 and express the constant with just a letter. Thus $3C$ is replaced by K.

If step 1 cannot be done, so that the equation cannot be expressed in the proper form, then the technique of separation of variables cannot be used and another method must be tried. For example, the equation $y' + xy = x^3$ cannot be expressed in the form $f(y)y' = g(x)$, so it cannot be solved by separation of variables. The next section studies a technique that can be used to solve this equation.

Constant Function Solutions

A subtle but important point has thus far been overlooked. In Example 1 the solution of the differential equation was stated to be $y = \sqrt[3]{3x^2 + K}$. However, this does not represent all solutions. There is one additional function that is a solution. It is the constant function $y = f(x) = 0$. This function satisfies the equation $y^3y' = 2xy$ because both sides are 0 when it is substituted for y. Why did the procedure not uncover this solution? It is because whenever an equation is divided by a variable, it must be assumed that the variable is not 0. In Example 1, the first step requires division by y. It is tacitly assumed that no solution $y = f(x)$ can equal 0 at any value of x. Notice also that by assuming y is not 0, we ignore the possibility that the function $f(x) = 0$ is a solution. This means that this specific function must be tested separately. By substituting $y = f(x) = 0$ into the original equation, we see that it indeed a solution. Therefore the solution of the differential equation $y^3y' = 2xy$ must be expressed as

$$y = \sqrt[3]{3x^2 + K}, \qquad y = 0$$

Sometimes this special case is already represented in the expression for the general solution. For instance, in the beginning of this section the general solution of the differential equation $y' + 3y = 0$ was found to be $y = Ke^{-3x}$. Its solution entailed dividing by y, so the constant function $y = 0$ should be checked to see if it is a solution. It does satisfy the equation, but notice that $y = 0$ is already represented in the general solution by the particular solution corresponding to $K = 0$. This means that it does not have to be singled out in the statement of the solution, as is the case for the equation in Example 1.

The method of separation of variables often requires division by a variable or an expression containing the variable. It is recommended that whenever this occurs set the expression equal to 0 and solve for y. This results in a constant function $y = C$. Test this function in the original equation to see if it is a solution. The next example demonstrates how this can happen.

EXAMPLE 2 | **Problem**

Solve the differential equation $y' + 2xy = 2x$.

SOLUTION

1. To separate the variables, subtract $2xy$ from each side of the equation to get $y' = 2x - 2xy$. Then factor $2x$ from the two terms on the right-hand side to get $y' = 2x(1 - y)$. Now divide the equation by $1 - y$, which yields

$$\frac{y'}{1 - y} = 2x$$

2. Replacing y' by $\dfrac{dy}{dx}$ and multiplying by dx results in

$$\frac{dy}{1 - y} = 2x \, dx$$

3. Integrating, using the substitution $u = 1 - y$ on the left-hand side, yields

$$-\ln(1 - y) = x^2 + C$$

4. To solve for y, multiply by -1 and then let each side of the equation be the exponent of e.

$$e^{\ln(1 - y)} = 1 - y = e^{-x^2 - C}$$
$$y = 1 - Ke^{-x^2} \qquad \text{(where } K = e^{-C}\text{)} \qquad\qquad (4)$$

Now investigate whether division by 0 resulted in a "lost" solution. Step 1 required division by $1 - y$, so the remainder of the steps assume that $1 - y$ is not 0. Setting $1 - y = 0$ yields $y = 1$. Test this constant function to see if it is a solution of the original equation, $y' + 2xy = 2x$. If $y = 1$, then $y' = 0$ (the derivative of a constant function is 0), so the equation becomes $0 + 2x(1) = 2x$, which is true for all values of x. Therefore $y = 1$ is a solution. We do not have to list the solution $y = 1$ separately because it is included in the general solution with $K = 0$. But it still had to be tested separately, because in equation (3) $K = e^C$ cannot be 0 for any value of C. Thus the general solution is

$$y = 1 - Ke^{-x^2}$$

The next example solves an initial value problem via separation of variables.

EXAMPLE 3 | **Problem**

Solve the initial value problem $y' = 4x^3y^2, \qquad y(0) = 10$

SOLUTION First find the general solution by the procedure.

1. Separate the variables.

$$\frac{y'}{y^2} = 4x^3$$

2. Replace y' by $\dfrac{dy}{dx}$ and multiply by dx.

$$y^{-2}\, dy = 4x^3\, dx$$

3. Integrate each side.

$$-y^{-1} = x^4 + C$$

4. Solve for y.

$$y = \frac{-1}{x^4 + C}$$

Set $y^2 = 0$, which gives $y = 0$. This is a solution but it is not the particular solution desired because $y(0) = 0$, not 10. To find the desired particular solution, set $y(0) = 10$.

$$10 = y(0) = \frac{-1}{0^4 + C}$$

Thus $10 = -1/C$ and $C = -1/10 = -.1$. Therefore the desired particular solution is

$$y = \frac{-1}{x^4 - .1}$$

World Oil Revisited

Let us return to the differential equation in Odell and Rosing's model of world oil supplies. In one run of their model they let $r = .05$ and $T = 100$ hundred million barrels annually. They let $x = 0$ correspond to the year 1940, and they measured x in 10-year intervals. The equation is

$$y' = .05y\left(1 - \frac{y}{100}\right) = .05y(1 - .01y)$$

Use separation of variables to solve the equation. Express y' as $\dfrac{dy}{dx}$, divide through the equation by $y\,(1 - .01y)$, and multiply by dx. The equation becomes

$$\frac{dy}{y(1 - .01y)} = .05\, dx$$

Express the fraction on the left as the sum of two fractions. (The details of this algebraic technique are outlined in the Case Study.) This yields

$$\frac{1}{y(1 - .01y)} = \frac{1}{y} + \frac{.01}{1 - .01y}$$

Integrating the equation yields

$$\int \frac{1}{y(1 - .01y)} = \int \frac{1}{y} + \int \frac{.01}{1 - .01y} = \int .05 \, dx$$

$$\ln y - \ln[1 - .01y] = .05x + C$$

$$\ln \left[\frac{y}{1 - .01y} \right] = .05x + C$$

Consider each side as an exponent of e.

$$e^{\ln [y/(1 - .01y)]} = e^{.05x + C}$$

$$\frac{y}{1 - .01y} = Ke^{.05x}$$

where $K = e^C$. Cross multiply and solve for y. This gives

$$y = \frac{Ke^{.05x}}{1 + .01 \, Ke^{.05x}}$$

To find the particular solution of the equation that Odell and Rosing use, let $y(0) = 50$, that is, assume that in 1940 the supply of oil was 50 million barrels. Then

$$y(0) = \frac{K}{1 + .01K} = 50$$

$$K = \frac{50}{.5} = 100$$

The particular solution is

$$y = \frac{100e^{.05x}}{1 + e^{.05x}}$$

This is an example of a logistic equation. It has an S-shaped curve that is common in many applications of calculus. Various applications from different disciplines are mentioned in the Case Study.

An Application from Sociology

Sociologists use the same type of differential equation to model various aspects of social process, for example, Arminger's* study of panels, Neilsen's† analysis of

*Gerhard Arminger, "Linear Stochastic Differential Equation Models for Panel Data with Unobserved Variables," *Sociological Methodology,* Vol. 16, 1986, pp. 187–212.

†François Neilsen, "The Flemish Movement in Belgium After World War II: A Dynamic Analysis," *American Sociological Review,* Vol. 45, 1980, pp. 76–84.

the Flemish movement in Belgium and Rosenfeld's* analysis of career dynamics. The next example illustrates one such use of the logistic equation in the social sciences.

EXAMPLE 4

In sociology the diffusion of information is of fundamental importance. Sociologists assume that the spread of rumors and messages throughout a population is predictable. It is assumed that the rate of diffusion is proportional to the frequency of contact between those who have the message and those who have not. If N represents the proportion of the population that has received the message at time x, and so $1 - N$ is the proportion that has not, then

$$N' = rN(1 - N)$$

The constant r is the growth constant of N.

Problem

Find the general solution of the differential equation.

SOLUTION Expressing N' as $\dfrac{dN}{dx}$ and separating variables yields

$$\frac{dN}{N(1 - N)} = r\, dx$$

Express the fraction on the left as the sum of two fractions.

$$\frac{1}{N(1 - N)} = \frac{1}{N} + \frac{1}{1 - N}$$

Integrating the equation yields

$$\int \frac{1}{N(1 - N)} = \int \frac{1}{N} + \int \frac{1}{1 - N} = \int r\, dx$$

$$\ln N - \ln [1 - N] = rx + C$$

$$\ln \left[\frac{N}{1 - N} \right] = rx + C$$

Consider each side as an exponent of e.

$$e^{\ln [N/(1 - N)]} = e^{rx + C}$$

$$\frac{N}{1 - N} = Ke^{rx}$$

where $K = e^C$. Cross multiply and solve for N. This gives

$$N = \frac{Ke^{rx}}{1 + Ke^{rx}}$$

*Rachel Rosenfeld, "Race and Sex Differences in Career Dynamics," *American Sociological Review*, Vol. 45, 1980, pp. 583–609.

EXERCISE SET 9–2

• In Problems 1 to 6 determine whether the differential equation can be expressed in the proper form to be solved by separation of variables.

1. $yy' = y^2 x^3$
2. $y^2 y' = y^{-3} x$
3. $yy' - y^2 = x^2$
4. $y' - xy = x^2$
5. $y' - x^2 y = x^2$
6. $y' = x^3 y + e^y$

In Problems 7 to 16 solve the differential equation by separation of variables.

7. $y' = xy^2$
8. $y' = x^3 y^4$
9. $y' = x - xy$
10. $y' = xy^2 + x^2 y^2$
11. $xyy' - x^2 = 4x^3$
12. $x^2 y' + x = \ln x$
13. $\dfrac{y'}{\ln x} = (y - 3)^3$
14. $x^2 y' = 4 - y$
15. $y' = e^{x+y}$
16. $e^x y' = y^2 e^{-x}$

In Problems 17 to 22 solve the initial value problem.

17. $y' = x^2 y^2$, $y(0) = 1$
18. $y' = x^4 y^4$, $y(0) = 3$
19. $y' - 3xy = 3x$, $y(0) = 1$
20. $xy' = x^2 y + x^4 y$, $y(1) = 0$
21. $xy' = -y$, $y(1) = 2$
22. $x^2 y' = y$, $y(1) = -1$

•• In Problems 23 to 28 solve the differential equation.

23. $xyy' = \dfrac{(y^2 - 4)}{x}$
24. $ye^x y' = xe^y$
25. $xy' = y \ln x$
26. $(1 + x)^2 y' = (1 + y)^2$
27. $yy' = x^2 e^y$
28. $y' \ln x = \dfrac{y}{x}$

••• 29. Show that the solution of the differential equation $y' + ky = 0$ is $y = Ce^{-kx}$ for any constant k.

30. Learning curves are used to describe how the cost of producing items decreases as the production process becomes more efficient. Lieberman studied 37 chemical products to learn how learning affects the chemical industry.* Assume that on the first day of production the plant can make A units of a product. As the proficiency of the process increases, the volume of production increases until a maximum of M units is made daily. Let y represent the number of units produced on day x and suppose the rate of production is proportional to $M - y$. Find a general formula for y.

31. When new technology is introduced to the marketplace, the increase in the number of firms adopting it is called the diffusion of the innovation. The principal question is how does the new technology affect market share. In an article studying the effect of the introduction of automatic teller machines on the banking industry, Hannan and McDowell employ a diffusion model based on the premise that the increase in the proportion of firms that adopt the innovation is proportional to the proportion of firms that have already adopted the innovation as well as the proportion of firms that have yet to adopt it.† Let $P(t)$ represent the proportion of firms that have adopted the innovation. Then $P(t)$ satisfies the differential equation

$$D_t P = aP(1 - P)$$

Solve the differential equation for $a = .2$.

32. Activated carbon is one of the most commonly used treatments for removing organic contaminants in drinking water. In a study to evaluate the performance of various techniques of using

---

*Marvin B. Lieberman, "The Learning Curve and Pricing in the Chemical Processing Industries," *Rand Journal of Economics*, Vol. 15, No. 2, 1984, pp. 213–227.

†Timothy H. Hannan and John M. McDowell, "Market Concentration and the Diffusion of New Technology in the Banking Industry," *Review of Economics and Statistics*, Vol. 66, 1984, pp. 686–691.

activated carbon in water treatment. Clark studied utilities from eight cities, including Miami and Cincinnati.‡ Water was treated by activated carbon in various ways and contaminants were adsorbed by the carbon. Let the amount of contaminant adsorbed at time t be given by $C(t)$.

In one part of the study it was assumed that the rate of change of $C(t)$ was proportional to $C - K$, where K is a constant depending on the type of treatment being used. Express this relationship as a differential equation, and solve the equation.

‡Robert M. Clark, "Evaluating the Cost and Performance of Field-scale Granular Activated Carbon Systems," *Environmental Science Technology,* Vol. 21, 1987, pp. 573–581.

9–3 INTEGRATING FACTORS

The theory of differential equations involves classifying equations into sets that have their own technique of solution. Usually the search for the appropriate technique is a trial-and-error method. In this section we study a technique called solution by integrating factors. It applies to equations whose form is similar to those that can be solved by separation of variables. For this reason it is often the method attempted if an equation cannot be solved by separation of variables.

Integrating Factors

Consider a differential equation in the form

$$y' + p(x)y = q(x)$$

This is called a **first-order linear differential equation.** This type of equation can be solved by a technique called **integrating factors.** The method of solution by integrating factors involves multiplying the equation by a function that transforms the equation into one that is readily integrated. This function is called the **integrating factor.** Given the equation in the preceding form, the integrating factor is the function

$$f(x) = e^{\int p(x)\, dx}$$

Example 1 illustrates the technique. Once it is shown how to apply the method we can demonstrate why it works.

EXAMPLE 1

Problem

Solve the differential equation $y' + 2xy = x$.

SOLUTION We have $p(x) = 2x$ and $q(x) = x$. The integrating factor is $f(x) = e^{\int 2x\, dx} = e^{x^2}$. Multiply the equation by the integrating factor.

$$(y' + 2xy)e^{x^2} = xe^{x^2}$$
$$y'e^{x^2} + 2xye^{x^2} = xe^{x^2} \tag{1}$$

The key step in the technique is to use implicit differentiation to obtain

$$\frac{d}{dx}(ye^{x^2}) = y'e^{x^2} + 2xye^{x^2}$$

Substituting this into equation 1 yields

$$\frac{d}{dx}(ye^{x^2}) = xe^{x^2}$$

Once the equation is in this form both sides can be integrated with respect to x. To integrate the left-hand side, note that the integral is merely the antiderivative of the derivative, which is the function itself. To integrate the right-hand side, use the substitution $u = x^2$ so $du = 2x\,dx$ and $(1/2)du = dx$. We get

$$\int \frac{d}{dx}(ye^{x^2})\,dx = \int xe^{x^2}\,dx$$

$$ye^{x^2} = \frac{1}{2}e^{x^2} + C$$

Dividing by e^{x^2} or, equivalently, multiplying by e^{-x^2} yields

$$y = \frac{1}{2} + Ce^{-x^2}$$

Justifying the Method

We can use Example 1 as a model to derive a formula for the solution by integrating factors. Consider the differential equation

$$y' + p(x)y = q(x)$$

The integrating factor is the function

$$f(x) = e^{\int p(x)\,dx}$$

Multiplying the equation by $f(x)$, we get

$$y'f(x) + p(x)yf(x) = q(x)f(x) \tag{2}$$

The left-hand side can be expressed as

$$\frac{d}{dx}[yf(x)]$$

Substituting this expression into equation (2) yields

$$\frac{d}{dx}[yf(x)] = q(x)f(x)$$

Integrating this equation gives

$$\int \frac{d}{dx}[yf(x)]\ dx = \int q(x)f(x)\ dx$$

The left-hand side of this equation is equal to $yf(x)$, so we get

$$yf(x) = \int q(x)f(x)\ dx$$

Dividing by $f(x)$ yields the formula for the general solution of a differential equation by integrating factors.

The formula for the general solution of the first-order linear differential equation

$$y' + p(x)y = q(x)$$

by the method of integrating factors is

$$y = \frac{1}{f(x)} \int q(x)f(x)\ dx$$

where $f(x) = e^{\int p(x)\ dx}$.

Notice that the factor $1/f(x)$ in the solution is not part of the integrand. The integration must be done first and the result is divided by $f(x)$.

The next few examples illustrate how to apply the method.

EXAMPLE 2 | **Problem**

Solve the differential equation $y' + y = x$.

SOLUTION We have $p(x) = 1$ and $q(x) = x$. The integrating factor is $f(x) = e^{\int dx} = e^x$. By the formula the general solution is

$$y = \frac{1}{e^x} \int xe^x\ dx \tag{3}$$

To integrate xe^x, use integration by parts: the formula is $\int u\ dv = uv - \int v\ du$. Let $u = x$ and $dv = e^x\ dx$. Then $du = dx$ and $v = \int dv = \int e^x\ dx = e^x$. Thus

$$\int xe^x\ dx = xe^x - \int e^x\ dx$$
$$= xe^x - e^x + C$$

Substituting this into equation (3) yields

$$y = \frac{1}{e^x}(xe^x - e^x + C)$$

Hence the general solution is

$$y = x - 1 + Ce^{-x}$$

Sometimes it is not obvious that the equation is in the proper form for solution by integrating factors. Example 3 gives an illustration of this possibility.

EXAMPLE 3

Problem

Solve the differential equation $xy' = 4x^2 + x - 2y$.

SOLUTION To see if the equation can be expressed in proper form, divide by x, or, equivalently, multiply by x^{-1} to get a term containing just y'. This gives

$$y' = 4x + 1 - 2x^{-1}y$$

Next isolate the terms containing just powers of x and the constant term. This is done by adding $2x^{-1}y$ to each side of the equation. This yields

$$y' + 2x^{-1}y = 4x + 1$$

The equation is now in proper form. We have $p(x) = 2x^{-1}$ and $q(x) = 4x + 1$. The integrating factor is

$$f(x) = e^{\int 2x^{-1} \, dx} = e^{2 \ln x} = e^{\ln x^2} = x^2$$

By the formula the general solution is

$$y = \frac{1}{x^2} \int (4x + 1)x^2 \, dx \tag{4}$$

To integrate, first express the integrand in two terms. That is, write $(4x + 1)x^2 = 4x^3 + x^2$. Then

$$\int (4x + 1)x^2 \, dx = \int (4x^3 + x^2) \, dx = x^4 + \frac{x^3}{3} + C$$

Substituting this into equation 4 yields

$$y = \frac{1}{x^2}\left(x^4 + \frac{x^3}{3} + C\right)$$

Hence the general solution is

$$y = x^2 + \frac{x}{3} + Cx^{-2}$$

An Application: New Product Impact on the Market

Differential equations are used to describe the impact that new technologies have on the marketplace. T. Fisher and R. Pry studied a variety of products and the effect they had on their industries when the products were first introduced.* If a new product or process excites the industry sufficiently to gain 15% of the market, most likely it will continue to gain a larger share until it dominates the market or another competitor takes its place. As examples, Fisher and Pry considered the introduction of water-based paints versus oil-based brands, sulfate versus tree-tapped turpentine, and detergents versus soap. Each product showed the same type of market share growth. The logistic equation is used by Fisher and Pry to model the market share of the new product at time t. The market share y of a competitor can be described by a differential equation of the following form:

$$y' - ky = c$$

The constant k is determined by the new products' potential impact and c is determined by the degree of confidence consumers have in the competitors' products.

EXAMPLE 4

Problem

Solve the differential equation $y' - ky = c$.

SOLUTION To find the general solution, use the integrating factor $e^{\int -k\, dx}$, which is $f(x) = e^{-kx}$. By the formula the general solution is

$$y = \frac{1}{e^{-kx}} \int ce^{-kx}\, dx$$

$$= \frac{1}{e^{-kx}} \left(\frac{-ce^{-kx}}{k} + C \right)$$

Hence the general solution is

$$y = -\frac{c}{k} + Ce^{kx}$$

The next example demonstrates another application of this type of equation in the field of investment analysis.

EXAMPLE 5

Problem

The manager of an investment portfolio wants to earn 5% per year, compounded continuously, with an investment of $1000 per year put into the account. Let y represent the amount of money in the account at time x. (a) Find a differential equation that satisfies these criteria. (b) Find the particular solution when the account starts with $1000 worth of stocks.

*T. C. Fisher and R. H. Pry, *Technological Forecasting and Social Change*, Vol. 3, 1971, pp. 75–91.

SOLUTION (a) The rate of increase of the amount of money in the account increases continuously by .05y and in constant increments of 1000. Thus the equation satisfying these criteria is

$$y' = .05y + 1000$$

This equation can be put into the form of the equation in Example 4 as follows:

$$y' - .05y = 1000$$

Thus $k = .05$ and $c = 1000$. Hence the solution of the equation is

$$y = -\frac{1000}{.05} + Ce^{.05x}$$

$$= -20{,}000 + Ce^{.05x}$$

(b) We are given that $y(0) = 1000$. Thus

$$y(0) = -20{,}000 + C = 1000$$
$$C = 21{,}000$$

Therefore the particular solution is

$$y = -20{,}000 + 21{,}000e^{.05x}$$

EXERCISE SET 9–3

● In Problems 1 to 6 determine whether the differential equation can be expressed in the proper form to be solved by integrating factors.

1. $y' = 2y + 1$
2. $y^2y' = y^3 + y^2$
3. $yy' - y^2 = yx^2$
4. $y' - xy = x^2$
5. $y' - xy^2 = x^2y$
6. $y' = x^3y + e^y$

In Problems 7 to 16 solve the differential equation by integrating factors.

7. $y' + 2y + 1 = 0$
8. $y' - y = x$
9. $y' - y = e^{-x}$
10. $y' + xy = 2x$
11. $y' + x^2y = 3x^2$
12. $x^2y' + y = 3$
13. $y' - y/x = x^3 - 5x$
14. $y' = y + 5x$
15. $y' = x^2(y + 2)$
16. $y' = ye^{-x} + 1$

In Problems 17 to 22 solve the initial value problem.

17. $y' = 2 - y$, $y(1) = 1$
18. $y' = x^3 - yx$, $y(1) = 1$
19. $y' - y/x = x^2 - 1$, $y(1) = 1$
20. $y' - y/x^2 = 5x^{-2}$, $y(1) = 3$
21. $y' = xe^{-x^2} - 2xy$, $y(0) = 1$
22. $y' - y/x^3 = x^{-3}$, $y(1) = 1$

●● In Problems 23 to 28 solve the differential equation.

23. $xyy' = xy^2 - x^2y$
24. $e^xy' = x - ye^x$
25. $xy' = y + x$
26. $(1 + x)y' = 1 + y$
27. $xy' = y - x - x^2e^x$
28. $xy' + 2y + 3xy = xe^{-3x}$

●●● 29. At time t a tank contains 10 pounds (lb) of salt dissolved in 80 gallons (gal) of water. Suppose water containing 1 lb of salt per gallon is being added to the tank at a rate of 5 gal/min, and that the mixed solution is being drained from the tank at the same rate. Find a formula for the amount y of salt in the tank at time t.

30. In an article* describing the growth rate of several types of populations of organisms, it was discovered that the rate of growth was proportional to the number of organisms present plus a constant that depended on the species. Find a differential equation satisfying the criteria.

31. In psychology an important model of learning, called the Hullian model, assumes that on each trial of an experiment the strength of the correct response is increased.† This is because the subject has learned more about the experiment from previous trials. The probability P that a correct response will be given at time x increases as x increases. It is assumed that P' depends not only on P itself, but also on a certain threshhold C, which P' cannot exceed. The Hullian model of learning is described by the equation $P' =$

$C - aP$ for constants a and C. Find the general solution of this equation.

32. Refer to Problem 34 in Section 9–1 which defined a differential equation used in radiology relating the rate of growth of tumors and the diffusion or spread of oxygen throughout the tumor.‡ Let $v(r)$ represent the concentration of oxygen r mm from the center of the spherical tumor. Then $v(r)$ satisfies the differential equation

$$rv'' + 2v' - Q = 0$$

where Q is a constant depending on the consumption of the oxygen in the tissue. Solve the differential equation using integrating factors by first substituting $y = v'$ and dividing through by r.

*Martin H. Kingsley, ''Models of Population Growth of Enzymes,'' *Population Studies*, Vol. 34, 1983.

†William P. Crane, ''Hullian Models of Learning in Education,'' *Psychological Review*, Vol. 81, 1977, pp. 56–58.

‡A. J. Franko and H. I. Freedman, ''Model of Diffusion of Oxygen to Spheroids Grown in Stationary Medium—I. Complete Spherical Symmetry,'' *Bulletin of Mathematical Biology*, Vol. 46, 1984, pp. 205–217.

CASE STUDY   Population Growth

Two facts about population growth are clear: there is a fixed, finite amount of land on this planet, and the human population continues to expand. Have we reached a critical point? Will overpopulation render our species extinct? These questions have a familiar ring today. But are they new to our generation? As early as 1890 politicians, scientists, and social scientists were concerned with the alarmingly sharp rise of the human population. The first attempts to obtain an accurate mathematical model of population growth utilized differential equations. Today's population analysis techniques use more complicated differential equations, but the basic foundation is very similar. We outline these early models and then see how the same models are used today in a variety of ways in order to understand diverse types of populations.

All life starts as a single cell. In higher forms of life this cell divides into two, these divide into four, then eight, and so on. The cells remain attached to each other and over a period of time the process results in a mature individual of the species. Counting the number of cells becomes impossible early in the process. Another way to measure the development of the growth process is to measure weight at periodic intervals. Early researchers, such as Malthus, Pearl, and Reed, found that the curve produced by measuring weight over time had a peculiar shape.* And the shape remained the same for different species. Figure 1 shows the weight curve for two different organisms. Figure 1a is the curve for the white male rat from the data of H. H. Donaldson, and Figure 1b is the curve for the growth of a pumpkin, taken from T. B. Robinson. Pearl and Reed noticed the similarity of the shapes of the curves. The increase is slow at first but then shows a dramatic rise. After a while the increase levels off. They referred to the curve as having a stretched string shape.

The equations of the curves were also similar. The data for the weight y of the rat, measured in grams, with time t measured in days, are approximated by the function

$$y = \frac{273}{1 + 75.2e^{-7.2x}}$$

The data for the pumpkin are approximated by

$$y = \frac{5190}{1 + 29732e^{-16.3x}}$$

Additional data were studied, including the regeneration of a tadpole's tail and the increase in the population of yeast cells. Each set of data produced the same type of S-shaped curve.

When Pearl and Reed were publishing their results they discovered that as early as 1838 a Belgian mathematician, P. F. Verhulst, had used the same type of curve in studying population growth, but his research did not lead very far. They called this type of graph a "logistic curve" because Verhulst first coined the term.

Could the growth of human population exhibit similar characteristics? To

*Raymond Pearl, *The Biology of Population Growth,* New York, Knopf, 1925.

FIGURE 1

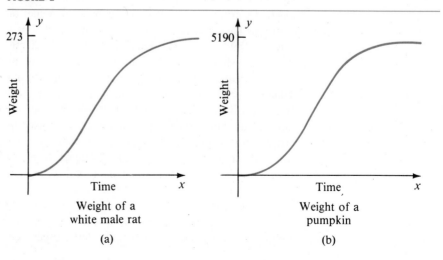

(a) Weight of a white male rat

(b) Weight of a pumpkin

answer this question, they had to discover the underlying principle of such growth. Let N represent the size of a species' population at time x. Pearl and Reed reasoned that if there was no competition from outside forces, N would increase at a constant rate r. This results in the growth equation $N'(x) = rN(x)$. Here r is called the *growth constant*. In a more realistic situation, competition for resources will be present. This competition will have a greater effect on the species as N gets larger. They assumed that the population cannot exceed a certain maximum size K. So the rate of increase N' is proportional to the term $1 - N/K$. When N is small this term is close to 1, so it has little effect on N'. But when it is close to K this term is small, so N' decreases. They assumed that N' satisfied the differential equation, called the *logistic equation*.

$$N' = rN\left(1 - \frac{N}{K}\right) = rN(1 - bN)$$

We use $b = 1/K$ to ease the computation. To solve this equation, we use separation of variables. First divide the equation by $N(1 - bN)$ and express N' as $\dfrac{dN}{dx}$. This results in the equation

$$\frac{1}{N(1 - bN)} \cdot \frac{dN}{dx} = r$$

Now multiply by dx.

$$\frac{dN}{N(1 - bN)} = r\,dx \tag{1}$$

To integrate the left-hand side, express the integrand as the sum of two fractions whose numerators have yet to be found.

$$\frac{1}{N(1 - bN)} = \frac{A}{N} + \frac{B}{1 - bN} \tag{2}$$

Adding the fraction on the right and equating the two numerators yields

$$1 = A(1 - bN) + BN = A + N(B - bA)$$

The right-hand side is a polynomial in N, which is equal to 1. This means that the term containing N must be 0 and the other term is 1. Therefore $A = 1$ and $B - bA = 0$. Substituting $A = 1$ into the latter equation yields $B = b$. Hence equation 2 becomes

$$\frac{1}{N(1 - bN)} = \frac{1}{N} + \frac{b}{1 - bN}$$

Using this in equation (1) enables us to integrate the left-hand side.

$$\int \frac{dx}{N(1 - bN)} = \int \frac{dx}{N} + \int \frac{b\, dx}{1 - bN}$$

$$= \ln N - \ln(1 - bN)$$

$$= \ln \left[\frac{N}{1 - bN} \right]$$

The integral of the right-hand side of equation (1) is $rx + C$. Therefore the integration of equation 1 yields

$$\ln \left[\frac{N}{1 - bN} \right] = rx + C$$

Consider each side as an exponent of e. This gives

$$\frac{N}{1 - bN} = e^{rx+C} = e^{rx}e^{C} = De^{rx}$$

We let $D = e^{C}$. Solving this equation for N yields

$$N = (1 - bN)De^{rx}$$

Expand the right-hand side and isolate N on the left-hand side of the equation. This yields

$$N = \frac{De^{rx}}{1 + bDe^{rx}} \tag{3}$$

How does this function compare with the weight functions for the rat and the pumpkin? To express this solution in the same way as earlier, divide the numerator and denominator by bDe^{rx}. To get a simpler expression, replace the constant of integration bD by c. The numerator becomes $1/b$, but earlier we let $b = 1/K$, where K is the maximum population of the species. Hence the numerator becomes K. Reverse the order of the terms in the denominator and express $1/e^{rx}$ as e^{-rx}. Then the solution to the logistic equation becomes

$$N = \frac{K}{1 + ce^{-rx}} \tag{4}$$

This is the form of the logistic equation for the Donaldson data on the weight of the rat with $K = 273$, $c = 75.2$, and $r = 7.2$. The equation given at the beginning of the Case Study was

$$y = \frac{273}{1 + 75.2e^{-7.2x}}$$

The first example demonstrates how Pearl used the logistic equation to model world population.

EXAMPLE 1   Pearl determined empirically that the maximum population that the earth could sustain would be 2026 million people. He also assumed a rate of increase of $r = .024$. His model therefore was the logistic equation

$$N' = .024N\left(1 - \frac{N}{2026}\right)$$

Problem

Find the general solution of this differential equation.

SOLUTION The general solution is the type of function in equation 4. Let $K = 2026$ and $r = .024$. Then the general solution is

$$N = \frac{2026}{1 + ce^{-.024x}}$$

EXAMPLE 2

Pearl let $x = 0$ represent the year 1600 and he assumed the population at that time was 445 million people.

Problem

Find the particular solution of Pearl's logistic equation.

SOLUTION Let $N(0) = 445$ in the solution in Example 1. This gives

$$N(0) = 445 = \frac{2026}{1 + ce^0} = \frac{2026}{1 + c}$$

$$445 + 445c = 2026$$

$$445c = 1581$$

$$c = 3.55$$

The particular solution is

$$N = \frac{2026}{1 + 3.55e^{-.024x}}$$

The graph of N is given in Figure 2.

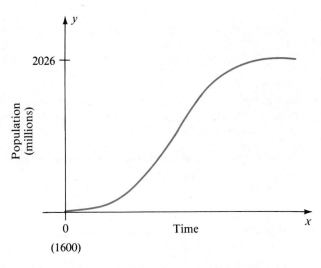

FIGURE 2

The logistic equation was the first realistic mathematical model for human population growth. Even though most modern studies of human population use more refined and complicated models, the logistic equation is still used today to model diverse patterns of growth over short periods. Here are several examples of the varied applications of the logistic equation.

1. There is a plethora of applications in the journal *Population Studies* that use the logistic equation. For instance, in the article "Mortality Crossovers: Reality or Bad Data?" researchers compare the yearly population growth of pairs of selected countries, such as Sri Lanka versus Sweden, Honduras versus France, and Hong Kong versus Denmark, to determine whether the population data are accurate.*

2. Several papers in the book *Population Growth and Economic Development: Issues and Evidence,* use the logistic equation. One article, "School Expenditures and Enrollments," by R. Schultz, studies the economic impact of national educational budgets of various countries. Economic indicators, such as investment, rate of return, supply, demand, and Cobb–Douglas production functions, are used to measure the effects of educational matters such as teacher training and demographic trends.†

3. Two interesting examples of where the logistic curve is applied in animal population growth are predator-prey models (the Hudson Bay Company on lynx and hare populations) in *Mathematical Models* by Richard Haberman,‡ and "Statistical Husbandry: Chance, Probability, and Choice in a Reindeer Management Economy" by T. Ingold.§

4. Two articles, one by Gerhard Arminger‖ and the other by Ross L. Matsueda and William T. Bielby,# in *Sociological Methodology* in 1986 provide examples of the use of the logistic curve in the social sciences.

5. In the business world forecasting demand, production, and sales are extremely important. A good example of how the logistic curve is used in various aspects of business is contained in a paper by T. C. Fisher and R. H. Pry.**

*Ansley J. Coale and Ellen E. Kisker, "Mortality Crossovers: Reality or Bad Data?" *Population Studies,* Vol. 40, 1986, pp. 389–402.

†R. Schultz, "School Expenditures and Enrollments," in D. Gale Johnson and Ronald D. Lee (eds.), *Population Growth and Economic Development: Issues and Evidence,* Madison, University of Wisconsin Press, 1987.

‡Richard Haberman, *Mathematical Models,* Englewood Cliffs, N.J., Prentice-Hall, 1977.

§T. Ingold, "Statistical Husbandry: Chance, Probability, and Choice in a Reindeer Management Economy," *Operations Research,* Vol. 54, 1982, pp. 311–322.

‖Gerhard Arminger, "Linear Stochastic Differential Equal Models for Panel Data with Unobserved Variables," *Sociological Methodology,* Vol. 16, 1986, pp. 187–212.

#Ross L. Matsueda and William T. Bielby, "Statistical Power in Covariance Structure Models." *Sociology Methodology,* Vol. 16, 1986, pp. 120–158.

**T. C. Fisher and R. H. Pry, "A Simple Substitution Model of Technological Change," *Technological Forecasting and Social Change,* Vol. 3, 1971, pp. 75–91.

CHAPTER REVIEW

Key Terms

9–1 **Introduction to Differential Equations**
Differential Equation
Order
Solution
General Solution

Arbitrary Solution
Particular Solution
Initial Condition
Initial Value Problem

9–2 **Separation of Variables**
Logistic Equation

Separable

9–3 **Integrating Factors**
First-Order Linear Differential Equation
Integrating Factors

Integrating Factor

Summary of Important Concepts

The growth equation is $y' = ry$

The logistic equation is $N' = rN(1 - N)$

The solution of $y' + p(x)y = q(x)$ is

$$y = \frac{1}{f(x)} \int q(x)f(x)d(x)$$

where $f(x) = e^{\int p(x)\, dx}$

REVIEW PROBLEMS

1. Show that the function $f(x)$ is a solution of the differential equation.

$$y'' - 2y' + y = 0, \qquad f(x) = 3e^x - 2xe^x$$

In Problems 2 and 3 find the general solution and a particular solution of the differential equation.

2. $y' = 2x^{-3} - e^x$ **3.** $y'' = e^{-3x} - 2x^{-1/3}$

4. Solve the initial value problem.

$$y'' = 2e^{-3x + x^{1/3}}, \qquad y(0) = 2, y'(0) = 1$$

5. Find the order of the differential equation.

$$y''' + e^x y + y'' = x^{1/2}$$

In Problems 6 and 7 solve the initial value problem.

6. $y' = x^2 e^{2x}, \qquad y(0) = 0$

7. $y'' = -xe^x, \qquad y(0) = 0, y'(0) = 1$

In Problems 8 and 9 determine whether the differential equation can be expressed in the proper form to be solved by separation of variables.

8. $y' - x^4 y = x^{1/2}$ **9.** $y' = x^3 y + xe^y$

In Problems 10 to 13 solve the differential equation by separation of variables.

10. $\dfrac{y'}{\ln x} = (y - 1)^2$ **11.** $x^3 y' = 2 - 3y$

12. $xyy' = \dfrac{(y^2 - 4)}{x}$ **13.** $y' \ln x = \dfrac{y}{x}$

In Problems 14 and 15 determine whether the differential equation can be expressed in the proper form to be solved by integrating factors.

14. $y' - xy^3 = x^4y$ **15.** $y' = x^3y + e^{2x}$

In Problems 16 and 17 solve the differential equation by integrating factors.

16. $y' - 2y = e^x$ **17.** $y' = y - 5x$

In Problems 18 to 20 solve the initial value problem.

18. $y' = x^3 - yx,$ $y(1) = 1$

19. $y' - \dfrac{y}{x^3} = x^{-3},$ $y(0) = 1$

20. $xy' = y - x - e^x,$ $y(0) = 0$

10 Trigonometric Functions

CHAPTER OVERVIEW   The study of circles, triangles, and angles dates back to the dawn of civilization. The trigonometric functions, which are defined in terms of these concepts, allow us to create models in such diverse disciplines as astronomy, ecology, navigation, and music. For example, to study sound from a vibrating tuning fork, we must consider "periodic" functions, which are functions that repeat their functional values periodically. These types of functions are utilized in such devices as telephones, stereo systems, radios, and television sets. The trigonometric functions best model this type of phenomenon.

In Section 10–1 we study radian measure of an angle and compare it with the more familiar degree measure. In Section 10–2 we consider the two fundamental trigonometric functions, the sine and cosine functions. Their definitions, graphs, and several relationships between them are presented. The next section looks at the calculus of the sine and cosine functions. Their derivatives and integrals are studied. Besides the sine and cosine, there are four other trigonometric functions: the tangent, cotangent, secant, and cosecant functions. They are presented in Section 10–4.

CASE STUDY PREVIEW

In the early 1980s the Melbourne, Australia, government started an extensive, controversial plan to build a complex underground system of tunnels. Part of the tunnel project would serve as a railway and another part would carry sewerage. The tunnel-boring operations revealed a lack of knowledge of the forces that were generated when cutting through diverse rock formations. Mining engineers in charge of the project contacted the University of Melbourne to start a research project to

419

design rock-boring tools with the proper shaft. Because such a machine uses rotating cutter heads, trigonometric functions must be used to model the operation of the shaft. The Case Study shows how several design problems were overcome.

10−1 RADIAN MEASURE

The practice of measuring an angle by degree dates back to the ancient Babylonians around the third century B.C. One full revolution is divided into 360 equal parts, each representing 1 degree. A degree is further divided into minutes and seconds. Although this form of angular measurement is still common, degree measure is being widely replaced by radian measure. In mathematics it is especially convenient to use radians because degrees are geometric quantities whereas radians are pure numbers. This allows us to perform operations on radians that would be impossible to do with degrees. Other branches of science are using radian measure more, and the international metric system has officially adopted radian measure. In this section we study radian measure and its connection to degree measure.

Radian Measure of Angles

In degree measure a right angle is 90 degrees (90°), one-half of a revolution is 180°, and a full revolution is 360°. To define radian measure, we also refer to revolutions but a reference circle is also needed. Consider a circle of radius 1 whose center is at the origin. It is called the unit circle. An angle with its vertex at the origin is called a central angle. Consider the central angle with its initial side on the positively directed x-axis and rotate the terminal side counterclockwise. As the terminal side rotates it traverses the circumference of the circle. When the terminal side traverses an arc of length 1 of the circumference the central angle is said to have a measure of one **radian.** (See Figure 1.)

FIGURE 1   x is an angle with a measure of 1 radian.

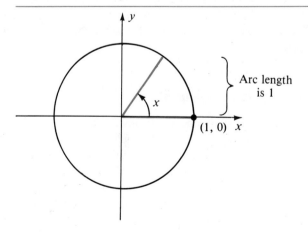

The circumference of the unit circle is 2π units. Since one full revolution of the terminal side traverses the circumference of the circle, there are 2π radians in one full revolution of the circle. Hence the angular measurement of 2π radians equals $360°$. We write

$$2\pi \text{ radians} = 360° \qquad (1)$$

Similarly, one-half of a revolution, or $180°$, is π radians. Expressed as an equation this is

$$\frac{1}{2} 2\pi \text{ radians} = 180°$$

$$\pi \text{ radians} = 180°$$

A right angle, or $90°$, is one-quarter of a revolution, so

$$\frac{1}{4} 2\pi \text{ radians} = 90°$$

$$\frac{\pi}{2} \text{ radians} = 90°$$

Figure 2 shows these two angles.

These particular instances suggest the general relationship between angular measurement in radians and degrees. From equation (1) we can write

$$1° = \frac{2\pi}{360} \text{ radians} = \frac{\pi}{180} \text{ radians}$$

This shows that to convert from degrees to radians, multiply the number of degrees by $\pi/180$.

Let r be the measurement of an angle in radians and d be its measurement in degrees. Then to convert from d^0 to r radians, use the formula

$$r = d^0 = d \cdot \frac{\pi}{180} \text{ radians}$$

To convert from r radians to d^0, solve this formula for d to get

$$d^0 = r = r \cdot \frac{180}{\pi} \text{ degrees}$$

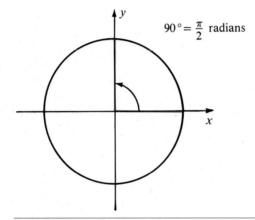

FIGURE 2

From now on we will assume that an angle is measured in radians unless otherwise stated. Examples 1 and 2 illustrate how to use these formulas.

EXAMPLE 1 | **Problem**

Convert the angles from degrees to radians: (a) 60°, (b) 30°, and (c) 135°.

SOLUTION Use the first formula in the display.

(a) $60° = 60 \cdot \dfrac{\pi}{180}$ radians $= \dfrac{\pi}{3}$ radians

(b) $30° = 30 \cdot \dfrac{\pi}{180}$ radians $= \dfrac{\pi}{6}$ radians

(c) $135° = 135 \cdot \dfrac{\pi}{180}$ radians $= \dfrac{3\pi}{4}$ radians

These angles are shown in Figure 3.

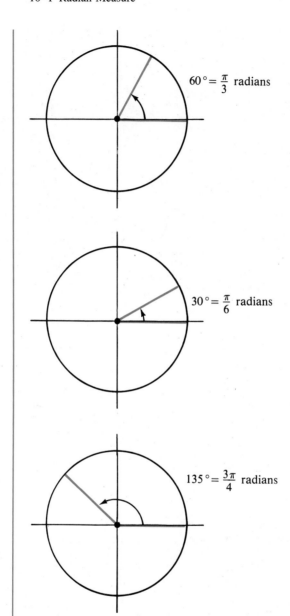

$60° = \frac{\pi}{3}$ radians

$30° = \frac{\pi}{6}$ radians

$135° = \frac{3\pi}{4}$ radians

FIGURE 3

EXAMPLE 2 | **Problem**

Convert the angles from radians to degrees: (a) $\frac{\pi}{4}$, (b) $\frac{2\pi}{3}$, and (c) $\frac{3\pi}{2}$.

SOLUTION Use the second formula in the display.

(a) $\dfrac{\pi}{4} = \dfrac{\pi}{4} \cdot \dfrac{180}{\pi}$ degrees $= 45°$

(b) $\dfrac{2\pi}{3} = \dfrac{2\pi}{3} \cdot \dfrac{180}{\pi}$ degrees $= 120°$

FIGURE 4

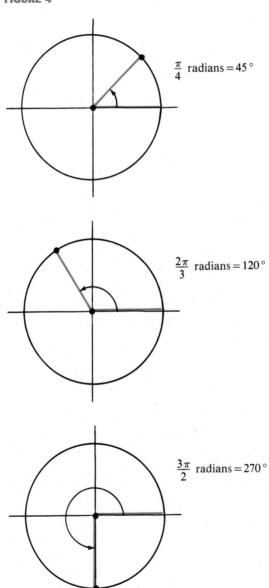

$\dfrac{\pi}{4}$ radians $= 45°$

$\dfrac{2\pi}{3}$ radians $= 120°$

$\dfrac{3\pi}{2}$ radians $= 270°$

(c) $\dfrac{3\pi}{2} = \dfrac{3\pi}{2} \cdot \dfrac{180}{\pi}$ degrees $= 270°$

These angles are shown in Figure 4.

Extending Radian Measure

Often applications require more than one revolution. For example, a combination lock often requires more than one revolution, and a lathe needs many revolutions to form a piece of wood or metal. If t is the radian measure of an angle, then the geometric interpretation of t is the distance traversed on the unit circle by the point P, starting from the point $(1, 0)$ and proceeding in a counterclockwise direction. (See Figure 5.) The radius of the unit circle turns through $t/2\pi$ revolutions. For $t > 2\pi$ the angle goes through more than one revolution. To illustrate:

$t = 4\pi$ radians   corresponds to $\dfrac{4\pi}{2\pi} = 2$ revolutions

$t = 6\pi$ radians   corresponds to $\dfrac{6\pi}{2\pi} = 3$ revolutions

$t = 7\pi$ radians   corresponds to $\dfrac{7\pi}{2\pi} = 3.5$ revolutions

Example 3 shows the connection between degree and radian measure for angles of more than one revolution.

EXAMPLE 3

Problem

Convert the angles from radians to degrees and express the angle in terms of the revolutions that it makes: (a) $\dfrac{9\pi}{4}$, (b) $\dfrac{10\pi}{3}$, and (c) $\dfrac{9\pi}{2}$

FIGURE 5   Radian measure t is the length traversed by P on the circumference of the unit circle.

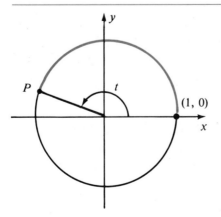

SOLUTION Use the second formula in the display.

(a) $\dfrac{9\pi}{4} = \dfrac{9\pi}{4} \cdot \dfrac{180}{\pi}$ degrees $= 405°$

The angle makes $9\pi/4(2\pi) = 9/8$ revolutions.

(b) $\dfrac{10\pi}{3} = \dfrac{10\pi}{3} \cdot \dfrac{180}{\pi}$ degrees $= 600°$

The angle makes $10\pi/3(2\pi) = 5/3$ revolutions.

(c) $\dfrac{9\pi}{2} = \dfrac{9\pi}{2} \cdot \dfrac{180}{\pi}$ degrees $= 810°$

The angle makes $9\pi/2(2\pi) = 9/4$ revolutions. (See Figure 6.)

It is also important in many applications to distinguish clockwise motion from counterclockwise motion. If the terminal side of the angle is rotated in the clockwise direction, we say the distance traversed on the unit circle is negative. Hence the radian measure is negative if the rotation is in the clockwise direction. For instance, the angle of measure $-\pi/4$ has its terminal side in the fourth quadrant. The angle it makes with the initial side is $45°$, but since it is rotated in the clockwise direction, the angle is negative. Example 4 illustrates several negative angles.

EXAMPLE 4

Problem

Convert the angles from radians to degrees and express the angle in terms of the revolutions that it makes: (a) $-\dfrac{\pi}{6}$, (b) $-\dfrac{4\pi}{3}$, and (c) $-\dfrac{9\pi}{2}$.

SOLUTION (a) $-\dfrac{\pi}{6} = -\dfrac{\pi}{6} \cdot \dfrac{180}{\pi}$ degrees $= -30°$

The angle makes $\pi/6(2\pi) = 1/12$ revolution in the clockwise direction.

(b) $-\dfrac{4\pi}{3} = -\dfrac{4\pi}{3} \cdot \dfrac{180}{\pi}$ degrees $= -240°$

The angle makes $4\pi/3(2\pi) = 2/3$ revolution in the clockwise direction.

(c) $-\dfrac{9\pi}{2} = -\dfrac{9\pi}{2} \cdot \dfrac{180}{\pi}$ degrees $= -810°$

The angle makes $9\pi/2(2\pi) = 9/4$ revolutions in the clockwise direction. (See Figure 7.)

Angular Speed

Consider a wheel rotating around its center at a constant rate. As a point on the circumference travels a certain distance it sweeps an angle x of the circle. The **angular speed** w of A is the rate at which the angle x changes per unit of time. We express w in terms of x and time t by

$$w = \frac{x}{t}$$

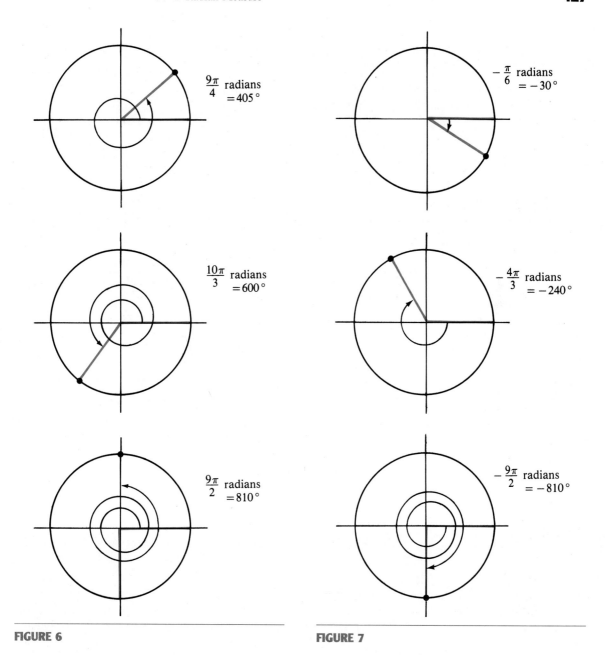

$\dfrac{9\pi}{4}$ radians $= 405°$

$-\dfrac{\pi}{6}$ radians $= -30°$

$\dfrac{10\pi}{3}$ radians $= 600°$

$-\dfrac{4\pi}{3}$ radians $= -240°$

$\dfrac{9\pi}{2}$ radians $= 810°$

$-\dfrac{9\pi}{2}$ radians $= -810°$

FIGURE 6

FIGURE 7

Angular speed is usually measured in radians per second (rad/s). For example, if a wheel is rotating at a rate of 1 revolution per second, then its angular speed is 2π rad/s because

$$w = \frac{2\pi}{1} = 2\pi$$

EXAMPLE 5 | The Case Study mentions a shaft of a rock-cutting machine that rotates at 90 revolutions per minute. Find the angular speed of the shaft.

SOLUTION The rate of 90 rpm corresponds to $90/60 = 3/2$ revolutions per second. Since the angular speed of 1 revolution per second is 2π rad/s, the angular speed of the shaft is

$$w = \frac{3}{2} \cdot 2\pi = 3\pi \text{ rad/s}$$

EXERCISE SET 10–1

• In Problems 1 to 6 convert the angles from degrees to radians.

1. 90°   2. 45°   3. 210°

4. 150°   5. 270°   6. 330°

In Problems 7 to 12 convert the angles from radians to degrees.

7. $\dfrac{\pi}{6}$ 8. $\dfrac{5\pi}{6}$ 9. $\dfrac{5\pi}{2}$

10. $\dfrac{3\pi}{4}$ 11. $\dfrac{7\pi}{4}$ 12. $\dfrac{7\pi}{3}$

In Problems 13 to 18 convert the angles from radians to degrees and express the angle in terms of the revolutions that it makes.

13. $\dfrac{11\pi}{4}$ 14. $\dfrac{7\pi}{3}$ 15. $\dfrac{7\pi}{2}$

16. $\dfrac{17\pi}{4}$ 17. $\dfrac{14\pi}{3}$ 18. $\dfrac{25\pi}{6}$

In Problems 19 to 24 convert the angles from radians to degrees and express the angle in terms of the revolutions that it makes.

19. $-\dfrac{5\pi}{6}$ 20. $-\dfrac{5\pi}{3}$ 21. $-\dfrac{11\pi}{2}$

22. $-\dfrac{5\pi}{4}$ 23. $-\dfrac{14\pi}{3}$ 24. $-\dfrac{29\pi}{6}$

•• In Problems 25 to 28 find the radian measure of the angle whose measure is given in degrees.

25. 390°   26. 540°

27. 1080°   28. 1110°

In Problems 29 to 32 find the radian measure of the angle that makes the given number of revolutions in a clockwise direction.

29. two complete revolutions

30. four complete revolutions

31. two and one-third revolutions

32. three and one-sixth revolutions

In Problems 33 to 36 find the radian measure of the angle that makes the given number of revolutions in a counterclockwise direction.

33. 2 complete revolutions

34. 4.5 revolutions

35. 3.25 revolutions

36. 5.75 revolutions

••• 37. Turning a combination lock two and one-sixth revolutions corresponds to an angle of what radian measure?

38. Find the radian measure of the angle that the hour hand of a clock defines as it advances from 3 P.M. to the given time: (a) 6 P.M., (b) 9 P.M., and (c) 6 A.M.

39. The shaft of a rock-cutting machine rotates at 60 rpm. Find the angular speed of the shaft.

40. The shaft of a rock-cutting machine rotates at 100 rpm. Find the angular speed of the shaft.

41. The National Safety Council has developed a method for determining safe approach speeds

for vehicles at intersections where the view is obstructed.* The method uses a sight, or visibility, triangle whose legs are the perpendicular paths of two vehicles approaching the intersection at right angles. The "sight line" is the line that connects the eyes of the drivers of the two vehicles. The obstruction may be a building or trees or something similar. If the sight line passes through the obstruction, the drivers cannot see each other. The angles made by the sight line and the vehicle paths are critical in computing safe approach speeds. Two common angles made by sight lines are 60° and 30°, respectively. Convert these angles to radians.

42. The transmission of sound waves can be visualized by picturing the propagation of waves on the surface of water. For example, when a tuning fork is struck, air molecules vibrate between the prongs, and this motion causes a wave to move away from the fork in much the same way a pebble dropped in an pond causes a wave on water. Trigonometry is used to model this phenomenon.† The stage that a particular air molecule has reached in its vibration is called its "phase," which is measured in radians. The phase p is related to the angular speed x of the particle by $p = tx + b$. Find p in radians when $t = .2$ seconds, $x = 50$ rad/s, and b is a one-quarter revolution.

43. Beech Aircraft, with headquarters in Wichita, Kansas, is a large producer of general aviation aircraft. In the late 1970s a customer requested that the Beech Super King Air, a twin turboprop aircraft, be modified to increase flying time from 9 hours to 11 hours.‡ This required an increase in fuel capacity of about 100 gallons. Beech decided to install new fuel tanks inside the wing tips. In order to avoid increasing wing stress too much, they decided to position the new tanks so they would drain by gravity into the main tanks. This avoided using heavy pumps. The angle of declination of the tanks was important. Several prototypes were built. They had angles of (a) 2°, (b) 3° and (c) 3.5°. Convert these angles to radians.

*Louis J. Pignataro, *Traffic Engineering: Theory and Practice,* Englewood Cliffs, NJ, Prentice-Hall, 1973, pp. 338–339.

†Robert Erickson, *Sound Structure in Music,* Berkeley, CA, University of California Press, 1975, pp. 1–60.

‡R. D. Marwill, "Tip Tanks for the Beech Super King Air," *Engineering Case Library,* American Society for Engineering Education, ECL 246, 1982, pp. 1–41.

10–2 THE SINE AND THE COSINE

In many applications the values of a variable quantity repeat themselves at regular intervals. For example, in the study of ecology the size of a particular population increases and decreases periodically. The functions used to model this type of behavior are the trigonometric functions. In this section we examine two trigonometric functions, the sine and cosine functions. In Section 10–4 the other four trigonometric functions are defined in terms of these two functions.

There are two common definitions of the sine and cosine. One depends on right triangles, while the other definition depends on the unit circle. Each type of definition has its own advantage. The right triangle definition makes it easier to compute the functional values of certain common angles. The circular definition allows us to extend the domain of the functions. We start with the right triangle definition and study how to compute specific functional values. Then we incorporate this definition with the more general definition.

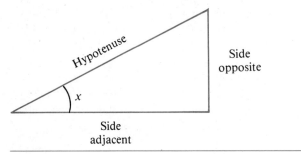

FIGURE 1

Right Triangle Definition

Let the number x represent the radian measure of an angle, and picture the angle in a right triangle as in Figure 1. Then the two functions, sine of x and cosine of x, denoted by **sin x** and **cos x**, are defined by the following ratios.

_____ **Definition** _____

$$\sin x = \frac{\text{length of side opposite}}{\text{length of hypotenuse}}$$

$$\cos x = \frac{\text{length of side adjacent}}{\text{length of hypotenuse}}$$

For most values of x the functional values of sin x and cos x are difficult to compute. A calculator with sin and cos keys is a great help. The values can also be taken from a table of trigonometric functions, such as the table in the Appendix. There are a few values of x, however, whose functional values can be derived from specific triangles and are easy to remember. Two such triangles are the 30-60-90 degree and the 45-45-90 degree triangles shown in Figure 2. The Pythagorean Theorem can be used to calculate the lengths of the sides of these triangles. Then the definition can be used to calculate the sine and cosine of the angles. For instance,

FIGURE 2a

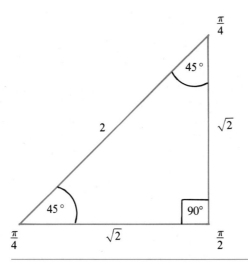

FIGURE 2b

in the 30-60-90 triangle the 30° angle has a radian measure of $\pi/6$. The length of the side opposite this angle is 1, while the length of the hypotenuse is 2. By the definition $\sin \pi/6 = 1/2$. In Example 1 we compute several additional values from these triangles.

EXAMPLE 1

Problem

Use the triangles in Figure 2 to compute (a) $\cos \pi/6$, (b) $\sin \pi/3$, (c) $\cos \pi/3$, (d) $\sin \pi/4$, and (e) $\cos \pi/4$.

SOLUTION (a) To compute $\cos \pi/6$, use the 30-60-90 degree triangle, where $\pi/6$ corresponds to 30°. The length of the side adjacent to this angle is $\sqrt{3}$, while the length of the hypotenuse is 2. By the definition $\cos \pi/6 = \sqrt{3}/2$.
(b) To compute $\sin \pi/3$, use the 30-60-90 triangle where $\pi/3$ corresponds to 60°. Then $\sin \pi/3 = \sqrt{3}/2$.
(c) Using the 30-60-90 triangle, we get $\cos \pi/3 = 1/2$.
For $x = \pi/4$ use the 45-45-90 triangle.
(d) $\sin \pi/4 = \sqrt{2}/2$; (e) $\cos \pi/4 = \sqrt{2}/2$.

If the lengths of the sides of a right triangle are known, then the sine and cosine of the angles can be computed without knowing the measure of the angle. Example 2 illustrates this.

EXAMPLE 2

Consider the right triangle in Figure 3 whose sides are of length 3, 4, and 5, respectively.

Problem

Compute (a) sin x and (b) cos x.

SOLUTION (a) To compute sin x, note that the length of the side opposite is 3 while the length of the hypotenuse is 5. By the definition sin x = 3/5.
(b) To compute cos x, note that the length of the side adjacent is 4, while the length of the hypotenuse is 5. By the definition cos x = 4/5.

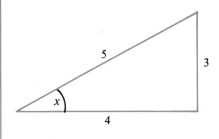

FIGURE 3

The Unit Circle Definition

A major obstacle in defining sin x and cos x in terms of lengths of sides of a triangle is that these functions are then defined only for angles between 0 and $\pi/2$ radians. To extend these definitions, we first view them in a slightly different way. This alternate definition of sin x and cos x can easily be extended.

　　Another way to describe sin x and cos x for x between 0 and $\pi/2$ is to place the angle in standard position in a unit circle. Then the angle is a central angle whose terminal side intersects the unit circle at a point P. Drop a perpendicular from P to the x-axis. This forms a right triangle whose hypotenuse is 1 since it is a radius of the unit circle. See Figure 4. The length of the side opposite the angle is the second, or y, coordinate of P. The length of the side adjacent to the angle is the first, or x, coordinate of P. Thus the coordinates of P are (cos x, sin x).

　　This view of sin x and cos x agrees with the right triangle definition for x between 0 and $\pi/2$. The advantage of this view is that we do not have to restrict x to this interval.

—————————— Unit Circle Definition of Sine and Cosine ——————————

Let x be any angle in standard position. Its terminal side intersects the unit circle at a point P. Define the first coordinate of P to be cos x and its second coordinate to be sin x. (See Figure 5.)

Example 3 shows how to compute sine and cosine for some angles that are not between 0 and $\pi/2$. The computation relies on Example 1.

Unit circle

FIGURE 4

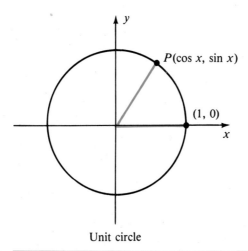

Unit circle

FIGURE 5

EXAMPLE 3 | Problem

Use the triangles in Figure 6 to compute (a) $\cos -\pi/6$, (b) $\sin -\pi/6$, (c) $\sin 2\pi/3$, and (d) $\sin 13\pi/4$.

SOLUTION (a) To compute $\cos -\pi/6$, note that the terminal side of the angle $-\pi/6$ is in the fourth quadrant. To find the coordinates of P, notice that the perpendicular from P to the x-axis forms a triangle that is congruent to the 30-60-90 triangle in Figure 2. Thus the lengths of the sides are the same, except that the y-coordinate is negative. Therefore P has the coordinates $(\sqrt{3}/2, -1/2)$. From the definition $\cos -\pi/6 = \sqrt{3}/2$.

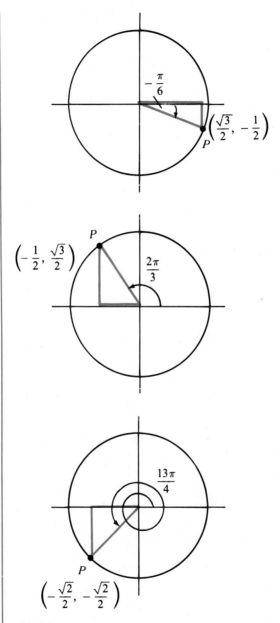

FIGURE 6

(b) To compute $\sin -\pi/6$, use the same point P in (a). Then $\sin -\pi/6 = -1/2$.

(c) To compute $\sin 2\pi/3$, use the 30-60-90 triangle formed in the second quadrant. The coordinates of P are $(-1/2, \sqrt{3}/2)$. Therefore $\sin 2\pi/3 = \sqrt{3}/2$.

(d) To compute $\sin 13\pi/4$, use the 45-45-90 triangle formed in the third quadrant. The coordinates of P are $(-\sqrt{2}/2,\ -\sqrt{2}/2)$. Therefore $\sin 13\pi/4 = -\sqrt{2}/2$.

Table 1 gives values of $\sin x$ and $\cos x$ for angles x that are multiples of $\pi/6$ and $\pi/4$ up to π. Each value can be verified by calculations similar to those in Example 3. For other angles use a calculator or a trigonometric table such as the one found in the Appendix. Table 1 also gives the values of three quadrantal angles, which are angles whose terminal sides lie on an axis. For example, for $x = 0$, the point P, which is the intersection of the terminal side and the unit circle is $(1, 0)$. Hence $\sin 0 = 0$ and $\cos 0 = 1$.

x	0	$\dfrac{\pi}{6}$	$\dfrac{\pi}{4}$	$\dfrac{\pi}{3}$	$\dfrac{\pi}{2}$	$\dfrac{2\pi}{3}$	$\dfrac{3\pi}{4}$	$\dfrac{5\pi}{6}$	π
$\sin x$	0	$\dfrac{1}{2}$	$\dfrac{\sqrt{2}}{2}$	$\dfrac{\sqrt{3}}{2}$	1	$\dfrac{\sqrt{3}}{2}$	$\dfrac{\sqrt{2}}{2}$	$\dfrac{1}{2}$	0
$\cos x$	1	$\dfrac{\sqrt{3}}{2}$	$\dfrac{\sqrt{2}}{2}$	$\dfrac{1}{2}$	0	$-\dfrac{1}{2}$	$-\dfrac{\sqrt{2}}{2}$	$-\dfrac{\sqrt{3}}{2}$	0

Properties of Sine and Cosine

There are several important relationships between $\sin x$ and $\cos x$. For every value x the point $(\cos x, \sin x)$ is the intersection of the the terminal side and the unit circle. The equation of the unit circle is $X^2 + Y^2 = 1$. This means that every point on the unit circle satisfies this equation. Therefore, replacing X with $\cos x$ and Y with $\sin x$ in the equation yields

$$\cos^2 x + \sin^2 x = 1$$

This relationship holds for every value of x. For this reason it is called an **identity.** The notation used for raising a trigonometric function to a power is illustrated in this identity. We write $\cos^2 x$ to represent $(\cos x)^2$ to simplify the notation. We now present several additional identities.

If x is any angle, then $x + 2\pi$ represents the angle x plus 1 revolution and $x + 2n\pi$ represents x plus π revolutions. Each of these angles has the same point of intersection of its terminal side with the unit circle, so that the sines of the angles are equal and the cosines of the angles are equal. This is also true of $x - 2n\pi$. This yields the following result.

$$\sin(x \pm 2n\pi) = \sin x \qquad \cos(x \pm 2n\pi) = \cos x$$

In Example 2 it can be seen that $\cos -\pi/6 = \cos \pi/6$. This is because the point P on the unit circle corresponding to $\pi/6$ has the same x coordinate as the point corresponding to $-\pi/6$. This is true for all angles. Example 2 also shows that $\sin -\pi/6 = -\sin \pi/6$. This is because the y coordinates of the corresponding points P are the negative of each other. See Figure 7. This is recorded in the following display:

$$\sin(-x) = -\sin x \qquad \cos(-x) = \cos x$$

Another identity that is true for all numbers a and b is

$$\sin(a + b) = \sin a \cos b + \cos a \sin b$$

This is called the "sine of a sum of two angles" formula. Its proof can be found in most trigonometry textbooks. Since it holds for any two angles, we can substitute $a = \pi/2$ and $b = -x$ to derive another important identity. We get

$$\sin\left(\frac{\pi}{2} - x\right) = \sin\frac{\pi}{2}\cos(-x) + \cos\frac{\pi}{2}\sin(-x)$$

$$= (1)\cos(-x) + (0)\sin(-x)$$

$$= \cos(-x) = \cos x$$

FIGURE 7

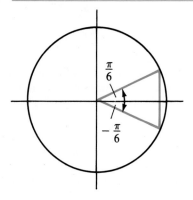

Similarly, we can show that $\cos(\pi/2 - x) = \sin x$.

$$\sin\left(\frac{\pi}{2} - x\right) = \cos x \qquad \cos\left(\frac{\pi}{2} - x\right) = \sin x$$

Other identities include

$$\sin(x - y) = \sin x \cos y - \cos x \sin y$$
$$\sin 2x = 2 \sin x \cos x$$
$$\cos 2x = \cos^2 x - \sin^2 x$$
$$\sin^2 x = (1/2)(1 - \cos 2x)$$
$$\cos^2 x = (1/2)(1 + \cos 2x)$$

Graphs of Sine and Cosine

The sine and cosine functions are used in applications to measure quantities that repeat their values over certain periods. The graphs of the sine and cosine display this periodicity property of the functions.

To see the behavior of $\sin x$ as x varies, follow the second coordinate of the point P as angle x rotates, starting from 0 and going through one revolution. As x increases from 0 to $\pi/2$, $\sin x$ increases from 0 to 1. The largest value that $\sin x$ takes on is 1. As x increases from $\pi/2$ to π, the second coordinate of P decreases to 0, so $\sin x$ decreases to 0. As x increases to $3\pi/2$, $\sin x$ decreases to -1. The smallest value that $\sin x$ takes on is -1. Then as x increases to 2π, $\sin x$ increases to 0. This completes one revolution of the circle. All these values will repeat as x continues through another revolution, that is, as x increases from 2π to 4π. This behavior is summarized in the following table.

As x Increases		sin x Varies	
From	To	From	To
0	$\dfrac{\pi}{2}$	0	1
$\dfrac{\pi}{2}$	π	1	0
π	$\dfrac{3\pi}{2}$	0	-1
$\dfrac{3\pi}{2}$	2π	-1	0

From this information a good sketch of $\sin x$ can be given (Figure 8). Because the graph of $\sin x$ repeats itself every 2π units, we say that the sine function is periodic with period 2π.

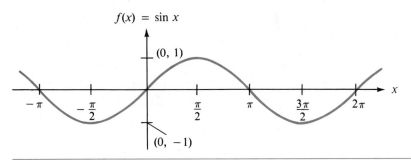

FIGURE 8

In a similar manner, the first coordinate of the point P can be studied as x goes through one revolution. The next table summarizes the behavior of the cosine function.

| As x Increases | | cos x Varies | |
From	To	From	To
0	$\dfrac{\pi}{2}$	1	0
$\dfrac{\pi}{2}$	π	0	-1
π	$\dfrac{3\pi}{2}$	-1	0
$\dfrac{3\pi}{2}$	2π	0	1

From this information a good sketch of cos x can be given (Figure 9). The cosine function is also periodic with period 2π.

The graph of the sine function occurs naturally in many settings. Trigonometry is especially useful in music. Figure 10 is a picture of a vibrating string on a violin.* It shows the string displacement on the vertical axis versus time on the horizontal axis. As we will see in the next section, there is a close connection between trigonometry and sound.

*Thanks to Dr. Robert Weinberg and Mr. David Nairns, Temple University, Department of Physics.

FIGURE 9

FIGURE 10

EXERCISE SET 10–2

• In Problems 1 to 4 use the triangles in Figure 2 to compute the given value.

1. $2 \cos \dfrac{\pi}{6}$

2. $3 \sin \dfrac{\pi}{3}$

3. $1 + \cos \dfrac{\pi}{3}$

4. $4 - \sin \dfrac{\pi}{4}$

In Problems 5 to 8 the right triangle in the accompanying figure has sides of length 5, 12 and 13, respectively. Compute the given value.

5. $\cos a$

6. $\sin a$

7. $\sin b$

8. $\cos b$

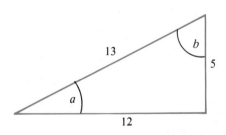

In Problems 9 to 20 compute the functional value.

9. $\cos \left(-\dfrac{\pi}{3} \right)$

10. $\sin \left(-\dfrac{\pi}{3} \right)$

11. $\sin \dfrac{4\pi}{3}$

12. $\cos \dfrac{4\pi}{3}$

13. $\sin \left(-\dfrac{4\pi}{3} \right)$

14. $\cos \left(-\dfrac{4\pi}{3} \right)$

15. $\cos \dfrac{5\pi}{4}$

16. $\sin \left(-\dfrac{5\pi}{4} \right)$

17. $\sin \left(\dfrac{15\pi}{4} \right)$

18. $\cos \dfrac{15\pi}{4}$

19. $\sin \left(-\dfrac{17\pi}{6} \right)$

20. $\cos \dfrac{19\pi}{6}$

•• In Problems 21 to 24 use a calculator or the table in the Appendix to find the functional values.

21. $\sin .5$

22. $\cos .6$

23. $\sin .1$

24. $\cos 1/10$

In Problems 25 to 28 find a number x such that $0 \le x \le \pi/2$ and x satisfies the given property.

25. $\sin x = \cos \dfrac{3\pi}{2}$

26. $\sin x = \sin \left(-\dfrac{3\pi}{2} \right)$

27. $\cos x = \sin 3\pi$

28. $\sin x = \sin 5\pi$

In Problems 29 to 32 sketch the graph of the function.

29. $f(x) = 2 \sin x$

30. $f(x) = 3 \sin x$

31. $f(x) = 1 + 2 \cos x$

32. $f(x) = -2 + 3 \sin x$

••• In Problems 33 to 36 sketch the graph of the function.

33. $f(x) = \sin 2x$ 34. $f(x) = 2 \sin 3x$

35. $f(x) = 1 - 2 \cos 2x$

36. $f(x) = -2 + 3 \sin 4x$

37. Why are $\sin x$ and $\cos x$ called "periodic" functions?

38. Find the largest and the smallest values of $f(x)$ and the value of x where they occur for the following functions: (a) $f(x) = \sin 2x$, (b) $f(x) = 2 \sin x$, and (c) $f(x) = 3 + 2 \cos 4x$.

39. Consider the formula for the movement of a typical air molecule of a simple sound

$$y(x) = 0.002 \cos(440\pi x)$$

Find the amplitude which is the largest value the function takes on.

40. In an article on how to make violins, Itokawa and Kumagai use the following formula to calculate the conductivity $C(x)$ of the f-holes of a violin for the angle x between the f-holes.* The f-holes of a violin amplify the vibration of the strings. Conductivity measures the speed of the sound wave through the f-holes.

$$C(x) = \frac{\pi}{2.9\sqrt{\cos x}}$$

For a given violin, $x = 84°$. Compute $C(x)$ for this value of x.

41. Kudzu is a weedy vine that forms a dense canopy over bare ground. It has been used as cattle fodder, for erosion control, and as a means to reclaim mine spoils. It may eventually be used as a source of sugar, and fiber for clothing, wallpaper, and paper. In an article studying the growth patterns of kudzu in central Maryland, Forseth and Teramura measured how the sun affected the leaf orientation by computing the cosine of incidence, defined as the cosine of the angle between the leaf lamina and a line perpendicular to the sun's direct beam.† The cosine of incidence directly affects the amount of irradiation that reaches the ground. For one type of kudzu leaf, the cosine of incidence was .9, whereas it was .8 for another type. Assuming the angles of incidence were between 0 and $\pi/2$ radians, find the angles corresponding to these values.

42. In his book showing the influence of mathematics in western culture, Morris Kline gives the following formula of the sound of a violin, which relates the position $y(t)$ at time t of a particular air molecule vibrating because of the violin strings:‡

$$y(t) = .06 \sin 18t + .02 \sin 36t + .01 \sin 54t$$

For each term, find the amplitude, which is the largest value that the function takes on.

*Hideo Itokawa and Chihiro Kumagai, "On the Study of Violin and Its Making," *Musical Accoustics: Piano and Wood Instruments*, Earle Kent (ed.), Stroudsburg, PA, Dowden, Hutchinson and Ross, 1979, pp. 55–84.

†Irwin N. Forseth and Alan H. Teramura, "Kudzu Leaf Energy Budget and Calculated Transpiration: At the Influence of Leaflet Orientation," *Ecology*, Vol. 67, 1987, pp. 564–570.

‡Morris Kline, *Mathematics in Western Culture*, London, Oxford University Press, 1953, pp. 287–303.

10–3 DIFFERENTIATION AND INTEGRATION

The graphs of the sine and cosine show the close connection between the two functions. The graph of the sine is identical to the graph of the cosine when moved to the left $\pi/2$ units. This is because of the identity $\sin(\pi/2 - x) = \cos x$. In this section we see another similarity between these functions when we calculate their derivatives. The rules governing their derivatives are straightforward but the proofs of the rules are not. The proofs are put off until the end of the section.

The Derivative of the Sine

To get an intuitive grasp of the derivative of the sine, superimpose the graphs of the sine and the cosine as in Figure 1. Then identify the slope of the tangent line at several points of the sine curve. For instance, the slope at $x = 0$ is 1; the slope at $x = \pi/2$ is 0; the slope at $x = \pi$ is -1. In each case the slope of the sine function at $x = a$ is cos a. This is true for all x, meaning that the derivative of sin x is cos x for all x. This is recorded in the following rule.

$$D_x \sin x = \cos x$$

A more rigorous verification of this rule is given at the end of this section.

If u is a function of x, we can apply the chain rule to get the sine version of the chain rule.

If u is a function of x, then

$$D_x \sin u = \cos u \cdot D_x u$$

FIGURE 1

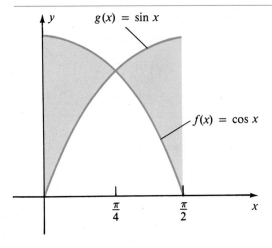

Example 1 illustrates how to apply the sine version of the chain rule. It should be noted that $\sin x^2 \neq (\sin x)^2$ where $(\sin x)^2$ is written $\sin^2 x$. To see this, let x be a specific value, say $x = \dfrac{\pi}{4}$, and compute each expression.

EXAMPLE 1 | **Problem**

Compute the derivative of the functions: (a) $f(x) = \sin 5x$, (b) $f(x) = \sin x^2$, and (c) $f(x) = \sin(x^3 - 5x)$.

SOLUTION (a) Let $u = 5x$. Then $D_x u = D_x 5x = 5$. Thus

$$f'(x) = D_x \sin 5x = (\cos 5x)5 = 5 \cos 5x$$

(b) Let $u = x^2$. Then $D_x u = D_x x^2 = 2x$. Thus

$$f'(x) = D_x \sin x^2 = (\cos x^2)2x = 2x \cos x^2$$

(c) Let $u = x^3 - 5x$. Then $D_x u = D_x(x^3 - 5x) = 3x^2 - 5$.

Thus

$$f'(x) = D_x \sin(x^3 - 5x) = [\cos(x^3 - 5x)](3x^2 - 5)$$
$$= (3x^2 - 5)\,3\cos(x^3 - 5x)$$

The Derivative of the Cosine

Once the derivative of $\sin x$ is known the derivative of $\cos x$ can be derived by using the chain rule and the two trigonometric identities.

$$\sin\left(\frac{\pi}{2} - x\right) = \cos x \qquad \text{and} \qquad \cos\left(\frac{\pi}{2} - x\right) = \sin x$$

Start with the first identity and take its derivative, using the sine version of the chain rule with $u = \pi/2 - x$.

$$D_x \cos x = D_x \sin\left(\frac{\pi}{2} - x\right)$$

$$= \cos\left(\frac{\pi}{2} - x\right) D_x\left(\frac{\pi}{2} - x\right)$$

$$= \cos\left(\frac{\pi}{2} - x\right)(-1)$$

$$= -\cos\left(\frac{\pi}{2} - x\right)$$

$$= -\sin x$$

The last conclusion is drawn from the second trigonometric identity above. This result is recorded in the display together with the cosine version of the chain rule.

$$D_x \cos x = -\sin x$$

If u is a function of x, then

$$D_x \cos u = -\sin u \cdot D_x u$$

Example 2 illustrates how to apply the cosine version of the chain rule.

EXAMPLE 2

Problem

Compute the derivative of the functions: (a) $f(x) = x^2 \cos 5x$ and (b) $f(x) = (5 + 3 \cos x^2)^3$.

SOLUTION (a) Use the product rule and the cosine version of the chain rule with $u = 5x$. Then $D_x u = D_x 5x = 5$. Thus

$$\begin{aligned}
f'(x) = D_x x^2 \cos 5x &= x^2 D_x \cos 5x + (\cos 5x) D_x x^2 \\
&= x^2(-\sin 5x)5 + (\cos 5x)2x \\
&= -5x^2 \sin 5x + 2x \cos 5x
\end{aligned}$$

(b) Two versions of the chain rule must be used. First $f(x)$ is of the form $g(x)^n$, where $g(x) = 5 + 3 \cos x^2$. Thus the extended power rule is used to find $f'(x)$. That is, $f'(x) = ng(x)^{n-1}g'(x)$. To compute $g'(x)$, we must use the cosine version of the chain rule with $u = x^2$. This results in

$$\begin{aligned}
f'(x) = ng(x)^{n-1}g'(x) \\
&= 3(5 + 3 \cos x^2)^2 D_x(5 + 3 \cos x^2) \\
&= 3(5 + 3 \cos x^2)^2(-3 \sin x^2)(D_x x^2) \\
&= 3(5 + 3 \cos x^2)^2(-3 \sin x^2)(2x) \\
&= -18x(5 + 3 \cos x^2)^2(\sin x^2)
\end{aligned}$$

Integration

The differentiation rules for the sine and cosine functions can be expressed as antidifferentiation formulas.

$$\int \sin x \, dx = -\cos x + C$$

$$\int \cos x \, dx = \sin x + C$$

These formulas can be used with integration techniques to evaluate a variety of integrals involving the sine and cosine functions. Example 3 uses substitution to compute two such integrals.

EXAMPLE 3 | **Problem**

Compute the integrals: (a) $\int 5 \cos 5x \, dx$ and (b) $\int x \sin x^2 \, dx$.

SOLUTION (a) Use the substitution $u = 5x$. Then $du = 5 \, dx$. Thus

$$\int 5 \cos 5x \, dx = \int (\cos 5x)(5 \, dx)$$

$$= \int \cos u \, du$$

$$= \sin u + C$$

$$= \sin 5x + C$$

(b) Use the substitution $u = x^2$. Then $du = 2x \, dx$ and $(1/2)du = x \, dx$. Thus

$$\int x \sin x^2 \, dx = \int (\sin x^2)(x \, dx)$$

$$= \int (\sin u) \frac{1}{2} \, du = \frac{1}{2} \int \sin u \, du$$

$$= -\frac{1}{2} \cos u + C$$

$$= -\frac{1}{2} \cos x^2 + C$$

The next example requires a more complicated substitution to compute a definite integral.

EXAMPLE 4 | **Problem**

Compute the definite integral.

$$\int_0^{\pi/2} 4 \sin^3 x \cos x \, dx$$

SOLUTION First find the indefinite integral

$$\int 4 \sin^3 x \cos x \, dx$$

Use the substitution $u = \sin x$. Then $du = \cos x \, dx$. Thus

$$\int 4 \sin^3 x \cos x \, dx = \int 4u^3 \, du = u^4 + C = \sin^4 x + C$$

The definite integral is then evaluated as follows:

$$\int_0^{\pi/2} 4 \sin^3 x \cos x \, dx = \sin^4 x \, \Big|_0^{\pi/2}$$

$$= \sin^4 \left(\frac{\pi}{2}\right) - \sin^4(0)$$

$$= 1^4 - 0^4 = 1$$

The next example shows how to find areas of regions bounded by trigonometric functions.

EXAMPLE 5 | **Problem**

Find the area of the region bounded by the graphs of $f(x) = \sin x$ and $g(x) = \cos x$ from $x = 0$ to $x = \dfrac{\pi}{2}$. See Figure 1.

SOLUTION To calculate the area of a region bounded by two curves, we need to determine the intervals where $f(x) \geq g(x)$. In those intervals we integrate $f(x) - g(x)$. Otherwise we integrate $g(x) - f(x)$. First determine where the graphs intersect by setting $f(x) = g(x)$. This yields

$$\sin x = \cos x \qquad \text{for } 0 \leq x \leq \frac{\pi}{2}$$

From Figure 1 the only value of x satisfying these properties is $x = \pi/4$. Also, from Figure 1 we have

$$\cos x \geq \sin x \qquad \text{for } 0 \leq x \leq \frac{\pi}{4}$$

$$\sin x \geq \cos x \qquad \text{for } \frac{\pi}{4} \leq x \leq \frac{\pi}{2}$$

To compute the area, two integrals are required:

$$\int_0^{\pi/4} (\cos x - \sin x) \, dx + \int_{\pi/4}^{\pi/2} (\sin x - \cos x) \, dx$$

$$= (\sin x + \cos x) \, \Big|_0^{\pi/4} + (-\cos x - \sin x) \, \Big|_{\pi/4}^{\pi/2}$$

$$= \left[\frac{\sqrt{2}}{2} + \frac{\sqrt{2}}{2} - (0 + 1)\right] + \left[-0 - 1 - \left(-\frac{\sqrt{2}}{2} - \frac{\sqrt{2}}{2}\right)\right]$$

$$= 2\sqrt{2} - 2$$

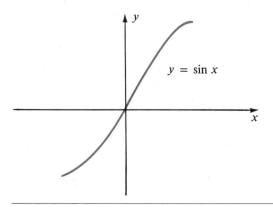

FIGURE 2   The tangent line at $x = 0$ appears to have slope 1.

Verification of the Derivative of the Sine

To verify that the derivative of the sine is the cosine, we need the following result:

$$\lim_{h \to 0} \frac{\sin h}{h} = 1 \tag{1}$$

An intuitive geometric argument of this result utilizes the graph in Figure 2. The slope of the tangent line of $y = \sin x$ at $x = 0$ appears to be 1. The secant lines that pass through $(0, 0)$ and $(h, \sin h)$ have slope $(\sin h)/h$. Since these slopes approximate the slope of the tangent line, they must approach 1. This means that the limit of $(\sin h)/h$ as h approaches 0 is 1.

Equation (1) can also be verified by using a calculator. Choose values of h that approach 0 and compute successive values of $(\sin h)/h$. Table 1 has several such calculations.

TABLE 1

h	.1	.01	.001	.0001	.00001	$-.001$	$-.0001$
$\sin h$	.09983	.01000	.00100	.00010	.00001	$-.00100$	$-.00010$
$\dfrac{\sin h}{h}$	.99833	.99998	.99999	.99999	.99999	$-.99999$	$-.99999$

In a similar way, the following result can be verified:

$$\lim_{h \to 0} \left[\frac{\cos h - 1}{h} \right] = 0 \tag{2}$$

We can verify the formula for the derivative of the sine function by using equations (1) and (2) along with the formula for the sine of a sum.

$$\sin(x + h) = \sin x \cos h + \cos x \sin h$$

From the definition of the derivative we have

$$f'(x) = \lim_{h \to 0} \frac{f(x + h) - f(x)}{h}$$

$$= \lim_{h \to 0} \left[\frac{\sin(x + h) - \sin x}{h} \right]$$

$$= \lim_{h \to 0} \left[\frac{\sin x \cos h + \cos x \sin h - \sin x}{h} \right]$$

Factor sin x from the first and third terms and split the fraction into two fractions. This yields

$$f'(x) = \lim_{h \to 0} \left[\sin x \frac{(\cos h - 1)}{h} + \cos x \frac{(\sin h)}{h} \right]$$

Use the fact that the limit of a sum is the sum of the limits. Also, since sin x and cos x are not affected as h approaches 0 they can be passed by the limit sign.

$$f'(x) = \sin x \left[\lim_{h \to 0} \frac{(\cos h - 1)}{h} \right] + \cos x \left[\lim_{h \to 0} \frac{(\sin h)}{h} \right]$$

Apply equations (1) and (2).

$$f'(x) = \sin x \cdot (0) + \cos x \cdot (1)$$
$$= \cos x$$

This verifies that $D_x \sin x = \cos x$.

Periodic Motion

Periodic motion is a natural phenomenon that occurs frequently in various settings. Simple examples are the oscillating motion of a pendulum and the rise and fall of the population of an animal species. To see how trigonometric functions model periodic motion, consider a vertical spring of natural length L. If a body of mass m is attached to the end of the spring, the spring is stretched to a new length, $L + 1$, as shown in Figure 3. If the mass is pulled beyond its equilibrium (resting) position, then the spring will exert a force F to restore itself to the resting position.

FIGURE 3

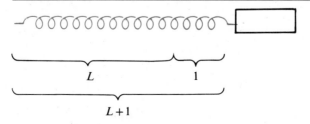

This force is proportional to the distance pulled y and is in the opposite direction. With k the constant of proportionality, this can be expressed as

$$F = -ky$$

By Newton's second law of motion the sum of the forces acting on the mass is given by $F = ma$, where a is the acceleration of the particle. By definition the acceleration is the second derivative of y: $a = D_x^2 y = y''(x)$. This yields the differential equation

$$my''(x) = -ky \qquad (3)$$

The solution of this equation can be shown to be

$$y(x) = A \cos(ax + b)$$

for constants A, a, and b, with $a = \sqrt{k/m}$. To see that this equation is a solution of the differential equation, compute y''.

$$y'(x) = -Aa \sin(ax + b) \qquad y''(x) = -Aa^2 \cos(ax + b)$$

Substituting this result into equation (3) yields

$$my''(x) = -mAa^2 \cos(ax + b) = -mA \left(\frac{k}{m}\right) \cos(ax + b)$$

$$= -kA \cos(ax + b)$$

$$= -ky(x)$$

Let us apply this concept of periodic motion to sound. The variety of sounds produced by musical instruments, voices, radios, and television sets is too great to study in one model. But a simple sound, that of a tuning fork, can be modeled by an elementary trigonometric function. When a prong of a tuning fork is struck, both prongs oscillate inward and outward very rapidly. This motion is similar to a stretched spring that moves according to a restoring force. The movement of the prongs causes the air around them to move back and forth, which in turn causes pressure to build. This results in waves, called sound waves, spreading from the fork. The phenomenon is similar to moving a stick rapidly in a pool of water. The motion of the stick causes waves to spread through the pool. Just as the wave in the water is made up of the oscillating movement of individual water molecules, so too does a sound wave consist of the oscillating movement of individual air molecules.

The motion of a single molecule of air can be modeled after the motion of a stretched string. Its motion is given by the formula

$$y(x) = A \cos(2\pi ax + b)$$

The variable x is the time (measured in seconds) that the molecule is in motion and y is the **displacement,** or the molecule's distance (measured in inches) from its rest position. The constant A is the **amplitude,** which is the maximum displacement and the constant a is the frequency, or the number of oscillations that the molecule makes per second. A typical formula for a simple sound is

$$y(x) = 0.001 \cos(2\pi 440x)$$
$$= 0.001 \cos(880\pi x)$$

This means that a typical air molecule oscillating in accordance with this formula moves back and forth about its rest position 440 times per second, and the farthest point it reaches from its rest position is 0.001 inches.

  More complex sounds are made up of oscillating air molecules that move according to formulas such as

$$y(x) = A \cos(ax + b) + B \sin(cx + d)$$

For example, in the article "On the Study of Violin and Its Making," Itokawa and Kumagai consider the frequency characteristics of the violin.* The tone quality at particular pitches is studied and various graphs are generated. The data are approximated by different combinations of sine and cosine functions. For instance, one particular violin at the "Wolfe-note" pitch exhibited the following behavior:

$$y(x) = .003 \cos(450\pi x) + .013 \sin(410\pi x)$$

*Hideo Itokowa and Chihiro Kumagai, "On the Study of Violin and Its Making," *Musical Acoustics: Piano and Wood Instruments*, Earle Kent (ed.), Stroudsburg, PA, Dowden, Hutchinson and Ross, 1977.

EXERCISE SET 10–3

• In Problems 1 to 6 compute the derivative of the function.

1. $f(x) = \sin 6x$

2. $f(x) = 4 \sin 6x$

3. $f(x) = \sin x^{-4}$,

4. $f(x) = -2 \sin x^{-4}$,

5. $f(x) = \sin(2x^4 - 5x^2)$

6. $f(x) = 3 \sin(x^{-1} - x^{1/2})$

In Problems 7 to 10 compute the derivative of the function.

7. $f(x) = x^2 \cos 6x$

8. $f(x) = 4x^3 \cos(-6x)$

9. $f(x) = (x + \cos 5x)^{-2}$

10. $f(x) = (x^2 + 2 \cos 3x)^{1/2}$

In Problems 11 to 16 compute the integral.

11. $\int 9 \cos 3x \, dx$

12. $\int 4 \cos(2x - 1) \, dx$

13. $\int x^2 \sin x^3 \, dx$

14. $\int x^{-2} \cos x^{-1} \, dx$

15. $\int 5 \cos^4 x \sin x \, dx$

16. $\int \dfrac{\cos x}{\sin x} \, dx$

In Problems 17 and 18 compute the definite integral.

17. $\int_0^{\pi/6} \sin x \cos^3 x \, dx$

18. $\int_0^{\pi/4} \cos x \sin^2 x \, dx$

In Problems 19 and 20 find the area of the region bounded by the graphs of functions.

19. $f(x) = \sin x$ and $g(x) = \cos x$ from $x = -\dfrac{\pi}{4}$ to $x = \dfrac{\pi}{2}$.

20. $f(x) = \sin x$ and $g(x) = \cos x$ from $x = -\dfrac{\pi}{4}$ to $x = \pi$.

•• In Problems 21 to 26 find $f'(x)$.

21. $f(x) = \sin(e^x) + e^{\sin x}$

22. $f(x) = \ln(\sin x) + \sin(\ln x)$

23. $f(x) = (\sin 3x + e^{3x})^{1/3}$

24. $f(x) = \sin(\cos 3x)$ 25. $f(x) = \cos^2 x^3$

26. $f(x) = x \sin^2(4x + 1)$

In Problems 27 to 30 evaluate the integrals.

27. $\displaystyle\int e^x \cos(e^x)\, dx$ 28. $\displaystyle\int \frac{\sin(\ln x)}{x}\, dx$

29. $\displaystyle\int (\cos x)e^{\sin x}\, dx$ 30. $\displaystyle\int \frac{\sin 3x}{\cos 3x}\, dx$

••• In Problems 31 and 32 find $f'(x)$.

31. $f(x) = \sin(e^{3x}) + e^{\sin 3x}$

32. $f(x) = \sin^2(\cos 3x)$

33. Evaluate $\displaystyle\int \sin^3 x\, dx$

34. Find $f''(x)$ for $f(x) = \sin x \cos x$.

35. Compute $f'(x)$ where $f(x) = \sin^2 x + \cos^2 x$. Give a reason why $f'(x)$ is such a simple function.

36. A water trough is to be constructed from three metal sheets each 1 meter wide and 5 meters long plus end panels in the shape of trapezoids. Find the angle at which the long panels should be joined so as to provide a trough of maximum volume.

37. In describing how sound waves travel, Beranek shows sound pressure $p(x)$ when the air molecule is at position x satisfies the differential equation*

$$p''(x) = -w^2 p(x)$$

where w is a constant. Verify that a solution of this differential equation is $p(x) = \cos wt$.

38. When the read-write mechanism of one of IBM's first large mainframe computers developed failure problems, engineers had to solve the problem quickly or IBM would lose the trust and credibility of its customers. In an article describing how IBM uncovered and solved the problem, Piziali shows that the difficulty was in the failure of a part called the pickup-head link.† This is the device that picks up information stored magnetically and moves the information to another position in the computer; it might, for instance, be moved to be processed by the printer. The pickup-head link was failing because it was put under too much stress over long periods as it came into contact with the drum on which the information was stored. Upon contact, the link was deflected, or bent. The deflection $D(x)$ depended on the amount of pressure x, also referred to as shock, measured in pounds per square inch. The formula for $D(x)$ was

$$D(x) = \frac{.125}{\cos .04x^{1/2}}$$

The rate of change of $D(x)$ as x increased was critical. Compute $D'(x)$.

39. Use a calculator to verify that

$$\lim_{h \to 0} \left[\frac{\cos h - 1}{h} \right] = 0$$

*Leo L. Beranck, *Noise Reduction,* New York, McGraw-Hill, 1980, pp. 20–25.
†R. Piziali, "The Pickup-head Link Failure," *Engineering Case Library,* American Society for Engineering Education, No. 186, 1972, pp. 1–16.

10–4 THE OTHER TRIGONOMETRIC FUNCTIONS

There are six trigonometric functions, so there are four trigonometric functions in addition to the sine and cosine functions. In this section we define the other four functions and their derivatives. Of the remaining four functions the tangent is the most important, in part because each of the other three functions, the cosecant, secant, and cotangent, is the reciprocal of the sine, cosine, and tangent, respectively. For this reason most of our attention in this section will be spent on the tangent.

Definitions

Many applications of calculus require the product and quotient of the sine and cosine functions. They appear so often that some of these products and quotients are given specific names for easy reference. Together with their abbreviations, these functions are the **tangent** function, expressed as tan x; the **cotangent** function, or cot x; the **secant** function, or sec x; and the **cosecant** function, or csc x. They are defined as follows:

$$\tan x = \frac{\sin x}{\cos x} \qquad \cot x = \frac{\cos x}{\sin x}$$

$$\sec x = \frac{1}{\cos x} \qquad \csc x = \frac{1}{\sin x}$$

Each function is defined for all x except those values where the denominator equals 0. For instance, cot x is defined for all x except where sin $x = 0$. That is, cot x is not defined for $x = 0$, π, 2π, and, in general, all $x = n\pi$.

A number of identities involving the trigonometric functions are studied in a course of trigonometry. In this section we cover only one. Several additional identities are mentioned in the exercises. To derive the identity, start with the identity derived in the previous section.

$$\sin^2 x + \cos^2 x = 1$$

Divide each term in this equation by $\cos^2 x$.

$$\frac{\sin^2 x}{\cos^2 x} + \frac{\cos^2 x}{\cos^2 x} = \frac{1}{\cos^2 x}$$

The first term equals $\tan^2 x$, the second term equals 1, and the third term equals $\sec^2 x$. This yields

$$\tan^2 x + 1 = \sec^2 x$$

The right triangle interpretation of the **tangent** is derived from the definitions of the sine and cosine. Recall that the sine is the ratio of the length of the side opposite and the length of the hypotenuse. The cosine is the ratio of the side adjacent and the length of the hypotenuse. From Figure 1 we get

$$\tan x = \frac{\sin x}{\cos x} = \frac{\text{(side opposite)/(hypotenuse)}}{\text{(side adjacent)/(hypotenuse)}}$$

Therefore when x is restricted to $0 < x < \pi/2$, we have the following formula:

$$\tan x = \frac{\text{side opposite}}{\text{side adjacent}}$$

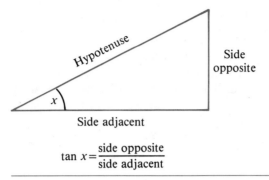

$$\tan x = \frac{\text{side opposite}}{\text{side adjacent}}$$

FIGURE 1

Example 1 illustrates how to evaluate the tangent function at some values of x.

EXAMPLE 1 | **Problem**

For the angle x in Figure 2 find (a) tan x, (b) cot x, (c) sec x, and (d) csc x.

SOLUTION (a) Using the formula with the length of the side opposite 3 units and the length of the side adjacent 4 units, tan $x = 3/4$.
(b) Since cot $x = (\cos x)/(\sin x)$, we see that cot $x = 1/\tan x$. Thus cot $x = 1/(3/4) = 4/3$.
(c) By definition sec $x = 1/\cos x$. Since cos $x = 4/5$, sec $x = 5/4$.
(d) By definition csc $x = 1/\sin x$. Since sin $x = 3/5$, csc $x = 5/3$.

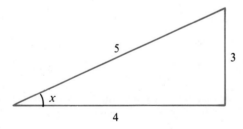

FIGURE 2

The Graph of the Tangent

The tangent is defined for all x where cos $x \neq 0$. That is, tan x is not defined for $x = \pi/2 + 2n\pi$. At these values tan x has a vertical asymptote because the denominator approaches 0 and the numerator approaches 1 as x approaches any of these values. The tangent is a periodic function with period $\pi/2$. The graph of tan x is given in Figure 3.

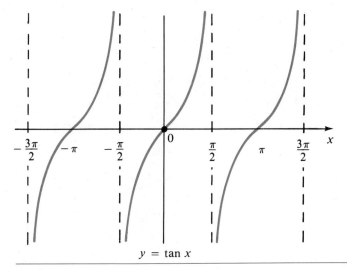

$y = \tan x$

FIGURE 3

The Derivative of the Tangent

The formula for the derivative of tan x can be derived from its definition by using the quotient rule.

$$D_x \tan x = D_x \left[\frac{\sin x}{\cos x} \right]$$

$$= \frac{(\cos x)(\cos x) - (\sin x)(-\sin x)}{(\cos x)^2}$$

$$= \frac{\cos^2 x + \sin^2 x}{\cos^2 x}$$

$$= \frac{1}{\cos^2 x} = \sec^2 x$$

Thus the derivative of tan x is $\sec^2 x$. This fact and the tangent version of the chain rule are recorded in the next display.

$$D_x \tan x = \sec^2 x$$

If u is a function of x, then

$$D_x \tan u = \sec^2 u \, D_x u$$

Example 2 illustrates how to apply these formulas.

EXAMPLE 2 | **Problem**

Compute the derivative of the functions: (a) $f(x) = \tan 5x$, (b) $f(x) = \tan x^2$, and (c) $f(x) = \tan(x^3 - 5x)$.

SOLUTION (a) Let $u = 5x$. Then $D_x u = D_x 5x = 5$. Thus

$$f'(x) = D_x \tan 5x = (\sec^2 5x)5 = 5 \sec^2 5x$$

(b) Let $u = x^2$. Then $D_x u = D_x x^2 = 2x$. Thus

$$f'(x) = D_x \tan x^2 = (\sec^2 x^2)2x = 2x \sec^2 x^2$$

(c) Let $u = x^3 - 5x$. Then $D_x u = D_x(x^3 - 5x) = 3x^2 - 5$. Thus

$$\begin{aligned} f'(x) &= D_x \tan(x^3 - 5x) \\ &= [\sec^2(x^3 - 5x)](3x^2 - 5) \\ &= (3x^2 - 5) \sec^2(x^3 - 5x) \end{aligned}$$

The next example shows how to use the derivative of the tangent with other differentiation formulas.

EXAMPLE 3 | **Problem**

Compute the derivative of the functions: (a) $f(x) = \tan^3 5x$ and (b) $f(x) = (1 - \tan x^2)^{-4}$

SOLUTION (a) Use the extended power rule with $u = \tan 5x$ and $n = 3$. The derivative of $\tan 5x$ was computed in Example 2. Thus

$$\begin{aligned} f'(x) &= D_x \tan^3 5x = 3 (\tan^2 5x)(D_x \tan 5x) \\ &= 3 (\tan^2 5x)(5 \sec^2 5x) \\ &= 15 (\tan^2 5x)(\sec^2 5x) \end{aligned}$$

(b) Use the extended power rule with $u = 1 - \tan x^2$ and $n = -4$. The derivative of $\tan x^2$ was computed in Example 2. Thus

$$\begin{aligned} f'(x) &= D_x(1 + \tan x^2)^{-4} \\ &= -4(1 + \tan x^2)^{-5} D_x(\tan x^2) \\ &= -4(1 + \tan x^2)^{-5}(2x \sec^2 x^2) \\ &= -8x(1 + \tan x^2)^{-5}(\sec^2 x^2) \end{aligned}$$

Derivatives of the Other Functions

The derivatives of the other trigonometric functions can also be derived by the quotient rules. The exercises will ask for their derivations. The formulas are

$$D_x \cot x = -\csc^2 x$$
$$D_x \sec x = \sec x \tan x$$
$$D_x \csc x = -\csc x \cot x$$

Incorporating these formulas into the chain rule yields the following, where u is a function of x.

$$D_x \cot u = -\csc^2 u \, D_x u$$
$$D_x \sec u = \sec u \tan u \, D_x u$$
$$D_x \csc u = -\csc u \cot u \, D_x u$$

The next example illustrates how to apply these formulas.

EXAMPLE 4 | **Problem**

Compute the derivative of the functions: (a) $f(x) = \cot 5x$, (b) $f(x) = \cot^4 5x$, and (c) $f(x) = \sec 2x + \csc x^2$.

SOLUTION (a) Use the formula for $D_x \cot u$ with $u = 5x$. Then $D_x u = 5$. Thus

$$f'(x) = D_x \cot 5x = -\csc^2 5x \, D_x 5x$$
$$= -5 \csc^2 5x$$

(b) Use the extended power rule with $u = \cot 5x$ and $n = 4$. The derivative of $\cot 5x$ was computed in part (a). Thus

$$f'(x) = D_x \cot^4 5x = 4(\cot^3 5x)(D_x \cot 5x)$$
$$= 4(\cot^3 5x)(-5 \csc^2 5x)$$
$$= -20(\cot^3 5x)(\csc^2 5x)$$

(c) Use the derivative of a sum rule.

$$f'(x) = D_x(\sec 2x + \csc x^2)$$
$$= D_x \sec 2x + D_x \csc x^2$$
$$= (\sec 2x \tan 2x)D_x 2x + (-\csc x^2 \cot x^2)D_x x^2$$
$$= 2 \sec 2x \tan 2x - 2x \csc x^2 \cot x^2$$

EXERCISE SET 10–4

• In Problems 1 to 4 find the given functional value of angle a where a is the angle in the accompanying figure.

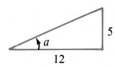

1. $\tan a$

2. $\cot a$

3. $\sec a$

4. $\csc a$

In Problems 5 to 12 compute the derivative of the function.

5. $f(x) = \tan(5x + 1)$ 6. $f(x) = \tan(x^2 + 5x)$

7. $f(x) = \tan(x^3 - 5x)$

8. $f(x) = \tan(x^{-3} - e^x)$

9. $f(x) = \tan^5 4x$ 10. $f(x) = \tan^{1/5} 5x$

11. $f(x) = (3x + \tan x^2)^{1/2}$

12. $f(x) = (\sin x + \tan 2x)^{-2}$

In Problems 13 to 15 derive the formula for the derivative of the given trigonometric function.

13. $f(x) = \cot x$ 14. $f(x) = \sec x$

15. $f(x) = \csc x$

In Problems 16 to 30 compute the derivative of the function.

16. $f(x) = \cot(5x + 3)$ 17. $f(x) = \sec(x^2 + 3x)$

18. $f(x) = \csc(x^2 - 5)$

19. $f(x) = \cot(x^{-3} - e^x)$

20. $f(x) = \sec^3 2x$ 21. $f(x) = \csc^{1/2} 3x$

22. $f(x) = (3x + \cot x^3)^{-2}$

•• 23. $f(x) = \ln \sec x$ 24. $f(x) = e^{\csc x}$

25. $f(x) = e^{\tan 6x}$

26. $f(x) = \ln \cot(4x - 1)$

27. $f(x) = \cot e^{3x}$

28. $f(x) = \tan \ln(3x + 1)$

29. $f(x) = \tan 2x \cot 3x$

30. $f(x) = \sec 4x \tan(4x + 1)$

••• 31. Express the differentiation formulas in this section as antidifferentiation formulas.

In Problems 32 to 38 evaluate the integral $\int f(x) \, dx$.

32. $f(x) = \sec^2 3x$

33. $f(x) = \sec 4x \tan 4x$

34. $f(x) = x \sec^2 x^2$ 35. $f(x) = x^2 \csc^2 x^3$

36. $f(x) = \sec^2 x \tan x$

37. $f(x) = \csc^2 x \cot^2 x$

38. $f(x) = \tan 2x$

39. Sound propagation in water has many applications, from sonar to fishing techniques. In an article studying the individual sound characteristics of the Mediterranean Sea, Leroy measured the relationship between the range $R(x)$ of a ray of sound when emitted into the sea at an angle x with the surface.* The relationship was found to be

$$R(x) = a \tan x + b \tan x^{1/2}$$

where a and b are constants depending on the time of year and the position on the sea. The rate of change of $R(x)$ is an important measure. Compute $R'(x)$.

40. There are many ways to describe how musical sounds are transmitted. In one such model of the sound of a piano, the impact of the hammer on a string causing it to vibrate is regarded as an electrical transmission.† The imparting of energy to the string by the hammer blow is analogous to connecting an inductance across a transmission line between its terminals. The impedance $Z(x)$ is a function of length of the string x given by the relationship

$$Z(x) = a \cot(2\pi bx)$$

Compute $Z'(x)$.

*C. C. Leroy, "Sound Propagation in the Mediterranean Sea," *Underwater Accoustics,* Vol. 2, V. M. Albers, (ed.), New York, Plenum Press, 1977, pp. 203–234.

†N. H. Fletcher, "A Vibrating String as an Electrical Transmission," *Musical Accoustics: Piano and Wind Instruments,* Earle L. Kent (ed.), Stroudsburg, PA, Dowden, Hutchinson and Ross, 1977, pp. 31–39.

CASE STUDY   Shaft Design*

All too often students equate mathematics with a magic wand—if you have a problem, fetch a math formula and "presto!"—out pops the solution. Real-world applications could not be further from this scenario. The development of most models involves trial-and-error techniques, mixed with a large dose of

*Adapted from W. P. Lewis, "A Case Study in Shaft Design," *Engineering Case Library,* Case number ECL 240, Washington, DC, American Society for Engineering Education, 1984.

frustration. This Case Study will give you a glimpse of what problem solving in the real world is like.

Melbourne's tunnel system   In the late 1970s the city government of Melbourne, Australia, decided to construct a large tunnel system beneath the city. One part of the system would consist of an underground railway and another would carry sewerage to a new treatment plant. The tunnel-boring operations revealed a lack of knowledge about the forces that are generated when cutting through different types of rock strata. The escalating costs and incessant delays were the cause of politicans' heated rhetoric and taxpayers' lament. A research program on rock-cutting had to be started immediately to design a special machine for the tunnel-boring task.

   To keep expenditures in line with projected budgets, tunneling rates through the various types of rock strata had to be predicted accurately. An important variable was the wear of the cutting tools, because there would be great stress on them. Accurate predictions of wear would enable researchers to plan efficiently for maintenance shutdowns. Exacerbating the problem was the fact that relevant information for local operating conditions was nonexistent. So an experimental test machine would have to be designed and tested. A young master's degree student, Gray Bailey, was given the task of designing and constructing the experimental machine.

   Bailey had confidence in most aspects of his design, but he wanted an expert opinion on the major design feature, the shaft that operates the cutting head. Since most of the stress of the cutting process would be on the shaft, the force on the shaft would have to be accurately determined. Bailey hired a University of Melbourne researcher, Ian McPherson, to help in the shaft design. The key decision was the diameter of the shaft. If it was too small, it would not wear well. But too large a shaft would decrease the efficiency of the motor and adversely affect the tunneling rate as well as the wear on the motor.

   In their first meeting McPherson asked Bailey for as many details of the project as he could relate. Bailey described the type of machine they were considering. "We have an electric motor available," said Bailey, "and based on research the maximum tangential force on the tool during cutting is most unlikely to exceed 50 kN (kiloNewtons). If you allow another 5% margin on this, I am sure your calculations would be very safe."

   Realizing that clients are frequently apt to forget or overlook relevant pieces of information, McPherson often extended the meeting with questions such as "Do you mind if I ask once again if that is all you can tell me?"

   After several of these prompts, Bailey assured him, "Now you know as much as I do!" They parted, and McPherson started the design. He felt confident he could solve the problem. Figures 1 and 2 show McPherson's schematic of the shaft layout. As the shaft turns, the cutting edges of the tool bore into the rock and essentially scrape it away. This puts tangential force F_t on the cutting edge of the tool. To calculate the maximum torque F, he used the formula

$$F = F_t r$$

FIGURE 1

where r is the radius of the tool. Since $r = 1.09$ and Bailey said to take F_t as $50 + .05(50) = 52.5$, McPherson calculated the maximum torque to be

$$F = 52.5(1.09) = 57.2$$

As he was about to turn his attention to the force on the shaft, McPherson realized he did not know the direction of rotation with respect to the motor. Should the rotation be through a negative (clockwise) angle of rotation or a positive (counterclockwise) angle? A phone call to Gray Bailey determined that Bailey had given no thought to the position of the drive motor. McPherson decided to base his design on a positive angle of rotation because it represented a more adverse load on the motor. If the client decided on the alternate design, there would be less load, so the design would still have ample support.

Next McPherson had to determine the force directly on the shaft. There were several points on the shaft that were subject to stress. He started with the force on the sprocket. Since F is the maximum torque, the force on the shaft would be the sum of the vertical components of the force F at each point of the circumference. Suppose the cutting edge of the tool makes an angle x with the vertical axis. He represented the tangential force per unit of length of circumference by f. Assume f is constant over the entire circumference. Let dF represent the force on a small part of the sprocket that causes the tool to rotate. If this small part of the sprocket on the circumference subtends an angle dx, then

FIGURE 2

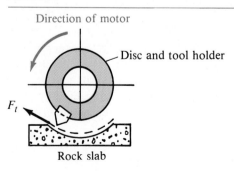

it has length $r\,dx$. Then $dF = fr\,dx$. Represent the vertical component of the force by F_v and the horizontal component by F_h. The force on the shaft at the sprocket is F_v. Then

$$dF_v = fr\,dx\,\sin x \qquad \text{and} \qquad F_h = fr\,dx\,\cos x$$

The sum of the individual forces dF is F. Since the force acts when the sprockets are engaged, it acts on the shaft from $x = 0$ to $x = \pi$, that is, from the point where the edge is facing down to where it is vertical as it rotates clockwise. This sum is given by

$$F_v = \int dF_v = \int_0^\pi fr\,\sin x\,dx = fr \int_0^\pi \sin x\,dx = fr(\cos x)\Big|_0^\pi = 2fr$$

$$F_h = \int dF_h = \int_0^\pi fr\,\cos x\,dx = fr \int_0^\pi \cos x\,dx = fr(\sin x)\Big|_0^\pi = 0$$

The maximum torque is $F = \pi rf \approx 57.2$. Solving this equation for f yields

$$f = \frac{57.2}{\pi r}$$

The first integral above states that the force on the shaft is $F_v = 2fr$. Substituting the preceding equation into this equation yields

$$F_v = 2fr = 2r\left(\frac{57.2}{\pi r}\right) = \frac{114.4}{\pi} \approx 36.5$$

This is the value of the force on the shaft at the sprocket.

McPherson then isolated three additional points on the shaft where there would be significant forces. He computed these forces in a similar manner. The equation for the shaft diameter was reduced to

$$d = 8.58n^{1/3} \text{ cm}$$

where n is a safety factor. He was unsure what value of n to use because the local conditions of rock strata were unknown. He decided to graph d as a function of n (Figure 3). From the graph he felt that a diameter of 14 cm would be reasonable because it would give a safety factor of just greater than 4.

McPherson called a meeting with Bailey and presented his design, including a shaft diameter of 14 cm. Ever cautious, McPherson kept asking himself, even during this "final" conversation, "What else could go wrong?" Since the shaft cutting edge scrapes along the rock, the shaft will tend to bend in the direction opposite from the scraping. This is called deflection.

"I just realized I forgot to ask you to specify the maximum allowable deflection," said McPherson.

Bailey was jolted. "Will the shaft deflect?"

"Of course it will! You've got high loads and a long bearing span. I feel deflection will be significant."

"Will it? Oh well, I'll leave it to you. You're the expert."

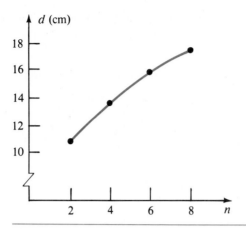

FIGURE 3

McPherson muttered, "There's more to this shaft design problem than meets the eye."

Back to the drawing boards. The calculations had to be reworked with deflection parameters included. McPherson's final calculation showed that to account for the maximum allowable deflection, d would have to be greater than 16.7. To allow a little margin of additional safety, he recommended a shaft of 18 cm. Bailey, however, felt that this would make the shaft too heavy. McPherson analyzed whether the design could be amended with support for the shaft. This led to a radical new design, which not only supported the shaft but also essentially eliminated the torque on the shaft, thus making it a much more efficient and reliable machine.

Bailey's experimental machine was built with this new design and it worked well. Construction began soon after tests showed the machine would run efficiently. If you ever visit Melbourne, take a ride on the underground railway, built in part by the McPherson–Bailey tunnel-boring machine.

Case Study Exercises

In Problems 1 to 4 compute the maximum torque on the shaft for the given values of r and F_t.

1. $r = 1.1,$ $F_t = 52$

2. $r = 1.5,$ $F_t = 55$

3. $r = 1.25,$ $F_t = 51.5$

4. $r = 1.15,$ $F_t = 50.5$

In Problems 5 to 8 compute the vertical component F_v of F for the given values of r and F_t.

5. $r = 1.1,$ $F_t = 52$

6. $r = 1.5,$ $F_t = 55$

7. $r = 2.5,$ $F_t = 50$

8. $r = 2.15,$ $F_t = 50.5$

CHAPTER REVIEW

Key Terms

10-1 Radian Measure
Radian   Angular Speed

10-2 The Sine and Cosine Functions
Sine   Identity
Cosine

10-3 Differentiation and Integration
Periodic Motion   Amplitude
Displacement

10-4 The Other Trigonometric Functions
Tangent   Secant
Cotangent   Cosecant

Summary of Important Concepts

$$r \text{ (radians)} = \frac{\pi}{180} d \text{ (degrees)}$$

$$\sin^2 + \cos^2 x = 1$$

$$\sin(x + 2n\pi) = \sin x$$

$$\cos(x + 2n\pi) = \cos x$$

$$\sin(-x) = \sin x$$

$$\cos(-x) = \cos x$$

$$\sin(a + b) = \sin a \cos b + \cos a \sin b$$

$$\sin\left(\frac{\pi}{2} - x\right) = \cos x$$

$$\cos\left(\frac{\pi}{2} - x\right) = \sin x$$

$$D_x \sin u = \cos u \, D_x u$$

$$D_x \cos u = -\sin u \, D_x u$$

$$D_x \tan u = \sec^2 u \, D_x u$$

$$D_x \cot u = -\csc^2 u \, D_x u$$

$$D_x \sec u = \sec u \tan u \, D_x u$$

$$D_x \csc u = \csc u \cot u \, D_x u$$

REVIEW PROBLEMS

1. Convert the angles from degrees to radians:
(a) 135°, (b) 210°.

2. Convert the angles from radians to degrees:
(a) $\dfrac{8\pi}{3}$, (b) $\dfrac{11\pi}{6}$

3. Convert the angles from radians to degrees and express the angle in terms of the revolutions that it makes: (a) $\dfrac{11\pi}{3}$, (b) $\dfrac{25\pi}{6}$

4. Find the radian measure of the angle that makes two and one-third revolutions in a clockwise direction.

5. Turning a combination lock two and one-sixth revolutions corresponds to an angle of what radian measure?

In Problems 6 to 9 find the functional value.

6. $\cos\left(-\dfrac{4\pi}{3}\right)$ **7.** $\sin\dfrac{15\pi}{4}$

8. $\sin .7$ **9.** $\cos\dfrac{1}{10}$

10. Find a number x such that $0 \le x \le \dfrac{\pi}{2}$ and satisfies the given property: $\sin x = \sin 5\pi$

In Problems 11 and 12 compute the derivative of the function.

11. $f(x) = 4 \sin (x^{-2} - x^{1/3})$

12. $f(x) = (x + 2 \cos 3x)^{1/4}$

In Problems 13 and 14 compute the integral.

13. $\displaystyle\int x^3 \sin x^4 \, dx$ **14.** $\displaystyle\int \dfrac{\sin x}{\cos x} \, dx$

15. Find $f'(x)$: $f(x) = \ln(\cos x) + \sin(\ln x)$

In Problems 16 to 20 compute the derivative of the function.

16. $f(x) = \tan(x^2 + 5x)$

17. $f(x) = (\sec x + \tan 2x)^{-2}$

18. $f(x) = \sec^4 2x$ **19.** $f(x) = e^{\csc x}$

20. $f(x) = \sec 2x \tan(3x + 2)$

11 Infinite Series

CHAPTER OVERVIEW   The two most widely used methods of finding a functional value of functions such as e^x, ln x, and sin x are to look up the value in a table and to use a calculator. But how do these numbers get into the table, and how does a calculator compute these values? For example, if you enter 1.1 in a calculator and press the e^x button, the number 3.004166 appears. Different calculators may vary in the last digit due to rounding off. In many calculators there is a slight delay during the calculation. In this chapter we see what steps the calculator uses to compute this number. The calculator approximates the number $e^{1.1}$ by substituting 1.1 into a particular polynomial.

The first section defines those types of polynomials that are associated with specific functions. They are called Taylor polynomials. Section 11–2 deals with infinite sums. In particular, we will see what it means to state that $1/3 = .3333. \ldots$. Section 11–3 explains further what it means to say that an infinite sum has a limit and how to determine it when such a limit exists. The last section ties together the ideas of Taylor polynomials and infinite series. An infinite series defined from a function in the manner of Section 11–1 is called a Taylor series.

CASE STUDY PREVIEW

A single change in demand can have far-reaching effects in various segments of an economy. This ripple effect has been known to economists for almost two centuries, but it was not until 1931, during the Great Depression, that a mathematical formulation was created allowing economists to measure these effects. The Case Study outlines the evolution of this measure, called the *multiplier principle*, and then it looks at a modern application of the principle to the control of the U.S. money supply by the Federal Reserve System.

11−1 TAYLOR POLYNOMIALS

The invention of hand-held calculators has proved to be a major scientific contribution. The time-consuming drudgery of elementary computation has been replaced by instantaneous and accurate calculation. Calculators are especially useful when evaluating functions such as e^x, $\ln x$, and $\sin x$. Instead of looking up values in tables, we can use a calculator, which gives immediate results that are accurate to many decimal places. How does a calculator evaluate these functions? Usually it is done by using a polynomial to approximate the function. This is because the circuits inside an electronic calculator are capable of performing only the basic operations of addition, subtraction, multiplication, and division. Polynomials can be evaluated with these operations, but many nonpolynomial functions cannot. The key to obtaining an accurate approximation of $f(a)$ is to select a polynomial $P(x)$ so that $P(a)$ is close to $f(a)$.

Taylor Approximations

We start by approximating $f(x) = e^x$ for values of x close to 0. For instance, what is the decimal equivalent of $e^{0.1}$? Our first approximation of $e^{0.1}$ will be with a first-degree polynomial. Then higher-degree polynomials will be used to generate more accurate approximations.

Our first goal is to approximate e^x by a polynomial of the form

$$P_1(x) = a + bx$$

The object is to find constants a and b so that $P_1(x)$ is close to e^x for values of x close to 0. $P_1(x)$ is a linear function whose graph is a straight line.

In Figure 1 we sketch the graph of $f(x) = e^x$ together with its tangent line at $x = 0$. The shape of the graph of e^x is close to its tangent line for values of x close to 0. There are two properties of $P_1(x)$ that cause their shapes to be similar: (1) each graph passes through $(0, 1)$, meaning that each has the same functional value at $x = 0$, and (2) the slope of each function is the same, meaning that their derivatives at $x = 0$ are equal. These two properties allow us to calculate the constants a and b by requiring that (1) $P_1(0) = f(0)$ and (2) $P'_1(0) = f'(0)$. For the second property remember that $P'_1(x) = b$ and $f'(x) = D_x(e^x) = e^x$.

1. $P_1(0) = a + b(0) = a = f(0) = e^0 = 1$
2. $P'_1(0) = b = f'(0) = e^0 = 1$

Therefore $a = 1$ and $b = 1$. Hence

$$P_1(x) = 1 + x$$

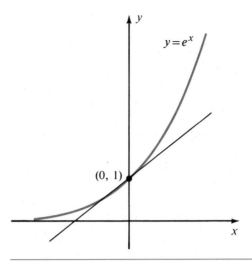

FIGURE 1

In general any differentiable function $f(x)$ can be approximated for values of x close to 0 by the first-degree polynomial

$$P_1(x) = f(0) + f'(0)x$$

This is called the *first Taylor polynomial of* $f(x)$ *at* $x = 0$. The graph of $P_1(x)$ is the tangent line of the graph of $y = f(x)$ at $x = 0$. The graphs are similar for values of x close to 0 because the following two properties are true:

1. $P_1(0) = f(0)$
2. $P_1'(0) = f'(0)$

From Figure 1 it is clear that the accuracy of the approximation of e^x by $P_1(x) = 1 + x$ decreases as x gets larger. The following table gives some values of each function:

TABLE 1

x	$P_1(x) = 1 + x$	e^x
0.0001	1.0001	1.0001
0.001	1.001	1.0010005
0.01	1.01	1.01005
0.1	1.1	1.10517
0.2	1.2	1.2214
0.5	1.5	1.64872
1	2	2.71828

The accuracy can be improved by using higher-degree polynomials. To obtain a second-degree polynomial $P_2(x)$ that will approximate $f(x) = e^x$, start with the general second-degree polynomial.

$$P_2(x) = a + bx + cx^2$$

To get a good approximation, we still need the previous two properties: (1) $P_2(0) = f(0)$ and (2) $P'_2(0) = f'(0)$. In addition, we require that (3) $P''_2(0) = f''(0)$. These three properties allow us to compute the constants a, b, and c.

1. $P_2(0) = a + b(0) + c(0) = a$
   $f(0) = e^0 = 1$
   Thus $a = 1$
2. $P'_2(0) = b + 2cx = b + 2c(0) = b$
   $f'(0) = e^0 = 1$
   Thus $b = 1$
3. $P''_2(0) = 2c$
   $f''(0) = e^0 = 1$
   Thus $2c = 1$, $c = 1/2$

Substitute the values $a = 1$, $b = 1$, and $c = 1/2$ into the expression for $P_2(x)$.

$$P_2(x) = 1 + x + \frac{1}{2} x^2$$

In general any function $f(x)$ whose first two derivatives exist can be approximated for values of x close to 0 by the second-degree polynomial

$$P_1(x) = f(0) + f'(0)x + f''(0)x^2$$

This is called the *second Taylor polynomial of $f(x)$ at $x = 0$*. Table 2 compares the approximations of $f(x) = e^x$ by the first and second Taylor polynomials. More accurate approximations can be derived by using a third-degree polynomial, $P_3(x)$. The three requirements above would remain the same and a fourth would be needed, namely, (4) $P'''_3(0) = f'''(0)$. Using computations similar to the preceding, we can

TABLE 2

x	$P_1(x) = 1 + x$	$P_2(x) = 1 + x + \frac{1}{2} x^2$	e^x
0.0001	1.0001	1.0001	1.0001
0.001	1.001	1.0010005	1.0010005
0.01	1.01	1.01005	1.0100502
0.1	1.1	1.105	1.10517
0.2	1.2	1.22	1.2214
0.5	1.5	1.6250	1.64872
1	2	2.5	2.71828

show that

$$P_3(x) = 1 + x + \frac{1}{2}x^2 + \frac{1}{6}x^3$$

The reason for using polynomials to approximate values of functions is that polynomials are easy to evaluate. The approximations for $f(x) = e^x$ at $x = 0.5$ were not especially accurate when we used $P_1(x)$ and $P_2(x)$. The next example shows how to obtain a more accurate approximation by using $P_3(x)$.

EXAMPLE 1

Problem

Use $P_3(x)$ to approximate $\sqrt{e} = e^{1/2}$.

SOLUTION In the expression for $P_3(x)$ let $x = 1/2$.

$$P_3(0.5) = 1 + \frac{1}{2} + \left(\frac{1}{2}\right)\left(\frac{1}{2}\right)^2 + \left(\frac{1}{6}\right)\left(\frac{1}{2}\right)^3$$

$$= 1 + \frac{1}{2} + \frac{1}{8} + \frac{1}{48} = \frac{79}{48}$$

$$\approx 1.64583$$

The value of $e^{1/2}$ to five decimal places is 1.64872.

Taylor Polynomials

The accuracy of the approximations of $y = f(x)$ for values close to $x = 0$ can be improved by forming polynomials of higher degree. The polynomials are found by using similar properties. Each polynomial evaluated at $x = 0$ is equal to $f(0)$. The polynomial of degree n has each of its first n derivatives evaluated at $x = 0$ equal to the corresponding derivative of $f(x)$ at $x = 0$. That is, $P_n(0) = f(0)$ and $P_n^{(k)}(0) = f^{(k)}(0)$. This is called the **nth Taylor polynomial** of $f(x)$ at $x = 0$.

The denominator in the term containing x^3 is 6. It is computed from the general polynomial by taking the first two derivatives of x^3, which are $3x^2$ and $2(3)x = 6x$. The denominator in the term containing x^4 is computed by taking the first three derivatives of x^4, which are $4x^3$, $3(4)x^2$, and $2(3)(4)x = 24x$. This suggests that the denominators of the Taylor polynomials are of the form $n!$, called n factorial. Recall the definition

$$0! = 1! = 1$$

and for all positive integers $n > 1$

$$n! = n[(n - 1)!] = n(n - 1)(n - 2) \cdots 2(1)$$

For example

$$2! = 2(1) = 2$$
$$3! = 3(2!) = 3(2) = 6$$
$$4! = 4(3!) = 4(3)(2)(1) = 24$$
$$5! = 5(4!) = 5(4)(3)(2)(1) = 120$$

If $f(0)$, $f'(0)$, $f''(0)$, . . . , $f^{(n)}(0)$ exist, then the **nth Taylor polynomial of** $f(x)$ **at** $x = 0$ is defined by

$$P_n(x) = f(0) + f'(0)x + \frac{f''(0)}{2} x^2 + \frac{f'''(0)}{3!} x^3 + \cdots + \frac{f^{(n)}(0)}{n!} x^n$$

EXAMPLE 2 | **Problem**

Find the first four Taylor polynomials for the function $f(x) = \ln(1 + x)$ at $x = 0$.

SOLUTION Compute the first four derivatives of $f(x)$ and evaluate each at $x = 0$.

$$f(x) = \ln(1 + x) \qquad\qquad f(0) = \ln 1 = 0$$
$$f'(x) = \frac{1}{1 + x} = (1 + x)^{-1} \qquad f'(0) = 1$$
$$f''(x) = -(1 + x)^{-2} \qquad\qquad f''(0) = -1$$
$$f'''(x) = 2(1 + x)^{-3} \qquad\qquad f'''(0) = 2$$
$$f^{(4)}(x) = -6(1 + x)^{-4} \qquad\qquad f^{(4)}(0) = -6$$

The first four Taylor polynomials for $f(x) = \ln(1 + x)$ at $x = 0$ are

$$P_1(x) = 0 + (1)x = x$$

$$P_2(x) = 0 + (1)x + \frac{(-1)}{2} x^2 = x - \frac{x^2}{2}$$

$$P_3(x) = 0 + (1)x + \frac{(-1)}{2} x^2 + \frac{(2)}{6} x^3 = x - \frac{x^2}{2} + \frac{x^3}{3}$$

$$P_4(x) = x - \frac{x^2}{2} + \frac{x^3}{3} - \frac{6x^4}{24} = x - \frac{x^2}{2} + \frac{x^3}{3} - \frac{x^4}{4}$$

The next example shows how the accuracy of the approximations improves as successive Taylor polynomials are used.

EXAMPLE 3 | **Problem**

Approximate $\ln(1.2)$ using the first four Taylor polynomials for the function $\ln(1 + x)$ at $x = 0$.

SOLUTION Substitute $x = 0.2$ in the Taylor polynomials computed in Example 2.

$$P_1(0.2) = 0.2$$

$$P_2(0.2) = 0.2 - \frac{(0.2)^2}{2} = 0.18$$

$$P_3(0.2) = 0.2 - \frac{(0.2)^2}{2} + \frac{(0.2)^3}{3} \approx 0.18267$$

$$P_4(0.2) = 0.2 - \frac{(0.2)^2}{2} + \frac{(0.2)^3}{3} - \frac{(0.2)^4}{4} \approx 0.18223$$

The value of $\ln 1.2$ accurate to five decimal places is 0.18232. The difference between this value and the approximation $P_4(0.2)$ is less than 0.0001.

EXAMPLE 4 | **Problem**

Find the nth Taylor polynomial at $x = 0$ for the function $f(x) = \dfrac{1}{1 - x}$.

SOLUTION Compute the first four derivatives of $f(x)$ and evaluate each at $x = 0$. Then determine a pattern to find the nth derivative and evaluate it at $x = 0$.

$$f(x) = \frac{1}{1 - x} = (1 - x)^{-1} \qquad\qquad f(0) = 1$$
$$f'(x) = -(1 - x)^{-2}(-1) = (1 - x)^{-2} \qquad f'(0) = 1$$
$$f''(x) = 2(1 - x)^{-3} \qquad\qquad f''(0) = 2$$
$$f'''(x) = 6(1 - x)^{-4} \qquad\qquad f'''(0) = 6$$
$$f^{(4)}(x) = 24(1 - x)^{-5} \qquad\qquad f^{(4)}(0) = 24$$

The fourth Taylor polynomial for $f(x) = 1/(1 - x)$ at $x = 0$ is

$$P_4(x) = 1 + x + \frac{2}{2}x^2 + \frac{6}{6}x^3 + \frac{24}{24}x^4$$

$$P_4(x) = 1 + x + x^2 + x^3 + x^4$$

The pattern is that $f^{(k)}(0) = k!$ for each k. Therefore the nth Taylor polynomial is

$$P_n(x) = 1 + x + x^2 + x^3 + \cdots + x^n$$

Taylor Polynomials at $x = a$

Sometimes it is not convenient to approximate a function by Taylor polynomials at $x = 0$. For example, to approximate $\sqrt{4.1}$, we use Taylor polynomials of $f(x) = \sqrt{x}$. But 4.1 is not close enough to $x = 0$ to guarantee an accurate approximation. It is possible to define Taylor polynomials at any value $x = a$. For this example we choose $a = 4$ to get an accurate approximation. The definition is similar to that of Taylor polynomials at $x = 0$. In the latter formula replace x by $x - a$ and evaluate the function and its derivatives at $x = a$.

If $f(a)$, $f'(a)$, $f''(a)$, . . . , $f^{(n)}(a)$ exist, then the **nth Taylor polynomial of $f(x)$ at $x = a$** is defined by

$$P_n(x) = f(a) + f'(a)(x - a) + \frac{f''(a)}{2}(x - a)^2 + \frac{f'''(a)}{3!}(x - a)^3$$
$$+ \cdots + \frac{f^{(n)}(a)}{n!}(x - a)^n$$

EXAMPLE 5

Problem

Approximate $\sqrt{4.1}$ by using the third Taylor polynomial at $x = 4$.

SOLUTION Find the first three derivatives of $f(x) = \sqrt{x}$ and evaluate them at $x = 4$.

$$f(x) = x^{1/2} \qquad f(4) = 2$$

$$f'(x) = \frac{x^{-1/2}}{2} \qquad f'(4) = \frac{(4)^{-1/2}}{2} = \frac{1}{4}$$

$$f''(x) = \frac{-x^{-3/2}}{4} \qquad f''(4) = \frac{-(4)^{-3/2}}{4} = \frac{-1}{32}$$

$$f'''(x) = \frac{3x^{-5/2}}{8} \qquad f'''(4) = \frac{(3)(4)^{-5/2}}{8} = \frac{3}{256}$$

Then the third Taylor polynomial at $x = 4$ is

$$P_3(x) = 2 + \frac{1}{4}(x - 4) - \frac{1}{64}(x - 4)^2 + \frac{1}{512}(x - 4)^3$$

To approximate $\sqrt{4.1}$, evaluate $P_3(x)$ at $x = 4.1$. It is

$$P_3(4.1) = 2 + \frac{1}{4}(4.1 - 4) - \frac{1}{64}(4.1 - 4)^2 + \frac{1}{512}(4.1 - 4)^3$$

$$= 2 + \frac{.1}{4} - \frac{.01}{64} + \frac{.001}{512} \approx 2.024845703$$

The value of $\sqrt{4.1}$ accurate to seven decimal places is 2.0248457.

The Remainder Theorem

The accuracy of an approximation to a function $y = f(x)$ by its nth Taylor polynomial at $x = a$ is the difference between the values of $f(x)$ and $P_n(x)$. It is called the *nth remainder of $f(x)$ at $x = a$*. Its definition is

$$R_n(x) = f(x) - P_n(x)$$

Usually it is not possible to compute this value, but it is possible to find an upper bound for the remainder. This means that the error of the approximation is less than a certain value, which gives a measure of the accuracy of the approximation. The upper bound is given by the following.

—————————————————————— Remainder Theorem ——————————————————————

Suppose M is a number such that $\left|f^{(n+1)}(t)\right| \leq M$ for all t between a and x. Then

$$|R_n(x)| < \frac{M}{(n + 1)!}|x - a|^{n+1}$$

EXAMPLE 6

Problem

Determine the accuracy of the approximation in Example 5.

SOLUTION The approximation used the third Taylor polynomial so $n = 3$. Compute the absolute value of the fourth derivative of $f(x) = \sqrt{x}$ for x between 4 and 4.1.

$$\left|f^{(4)}(x)\right| = \frac{15x^{-7/2}}{16}$$

This is a decreasing function (because its derivative is negative) between 4 and 4.1, so that its maximum value occurs at $x = 4$. Thus

$$M = \left|f^{(4)}(4)\right| = \frac{15}{16}(4)^{-7/2} - \frac{15}{2048} \approx 0.0074$$

Therefore for this example

$$|R_3(x)| < \frac{M}{4!}|4.1 - 4|^4 \approx \frac{(0.0074)}{24}(0.1)^4 \approx 0.00000003$$

This shows that the approximation of $\sqrt{4.1}$ using $P_3(4.1)$ is accurate to within 0.00000003. In Example 5 we mentioned that the accuracy is even better than this, but we used a calculator to find the value of $\sqrt{4.1}$. The remainder theorem is useful when no such means of calculation is available.

EXERCISE SET 11–1

• In Problems 1 to 4 use the third Taylor polynomial $P_3(x)$ of the function $f(x) = e^x$ to approximate the given number.

1. $e^{1/4}$ 2. $e^{1/5}$

3. $e^{3/4}$ 4. $e^{3/5}$

In Problems 5 to 8 find the first four Taylor polynomials at $x = 0$ for the function.

5. $f(x) = \ln(1 + 2x)$ 6. $f(x) = \ln(1 - x)$

7. $f(x) = e^{2x}$ 8. $f(x) = e^{-2x}$

In Problems 9 to 12 approximate the number using the fourth Taylor polynomial for the function $f(x) = \ln(1 + x)$ at $x = 0$.

9. $\ln(1.01)$ 10. $\ln(1.11)$

11. $\ln(0.9)$ 12. $\ln(0.09)$

In Problems 13 to 16 find the nth Taylor polynomial at $x = 0$ for the function.

13. $f(x) = e^x$ 14. $f(x) = \ln(1 + x)$

15. $f(x) = \dfrac{1}{1 + x}$ 16. $f(x) = e^{2x}$

In Problems 17 to 20 find the nth Taylor polynomial at $x = a$ for the function and the given value of a.

17. $f(x) = e^{x-1}, \quad a = 1$

18. $f(x) = \ln(2 - x), \quad a = 1$

19. $f(x) = \dfrac{1}{4 - x}, \quad a = 3$

20. $f(x) = e^{2x}, \quad a = 1$

•• In Problems 21 and 22 fill in the table for the given Taylor polynomials at $x = 0$ for the function $f(x)$.

21. $f(x) = \sin x$

x	$P_3(x) = x - \dfrac{x^3}{6}$	$P_5(x) = x - \dfrac{x^3}{6} + \dfrac{x^5}{120}$
0.001		
0.01		
0.1		
0.2		

22. $f(x) = \cos x$

x	$P_2(x) = 1 - \dfrac{x^2}{2}$	$P_4(x) = 1 - \dfrac{x^2}{2} + \dfrac{x^4}{24}$
0.001		
0.01		
0.1		
0.2		

In Problems 23 to 30 find the fourth Taylor polynomial at $x = 0$.

23. $f(x) = \cos 2x$ 24. $f(x) = e^{x+1}$

25. $f(x) = xe^x$ 26. $f(x) = xe^{-x}$

27. $f(x) = x \sin x$ 28. $f(x) = \dfrac{1}{1 - 2x}$

29. $f(x) = (1 - x)^{1/2}$ 30. $f(x) = (4 - x)^{3/2}$

In Problems 31 and 32 find the nth Taylor polynomial at $x = 0$.

31. $f(x) = \cos x$ 32. $f(x) = \sin x$

33. Use the third Taylor polynomial at $x = 9$ of $f(x) = \sqrt{x}$ to approximate $\sqrt{9.1}$.

34. Use the third Taylor polynomial at $x = 2$ of $f(x) = \ln x$ to approximate $\ln 2.1$.

35. Use the third Taylor polynomial at $x = 1$ of $f(x) = 1/x$ to approximate $1/1.1$.

36. Use the third Taylor polynomial at $x = 1$ of $f(x) = \sin x$ to approximate $\sin 1.1$.

37. Determine the accuracy in using $P_3(x)$ of $f(x) = e^x$ at $x = 0$ to approximate $e^{0.1}$ by showing that the error of the approximation is at most $f^{(4)}(0.1)(0.1)^4/4!$.

38. Determine the accuracy in using $P_2(x)$ of $f(x) = \sqrt{x}$ at $x = 0$ to approximate $\sqrt{0.1}$ by showing that the error of the approximation is at most $f^{(3)}(0.1)(0.1)^3/3!$.

In Problems 39 to 42 use the second Taylor polynomial at $x = 0$ to approximate the given number.

39. The revenue $R(x)$ in thousands of dollars from selling x thousand units is

$$R(x) = 100 \ln(1 + 50x)$$

Find $R(0.01)$.

40. The cost $C(x)$ in millions of dollars of producing x thousand units is

$$C(x) = \frac{100x}{1 + 5x}$$

Find $C(0.2)$.

41. Psychologists measure the amount of learning $P(x)$ that a group acquires in x minutes when accomplishing a certain task by the function

$$P(x) = \frac{1}{2 - e^{-0.1x}}$$

Find $P(0.1)$.

42. Sociologists show that the percentage $P(x)$ of people who have heard a rumor after x days is given by the function

$$P(x) = \frac{e^{0.2x}}{2e^{0.2x} - 1}$$

Find $P(0.1)$.

●●● 43. For the function

$$f(x) = 2 + 4x + \frac{7x^2}{2!} + \frac{5x^3}{3!}$$

find (a) $f(0)$, (b) $f'(0)$, (c) $f''(0)$, and (d) $f'''(0)$.

44. For the function

$$f(x) = 1 - 2(x - 2) + \frac{5}{2!}(x - 2)^2$$

$$+ \frac{11}{3!}(x - 2)^3$$

find (a) $f(2)$, (b) $f'(2)$, (c) $f''(2)$, and (d) $f'''(2)$.

45. In an article investigating various models of how different industries evolve, Marchetti studied the substitution of cars for horses as a means of transportation in the years 1900 to 1930.[*] The number of cars in use, $N(x)$, at time x was determined to be

$$N(x) = \frac{1000}{1 + e^{-.02x}}$$

Approximate $N(1)$, the number of cars in the United States in 1901, using the fourth Taylor polynomial for the function $f(x) = e^{-.02x}$ and then substituting this value into the formula for $N(x)$.

46. In an article studying the effects of a society using part of its gross national product to fund research, Dresch and Janson use the following function to measure the likelihood $P(x)$ that an individual with IQ x will pursue a career in science:[†]

$$P(x) = \frac{.5}{1 + e^{-.01x}}$$

Approximate $P(100)$ using the fourth Taylor polynomial for the function $f(x) = e^{-.01x}$ and then substituting this value into the formula for $P(x)$.

[*]Cesare Marchetti, "Infrastructure for Movement," *Technical Forecasting and Social Change,* Vol. 32, 1987, pp. 373–393.

[†]Stephen P. Dresch and Kenneth R. Janson, "Giants, Pygmies and Social Costs of Funding Research or Price Revisited," *Technical Forecasting and Social Change,* Vol. 32, 1987, pp. 323–340.

11-2 INFINITE SERIES

An **infinite series** is the "addition" of an infinite number of terms. The addition is not done term by term as is done for a finite number of terms. Rather, a sequence of numbers is generated, and if this sequence has a limit, we say that the sum has a limit. If the terms are designated by a_i, the sum is indicated by

$$a_1 + a_2 + a_3 + \cdots + a_n + \cdots$$

The term a_n is called the nth term. If the pattern of the terms is evident, the nth term may be omitted. In some examples it is convenient to start with the 0th term; that is, the first term is derived from the nth term by letting $n = 0$. The context makes this clear.

Some Examples

The fraction 1/9 has a decimal equivalent that is a repeating decimal. It is

$$\frac{1}{9} = 0.111111 \ldots$$

This can be seen by dividing 9 into 1.

$$\frac{.1111 \ldots}{9)\overline{1.000}}$$

This decimal is written in expanded notation as

$$0.1111 \ldots = 0.1 + 0.01 + 0.001 + 0.0001 + \cdots$$

$$= \frac{1}{10} + \frac{1}{10^2} + \frac{1}{10^3} + \frac{1}{10^4} + \cdots + \frac{1}{10^n} + \cdots$$

This is an example of an infinite series that has a sum. That is, the infinite series $0.1 + 0.01 + 0.001 + \cdots$ has the sum 1/9. To illustrate how this sum is computed, form the **partial sums,** which are the first term, the sum of the first two terms, the sum of the first three terms, and so on. These partial sums are denoted by S_1, S_2, S_3, and, in general, S_n. For the previous infinite series the partial sums are

$$S_1 = 0.1$$
$$S_2 = 0.11$$
$$S_3 = 0.111$$
$$S_4 = 0.1111$$
$$\vdots \qquad \vdots$$
$$S_n = 0.111 \ldots 1$$

This is a sequence of numbers. A **sequence** is a function whose domain is the set of natural numbers, and so we can investigate whether it has a limit as n approaches infinity. As previously indicated, this sequence has 1/9 as its limit. Thus we say that the infinite series has the sum 1/9.

For an example of an infinite series that does not have a sum, consider the series

$$1 + 1 + 1 + \cdots$$

This "sum" can be made as large as possible by adding as many terms as desired. To illustrate this, form the partial sums

$$S_1 = 1$$
$$S_2 = 1 + 1 = 2$$
$$S_3 = 1 + 1 + 1 = 3$$
$$\vdots \qquad \qquad \vdots$$
$$S_n = 1 + 1 + 1 + \cdots + 1 = n$$

This sequence does not approach any number as n approaches infinity. Thus it has no limit. Therefore the series does not have a sum.

Convergence and Divergence

In general the sum S_n of the first n terms of an infinite series is called the **nth partial sum.** An infinite series whose partial sums approach a limit is called **convergent** and the limit is called its **sum.** An infinite series whose partial sums do not approach a limit is called **divergent.** The series

$$0.1 + 0.01 + 0.001 + \cdots$$

is convergent with sum 1/9. The series

$$1 + 1 + 1 + \cdots$$

is divergent. A divergent series does not need to have partial sums that increase without bound. The series

$$1 - 1 + 1 - 1 + 1 - \cdots$$

alternately adds 1 and subtracts 1. The first four partial sums are

$$S_1 = 1 \qquad \qquad = 1$$
$$S_2 = 1 - 1 \qquad \qquad = 0$$
$$S_3 = 1 - 1 + 1 \qquad = 1$$
$$S_4 = 1 - 1 + 1 - 1 = 0$$

The partial sums alternate between 1 and 0, so they do not approach a limit. Therefore the series is divergent.

Geometric Series

Determining whether an infinite series converges is not usually an easy task. Consider the following two series:

$$1 + \frac{1}{2} + \frac{1}{3} + \frac{1}{4} + \frac{1}{5} + \cdots \tag{1}$$

$$1 + \frac{1}{2} + \frac{1}{4} + \frac{1}{8} + \frac{1}{16} + \cdots \tag{2}$$

They look similar, but they are actually very different in that the first series is divergent and the second is convergent. To see that equation (2) is convergent, consider the first five partial sums.

$$S_1 = 1 \qquad\qquad\qquad\qquad = 1$$

$$S_2 = 1 + \frac{1}{2} \qquad\qquad\qquad = \frac{3}{2} = 2 - \frac{1}{2}$$

$$S_3 = 1 + \frac{1}{2} + \frac{1}{4} \qquad\qquad = \frac{7}{4} = 2 - \frac{1}{4}$$

$$S_4 = 1 + \frac{1}{2} + \frac{1}{4} + \frac{1}{8} \qquad = \frac{15}{8} = 2 - \frac{1}{8}$$

$$S_5 = 1 + \frac{1}{2} + \frac{1}{4} + \frac{1}{8} + \frac{1}{16} = \frac{31}{16} = 2 - \frac{1}{16}$$

The pattern shows that S_n is equal to 2 minus the last term, which is $1/2^{n-1}$. That is

$$S_n = 2 - \frac{1}{2^{n-1}}$$

As n approaches infinity $1/2^{n-1}$ approaches 0, so the limit of S_n is 2. This series is an example of a **geometric series.** It is defined by

$$a + ar + ar^2 + ar^3 + \cdots + ar^n + \cdots$$

where the constants a and r are not 0 and r is called the **ratio.** (The ratio of any two consecutive terms is r.) The infinite series (2) is a geometric series with $a = 1$ and $r = 1/2$. It is possible to derive a rule to determine when a geometric series is convergent. Consider the nth partial sum S_n of the geometric series $a + ar + ar^2 + \cdots$ and multiply it by r to get rS_n. Then align like terms.

$$S_n = a + ar + ar^2 + ar^3 + \cdots + ar^{n-1}$$
$$rS_n = \phantom{a + {}} ar + ar^2 + ar^3 + \cdots + ar^{n-1} + ar^n$$

To obtain an expression for S_n, subtract the second expression from the first.

$$S_n - rS_n = a - ar^n$$

$$S_n(1 - r) = a(1 - r^n)$$

$$S_n = \frac{a(1 - r^n)}{1 - r} = \frac{a}{1 - r} - \frac{ar^n}{1 - r}$$

If $|r| < 1$, then r^n approaches 0 as n approaches infinity. In this case the second term approaches 0 and the sum is $a/(1 - r)$. Thus the series is convergent.

If $|r| \geq 1$, then r^n approaches infinity as n approaches infinity and the series is divergent.

_____ **Geometric Series** _____

For the infinite series

$$a + ar + ar^2 + ar^3 + \cdots + ar^n + \cdots$$

(1) If $-1 < r < 1$, then the series converges and the sum of the series is

$$\frac{a}{1 - r}$$

(2) If $r \leq -1$ or $r \geq 1$, then the series diverges.

For example, the geometric series with $a = 0.1$ and $r = 0.1$ is $0.1 + 0.01 + 0.001 + \cdots$. We saw previously that the sum of this series is $1/9$. Using the formula, we get $0.1/(1 - 0.1) = 0.1/0.9 = 1/9$. Also, using the formula for the geometric series in (2) with $a = 1$ and $r = 1/2$ yields the sum $1/(1 - 1/2) = 2$.

EXAMPLE 1 | **Problem**

Determine whether the geometric series converge. If so, find the sum:

(a) $\dfrac{1}{3} + \left(\dfrac{1}{3}\right)^2 + \left(\dfrac{1}{3}\right)^3 + \cdots$,

(b) $1 + \dfrac{4}{3} + \left(\dfrac{4}{3}\right)^2 + \left(\dfrac{4}{3}\right)^3 + \cdots$,

(c) $0.6 + 0.06 + 0.006 + 0.0006 + \cdots$.

SOLUTION (a) For this series $a = 1/3$ and $r = 1/3$. Since r is between -1 and 1, the series is convergent. The sum is

$$\frac{1/3}{(1 - 1/3)} = \frac{1/3}{2/3} = \frac{1}{2}$$

(b) For this series $a = 1$ and $r = 4/3$. Since $r > 1$, the series is divergent.
(c) For this series $a = 0.6$ and $r = 0.1$. Since r is between -1 and 1, the series is convergent. The sum is $0.6/(1 - 0.1) = 0.6/0.9 = 2/3$. This series is the decimal equivalent of $2/3$, which is $0.6666 \ldots$. Since $2/3 = 6(1/9)$, it is not surprising to see that $0.6666 \ldots = 6(0.1111 \ldots)$.

The decimal equivalent of a rational number is either a terminating or repeating decimal. A repeating decimal can be expressed as a geometric series, as was done earlier for $0.1111 \ldots$ and $0.6666 \ldots$. Repeating decimals can have more than one repeating digit, and the first few terms may not be part of the series.

EXAMPLE 2 | **Problem**

Find the fraction that is equal to the repeating decimal: (a) 0.363636 . . . and (b) 0.4999 . . . .

SOLUTION (a) Express the decimal as a series.

$$0.363636 \ldots = 0.36 + 0.0036 + 0.000036 + \cdots$$

This is a geometric series with $a = 0.36$ and $r = 0.01$. The sum is

$$\frac{0.36}{(1 - 0.01)} = \frac{0.36}{0.99} = \frac{4}{11}$$

(b) Express the decimal as a series.

$$0.4999 \ldots = 0.4 + 0.09 + 0.009 + 0.0009 + \cdots$$

The first term, 0.4, is not part of the geometric series because the ratio of any other two terms is 0.1, but the ratio of the first two terms is $0.09/0.4 = 9/40$. Ignore the first term for the time being and find the sum of the remaining terms. The terms form a geometric series with $a = 0.09$ and $r = 0.1$.

$$0.09 + 0.009 + 0.0009 + \cdots$$

The sum is $0.09/(1 - 0.1) = 0.09/0.9 = 0.1$. Therefore the sum of the first term, 0.4, and the sum of the geometric series, 0.1, is equal to 0.5. This means that

$$0.4999 \ldots = 0.4 + 0.1 = 0.5$$

That is, this series has the same sum as the series in Example 1(a). The decimal equivalents of the terms in Example 1(a) can be shown to sum to 0.4999 . . . .

The Harmonic Series

The series (1) mentioned earlier is

$$1 + \frac{1}{2} + \frac{1}{3} + \frac{1}{4} + \frac{1}{5} + \cdots$$

This is called the **harmonic series.** It is not a geometric series because the ratios of consecutive terms are not equal. We stated that this series is not convergent. This is because the partial sums can be made as large as desired by taking enough terms. Hence they do not have a limit. Calculate several partial sums and gather the terms as follows:

$$S_1 = 1$$

$$S_2 = 1 + \frac{1}{2} = \frac{3}{2}$$

$$S_4 = 1 + \frac{1}{2} + \frac{1}{3} + \frac{1}{4} = S_2 + \frac{1}{3} + \frac{1}{4} > S_2 + \left(\frac{1}{4} + \frac{1}{4}\right) > \frac{3}{2} + \frac{1}{2}$$

$$S_8 = 1 + \frac{1}{2} + \frac{1}{3} + \frac{1}{4} + \frac{1}{5} + \frac{1}{6} + \frac{1}{7} + \frac{1}{8}$$

$$= S_4 + \frac{1}{5} + \frac{1}{6} + \frac{1}{7} + \frac{1}{8} > S_4 + \left(\frac{1}{8} + \frac{1}{8} + \frac{1}{8} + \frac{1}{8}\right)$$

$$> \frac{3}{2} + \frac{1}{2} + \frac{1}{2}$$

Following this pattern, we get

$$S_{16} = 1 + \frac{1}{2} + \frac{1}{3} + \frac{1}{4} + \frac{1}{5} + \frac{1}{6} + \frac{1}{7} + \frac{1}{8} + \frac{1}{9} + \cdots + \frac{1}{16}$$

$$> S_8 + \left(\frac{1}{16} + \frac{1}{16} + \frac{1}{16} + \frac{1}{16} + \frac{1}{16} + \frac{1}{16} + \frac{1}{16} + \frac{1}{16}\right)$$

$$> \frac{3}{2} + \frac{1}{2} + \frac{1}{2} + \frac{1}{2}$$

$$S_{32} > \frac{3}{2} + \frac{1}{2} + \frac{1}{2} + \frac{1}{2} + \frac{1}{2}$$

This means that by doubling the number of terms added at each step, we increase the sum by more than 1/2. Thus the sum can be made as large as desired by taking enough terms. Therefore the harmonic series diverges.

An Application from Economics

When a government or a firm makes an investment in a community, there is often a residual effect beyond the amount of money initially expended. For example, if a city government makes a business tax cut of $1,000,000, the end result may be more revenue for the city, because the "extra" money is invested in new industry, which produces more business revenue. This in turn produces tax revenue. Economists call this the *multiplier effect*.

EXAMPLE 3

A city enacts a $1,000,000 tax benefit for business. Suppose each year this money produces an 80% return on investment, indicating that the revenue generated by this money at the end of the year is 80% of the total. This amount is then reinvested and it generates a return of 80%.

Problem

With these assumptions, determine how much tax revenue is ultimately generated by the tax cut if the business tax rate is 30% per year.

SOLUTION First determine how much revenue is generated by the tax benefit in the long run. The tax revenue is 30% of this sum. At the end of the first year the investment of this money generates $800,000 in business income. So in the first year the taxable income generated from the previous year's tax cut is $1,000,000(0.8) = $800,000. At the end of the following year 80% of this amount is generated in business income, which is $800,000(0.80) = $640,000. In the next year 80% of this amount is generated as business income, and so on. The amounts of income that the city taxes are

$$800,000 + 640,000 + 512,000 + \cdots$$
$$= 800,000 + 800,000(0.8) + 800,000(0.8)^2 + \cdots$$

This is a geometric series with $a = 800,000$ and $r = 0.8$. Its sum is

$$\frac{800,000}{1 - 0.8} = \frac{800,000}{0.2} = 4,000,000$$

The amount of tax revenue on this amount based on a 30% tax rate is $4,000,000(0.30) = $1,200,000. Therefore the tax benefit will eventually generate more tax revenue for the city than it granted.

Radioactivity

A bitter debate about radiation hazards has led to numerous protests and angry rhetoric at nuclear power plants. The concern is twofold. First, there is the threat of release of radioactivity into the atmosphere due to reactor malfunction and second, the long-term threat of nuclear waste. The latter hazard is best explained by infinite sums. What does it mean to say that the nuclear by-product of a reactor will "never" disappear? A radioactive element is one that continually decays into new elements. High-energy radiation is released and when these emissions enter the body they injure cells by tearing electrons away from atoms. The *half-life* of a radioactive element is the time it takes for one-half of the element to decay. For example, iodine-131, denoted by ^{131}I, has a half-life of about eight days. This means that 1 pound (lb) of ^{131}I is reduced to 1/2 lb in eight days. In eight more days the amount is reduced to 1/4 lb, and so on. A common fallacy is that after two half-lives a radioactive material will decay entirely.

EXAMPLE 4

Problem

What fraction of a radioactive material will decay after (a) four half-lives? (b) after six half-lives?

SOLUTION (a) After the first half-life 1/2 of the material decays. One-half of this material, or $(1/2)(1/2) = 1/4$, decays during the second half-life. Therefore after two half-lives the amount of the original material that has decayed is $1/2 + 1/4$. During the third half-life the additional amount of material that decays is $(1/2)(1/4) = 1/8$. And after four half-lives the amount that decays is

$$\frac{1}{2} + \frac{1}{4} + \frac{1}{8} + \frac{1}{16} = \frac{15}{16}$$

(b) After two more half-lives, that is, after six half-lives, the amount that decays is

$$\frac{1}{2} + \frac{1}{4} + \frac{1}{8} + \frac{1}{16} + \frac{1}{32} + \frac{1}{64} = \frac{63}{64}$$

The two sums in Example 4 are the partial sums S_4 and S_6 of the infinite series

$$\frac{1}{2} + \frac{1}{4} + \frac{1}{8} + \cdots + \frac{1}{2^n} + \cdots$$

It is a geometric series with $a = r = 1/2$. Its sum is

$$\frac{1/2}{(1 - 1/2)} = \frac{1/2}{1/2} = 1$$

This implies that in an "infinite" amount of time all of the material will decay. There are, however, several practical considerations. On one hand, it makes no sense to speak of an infinite amount of time, meaning that in any finite amount of time radioactive material never completely decays. But when the material becomes very small, say, the size of an atom, for all practical purposes, it has completely decayed. On the other hand, it takes many half-lives for that amount to decay. This is the point that worries many people—radioactive material takes so long to decay that once it is produced, it is with us virtually forever.

EXERCISE SET 11–2

• In Problems 1 to 8 determine whether the geometric series converges. If so, find the sum.

1. $0.3 + 0.03 + 0.003 + 0.0003 + \cdots$

2. $0.2 + 0.02 + 0.002 + 0.0002 + \cdots$

3. $1 + \dfrac{5}{3} + \left(\dfrac{5}{3}\right)^2 + \left(\dfrac{5}{3}\right)^3 + \cdots$

4. $1 + \dfrac{5}{4} + \left(\dfrac{5}{4}\right)^2 + \left(\dfrac{5}{4}\right)^3 + \cdots$

5. $\dfrac{2}{3} + \left(\dfrac{2}{3}\right)^2 + \left(\dfrac{2}{3}\right)^3 + \cdots$

6. $\dfrac{3}{4} + \left(\dfrac{3}{4}\right)^2 + \left(\dfrac{3}{4}\right)^3 + \cdots$

7. $1 + \dfrac{1}{8} + \left(\dfrac{1}{8}\right)^2 + \left(\dfrac{1}{8}\right)^3 + \cdots$

8. $1 + \dfrac{2}{9} + \left(\dfrac{2}{9}\right)^2 + \left(\dfrac{2}{9}\right)^3 + \cdots$

In Problems 9 to 14 find the fraction that is equal to the repeating decimal.

9. $0.353535\ldots$ 10. $0.454545\ldots$

11. $0.717171\ldots$ 12. $0.838383\ldots$

13. $0.2999\ldots$ 14. $0.7999\ldots$

In Problems 15 to 18 a city enacts a $1,000,000 tax benefit for business and each year this money produces an $r\%$ return on investment. This amount is then reinvested and it generates a return of $r\%$, and so on. Determine how much tax revenue is ultimately generated by the tax cut if the business tax rate is 20% per year.

15. $r = 90$ 16. $r = 70$

17. $r = 75$ 18. $r = 50$

In Problems 19 to 22 determine what fraction of a radioactive material will decay after n half-lives.

19. $n = 5$ 20. $n = 7$

21. $n = 10$ 22. $n = 12$

•• In Problems 23 to 30 a geometric series is given. Find the ratio and determine if the series is convergent. If it is convergent, find the sum of the series.

23. $32 + 16 + 8 + 4 + 2 + \cdots$

24. $9 + 3 + 1 + \dfrac{1}{3} + \cdots$

25. $2 + 4 + 8 + 16 + 32 + \cdots$

26. $1 + 1.1 + 1.21 + 1.331 + \cdots$

27. $10 + 9 + 8.1 + 7.29 + \cdots$

28. $10 + 1 + .1 + .01 + \cdots$

29. $64 + 48 + 36 + 27 + \cdots$

30. $.1 + .101 + .10201 + .1030301 + \cdots$

In Problems 31 to 34 determine if the series converges.

31. $\dfrac{1}{2} + \dfrac{1}{3} + \dfrac{1}{4} + \dfrac{1}{5} + \cdots$

32. $2 + 1 + \dfrac{1}{2} + \dfrac{1}{3} + \dfrac{1}{4} + \dfrac{1}{5} + \cdots$

33. $3 + \dfrac{3}{2} + 1 + \dfrac{3}{4} + \dfrac{3}{5} + \dfrac{1}{2} + \dfrac{3}{7} + \cdots + \dfrac{3}{n} + \cdots$

34. $2 + \dfrac{3}{2} + \dfrac{4}{3} + \dfrac{5}{4} + \dfrac{6}{5} + \cdots + \dfrac{n+1}{n} + \cdots$

In Problems 35 to 38 the nth term of the series is given. Find the first four partial sums.

35. $a_n = \dfrac{1}{n}$ 36. $a_n = \dfrac{1}{2n+1}$

37. $a_n = \dfrac{2}{3n+1}$ 38. $a_n = \dfrac{1}{n^2}$

39. A salesperson is paid a commission on a contract in the following manner: the salesperson receives $1000 in the first year, $600 in the second year, $360 in the third year, and so on. How much is the person paid (a) in the first four years, (b) in perpetuity, that is, what is the sum of the geometric series?

40. A firm pays $1,000,000 in salaries to employees in a city. The employees then spend 80% of this amount in the city. This is income to the people selling the goods in the city. These people then spend 80% of this income in the city. As this continues, what is the net income to the people of the city generated from the initial payment of the firm? This is an example of the "multiplier effect" applied to salary.

41. A sociologist assumes that for every 100 people who are told a rumor, 50% will tell another person the rumor. Of these people, 50% will tell another person. How many people will eventually be told the rumor if 128 were initially told?

42. A patient receives 10 milligrams of a drug. Each hour 50% of the amount of the drug present in the body is eliminated. How much of the drug is present after (a) 3 hours, (b) 5 hours, (c) 10 hours?

●●● 43. For any nonzero number a, explain why the series

$$a + \frac{a}{2} + \frac{a}{3} + \frac{a}{4} + \frac{a}{5} + \cdots + \frac{a}{n} + \cdots$$

is divergent.

44. Explain why the series

$$\frac{1}{2} + \frac{1}{4} + \frac{1}{6} + \frac{1}{8} + \cdots + \frac{1}{2n} + \cdots$$

is divergent.

45. The term "multiplier" is used in economics to refer to the total effect over time that is produced by an initial investment. The term was first introduced by R. F. Kohn in an article showing how investment in housing can affect unemployment.* To illustrate the concept Kohn used a small community. Suppose the government spends an initial investment of $1,000,000 to produce new jobs in the building industry. This can be viewed as an investment of $1,000,000 in employment as it becomes salary to individuals. These individuals spend their income in the community, and 20% of their income produces more money for additional jobs. Then 20% of this amount is spent producing more jobs, and so on. What is the total amount that produces new income?

46. Planning for regional growth and economic development often involves analysis of the long range impact of investments. In an article comparing various types of measurement of long range impact, Kuehen, et al., studied various industries in the state of Missouri.† In one part of the study, they computed that if an initial investment is made in the agriculture industry, an income (in goods and services) equal to 75% of the investment would be returned to the industry in the first year. In the second year, 75% of the latter amount would return to the industry, and so on. What is the total effect of the initial investment?

*R. F. Kohn, "The Relation of Home Investment to Unemployment," *The Economic Journal,* Vol. 31, 1931, pp. 173–198.

†John A. Kuehen et al., "Comparisons of Multipliers from Input-output and Economic Base Models," *Land Economics,* Vol. 61, 1985, pp. 129–135.

11–3 TESTS FOR CONVERGENCE

We can test a geometric series for convergence by computing its ratio. If the series converges, then its sum can be calculated. For many other types of infinite series it is not actually possible to find the sum of the series if it converges. For these series we use various tests to determine whether the series converges. Even though the sum cannot be computed exactly, it is possible to approximate the limit of the partial sums with good accuracy. In the first part of this section we study infinite series whose *terms are positive*. The last part of the section deals with infinite series with negative terms.

The Comparison Test

Consider the two infinite series

$$a_1 + a_2 + a_3 + \cdots + a_n + \cdots \tag{1}$$

$$b_1 + b_2 + b_3 + \cdots + b_n + \cdots \tag{2}$$

Suppose each term of series (1) is less than or equal to the corresponding term of series (2); that is, $a_n \leq b_n$ for all n. If we know that series (2) converges, what can be said about series (1)? Since each term is less than its corresponding term, we have

$$a_1 + a_2 + \cdots a_n \leq b_1 + b_2 + \cdots + b_n$$

This means that the nth partial sum of series (1) is less than or equal to the nth partial sum of series (2). Since the latter sums approach a limit, the partial sums of series (1) must approach a number less than or equal to this limit. Therefore series (1) must also converge.

This result is part of the comparison test. The second part of the theorem states that if series (1) diverges and each term of series (2) is greater than or equal to its corresponding term in series (1), then series (2) also diverges.

_____ **The Comparison Test** _____

If the two series

$$a_1 + a_2 + a_3 + \cdots + a_n + \cdots \tag{1}$$

$$b_1 + b_2 + b_3 + \cdots + b_n + \cdots \tag{2}$$

have the property that

$$0 < a_n \leq b_n$$

then

i. If series (2) converges, then series (1) converges.
ii. If series (1) diverges, then series (2) diverges.

Figure 1 provides a geometric view of the comparison test. The width of each rectangle is 1 and the heights are a_n and b_n, respectively. This means that the areas of the rectangles are equal to the terms of the series. Each shaded rectangle corresponding to a_n lies within (or is equal to) the rectangle corresponding to b_n. If the sum of the areas b_n is finite, meaning that the corresponding series is convergent, then the sum of the rectangles a_n must also be finite and must sum to a number less than or equal to the sum of b_n. So the series corresponding to the a_n converges. Conversely, if the sums of areas a_n increase without bound, meaning the corresponding series is divergent, then the sum of areas b_n must also increase without bound, and so the corresponding series is divergent.

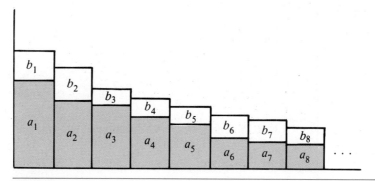

FIGURE 1

EXAMPLE 1

Problem

Determine whether the series

$$\frac{1}{2} + \frac{1}{3} + \frac{1}{5} + \frac{1}{9} + \cdots + \frac{1}{2^n + 1} + \cdots \tag{3}$$

converges.

SOLUTION The series can be compared with the geometric series

$$1 + \frac{1}{2} + \frac{1}{4} + \frac{1}{8} + \cdots + \frac{1}{2^n} + \cdots \tag{4}$$

For this geometric series $r = 1/2$, so the series converges with a sum of 2. Each term of series (3) is less than the corresponding term of series (4).

$$\frac{1}{2} < 1 \qquad \frac{1}{3} < \frac{1}{2} \qquad \frac{1}{5} < \frac{1}{4} \qquad \frac{1}{9} < \frac{1}{8}$$

In general for each n

$$\frac{1}{2^n + 1} < \frac{1}{2^n}$$

By the comparison test series 3 is convergent, and its sum is ≤ 2.

The comparison test states that if a series is to be compared with a known divergent series, each term of the divergent series must be less than or equal to the unknown series. The latter requirement can be relaxed some. The next example uses the following fact about infinite series. The convergence or divergence of an infinite series is not affected by the first few terms. In other words, if the first n terms were omitted, the remaining infinite series would converge if and only if the original series converges.

EXAMPLE 2 | **Problem**

Determine whether the series

$$\frac{1}{2} + \frac{2}{3} + \frac{3}{4} + \frac{4}{5} + \cdots + \frac{n}{n+1} + \cdots \tag{5}$$

converges.

SOLUTION The series can be compared with the harmonic series

$$1 + \frac{1}{2} + \frac{1}{3} + \frac{1}{4} + \frac{1}{5} + \cdots + \frac{1}{n} + \cdots \tag{6}$$

This is a known divergent series. The difficulty is that the first term does not satisfy the correct inequality in order to apply the comparison test. It is greater than, not less than, the first term of the series to be tested. Omit the first term of series (6). The resulting series is

$$\frac{1}{2} + \frac{1}{3} + \frac{1}{4} + \frac{1}{5} + \cdots + \frac{1}{n+1} + \cdots \tag{7}$$

Each term of series (7) is less than or equal to the corresponding term of series (5) because

$$\frac{1}{n+1} \leq \frac{n}{n+1}$$

By the comparison test series (5) is divergent.

Application of the comparison test is a trial-and-error process. If a known series does not work, meaning that the inequality is not satisfied by each pair of terms, another known series must be used. It might be enough to omit several of the first terms, as was done in Example 2. Or perhaps an entirely different series must be used.

Another pitfall to be avoided is applying the inequality incorrectly. If you suspect the series to be tested is convergent, a known convergent series must be used whose terms are greater than or equal to corresponding terms of the series under question. But if you suspect the series is divergent, the terms of the known divergent series must be less than the terms of the series under question.

The Ratio Test

If the comparison test does not produce results, another test, called the **ratio test,** can be used. The ratio test is sometimes more difficult to apply because it requires algebraic manipulation and the evaluation of a limit, but it can often be successfully applied to an infinite series where a comparison test is not clear. Rather than comparing each term of the series with the terms of another series, the ratio test requires that the ratio of successive terms of the series be computed. Then the limit of these ratios is used to test for convergence. Absolute value signs are used in the statement of the ratio test to allow us to apply it to series containing negative terms. This will be done at the end of the section. The absolute value signs are omitted in the examples in this section because all terms in the series are positive.

––––––––––––––––––––––– **The Ratio Test** –––––––––––––––––––––––

For the series

$$a_1 + a_2 + \cdots + a_n + \cdots$$

where $a_n > 0$ for all n, compute the ratio a_{n+1}/a_n for all n. If the following limit exists, let

$$R = \lim_{n \to \infty} \frac{|a_{n+1}|}{|a_n|}$$

Then

1. If $0 \leq R < 1$, the series converges.
2. If $R > 1$, the series diverges.
3. If $R = 1$, no conclusion can be made.

EXAMPLE 3

Problem

Determine whether the series

$$\frac{1}{3} + \frac{2}{3^2} + \frac{3}{3^3} + \frac{4}{3^4} + \cdots + \frac{n}{3^n} + \cdots \tag{8}$$

converges.

SOLUTION The general term a_n is

$$a_n = \frac{n}{3^n}$$

To find a_{n+1}, replace n by $n + 1$ in the expression for a_n.

$$a_{n+1} = \frac{n + 1}{3^{n+1}}$$

Form the ratio and simplify the expression.

$$\frac{a_{n+1}}{a_n} = \frac{(n + 1)/3^{n+1}}{n/3^n}$$

$$= \frac{(n + 1)}{n} \frac{3^n}{3^{n+1}}$$

$$= \frac{(n + 1)}{n} \frac{1}{3} = \frac{1}{3}\left(1 + \frac{1}{n}\right)$$

Take the limit as n approaches infinity.

$$\lim_{n \to \infty} \frac{a_{n+1}}{a_n} = \lim_{n \to \infty} \frac{1}{3} \left(1 + \frac{1}{n} \right)$$

$$= \frac{1}{3}$$

By the ratio test, with $R = 1/3 < 1$, series (8) is convergent.

The ratio test is useful when the series contains factorial expressions because the factorials cancel in the numerator and denominator of the ratio.

EXAMPLE 4 | **Problem**

Determine whether the series

$$2 + \frac{2^2}{2!} + \frac{2^3}{3!} + \frac{2^4}{4!} + \cdots + \frac{2^n}{n!} + \cdots \tag{9}$$

converges.

SOLUTION The general term a_n is

$$a_n = \frac{2^n}{n!}$$

Compute a_{n+1}.

$$a_{n+1} = \frac{2^{n+1}}{(n + 1)!}$$

Form the ratio and simplify the expression.

$$\frac{a_{n+1}}{a_n} = \frac{2^{n+1}/(n + 1)!}{2^n/n!}$$

$$= \frac{2^{n+1}}{2^n} \frac{n!}{(n + 1)!}$$

$$= \frac{2}{(n + 1)}$$

Take the limit as n approaches infinity.

$$\lim_{n \to \infty} \frac{a_{n+1}}{a_n} = \lim_{n \to \infty} \frac{2}{n + 1} = 0$$

By the ratio test, with $R = 0 < 1$, series (9) is convergent.

The next example shows that the size of the first few terms is sometimes not a good indication of the behavior of the remaining terms.

EXAMPLE 5 | **Problem**

Determine whether the series

$$\frac{1}{10} + \frac{2!}{100} + \frac{3!}{1000} + \cdots + \frac{n!}{10^n} + \cdots \qquad (10)$$

converges.

SOLUTION The general term a_n is

$$a_n = \frac{n!}{10^n}$$

Compute a_{n+1}.

$$a_{n+1} = \frac{(n+1)!}{10^{n+1}}$$

Form the ratio and simplify the expression.

$$\frac{a_{n+1}}{a_n} = \frac{(n+1)!/10^{n+1}}{n!/10^n}$$

$$= \frac{(n+1)!}{n!} \frac{10^n}{10^{n+1}}$$

$$= \frac{n+1}{10}$$

Take the limit as n approaches infinity.

$$\lim_{n\to\infty} \frac{a_{n+1}}{a_n} = \lim_{n\to\infty} \frac{n+1}{10} = \infty$$

By the ratio test, with $R = \infty > 1$, series (10) is divergent.

The first few terms in series (10) seem to indicate that the series is convergent because the terms are certainly getting very small. But eventually in later terms the numerators become much larger than the denominators. This causes the series to diverge.

For a series to converge it is necessary that the nth term approach zero as n approaches infinity. But the converse is not true. That is, knowing that a_n approaches zero as n approaches infinity does not guarantee that the series is convergent. For example, consider the harmonic series

$$1 + \frac{1}{2} + \frac{1}{3} + \cdots + \frac{1}{n} + \cdots$$

Its nth term, $a_n = 1/n$, approaches zero as n approaches infinity, but it is a divergent series.

Series with Negative Terms

A common type of series with negative terms that will be studied in the next section is an **alternating series.** It is a series in which the signs of successive terms alternate: suppose $a_n > 0$, then an alternating series is of the form

$$a_1 - a_2 + a_3 - a_4 + \cdots$$

An alternating series can also start with a negative term. Here are some examples of alternating series.

$$1 - \frac{1}{3} + \frac{1}{3^2} - \frac{1}{3^3} + \frac{1}{3^4} - \cdots + \frac{(-1)^n 1}{3^n} + \cdots$$

$$-\frac{1}{10} + \frac{2}{100} - \frac{3}{1000} + \cdots + \frac{(-1)^n n}{10^n} + \cdots$$

$$1 - \frac{1}{2} + \frac{1}{3} - \cdots + \frac{(-1)^{n+1} 1}{n} + \cdots$$

In the first series n starts with 0, and in the second and third series n starts with 1. There is a simple test for convergence for alternating series.

_____ Alternating Series Test _____

Let $a_n > 0$, then the alternating series

$$a_1 - a_2 + a_3 - a_4 + \cdots$$

converges if

1. $a_n > a_{n+1}$ for all n greater than a fixed integer N, and
2. a_n approaches 0 as n approaches infinity.

EXAMPLE 6

Problem

Determine whether the series

$$1 - \frac{1}{2} + \frac{1}{3} - \frac{1}{4} + \cdots + \frac{(-1)^{n+1} 1}{n} + \cdots$$

converges.

SOLUTION For the first part of the alternating series test note that for all n

$$a_n = \frac{1}{n} > \frac{1}{n + 1} = a_{n+1}$$

For the second part of the test it is clear that $a_n = 1/n$ approaches zero as n approaches infinity. Therefore the alternating series converges.

EXERCISE SET 11–3

● In Problems 1 to 8 use the comparison test to determine whether the series converges.

1. $\dfrac{1}{3} + \dfrac{1}{4} + \dfrac{1}{6} + \dfrac{1}{10} + \cdots + \dfrac{1}{2^n + 2} + \cdots$

2. $\dfrac{1}{6} + \dfrac{1}{7} + \dfrac{1}{9} + \dfrac{1}{13} + \cdots + \dfrac{1}{2^n + 5} + \cdots$

3. $\dfrac{1}{2} + \dfrac{1}{4} + \dfrac{1}{10} + \dfrac{1}{28} + \cdots + \dfrac{1}{3^n + 1} + \cdots$

4. $\dfrac{1}{2} + \dfrac{1}{5} + \dfrac{1}{17} + \dfrac{1}{65} + \cdots + \dfrac{1}{4^n + 1} + \cdots$

5. $\dfrac{1}{3} + \dfrac{2}{4} + \dfrac{3}{5} + \dfrac{4}{6} + \cdots + \dfrac{n}{n + 2} + \cdots$

6. $\dfrac{1}{6} + \dfrac{2}{7} + \dfrac{3}{8} + \dfrac{4}{9} + \cdots + \dfrac{n}{n + 5} + \cdots$

7. $1 + 1.1 + 1.2 + 1.3$
$+ \cdots + \left(1 + \dfrac{n}{10}\right) + \cdots$

8. $0.1 + 0.11 + 0.12 + 0.13$
$+ \cdots + \left(0.1 + \dfrac{n}{100}\right) + \cdots$

In Problems 9 to 16 use the ratio test to determine whether the series converges.

9. $\dfrac{1}{4} + \dfrac{2}{4^2} + \dfrac{3}{4^3} + \dfrac{4}{4^4} + \cdots + \dfrac{n}{4^n} + \cdots$

10. $\dfrac{1}{1.5} + \dfrac{2}{(1.5)^2} + \dfrac{3}{(1.5)^3}$
$+ \cdots + \dfrac{n}{(1.5)^n} + \cdots$

11. $1 + \dfrac{3}{2^2} + \dfrac{4}{2^3} + \dfrac{5}{2^4}$
$+ \cdots + \dfrac{n + 1}{2^n} + \cdots$

12. $1 + \dfrac{5}{3^2} + \dfrac{7}{3^3} + \dfrac{9}{3^4}$
$+ \cdots + \dfrac{2n + 1}{3^n} + \cdots$

13. $3 + \dfrac{3^2}{2!} + \dfrac{3^3}{3!} + \dfrac{3^4}{4!} + \cdots + \dfrac{3^n}{n!} + \cdots$

14. $e + \dfrac{e^2}{2!} + \dfrac{e^3}{3!} + \dfrac{e^4}{4!} + \cdots + \dfrac{e^n}{n!} + \cdots$

15. $\dfrac{1}{7} + \dfrac{2}{9} + \dfrac{3!}{11} + \dfrac{4!}{13} + \cdots + \dfrac{n!}{2n + 5} + \cdots$

16. $\dfrac{1}{100} + \dfrac{2}{10,000} + \dfrac{3!}{100^3} + \cdots + \dfrac{n!}{100^n} + \cdots$

In Problems 17 to 20 use the alternating series test to determine whether the series converges.

17. $1 - \dfrac{1}{2^2} + \dfrac{1}{2^3} - \dfrac{1}{2^4} + \cdots + \dfrac{(-1)^n 1}{2^n} + \cdots$

18. $\dfrac{1}{2} - \dfrac{1}{4} + \dfrac{1}{6} - \dfrac{1}{8} + \cdots + \dfrac{(-1)^{n+1} 1}{2n} + \cdots$

19. $\dfrac{1}{2} - \dfrac{2}{3} + \dfrac{3}{4} - \dfrac{4}{5} + \cdots + \dfrac{(-1)^{n+1} n}{n + 1} + \cdots$

20. $1 - \dfrac{3}{4} + \dfrac{4}{6} - \dfrac{5}{8}$
$+ \cdots + \dfrac{(-1)^{n+1}(n + 1)}{2n} + \cdots$

●● In Problems 21 to 38 use the comparison test or the ratio test to determine whether the series converges.

21. $\dfrac{1}{4} + \dfrac{2}{10} + \dfrac{3}{28} + \dfrac{4}{82} + \cdots + \dfrac{n}{3^n + 1} + \cdots$

22. $\dfrac{1}{7} + \dfrac{2}{9} + \dfrac{3}{13} + \dfrac{4}{21} + \cdots + \dfrac{n}{2^n + 5} + \cdots$

23. $\dfrac{1}{2} + \dfrac{4}{2^2} + \dfrac{9}{2^3} + \dfrac{16}{2^4} + \cdots + \dfrac{n^2}{2^n} + \cdots$

24. $\dfrac{1}{4} + \dfrac{8}{4^2} + \dfrac{27}{4^3} + \dfrac{64}{4^4} + \cdots + \dfrac{n^3}{4^n} + \cdots$

25. $1 + \dfrac{1}{7} + \dfrac{1}{25} + \dfrac{1}{79} + \cdots + \dfrac{1}{3^n - 2} + \cdots$

26. $1 + \dfrac{1}{13} + \dfrac{1}{61} + \dfrac{1}{253}$

$\qquad + \cdots + \dfrac{1}{4^n - 3} + \cdots$

27. $\dfrac{1}{2} + \dfrac{2}{5} + \dfrac{3}{10} + \dfrac{4}{17} + \cdots + \dfrac{n}{n^2 + 1} + \cdots$

28. $\dfrac{1}{4} + \dfrac{3}{7} + \dfrac{4}{12} + \dfrac{5}{19} + \cdots + \dfrac{n + 1}{n^2 + 3} + \cdots$

29. $1 + \dfrac{2}{1024} + \dfrac{3!}{3^{10}} + \dfrac{4!}{4^{10}} + \cdots + \dfrac{n!}{n^{10}} + \cdots$

30. $\dfrac{1}{5} + \dfrac{1}{4} + \dfrac{1}{6} + \dfrac{1}{10} + \cdots + \dfrac{n!}{2^n + 3} + \cdots$

31. $\dfrac{1}{11} + \dfrac{3}{12} + \dfrac{9}{14} + \dfrac{27}{18}$

$\qquad + \cdots + \dfrac{3^n}{2^n + 10} + \cdots$

32. $\dfrac{1}{11} + \dfrac{2}{13} + \dfrac{4}{18} + \dfrac{8}{37}$

$\qquad + \cdots + \dfrac{2^n}{3^n + 10} + \cdots$

33. $1 + 2 + \dfrac{9}{6} + \dfrac{16}{24} + \cdots + \dfrac{n^2}{n!} + \cdots$

34. $1 + 4 + \dfrac{27}{6} + \dfrac{64}{24} + \cdots + \dfrac{n^3}{n!} + \cdots$

••• 35. $2 + 8 + 12 + \dfrac{32}{3} + \cdots + \dfrac{n^2 2^n}{n!} + \cdots$

36. $3 + \dfrac{1}{36} + \dfrac{2}{243} + \cdots + \dfrac{n!}{n^3 3^n} + \cdots$

37. $\dfrac{2}{5} + \dfrac{8}{25} + \dfrac{27}{125} + \cdots + \dfrac{n^2 2^n}{5^n} + \cdots$

38. $\dfrac{3}{2} + \dfrac{9}{32} + \dfrac{1}{8} + \cdots + \dfrac{3^n}{n^3 2^n} + \cdots$

Problems 39 to 44 provide an intuitive verification of the ratio test.

39. Let $a_1 + a_2 + \cdots + a_n + \cdots$ be an infinite series with $\lim\limits_{n \to \infty} \dfrac{a_{n+1}}{a_n} = r < 1$ and $a_n > 0$. Let s be the midpoint between r and 1, so $s = (r + 1)/2$. Explain why there exists an integer N such that $a_{n+1}/a_n < s$ for all $n \geq N$.

40. Explain why $a_{N+1} < s a_N$, $a_{N+2} < s a_{N+1} < s^2 a_N$, $a_{N+3} < s a_{N+2} < s^3 a_N$, and in general $a_{N+i} < s^i a_N$.

41. Explain why the series

$$s + s^2 + s^3 + \cdots + s^n + \cdots$$

is a convergent series.

42. Explain why the series

$$s a_N + s^2 a_N + s^3 a_N + \cdots + s^n a_N + \cdots$$

is a convergent series.

43. Use the comparison test to show why the series

$$a_1 + a_2 + \cdots + a_n + \cdots$$

is convergent for $r < 1$.

44. Use steps similar to Problems 39 to 43 to show why the series

$$a_1 + a_2 + \cdots + a_n + \cdots$$

is divergent for $r \geq 1$.

45. In a discussion of how to determine the useful life of a system of machinery, such as a computer or part of an assembly line, Ascher and Feingold use the infinite series*

$$1 + t + t^2 + t^3 + \cdots$$

For which of the following values of t does this series converge: (a) 1/2, (b) 3/4 (c) 2.1?

46. When a bow strikes the string of a violin, the string vibrates causing an air wave through the

---

*Harold Ascher and Harry Feingold, *Repairable Systems Reliability*, New York, Marcel Dekker, 1984, pp. 70–88.

violin. The end result is the music of the violin. There are many models used to describe how this sound is created. In one such model, a vibrating string is modeled as an oscillating pendulum.* If x is the maximum angular displacement then the period of oscillation T is given

by

$$T = a\left(1 + \frac{1^2}{2^2}\frac{\sin^2 x}{2} + \frac{1^2 3^2}{2^2 4^2}\frac{\sin^4 x}{2} + \cdots\right)$$

Find the nth term of the infinite series. Does the series coverge for $x = 60°$?

*Alexander Wood, *The Physics of Music,* 7th edition, London, Chapman and Hall, 1975, pp. 1–85.

11–4 TAYLOR SERIES

In Section 11–1 we approximated functional values of $y = f(x)$ near $x = 0$ using the nth Taylor polynomial. It is defined as

$$P_n(x) = f(0) + f'(0)x + \frac{f''(0)}{2}x^2 + \frac{f'''(0)}{3!}x^3 + \cdots + \frac{f^{(n)}(0)}{n!}x^n$$

The nth Taylor polynomial is defined only when the first n derivatives of $f(x)$ are defined at $x = 0$. In Section 11–2 we studied infinite series. Combining these two ideas, the concept of **Taylor series** can be extended to an infinite series, provided that all the derivatives of $f(x)$ are defined at $x = 0$. This means that for each value of x the nth Taylor polynomial $P_n(x)$ at $x = 0$ is the nth partial sum of the Taylor series (expanded about $x = 0$). (Later in this section we will see how Taylor polynomials at $x = a$ give rise to Taylor series expanded about $x = a$. For now the term "Taylor series" means Taylor series expanded about $x = 0$.) The definition means that for a particular value of x the corresponding value of the Taylor series is the limit as n approaches infinity of the partial sums $P_n(x)$. For a given value of x, if the series converges, its sum is the value of the Taylor series; if it diverges, the Taylor series does not exist for that value of x.

Taylor Series

If all the derivatives of $f(x)$ are defined at $x = 0$, then the Taylor series for $f(x)$ is the infinite series

$$f(0) + f'(0)x + \frac{f''(0)}{2}x^2 + \cdots + \frac{f^{(n)}(0)}{n!}x^n + \cdots$$

EXAMPLE 1 | **Problem**

Find the Taylor series for $f(x) = e^x$.

SOLUTION First compute the derivatives of $f(x)$ and evaluate them at 0.

$$f'(x) = e^x \qquad f'(0) = e^0 = 1$$
$$f''(x) = e^x \qquad f''(0) = e^0 = 1$$
$$f'''(x) = e^x \qquad f'''(0) = e^0 = 1$$
$$f^{(n)}(x) = e^x \qquad f^{(n)}(0) = e^0 = 1$$

From the definition the Taylor series for $f(x) = e^x$ is

$$1 + x + \frac{x^2}{2} + \frac{x^3}{3!} + \cdots + \frac{x^n}{n!} + \cdots \tag{1}$$

The first $n + 1$ terms of the Taylor series in Example 1 is the nth Taylor polynomial $P_n(x)$ at $x = 0$ for $f(x) = e^x$. It was computed in Section 11–1. So the computation of the Taylor series is similar to finding the nth Taylor polynomial. But the nth Taylor polynomial can always be evaluated at x, assuming the function has n derivatives, while sometimes the Taylor series will not converge, meaning it will not have a value at x. The next question is, where does the series converge? The ratio test is used to answer this question.

EXAMPLE 2 | **Problem**

For what values of x does the Taylor series for $f(x) = e^x$ converge?

SOLUTION Let x be a particular number. Then the Taylor series (1) is an infinite series of numbers with the general term $a_n = x^n/n!$. To apply the ratio test, find $a_{n+1} = x^{n+1}/(n + 1)!$ and form the ratio

$$\frac{|a_{n+1}|}{|a_n|} = \frac{|x^{n+1}/(n + 1)!|}{|x^n/n!|}$$

$$= \frac{|x^{n+1}|}{|x^n|} \frac{n!}{(n + 1)!}$$

$$= \frac{|x|}{n + 1}$$

Find the limit as n approaches infinity. Remember that x is a given number, so it is not affected as n approaches infinity.

$$\lim_{n \to \infty} \frac{|a_{n+1}|}{|a_n|} = \lim_{n \to \infty} \frac{|x|}{n + 1} = (0)|x| = 0$$

This means that for any given value of x the Taylor series converges because the limit as n approaches infinity of the ratio is always 0, which is less than 1. Hence the Taylor series for $f(x) = e^x$ converges for all values of x.

The next example shows that the Taylor series for a function may converge for some values of x and diverge for others.

EXAMPLE 3 | **Problem**

Find (a) the Taylor series for the function $f(x) = 1/(1 - x)$ and (b) the values of x for which the series converges.

SOLUTION (a) Compute the first four derivatives of $f(x)$ and evaluate each at $x = 0$. Then determine a pattern to find the nth derivative and evaluate it at $x = 0$. These computations were carried out in Section 11–1, Example 4.

$$f(x) = \frac{1}{1 - x} = (1 - x)^{-1} \qquad\qquad f(0) = 1$$
$$f'(x) = -(1 - x)^{-2}(-1) = (1 - x)^{-2} \qquad f'(0) = 1$$
$$f''(x) = 2(1 - x)^{-3} \qquad\qquad f''(0) = 2$$
$$f'''(x) = 6(1 - x)^{-4} \qquad\qquad f'''(0) = 6$$
$$f^{(4)}(x) = 24(1 - x)^{-5} \qquad\qquad f^{(4)}(0) = 24$$

The fourth Taylor polynomial for $f(x) = 1/(1 - x)$ at $x = 0$ is

$$P_4(x) = 1 + x + \frac{(2)}{2}x^2 + \frac{(6)}{6}x^3 + \frac{(24)}{24}x^4$$

$$P_4(x) = 1 + x + x^2 + x^3 + x^4$$

The pattern is $f^{(k)}(0) = k!$ for each k. Therefore the Taylor series is

$$1 + x + x^2 + x^3 + \cdots + x^n + \cdots$$

(b) To apply the ratio test, compute $a_n = x^n$ and $a_{n+1} = x^{n+1}$ and form the ratio

$$\frac{|a_{n+1}|}{|a_n|} = \frac{|x^{n+1}|}{|x^n|} = |x|$$

Find the limit as n approaches infinity.

$$\lim_{n \to \infty} \frac{|a_{n+1}|}{|a_n|} = \lim_{n \to \infty} |x| = |x|$$

The ratio test states that the series converges when $|x| < 1$. The series diverges for $x > 1$ or $x < -1$. For $x = \pm 1$ the ratio test gives no conclusion. The cases when $x = 1$ and $x = -1$ must be determined separately. For $x = 1$ the series becomes

$$1 + 1 + 1 + 1 + \cdots$$

This series diverges. For $x = -1$ the series is

$$1 - 1 + 1 - 1 + \cdots$$

This series diverges. Therefore the Taylor series for $f(x) = 1/(x - 1)$ converges for $-1 < x < 1$ and diverges for other values of x.

The set of values of x where a Taylor series converges is always an interval. It is called the **interval of convergence** of the Taylor series. Example 3 states that the interval of convergence for $f(x) = 1/(1 - x)$ is $(-1, 1)$. The interval of convergence is found by using the ratio test, except that the end points must be tested separately.

Another way to view the Taylor series in Example 3 is to note that it is a geometric series with $a = 1$ and ratio $r = x$. A geometric series is convergent if $-1 < r < 1$, which agrees with the result in Example 3.

Properties of Taylor Series

A Taylor series itself can be regarded as a function. In fact for all the functions considered in this text, if the Taylor series of $f(x)$ converges when $x = c$, then the sum of the series is $f(c)$. For example, the Taylor series for $f(x) = e^x$ converges for all x, so in particular, it converges for $x = 1.5$. The sum of the Taylor series of e^x when $x = 1.5$ is $e^{1.5}$. That is

$$e^{1.5} = 1 + 1.5 + \frac{1.5^2}{2!} + \frac{1.5^3}{3!} + \cdots + \frac{1.5^n}{n!} + \cdots$$

In general suppose $f(x)$ has a Taylor series

$$a_0 + a_1 x + a_2 x^2 + a_3 x^3 + \cdots + a_n x^n + \cdots$$

where $a_n = f^{(n)}(0)/n!$. If the series converges for $x = c$, then

$$f(c) = a_0 + a_1 c + a_2 c^2 + a_3 c^3 + \cdots + a_n c^n + \cdots$$

Therefore for all values of x where the Taylor series converges

$$f(x) = a_0 + a_1 x + a_2 x^2 + a_3 x^3 + \cdots + a_n x^n + \cdots$$

where $a_n = f^{(n)}(0)/n!$. In this way a Taylor series can be regarded as a polynomial with ''infinite'' degree. The advantage of this view is that many operations on polynomials are valid for the Taylor series. For example, we can multiply a Taylor series by x to get the Taylor series for $xf(x)$. Thus

$$xf(x) = a_0 x + a_1 x^2 + a_2 x^3 + \cdots + a_n x^{n+1} + \cdots$$

Here are some additional operations that are valid for Taylor series.

1. If the Taylor series for $f(x)$ is multiplied by a constant or a constant times a power of x, cx^n, the result is the Taylor series for $cx^n f(x)$.
2. If x is replaced by a power of x or a constant times a power of x, cx^n, the result is the Taylor series for $f(cx^n)$.
3. If the Taylor series for two functions $f(x)$ and $g(x)$ are added term by term, the result is the Taylor series for $f(x) + g(x)$.
4. If the Taylor series for $f(x)$ is differentiated term by term, the result is the Taylor series for $f'(x)$.
5. If the Taylor series for $f(x)$ is integrated term by term, the result is the Taylor series for $\int f(x)\, dx$.

These operations make it possible to find the Taylor series of some functions without using the formal definition. If the function can be found by using one of these operations, then its Taylor series can be computed by using the same operation. For many of these operations the values of x where the new Taylor series is convergent remain the same as for $f(x)$. Sometimes, however, the ratio test must be applied to find the appropriate values where the series converges.

EXAMPLE 4

Problem

Use the Taylor series for $f(x) = \dfrac{1}{1 - x}$ computed in Example 3 to find the

Taylor series for the functions: (a) $g(x) = \dfrac{4x^3}{1 - x}$, (b) $h(x) = \dfrac{1}{1 - 2x^2}$,

(c) $k(x) = \dfrac{1}{(1 - x)^2}$.

SOLUTION (a) Since $g(x) = 4x^3 f(x)$, use operation (1) with $c = 4$ and $n = 3$. Then

$$g(x) = 4x^3 + 4x^4 + 4x^5 + \cdots + 4x^{n+3} + \cdots$$

(b) Since $h(x) = f(2x^2)$, use operation (2) with $c = 2$ and $n = 2$. Then

$$h(x) = 1 + 2x^2 + (2x^2)^2 + (2x^2)^3 + \cdots + (2x^2)^n + \cdots$$
$$= 1 + 2x^2 + 4x^4 + 8x^6 + \cdots + 2^n x^{2n} + \cdots$$

(c) Since $k(x) = f'(x)$, use operation (4).

$$k(x) = 1 + 2x + 3x^2 + 4x^3 + \cdots + nx^{n-1} + \cdots$$

The operations on Taylor series also provide a method to approximate definite integrals. For example, the standard normal curve of statistics has the equation

$$f(x) = \frac{1}{\sqrt{2\pi}} e^{-x^2/2}$$

The area under this curve is an important concept in statistics. But areas under the curve cannot be found by the fundamental theorem of calculus because there is no simple formula for an antiderivative of the integrand. Numerical integration techniques must be applied and Taylor series can be used to calculate these areas with good accuracy.

EXAMPLE 5

Problem

Find the area under the standard normal curve from $x = 0$ to $x = .5$, which is given by

$$\frac{1}{\sqrt{2\pi}} \int_0^{.5} e^{-x^2/2} \, dx$$

SOLUTION Use the Taylor series for e^x.

$$e^x = 1 + x + \frac{x^2}{2!} + \frac{x^3}{3!} + \cdots + \frac{x^n}{n!} + \cdots$$

Replace x by $-x^2/2$.

$$e^{-x^2/2} = 1 + \frac{-x^2}{2} + \frac{1}{2!}\left(-\frac{x^2}{2}\right)^2 + \frac{1}{3!}\left(-\frac{x^2}{2}\right)^3$$

$$+ \cdots + \frac{1}{n!}\left(-\frac{x^2}{2}\right)^n + \cdots$$

$$= 1 - \frac{x^2}{2} + \frac{x^4}{8} - \frac{x^6}{48} + \cdots + (-1)^n \frac{x^{2n}}{n!2^n} + \cdots$$

Integrating yields

$$\frac{1}{\sqrt{2\pi}}\int_0^{.5} e^{-x^2/2}\,dx = \frac{1}{\sqrt{2\pi}}\left[x - \frac{x^3}{6} + \frac{x^5}{40} - \frac{x^7}{336}\right.$$

$$\left. + \cdots + \frac{(-1)^n x^{2n+1}}{(2n+1)n!2^n} + \cdots \right]\Bigg|_0^{.5}$$

$$= \frac{1}{\sqrt{2\pi}}\left[.5 - \frac{.5^3}{6} + \frac{.5^5}{40} - \frac{.5^7}{336} + \cdots + \frac{(-1)^n .5^{2n+1}}{(2n+1)n!2^n} + \cdots \right]$$

The infinite series on the right is convergent and its sum is the definite integral on the left. Taking the sum of the first four terms yields the approximation .1915. This is accurate to four decimal places. A more accurate approximation can be made by summing a larger amount of terms.

Taylor Series Expanded About a

Thus far, by the term Taylor series we have meant Taylor series expanded about 0. Just as we have defined Taylor polynomials at a, we can also define **Taylor series expanded about $x = a$.** Recall that the nth Taylor polynomial at a is defined by

$$P_n(x) = f(a) + f'(a)(x - a) + \frac{f''(a)}{2}(x - a)^2 + \frac{f'''(a)}{3!}(x - a)^3$$

$$+ \cdots + \frac{f^{(n)}(a)}{n!}(x - a)^n$$

If all of the derivatives of $f(x)$ are defined at a, then the Taylor series of $f(x)$ expanded about a is defined by

$$f(a) + f'(a)(x - a) + \frac{f''(a)}{2}(x - a)^2 + \frac{f'''(a)}{3!}(x - a)^3$$

$$+ \cdots + \frac{f^{(n)}(a)}{n!}(x - a)^n + \cdots$$

EXERCISE SET 11–4

• In Problems 1 to 14 find the Taylor series for $f(x)$.

1. $f(x) = e^{2x}$

2. $f(x) = e^{3x}$

3. $f(x) = e^{-x}$

4. $f(x) = e^{-3x}$

5. $f(x) = \dfrac{1}{1 + x}$

6. $f(x) = \dfrac{1}{1 + 2x}$

7. $f(x) = \dfrac{1}{1 - 2x}$

8. $f(x) = \dfrac{2}{2 - x}$

9. $f(x) = (1 + x)^{1/2}$

10. $f(x) = (1 + x)^{-1/2}$

11. $f(x) = (1 + x)^{-2}$

12. $f(x) = (1 + 2x)^{-2}$

13. $f(x) = \ln(1 + x)$

14. $f(x) = \ln(1 + 3x)$

In Problems 15 to 22 find the interval of convergence of the Taylor series for the function in the given problem.

15. Problem 1

16. Problem 2

17. Problem 5

18. Problem 6

19. Problem 9

20. Problem 10

21. Problem 11

22. Problem 12

In Problems 23 to 26 use the Taylor series for $f(x) = \dfrac{1}{1 - x}$ computed in Example 3 to find the Taylor series for the given function.

23. $f(x) = \dfrac{2x^4}{1 - x}$

24. $f(x) = \dfrac{1}{1 - x^3}$

25. $f(x) = \dfrac{1}{(1 - x)^3}$

26. $f(x) = \dfrac{x}{1 + x}$

In Problems 27 to 30 use the Taylor series for $f(x) = e^x$ computed in Example 1 to find the Taylor series for the given function.

27. $f(x) = xe^x$

28. $f(x) = xe^{4x}$

29. $f(x) = e^{-x}$

30. $f(x) = x^2 e^{-x}$

In Problems 31 to 34 find the area under the standard normal curve from $x = 0$ to $x = a$.

31. $a = 1$

32. $a = 1.5$

33. $a = .25$

34. $a = 2$

•• In Problems 35 to 46 find the Taylor series expanded about a.

35. $f(x) = \sin x$, $\quad a = 0$

36. $f(x) = \cos x$, $\quad a = 0$

37. $f(x) = \sin x$, $\quad a = \dfrac{\pi}{2}$

38. $f(x) = \cos x$, $\quad a = \dfrac{\pi}{2}$

39. $f(x) = 1/x$, $\quad a = 1$

40. $f(x) = 2/x$, $\quad a = 2$

••• 41. $f(x) = x \sin x$, $\quad a = 0$

42. $f(x) = x \cos x$, $\quad a = 0$

43. $f(x) = \sin x^2$, $\quad a = 0$

44. $f(x) = \cos x^2$, $\quad a = 0$

45. $f(x) = x^2 \sin x^2$, $\quad a = \dfrac{\pi}{2}$

46. $f(x) = x^3 \cos x^2$, $\quad a = \dfrac{\pi}{2}$

47. An important ingredient in many economic models is the computation of supply and demand functions from specific data. In an article investigating various ways to compute such functions, Gallant studied the method of Taylor series as one means of producing demand functions.[*] In one part of the study the Taylor series for $f(x) = \ln(x + 1)$ at $x = 1$ is computed. Find this Taylor series.

[*]A. Ronald Gallant, "The Fourier Flexible Form," *American Journal of Agricultural Economics*, Vol. 66, 1984, pp. 204–208.

48. Psychologists use Taylor series to model various characteristics of tests. Wickens studied a psychological test in which the subject was asked to perform a task that required a trial-and-error approach to master the task.[†] He let T represent the number of errors the subject made before the task was learned. Thus T is defined as the sum of the errors made on the successive trials of the experiment, so $T = X_1 + X_2 + X_3 + \cdots$, where X_n is 1 if an error was made or 0 if the response was correct. The likelihood $E(x)$ that the subject masters the task is defined by a Taylor series where x is the probability of an error being made on a given trial. Under certain circumstances, it was shown that

$$E(x) = x + \frac{x^2}{2!} + \frac{x^3}{3!} + \cdots$$

For what function is this the Taylor series at $x = 0$?

[†]Thomas D. Wickens, *Models for Behavior: Stochastic Processes in Psychology*, San Francisco, W. H. Freeman, 1982, pp. 1–76.

CASE STUDY   **The Multiplier Principle in Economics**

Like a ripple in a pond, a single change in demand can begin a sequence of events with multiple cumulative effects in one industry or an entire economy. This ripple effect has been known to economists for a long time. Indeed, the idea of cumulative effects beyond the initial impact of an event was studied extensively by David Ricardo in the early 19th century. But it wasn't until the publication of R. F. Kohn's famous article in 1931, ''The Relation of Home Investment to Unemployment,'' that such long-term effects were actually calculated.[*] Kohn's study was the first deliberate attempt to use a formula to measure the effect of public works on unemployment and the subsequent increase in net expenditure. The paper deals with the highly controversial topic of the use by the government of public works to stem the debilitating effects of the Depression. The Depression was worldwide, so this topic had been widely debated in social and economic circles, but Kohn was the first to attempt any calculation of the effects on unemployment.

A key ingredient in Kohn's study is the ''multiplier principle.'' It had been used by earlier writers in several disciplines, but it had never been applied by economists. For example, Jul. Wulff, a member of the Danish parliament in 1886, computed the secondary effects of public works on employment by considering the effect of an increase in employment. He assumed that if wages were increased by a specific amount, say, I, then 60% of the increased income would be spent on Danish goods and services, while the remaining 40% would be spent on imports. The subsequent increase in receipts at home would be $60\%(I) = .6I$. Of this sum, again 60% would be spent at home, meaning that an additional $60\%(.6I) = .36I$ would be spent on Danish products and services. Continuing this reasoning, the total effect of the increase in wages I would be

$$I + .6I + .36I + \cdots + (.6)^n I + \cdots$$

[*]R. F. Kohn, ''The Relation of Home Investment to Unemployment,'' *Economic Journal*, Vol. 31, 1931, pp. 173–198.

This is a geometric series, with the first term I and ratio .6, so Wulff concluded that the total effect is

$$\frac{I}{1 - .6} = \frac{I}{.4} = 2.5I$$

Therefore the cumulative effect is two and a half times greater than the initial impact of the increase in wages. The ratio, 0.6, is referred to as the *multiplier*, and computing the total effect by a geometric series with the ratio equal to the multiplier is called the *multiplier principle.**

Kohn applied the multiplier principle to the problem of the effect of investment on employment, and this changed the way economists measure cumulative changes. Here is the essence of Kohn's argument. Suppose new employees are hired in an industry and their total amount of wages is W. These new employees produce an increase in profit P, and so still more employees can be hired. The number of new employees hired depends on W and P. Suppose that a certain fraction of the profit is to be used to hire new employees. Since P depends on W, this fraction can be thought of as a percentage of W. Let the fraction be p, meaning that the increase in profit from the initial expenditure of increased wages W produces new employees with wages pW, where p lies between 0 and 1. From these wages more profit is generated, resulting in new employees' earning $p(pW) = p^2W$ in wages. The long-term effect is

$$W + pW + p^2W + \cdots + p^nW + \cdots$$

Kohn concluded that the total cumulative effect from an initial expenditure of increased wages W is

$$\frac{W}{1 - p}$$

Calculating the value of the multiplier is usually a complex exercise. This was true for Kohn and it is true today. There are various models that are used, and many areas of economics that use the multiplier principle. For example, input-output models of various facets of the economy use multipliers to forecast trends in industries.†

Money stock   One specific economic application of the multiplier principle is in the control of the U.S. money supply by the Federal Reserve System (FRS). The FRS is the central banking organization of the U.S. government. Its most important function is to control the flow of money, including cash and credit. In general, if credit is plentiful, businesses can borrow money readily to expand, thereby improving the economy, and individuals can take out loans at lower interest rates to make large purchases, such as homes, automobiles, and college

*See Fr. Johannsen, "Tidens Krav," *Berlingske Tidende,* March 1927.

†See A. D. Sandoval, "Constant Relationship Between Input-Output Income Multipliers," *Review of Economics and Statistics,* Vol. 49, pp. 599–600; and U.S. Department of Commerce, *The Detailed Input-Output Structure of the U.S. Economy,* Washington, DC, 1979.

tuition. However, too much credit can result in businesses and individuals overextending their ability to repay loans. So too much credit as well as too little can be detrimental to the economy.

   Over 5,000 banks belong to the FRS. Each member bank operates in one of 12 Federal Reserve districts, and each district has a Federal Reserve Bank. Member banks hold stock in the Federal Reserve Bank. The FRS sets reserve requirements for member banks. This means that each member bank must set aside a certain percentage of its deposits in a reserve fund. They can use the remaining part of their deposits to make loans. In return, the banks can borrow money from the FRS. The rate of interest that the FRS charges the banks is called the *discount rate*. The FRS uses the discount rate to influence credit conditions. This is because the member banks adjust the rate that they charge their customers according to the discount rate.

   One method used by the FRS to determine the size of the money stock is to forecast how much money will be needed in the immediate future. A common forecasting tool is the multiplier principle. In this context the multiplier is called the *money growth rate, r*. If $r = .10$, then the money stock is to increase by 10% in the next time period, which is usually monthly or quarterly. This change then results in more capital being generated, and it is assumed that this 10% increase causes a further increase of 10% of this amount, or $.1(.1) = .01$. Continuing, the total effect of the initial increase of 10% results in a total increase of

$$1 + .1 + .01 + .001 + \cdots + (.1)^n + \cdots$$

The sum of this series is $1/(1 - .1) = 10/9$. In practice, the FRS computes the money multiplier each month to help determine how much to increase or decrease the money supply.*

Case Study Exercises

In Problems 1 to 4 find the total cumulative effect from an initial expenditure of wages W for the given value of p in Kohn's model of the effect of investment on employment.

1. $W = \$1,000,000$, $p = 25\%$
2. $W = \$1,000,000$, $p = 75\%$
3. $W = \$10,000,000$, $p = 10\%$
4. $W = \$10,000,000$, $p = 80\%$

In Problems 5 to 8 find the total cumulative effect on the money supply for the given money growth rate, r.

5. $r = 12\%$ 6. $r = 2\%$
7. $r = 20\%$ 8. $r = 25\%$

9. Give a practical reason why the fraction p mentioned in Kahn's model must be between 0 and 1.

10. Explain from a mathematical point of view why the fraction p mentioned in Kohn's model must between 0 and 1.

*R. W. Hafer et al., "Forecasting the Money Multiplier: Implications for Money Stock Control and Economic Activity," *Federal Reserve Bank of St. Louis Review*, October 1983, pp. 22–33.

CHAPTER REVIEW

Key Terms

11–1 Taylor Polynomials

Taylor Polynomials

Taylor Polynomial at $x = a$

nth Taylor Polynomial

Remainder Theorem

Taylor Polynomial at $x = 0$

11–2 Infinite Series

Infinite Series

Convergent

nth Term

Divergent

Partial Sums

Geometric Series

Sequence

Ratio

nth Partial Sum

Harmonic Series

Sum of a Series

11–3 Tests for Convergence

Comparison Test

Alternating Series

Ratio Test

11–4 Taylor Series

Taylor Series

Operations on Taylor Series

Interval of Convergence

Taylor Series Expanded About $x = a$

Summary of Important Concepts

nth Taylor polynomial of $f(x)$ at $x = 0$ is

$$P_n(x) = f(0) + f'(0)x + \frac{f''(0)}{2} x^2 + \cdots + \frac{f^{(n)}(0)}{n!} x^n$$

nth Taylor polynomial of $f(x)$ at $x = a$ is

$$P_n(x) = f(a) + f'(a)(x - a) + \frac{f''(a)}{2}(x - a)^2 + \cdots + \frac{f^{(n)}(a)}{n!}(x - a)^n$$

Suppose M is a number such that $|f^{(n+1)}(t)| \leq M$ for all t between a and x, then

$$|R_n(x)| < \frac{M}{(n + 1)!} |x - a|^{n+1}$$

The sum of the series $a + ar + ar^2 + \cdots + ar^n + \cdots$ is

$$\frac{a}{1 - r} \text{ for } -1 < r < 1.$$

The Taylor series of $f(x)$ at $x = 0$ is

$$f(0) + f'(0)x + \frac{f''(0)}{2} x^2 + \cdots + \frac{f^n(0)}{n!} x^n + \cdots$$

REVIEW PROBLEMS

1. Use the third Taylor polynomial $P_3(x)$ of the function $f(x) = e^x$ to approximate $e^{4/5}$.

2. Find the first four Taylor polynomials at $x = 0$ for the function $f(x) = e^{-3x}$.

3. Approximate the number $\ln(1.5)$ using the fourth Taylor polynomial for the function $f(x) = \ln(1 + x)$ at $x = 0$.

4. Find the nth Taylor polynomial at $x = 0$ for the function $f(x) = \dfrac{1}{1 + x}$.

5. Find the nth Taylor polynomial at $x = a$ for the function $f(x) = e^{x-1}$ at $a = 1$.

6. Find the fourth Taylor polynomial at $x = 0$ for $f(x) = xe^{-x}$.

7. Determine whether the geometric series converges. If so, find the sum

$$1 + \frac{4}{9} + \left(\frac{4}{9}\right)^2 + \left(\frac{4}{9}\right)^3 + \cdots$$

8. Determine whether the series converges. If so, find the sum

$$27 + 9 + 3 + 1 + \frac{1}{3} + \cdots$$

9. Find the fraction that is equal to the repeating decimal $.373737...$

10. A city enacts a $1,000,000 tax benefit for business, and each year this money produces a 75% return on investment. This amount is then reinvested and it generates a return of 75%, and so on. Determine how much tax revenue is ultimately generated by the tax cut if the business tax rate is 20% per year.

11. Determine what fraction of a radioactive material will decay after six half-lives.

12. Determine whether the series converges.

$$5 + \frac{5}{2} + \frac{5}{3} + \frac{5}{4} + \frac{5}{5} + \frac{5}{6} + \frac{5}{7}$$
$$+ \cdots + \frac{5}{n} + \cdots$$

13. Use the comparison test to determine whether the series converges.

$$\frac{1}{3} + \frac{1}{4} + \frac{1}{6} + \frac{1}{10} + \cdots + \frac{1}{2^n + 2} + \cdots$$

14. Use the ratio test to determine whether the series converges.

$$\frac{1}{4} + \frac{2}{4^2} + \frac{3}{4^3} + \frac{4}{4^4} + \cdots + \frac{n}{4^n} + \cdots$$

In Problems 15 and 16 use the comparison test or the ratio test to determine whether the series converges.

15. $\dfrac{1}{5} + \dfrac{2}{11} + \dfrac{3}{29} + \dfrac{4}{83} + \cdots + \dfrac{n}{3^n + 2} + \cdots$

16. $\dfrac{1}{3} + \dfrac{4}{3^2} + \dfrac{9}{3^3} + \dfrac{16}{3^4} + \cdots + \dfrac{n^2}{3^n} + \cdots$

17. Find the Taylor series for $f(x) = (1 + x)^{1/2}$ and find the values of x where the Taylor series converges.

18. Use the Taylor series for $f(x) = \dfrac{1}{1 - x}$ computed in Example 3 in Section 11–4 to find the Taylor series for the function $f(x) = \dfrac{2x^5}{1 - x}$.

19. Find the area under the standard normal curve from $x = 0$ to $x = 1.75$.

20. Find the Taylor series for $f(x) = \sin x$ expanded about $\dfrac{\pi}{2}$.

12 Probability and Calculus

Section 12–1
Discrete Probability Distributions

Section 12–2
Continuous Probability Distributions

Section 12–3
Expected Value and Standard Deviation: Continuous Random Variables

Section 12–4
Exponential and Normal Distributions

Case Study
Reliability

CHAPTER OVERVIEW   The theory of probability is the branch of mathematics that helps find patterns in seemingly haphazard occurrences. In the first section we study some fundamental principles of probability theory. It considers probability experiments that have a finite number of possible outcomes. They are called discrete probability distributions. Sometimes the number of outcomes in an experiment can be infinite. If the outcomes can be any number in an interval, the distribution is called continuous. This type of experiment is introduced in Section 12–2. The concepts of expected value, variance, and standard deviation for discrete random variables require the addition of numbers. To sum quantities over an interval, we need the definite integral. In Section 12–3 it is shown how these concepts are defined by using calculus. There are many types of probability distributions. In Section 12–4 we study two of the most common continuous probability distributions, the exponential distribution and the normal distribution.

CASE STUDY PREVIEW

When the space shuttle *Challenger* burst into flames on January 28, 1986 the entire space program suffered a setback of major proportions: the next shuttle flight would not take place until the National Aeronautics and Space Administration endured over two years of agonizing turmoil. Something went wrong. The most likely candidates are the O-rings—rubber gaskets used to seal the joints of the booster rockets. The Case Study shows how reliability theory is used to test parts and components to predict their probability of failure. In particular, reliability testing of the O-rings was suspect before that critical launch.

12–1 DISCRETE PROBABILITY DISTRIBUTIONS

There are many events in the real world that seem to happen by chance. By studying the behavior of some of these seemingly unpredictable events, we can often determine patterns. The theory of probability is the branch of mathematics that helps find these patterns. In this section we study some fundamental principles of probability theory. To introduce these ideas, let us look at an example.

Introduction: An Example

In the early 1970s the Massachusetts Department of Public Works conducted a study of speed limits in an effort to improve motor vehicle safety. Highway accident data were analyzed on a surburban stretch of Route 2, a major road entering Boston from the west. In an article* about the study William B. Fairley raised the critical question: In what season of the year did the most accidents occur? The year was divided into four quarters, the first being January to March, then April to June, and so forth. Data were gathered for each quarter from 1970 to 1972. Table 1 gives the quarterly count and the average for each quarter. For example, during the three years of the study there was a total of 302 accidents in the first quarter, 138 in 1970, 74 in 1971, and 90 in 1972. The averages are computed by dividing each quarterly count by the total number of accidents in all quarters, 1269. Thus the quarterly average for quarter 1 is 302/1269 = .24.

TABLE 1   Accidents on Route 2, 1970–1972

Quarter	Quarterly Count	Quarterly Average
1	302	.24
2	244	.19
3	283	.22
4	440	.35
Total	1269	

By looking at the averages, it is easy to determine that not only were there more accidents in the fourth quarter, but there were almost twice as many in quarter 4 as in quarter 2. The percentages in the Quarterly Average column are called **relative frequencies.** They arise from the fraction, m/N, where m is the number of occurrences of accidents in a quarter and N is the total number of accidents. If we select an accident at random from the total of 1269, then the **probability** that it occurred in quarter 4 is .35. This is expressed as Pr(quarter 4) = .35.

To say that the probability of this event is .35 means that, since the event has occurred 35% of the time in the past, it is reasonable to predict that it will happen 35% of the time in the future. In this way probability is used to predict

*William B. Fairley, "Accidents on Route 2: Two-way Structures for Data," Statistics and Public Policy, Boston, Addison-Wesley, 1977, pp. 23–50.

future trends. That is, under the same conditions 35% of the accidents on Route 2 will occur in quarter 4.

The phrase "under the same conditions" spells the difference between studies of real-world phenomena and games of chance, such as tossing coins or rolling dice, or playing cards or roulette. Games are repeated exactly, but it is much more difficult to repeat real-world experiments under exactly the same conditions. Therefore games are used as experiments to illustrate the concepts of probability because they are familiar and easy to demonstrate, and on each play they are repeated identically.

Sample Spaces

The basic idea in probability is to study outcomes of experiments. By an **experiment** we mean an action whose possible outcomes can be determined and recorded. For example, four simple experiments are

Experiment A: toss a coin and observe heads or tails.
Experiment B: roll a die and record the number on the top face.
Experiment C: an urn contains 2 red marbles, 3 white marbles, and 5 blue marbles; a marble is drawn at random and its color is recorded.
Experiment D: in the Fairley study of accidents on Route 2 a researcher chooses an accident at random and records the quarter in which it occurs.

Each time an experiment is performed an outcome is observed. The set of possible outcomes is called the **sample space** for the experiment. For instance, Experiment A has two outcomes, heads and tails. If we let "H" represent heads and "T" represent tails, then $S = \{H, T\}$ is the sample space for Experiment A.

EXAMPLE 1

Problem

Find the sample space for Experiments B, C, and D.

SOLUTION Experiment B has six possible outcomes: 1, 2, 3, 4, 5, 6. Hence the sample space for Experiment B is

$$S = \{1, 2, 3, 4, 5, 6\}$$

Experiment C has three outcomes. If we let R represent the outcome red, W represent white, and B represent blue, then its sample space is

$$S = \{R, W, B\}$$

Experiment D has four outcomes. Its sample space is

$$S = \{\text{quarter 1, quarter 2, quarter 3, quarter 4}\}$$

Events

Often we are interested in looking at the possibility that more than one outcome is observed. For example, in Experiment D, suppose we wanted to determine how many accidents occurred in the first six months of the year. The corresponding outcomes would be quarter 1 and quarter 2. In mathematical language this is the set {quarter 1, quarter 2}, which is a subset of the sample space. In general, a subset of the sample space is called an **event.** The next example describes events associated with Experiment B.

EXAMPLE 2 | **Problem**

Describe in terms of sets the two events E and F associated with Experiment B where

    E: an odd number occurs
    F: a number less than 5 occurs

SOLUTION Event E consists of the three outcomes, 1, 3, and 5. Thus

    $E = \{1, 3, 5\}$

Event F consists of the outcomes 1, 2, 3, and 4. Thus

    $F = \{1, 2, 3, 4\}$

Probabilities

The probability of an event is the likelihood that it will occur. For instance, if a fair coin is tossed, the likelihood of observing heads is the same as observing tails, so each will occur $\frac{1}{2}$ the time in the long run, that is, when the experiment is repeated often. For the event E its probability is expressed as $Pr(E)$. For the experiment of tossing a coin $Pr(\text{heads}) = \frac{1}{2}$ and $Pr(\text{tails}) = \frac{1}{2}$. For the Fairley study the probability of each quarter is its relative frequency. Thus $Pr(\text{quarter 1}) = .24$. If there are N equally likely outcomes in a sample space and event E consists of m outcomes, then $Pr(E) = m/N$. The next example shows how to assign probabilities to events for the experiment of tossing a die, whose six outcomes are equally likely.

EXAMPLE 3 | The experiment is to toss a die and record the number on the top face. Define the events E, F, and G as follows:

    $E = \{3, 4\}$
    $F = \{1, 3, 4, 5, 6\}$
    G: an even number occurs

Problem

Find (a) $Pr(E)$, (b) $Pr(F)$, and (c) $Pr(G)$.

SOLUTION (a) In the formula $m = 2$ and $N = 6$, so $Pr(E) = \frac{2}{6} = \frac{1}{3}$. (b) Here $m = 5$ and $N = 6$, so $Pr(F) = \frac{5}{6}$. (c) There are three outcomes that are even, so $G = \{2, 4, 6\}$. Thus $m = 3$ and so $Pr(G) = \frac{3}{6} = \frac{1}{2}$.

Random Variables and Probability Distributions

A notation used with the sample space $S = \{a_1, a_2, \ldots, a_n\}$ involves letting X represent the outcome of the experiment. Then X is a variable that can assume the values $a_1, a_2, \ldots, a_n$. It is called a **random variable** because the occurrence of the values of X depends on chance. Each outcome a_i is assigned a probability p_i. This is expressed as $Pr(a_i) = p_i$. For example, for the experiment of tossing a die the random variable X has six outcomes, 1 through 6. Each has probability $\frac{1}{6}$. In particular, to express that the probability of the outcome 2 is $\frac{1}{6}$, we write $Pr(X = 2) = \frac{1}{6}$. In general $Pr(X = a) = p$ means that the probability of outcome a is p.

When random variables are used the outcomes are assumed to be numbers. For example, for the experiment of tossing a coin, instead of observing H or T, we record the number of heads. The sample space is $\{0, 1\}$. For the Fairley study let the sample space be $\{1, 2, 3, 4\}$, where each outcome represents the quarter in which an accident occurs. From now on we use only sample spaces whose outcomes are numbers.

When the outcomes are listed in a table with their probabilities it is called a **probability distribution.** For example, the probability distributions for the experiments of tossing a coin (and recording the number of heads) and rolling a die are as follows:

Toss a Coin		Roll a Die	
Outcome	Probability	Outcome	Probability
0	$\frac{1}{2}$	1	$\frac{1}{6}$
1	$\frac{1}{2}$	2	$\frac{1}{6}$
		3	$\frac{1}{6}$
		4	$\frac{1}{6}$
		5	$\frac{1}{6}$
		6	$\frac{1}{6}$

Sometimes tables are written horizontally rather than vertically as in the next example.

EXAMPLE 4

Problem

Find the probability distribution for the random variable defined by the Fairley study.

SOLUTION Let the values of the random variable be 1, 2, 3, and 4. Their corresponding probabilities are the relative frequencies given in Table 1. The probability distribution is

quarter, $X = a$	1	2	3	4
probability, p	.24	.19	.22	.35

Another way to define a random variable X is to give a bar graph with the values of X on the horizontal axis and their corresponding probabilities on the vertical axis. This is called a **probability histogram.** Two examples are shown in Figure 1.

Expected Value

The mean, or arithmetic average, of a set of numbers is a measure of the central tendency of the numbers. Intuitively, the mean is a measure of the "center" of the numbers. In a similar way, the central tendency of a random variable is measured by the **expected value.** Consider the random variable X defined by the probability distribution

outcome, $X = a$	a_1	a_2	$\cdots$	a_n
probability, p	p_1	p_2	$\cdots$	p_n

The expected value of X, written $E(X)$, is defined by

$$E(X) = a_1 p_1 + a_2 p_2 + \cdots + a_n p_n$$

EXAMPLE 5 | **Problem**

Find $E(X)$ for (a) Experiment 2 and (b) Experiment 4, defined in the beginning of this section.

FIGURE 1

Probability histograms

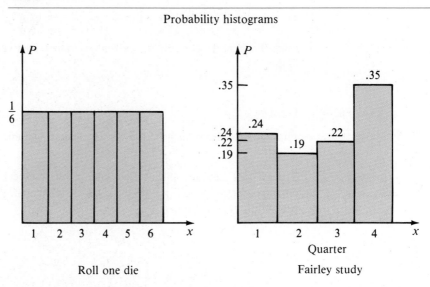

Roll one die   Fairley study

SOLUTION (a) The probability distribution is

outcome, $X = a$	1	2	3	4	5	6
probability, p	$\frac{1}{6}$	$\frac{1}{6}$	$\frac{1}{6}$	$\frac{1}{6}$	$\frac{1}{6}$	$\frac{1}{6}$

Therefore

$$E(X) = 1(\tfrac{1}{6}) + 2(\tfrac{1}{6}) + 3(\tfrac{1}{6}) + 4(\tfrac{1}{6}) + 5(\tfrac{1}{6}) + 6(\tfrac{1}{6})$$
$$= \tfrac{21}{6} = \tfrac{7}{2}$$

This shows that the $E(X)$ is not necessarily one of the values of X, just as the mean of a set of numbers is not necessarily one of the numbers. It means that if you were to roll a die repeatedly and record the outcome each time, then you would expect the arithmetic mean of these outcomes to be approximately $\frac{7}{2}$.
(b) The probability distribution is

outcome, $X = a$	1	2	3	4
probability, p	.24	.19	.22	.35

Therefore

$$E(X) = 1(.24) + 2(.19) + 3(.22) + 4(.35)$$
$$= .24 + .38 + .66 + 1.40 = 2.68$$

Standard Deviation

While the expected value of a random variable X measures the central tendency of X, the **variance** measures the spread of the outcomes. Just as $E(X)$ is a weighted average, so, too, the variance is the weighted sum of the squares of the differences between the outcomes and $E(X)$. Let X be a random variable with outcomes a_i and corresponding probabilities p_i. To simplify the notation, let E represent the expected value. The variance of X, denoted by Var(X), is defined by

$$\text{Var}(X) = (a_1 - E)^2 p_1 + (a_2 - E)^2 p_2 + \cdots + (a_n - E)^2 p_n$$

The **standard deviation** of X, denoted by the Greek letter σ, is defined to be the square root of Var(X). That is

$$\sigma = \sqrt{\text{Var}(X)}$$

The exact interpretation of the variance and the standard deviation is more difficult to explain than the expected value. Most of the outcomes will be within 3 standard deviations of the expected value. If the outcomes are spread out evenly about the expected value, then most of the outcomes will be within 2 standard deviations. In general, if the standard deviation of random variable X is larger than another random variable, then the distribution of random variable X is spread out more than the other's.

EXAMPLE 6 An ecologist investigates the reproductive behavior of water fowl to determine whether DDT is causing infertile eggs. A frequency distribution gives the number of nests that hold between zero and five eggs.

eggs in nest	0	1	2	3	4	5
frequency	15	25	25	20	10	5

Problem

Compute the (a) expected value, (b) variance, and (c) standard deviation.

SOLUTION The random variable X consists of the outcomes 0 through 5. To compute their probabilities, change the frequencies into relative frequencies. The sum of the frequencies is 100, so divide each frequency by 100. This yields the probability distribution.

outcome, $X = a$	0	1	2	3	4	5
probability	.15	.25	.25	.20	.10	.05

(a) $E(X) = 0(.15) + 1(.25) + 2(.25) + 3(.20) + 4(.10) + 5(.05)$
$= 0 + .25 + .50 + .60 + .40 + .25$
$= 2$

This indicates that the average number of eggs is 2.

(b) $\text{Var}(X) = (0 - 2)^2(.15) + (1 - 2)^2(.25) + (2 - 2)^2(.25)$
$+ (3 - 2)^2(.20) + (4 - 2)^2(.10) + (5 - 2)^2(.05)$
$= (4)(.15) + (1)(.25) + (0)(.25) + (1)(.20)$
$+ (4)(.10) + (9)(.05)$
$= .6 + .25 + 0 + .2 + .4 + .45$
$= 1.9$

(c) $\sigma = \sqrt{1.9} \approx 1.38$

Basic Properties of Distributions

Probability distributions have several important basic properties. When we assign a probability to an outcome, it must be a number between 0 and 1, inclusive. An event that is certain to happen, called a "certain event," has probability 1, while an "impossible event" has probability 0. For example, if a die is rolled, a certain event would be "observe a number less than 7" and an impossible event is "observe a number greater than 6."

The sample space must exhaust all possible outcomes. This means that the sum of the probabilities of the outcomes in the sample space must add up to 1. Finally, the outcomes in the sample space must be mutually exclusive.

_____ **Basic Properties of Probability Distributions** _____

Let S be the sample space for an experiment and let the probabilities assigned to the outcomes in S be $p_1, p_2, \ldots, p_n$. Then

1. For all $i = 1, \ldots, n,$ $0 \le p_i \le 1$.
2. $p_1 + p_2 + \cdots + p_n = 1$.
3. If E is the certain event, then $Pr(E) = 1$.
4. If E is the impossible event, then $Pr(E) = 0$.

EXERCISE SET 12–1

- In Problems 1 to 14 find the sample space of the experiment.

1. Roll a die and record whether the number on the top face is even or odd.

2. Roll a die and record whether the number on the top face is less than 3, equal to 3, or greater than 3.

3. Select a letter from the word "probability."

4. Select a letter from the word "dzo."

5. Select a card from an ordinary deck and record its color.

6. An ordinary deck of cards is shuffled and a card is selected at random and its suit is recorded.

7. A survey of voters is taken to determine their affiliation. They are classified as Democratic, Republican, or independent. A voter is selected at random and the voter's affiliation is recorded.

8. A person is selected and his or her birth month is recorded.

9. At a weekly meeting of Weight Watchers, members record whether they gained weight, lost weight, or stayed the same. Two members are selected at random and their weight changes are recorded.

10. A manufacturer chooses a sample of four items from a shipment and records the number of defective items in the sample.

11. A financial analyst keeps a record of whether the closing price of a particular stock increases,

decreases, or is unchanged from the previous day's close. The record is kept for four days.

12. The analyst in Problem 11 records the number of days of the four days that the closing price of the stock increases.

13. A farmer records over a three-year period whether corn, potatoes, or alfalfa is planted in a particular field in a given year. Each crop is planted once in the field in the three-year period.

14. The farmer in Problem 13 records whether the same three crops were planted in a field. In this field no crop was planted in successive years.

In Problems 15 and 16 describe in terms of sets the two events E and F associated with the given experiment.

E: an odd number occurs
F: a number greater than 5 occurs

15. A person is selected and his or her birth month is recorded.

16. A manufacturer chooses a sample of eight items from a shipment and records the number of defective items in the sample.

In Problems 17 to 20 the experiment is to toss a die and record the number on the top face. Find $Pr(E)$ for the given event E.

17. $E = \{2, 4\}$

18. $E = \{1, 2, 5, 6\}$

19. E: a prime number occurs

20. E: a number less than 5 occurs

In Problems 21 to 24, for the given random variable X, find (a) the probability histogram and (b) $E(X)$.

21.
outcome, $X = a$	1	2	3	4	5	6
probability, p	$\frac{1}{9}$	$\frac{1}{6}$	$\frac{1}{6}$	$\frac{1}{6}$	$\frac{1}{6}$	$\frac{2}{9}$

22.
outcome, $X = a$	1	2	3	4	5	6
probability, p	$\frac{1}{9}$	$\frac{1}{9}$	$\frac{1}{9}$	$\frac{1}{3}$	$\frac{1}{9}$	$\frac{2}{9}$

23.
outcome, $X = a$	1	2	3	4
probability, p	.26	.10	.28	.36

24.
outcome, $X = a$	1	2	3	4	5
probability, p	.20	.22	.14	.26	.18

In Problems 25 and 26 a frequency distribution is given for the ccological investigation given in Example 6. Compute the (a) expected value, (b) variance, and (c) standard deviation.

25.
eggs in nest	0	1	2	3	4	5
frequency	25	15	35	10	10	5

26.
eggs in nest	0	1	2	3	4	5	6
frequency	25	15	25	10	10	5	10

•• In Problems 27 to 30 find the sample space of the experiment.

27. A coin is tossed two times and the number of heads is recorded.

28. A die is tossed two times and the sum of the numbers on the top faces is recorded.

29. A coin is tossed three times and the number of heads is recorded.

30. A die and a coin are tossed and the number on the top face of the die and the number of heads are recorded.

In Problems 31 to 34, for the given random variable X, find (a) the probability histogram and (b) $E(X)$.

31.
outcome, $X = a$	10	11	12	13	14	15
probability, p	.1	.1	.2	.3	.2	.1

32.
outcome, $X = a$	20	21	22	23	24	25
probability, p	.2	.3	.1	.1	.2	.1

33.
outcome, $X = a$	10	20	30	40	50	60
probability, p	.11	.1	.38	.1	.21	.1

34.
outcome, $X = a$	110	120	130	140	150	160
probability, p	.21	.16	.32	.17	.04	.10

In Problems 35 to 40 a random variable is defined. Find the probability distribution and compute $E(X)$, $Var(X)$, and σ.

35. The number of accidents per month was recorded on a highway for a year. There were three months with no accidents, two months with one accident, three months with two accidents and four months with three accidents. A month is selected at random and the number of accidents is observed. Let X represent the number of accidents.

36. A lottery ticket is purchased for $10. The probability of winning $50 is .08 and the probability of winning $100 is .02. The probability of losing is .9. Let X represent the amount won or lost.

37. An automobile dealer has four types of cars selling for $21,000, $17,000, $14,000, and $9000. The probability that a shopper will buy each type of car is, respectively, .1, .2, .4 and .3. Let X represent the price of the car.

38. A student taking an economics exam has a probability of .7 of scoring 80%, .2 of scoring 85%, .05 of scoring 90%, and .05 of scoring 100%. Let X represent the student's score.

39. A biologist investigating airborne bacteria places a Petri plate containing glucose under a ventilation duct. The number of bacteria colonies that appear per square millimeter is tabulated as follows:

colonies per cell	0	1	2	3	4	5
cells	5	7	10	6	2	1

40. An economist determines trends in the stock market by comparing the average price per share of ten selected stocks from month to month. The prices for the first month are $15, $17, $16,

$17, $18, $15, $15, $16, $15, and $18. Let X represent the price of the stock.

41. Insurance companies and the legal system have several methods to compute an employee's potential earnings that are used as the basis of an award in the event of a wrongful death or disabling injury. But it is much harder to establish acceptable amounts to award housewives in the event of wrongful death or injury. Glennon and Lane investigate solutions to this type of problem by classifying housewives into categories, or professions, that they would most likely have entered given individual qualifications, education and experience.* These professions have established benefits, so a housewife's award would be commensurate with that occupation's award. The data are based on the 1980 Census Public Use samples and are given in the following table.

Occupation	Annual Income ($1000)	Percent of Women in this Occupation
Craft	12	22
Professional	17	29
Service	11	5
Technical	15	44

Find (a) the probability histogram, and (b) $E(X)$.

42. An insurance company must keep a certain amount of funds in reserve for payment of claims. If a company's reserve is too large, there are less funds for investment and earnings. On the other hand, if the reserve is too small and claims are made against the company that are larger than the reserve, the company is said to be in "ruin." So the size of the reserve is critical. In an article studying how to calculate the probability of ruin, Panjer uses the following table from several life insurance companies, giving the distribution of claims by size x, measured in 1000 dollars, and where $p(x)$ is the proportion of claims of that size.†

x	2	4	8	10	12	14	16	20	25
$p(x)$	.15	.08	.11	.10	.10	.11	.10	.18	.07

Find (a) the probability histogram, and (b) $E(X)$.

43. By the "optimal diet theory" ecologists mean that predators should attack the largest size prey available in order to have the best, or optimal, diet. In an article comparing this theory to the actual diet of a predator fish called the mosquitofish, Bence and Murdoch measured the size of the prey attacked and the percentage of successful attacks on that size prey per minute.‡ The data are shown in the following table.

x, Prey size (mm)	.9	1.76	2.35	3.41
$p(x)$, Proportion of successful attacks	.02	.06	.16	.76

Find (a) the probability histogram, and (b) $E(X)$.

*Dennis Glennon and Julia Lane, "Imputing a Housewife's Earnings in a Wrongful Death and Injury Case," *Journal of Risk and Insurance,* Vol. 53, 1986, pp. 734–743.

†Harry H. Panjer, "Direct Calculation of Ruin Probabilities," *Journal of Risk and Insurance,* Vol. 53, 1986, pp. 521–529.

‡James R. Bence and W. W. Murdoch, "Prey Selection by the Mosquitofish: Relation to Optimal Diet Theory," *Ecology,* Vol. 67, 1986, pp. 324–336.

12–2 CONTINUOUS PROBABILITY DISTRIBUTIONS

Each probability distribution studied in the previous section has a finite number of outcomes in the sample space. They are called discrete distributions. Sometimes the number of outcomes in an experiment can be infinite. If the outcomes can be any number in an interval, the distribution is called *continuous*. For example, (1) when a meteorologist measures the temperature of the ocean water in degrees Celsius the outcome is any number in [0, 100]; (2) when an investment manager measures the percent of the total value of a portfolio represented by one stock in the portfolio the outcome is any number in [0, 100]; (3) when a bank measures the time, in minutes per 8-hour day, that at least one customer was waiting in line, the outcome can be any number in [0, 480].

The concepts of expected value, variance, and standard deviation for discrete random variables require the addition of numbers. To sum quantities over an interval, we need the definite integral. This is how calculus is used in probability.

Continuous Probability Distributions

Consider the Fairley study of accidents on Route 2 in Massachusetts from the previous section. The experiment is to record the quarter in which an accident occurs. The sample space has four outcomes. Suppose the researchers were to use a larger sample space; that is, suppose they recorded the month in which the accident occurred. Then the sample space would have 12 outcomes. If they then recorded the date of the accident, there would be 365 outcomes. The corresponding histograms would have 12 and 365 rectangles, respectively. And if the researchers recorded the accident's time of day to the minute, the histogram would appear to be a continuous function. As the number of outcomes in the sample space increases, the histogram resembles a **continuous distribution.**

A random variable X that equals any number in an interval is called a **continuous random variable.** The important distinction between a discrete distribution and a continuous probability distribution centers on how probabilities are assigned to outcomes. In a discrete distribution each of the finite number of outcomes that the random variable can take on is assigned a probability. Since there are an infinite number of outcomes for a continuous random variable X, each outcome is not assigned a specific probability. Instead, probabilities are assigned to intervals of numbers. This is done by using the concept of a *probability density function*. It assigns probabilities of the form $Pr(a \leq X \leq b)$, which is read "the probability that an outcome in the interval $[a, b]$ is observed." The definite integral is used to define this type of probability.

Definition

Let X be a continuous random variable and let $f(x)$ be a continuous, nonnegative function with the property that

$$Pr(a \leq X \leq b) = \int_a^b f(x)\, dx$$

Then $f(x)$ is called a **probability density function** for X.

Figures 1 and 2 show how the graph of a probability density function is similar to a probability histogram for a discrete random variable with many outcomes. There are two important properties that a probability density function must satisfy. The probabilities of the outcomes for a discrete random variable must sum to 1. For a

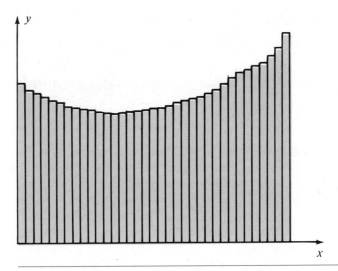

FIGURE 1 Probability histogram for a discrete random variable with many outcomes.

continuous random variable this means that the definite integral of the density function over the interval on which it is defined must be 1. Also, each probability of an outcome for a discrete random variable is greater than or equal to zero, indicating that the density function of a continuous random variable must be greater than or equal to zero.

——————————— **Properties of a Probability Density Function** ———————————
If $f(x)$ is a probability density function for the random variable X with sample space $[c, d]$, then

1. $\displaystyle\int_c^d f(x)\, dx = 1$

2. $f(x) \geq 0$ for all x in $[c, d]$

FIGURE 2 A probability density function.

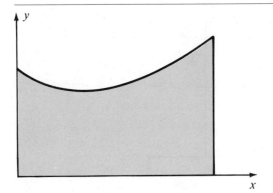

EXAMPLE 1 | **Problem**

Determine whether $f(x) = 2x$ is a probability density function on the interval $[0, 1]$.

SOLUTION To see if $f(x) = 2x$ satisfies property 1, compute

$$\int_0^1 2x \, dx = x^2 \Big|_0^1 = 1^2 - 0^2 = 1$$

This shows that the function satisfies property 1. In the interval $[0, 1]$, $f(x) = 2x$ is greater than or equal to 0 so it satisfies property 2. Therefore the $f(x) = 2x$ is a density function on $[0, 1]$.

The next example shows how to compute probabilities of a continuous random variable.

EXAMPLE 2 | **Problem**

For the probability density function $f(x) = 2x$ on the interval $[0, 1]$ find the probability that x is between (a) 0 and .5 and (b) .5 and 1.

SOLUTION (a) Compute

$$Pr(0 \leq X \leq .5) = \int_0^{.5} 2x \, dx = x^2 \Big|_0^{.5} = (.5)^2 - 0 = .25$$

This probability corresponds to the area of the region bounded by $f(x) = 2x$ and the x-axis from 0 to .5. (See Figure 3.)

FIGURE 3

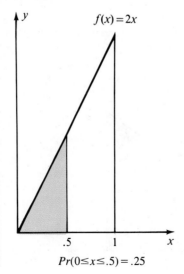

$Pr(0 \leq x \leq .5) = .25$

(b) Compute

$$Pr(.5 \leq X \leq 1) = \int_{.5}^{1} 2x \, dx = x^2 \Big|_{.5}^{1} = 1 - (.5)^2 = .75$$

This probability corresponds to the area of the region bounded by $f(x) = 2x$ and the x-axis from .5 to 1. (See Figure 4.)

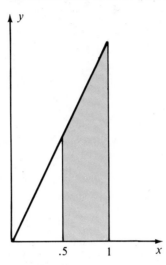

$$Pr(.5 \leq x \leq 1) = .75$$

FIGURE 4

EXAMPLE 3

Problem

Determine whether $f(x) = 6x - 3x^2$ is a probability density function on the interval $[0, 2]$. If it is not, find a number a such that $af(x)$ is a probability density function on the interval $[0, 2]$.

SOLUTION Test whether the function satisfies property 1.

$$\int_{0}^{2} (6x - 3x^2) \, dx = (3x^2 - x^3) \Big|_{0}^{2} = 12 - 8 = 4$$

This means that the function is not a probability density function on $[0, 2]$ because the integral is not 1. Since the function is greater than or equal to 0 on the $[0, 2]$, all that is necessary to find the desired density function is to divide $f(x)$ by 4. Let $a = \frac{1}{4}$. The function $(6x - 3x^2)/4$ satisfies both properties.

EXAMPLE 4

A battery manufacturing firm determines that a random variable X is the time x that a defective battery is produced, where x is measured in hours after a shift has begun production. The density function is

$$f(x) = \frac{1}{256} (24x - 3x^2)$$

Problem

Find the probability that a defective battery is produced between 0 and 2 hours into the shift.

SOLUTION Compute

$$Pr(0 \leq X \leq 2) = \int_0^2 \frac{1}{256} (24x - 3x^2)$$

$$= \frac{1}{256} (12x - x^3) \Big|_0^2 = \frac{16}{256} = \frac{1}{16}$$

EXERCISE SET 12–2

In Problems 1 to 8 determine whether $f(x)$ is a probability density function on the interval.

1. $f(x) = 2x - 1$, [1, 2]

2. $f(x) = x - \dfrac{1}{2}$, [1, 2]

3. $f(x) = \dfrac{5}{4} - \dfrac{x}{2}$, [0, 1]

4. $f(x) = 5 - 2x$, [0, 1]

5. $f(x) = x^2$, [0, 2]

6. $f(x) = \dfrac{x^2}{9}$, [0, 3]

7. $f(x) = \dfrac{3x(4 - x)}{32}$, [2, 4]

8. $f(x) = \dfrac{3x(4 - x)}{32}$, [0, 4]

In Problems 9 to 12 find the probability that x is between a and b for the probability density function $f(x) = 2x$ on the interval [0, 1].

9. $a = 0$, $b = .2$ 10. $a = 0$, $b = .8$

11. $a = .1$, $b = .2$ 12. $a = .5$, $b = .8$

In Problems 13 to 16 find the probability that x is between a and b for the probability density function $f(x) = 3x^2$ on the interval [0, 1].

13. $a = 0$, $b = .8$ 14. $a = .1$, $b = .2$

15. $a = .5$, $b = 1$ 16. $a = 0$, $b = 1$

In Problems 17 and 18 find the probability that x is between a and b for the probability density function

$$f(x) = \frac{12 - 3x^2}{32} \text{ on the interval } [-2, 2].$$

17. $a = -2$, $b = 1$

18. $a = -1$, $b = 0$

In Problems 19 to 28 find a number a such that $af(x)$ is a probability density function on the given interval.

19. $f(x) = x^2$, [0, 4]

20. $f(x) = 4 - x^2$, [0, 2]

21. $f(x) = 1 - x^2$, [-1, 1]

22. $f(x) = x - x^2$, [0, 1]

23. $f(x) = x(2 - x)$, [0, 2]

24. $f(x) = x^{1/2}$, [0, 2]

•• 25. $f(x) = e^{-x}$, [0, ln 2]

26. $f(x) = x^{-1}$, [1, 3]

27. $f(x) = 1 - e^{-x}$, [0, ln 2]

28. $f(x) = 4 - x^{-1}$, [1, 2]

In Problems 29 and 30 find the probability that x is between a and b for the probability density function $f(x) = 1 - x^{-1/2}$ on the interval [1, 4].

29. $a = 1$, $b = 2$ 30. $a = 2$, $b = 4$

In Problems 31 to 34 find the indicated probability for the probability density function $f(x) = 1 - x^{-1/2}$ on the interval [1, 4].

31. $Pr(X \leq 2)$ 32. $Pr(X \geq 2)$

33. $Pr(3 \leq X)$ 34. $Pr(3 \geq X)$

35. A computer chip manufacturing firm determines that a random variable X is the time x that a defective chip is produced, where x is measured in hours after a shift has begun production. The density function is $f(x) = \dfrac{3}{448}(10x - x^2)$, defined on the interval $[0, 8]$. Find the probability that a defective chip is produced between 0 and 4 hours into the shift.

36. An accounting firm uses the random variable X that measures the time x it takes for a late bill to be paid, where x is measured in months. The probability density function is

$$f(x) = \left(\frac{1}{12}\right)(x^2 - 6x + 10),$$

for x in the interval $[0, 3]$. Find the probability that a late bill is paid within the first two months.

37. When Baltimore Gas and Electric Company had to select a strategy for increasing its electricity-generating capacity, it had to choose between conventional means and several new technologies for its new site near Baltimore. The key issue was whether the perceived advantages of the new technologies were worth the risks of not having the generation capacity early enough to meet load growth. In an article discussing this choice, Keeney et al.* used the probability density function

$$f(x) = 1.2(1 - e^{.18(x-5)})$$

defined on the interval $[0, 1]$ to measure the probability that the new plant would meet demand with the new technologies where x is the percent of cost overrun, from 0 to 1 (hundred percent). Find the probability that demand will be met if the cost overrun is between 0 and .1 percent.

38. Hazardous wastes are often the by-product of other valuable products. For example, dangerous chemicals used in the dry cleaning process might reach water supplies if they are not disposed of properly. In an article discussing the government's role in determining policy for disposing of hazardous waste, Johnson and Ulen use a probability density function $q(x)$ which measures the likelihood that a firm will not have to pay a fine for improper disposal.† They define x to be the level of the firm's precaution in preventing an accident, and they let $q(x)$ be the density function measuring the likelihood that the firm will not have to pay a fine. Then the probability the firm will not have to pay a fine given that the firm supports a level of precaution between a and b units is given by

$$\int_a^b q(x)\,dx$$

Consider a firm that disposes of hazardous waste by using a mechanical cleaning process. Then the level of support x is the proportion of time the process is working efficiently. Let $q(x) = 6 - 6x^{1/2}$ be defined on the interval $[.25, 1]$. Find the probability that the firm will not have to pay a fine if it supports a level of precaution between .5 and 1.

*Ralph L. Keeney, "An Analysis of Baltimore Gas and Electric Company's Technology Choice," *Operations Research,* Vol. 34, 1986, pp. 18–41.

†Gary V. Johnson and Thomas S. Ulen, "Designing Public Policy Toward Hazardous Wastes: The Role of Administrative Regulations and Legal Liability Rules," *American Journal of Agricultural Economics,* Vol. 68, 1986, pp. 1266–1271.

12–3 EXPECTED VALUE AND STANDARD DEVIATION: CONTINUOUS RANDOM VARIABLES

The expected value of a discrete random variable X is the sum of the products of the values of X and their corresponding probabilities. A continuous random variable is defined on an interval, so it is not possible to define the expected value in the same way. It is not possible to sum the values in an interval. The definite integral is used to define the expected value and the standard deviation of continuous random variables.

Expected Value

The expected value of a continuous random variable X measures the central tendency of the values of X, in the same way as it does for discrete random variables. For a discrete random variable with values x and probabilities p the expected value is the sum of the products xp. For a continuous random variable the analog of the probabilities p is the probability density function $f(x)$. Extending the idea of the "sum of the terms xp" to continuous random variables yields the definite integral of $xf(x)$ over the interval in which $f(x)$ is defined.

_____ Definition _____

Let X be a continuous random variable with probability density function $f(x)$ defined on the interval $[a, b]$. The **expected value** $E(X)$ of X is

$$E(X) = \int_a^b xf(x)\, dx$$

_____

Example 1 shows how to compute the expected value for a **uniform** continuous random variable, which is one with a probability density function with a constant function.

EXAMPLE 1 | **Problem**

Compute the expected value of the uniform random variable X with probability density function $f(x) = .1$, defined on interval $[0, 10]$.

SOLUTION From the definition

$$E(X) = \int_0^{10} x(.1)\, dx = (.1) \int_0^{10} x\, dx$$

$$= (.1) \frac{x^2}{2} \bigg|_0^{10} = (.1)(50) = 5$$

The result in Example 1 is to be expected because the value 5 is the midpoint of the sample space $[0, 10]$ and the density function is constant. But if the density function is not uniform over the interval, $E(X)$ need not be the midpoint.

EXAMPLE 2 | **Problem**

Compute the expected value of the random variable X with probability density function $f(x) = .02x$, defined on $[0, 10]$.

SOLUTION From the definition

$$E(X) = \int_0^{10} x(.02x)\, dx = (.02) \int_0^{10} x^2\, dx$$

$$= (.02)\frac{x^3}{3}\Big|_0^{10} = (.02)\frac{(10)^3}{3} = \frac{20}{3}$$

Figure 1 illustrates the difference between the expected values of the random variables in Examples 1 and 2. If the area bounded by $f(x)$ and the x-axis from 0 to 10 were a weight, the fulcrum would be located at the expected value. In Figure 1a the weight is evenly distributed, so the fulcrum is in the center. In Figure 1b there is more weight to the right of center, so the fulcrum must be to the right of center.

Variance and Standard Deviation

For a discrete random variable the variance measures the spread of the outcomes. Just as $E(X)$ is a weighted average, so, too, the variance is the weighted sum of the squares of the differences between the outcomes and $E(X)$. For a continuous random variable the variance measures the spread of the density function. If X is a discrete random variable with values x and probabilities p, and expected value E, then the variance is the sum of the products $(x - E)^2 p$. Generalizing this idea for continuous random variables yields the definite integral of $(x - E)^2 f(x)$ over the interval that $f(x)$ is defined. We represent the expected value here by E to simplify notation.

FIGURE 1

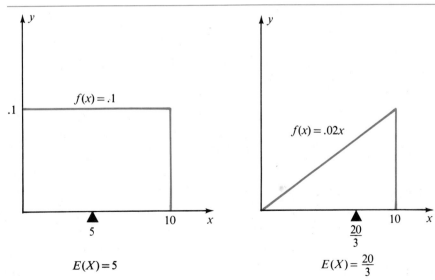

$f(x) = .1$

$E(X) = 5$

$f(x) = .02x$

$E(X) = \dfrac{20}{3}$

Definition

Let X be a continuous random variable with probability density function $f(x)$ defined on the interval $[a, b]$ and let E be the expected value of X. The **variance** $V(X)$ of X is

$$V(X) = \int_a^b (x - E)^2 f(x)\, dx$$

The **standard deviation** of X is $\sigma = \sqrt{V(X)}$.

EXAMPLE 3 | **Problem**

Compute the variance and the standard deviation of the random variables X with probability density function $f(x)$ defined on interval $[0, 10]$: (a) $f(x) = .1$ and (b) $f(x) = .02x$.

SOLUTION The expected values were computed in Examples 1 and 2. (a) From the definition with $E = 5$

$$V(X) = \int_0^{10} (x - 5)^2 (.1)\, dx = (.1) \int_0^{10} (x^2 - 10x + 25)\, dx$$

$$= (.1) \left(\frac{x^3}{3} - 5x^2 + 25x \right) \Big|_0^{10}$$

$$= (.1) \left(\frac{1000}{3} - 500 + 250 \right)$$

$$= \frac{25}{3} \simeq 8.33$$

$$\sigma = \sqrt{8.33} \simeq 2.89$$

(b) From the definition with $E = \dfrac{20}{3}$

$$V(X) = \int_0^{10} \left(x - \frac{20}{3} \right)^2 (.02x)\, dx$$

$$= (.02) \int_0^{10} \left(x^3 - \frac{40x^2}{3} + \frac{400x}{9} \right) dx$$

$$= (.02) \left(\frac{x^4}{4} - \frac{40x^3}{9} + \frac{200x^2}{9} \right) \Big|_0^{10}$$

$$= (.02) \left(\frac{10,000}{4} - \frac{40,000}{9} + \frac{20,000}{9} \right)$$

$$= \frac{100}{18} = \frac{50}{9} \simeq 5.56$$

$$\sigma = \sqrt{\frac{50}{9}} \simeq 2.36$$

Alternate Formula

The computation of the variance can be simplified by expanding the integrand and manipulating the result algebraically. This yields an **alternate formula.** To derive the new formula, start with the original expression $V(X)$.

$$V(X) = \int_a^b (x - E)^2 f(x) \, dx$$

$$= \int_a^b (x^2 - 2Ex + E^2) f(x) \, dx$$

$$= \int_a^b (x^2 f(x) - 2Exf(x) + E^2 f(x)) \, dx$$

$$= \int_a^b x^2 f(x) \, dx - 2E \int_a^b xf(x) \, dx + E^2 \int_a^b f(x) \, dx \qquad (1)$$

The second term of equation (1) contains an integral that is the definition of E.

$$E = \int_a^b xf(x) \, dx$$

The third term of equation (1) contains an integral that is equal to 1, because from the properties of probability density functions

$$\int_a^b f(x) \, dx = 1$$

Substituting these two results into equation (1) gives

$$V(X) = \int_a^b x^2 f(x) \, dx - 2E(E) + E^2(1)$$

$$= \int_a^b x^2 f(x) \, dx - E^2$$

Let X be a continuous random variable with probability density function $f(x)$ defined on an interval $[a, b]$ and let E be the expected value of X. Then the alternate formula for the variance $V(X)$ of X is

$$V(X) = \int_a^b x^2 f(x) \, dx - E^2$$

EXAMPLE 4

A department store determines that the proportion x of customers that pay for their purchases by credit card on a given day is given by the probability density function $f(x) = 3x^2 - 6x + 3$ for x in $[0, 1]$.

Problem

Compute (a) the expected value and (b) the variance.

SOLUTION (a) From the definition

$$E = \int_0^1 x(3x^2 - 6x + 3)\, dx$$

$$= \int_0^1 (3x^3 - 6x^2 + 3x)\, dx$$

$$= \left(\frac{3x^4}{4} - 2x^3 + \frac{3x^2}{2}\right)\Big|_0^1$$

$$= \frac{3}{4} - 2 + \frac{3}{2} = \frac{1}{4}$$

This means that the average number of customers that pay by credit card is $\frac{1}{4}$ or 25%.

(b) Use the alternate formula with $E = \frac{1}{4}$.

$$V(X) = \int_0^1 [x^2(3x^2 - 6x + 3)]\, dx - \left(\frac{1}{4}\right)^2$$

$$= \int_0^1 [(3x^4 - 6x^3 + 3x^2)]\, dx - \frac{1}{16}$$

$$= \left(\frac{3x^5}{5} - \frac{6x^4}{4} + x^3\right)\Big|_0^1 - \frac{1}{16}$$

$$= \frac{3}{5} - \frac{3}{2} + 1 - \frac{1}{16}$$

$$= \frac{3}{80} \approx .038$$

EXERCISE SET 12–3

• In Problems 1 to 10 compute the expected value of the random variable X with probability density function $f(x)$ defined on the given interval.

1. $f(x) = .2$, $[0, 5]$ 2. $f(x) = .5$, $[2, 4]$

3. $f(x) = .005x$, $[0, 20]$

4. $f(x) = .5x$, $[0, 2]$

5. $f(x) = x - \frac{1}{2}$, $[1, 2]$

6. $f(x) = \frac{5}{4} - \frac{x}{2}$, $[0, 1]$

7. $f(x) = \frac{5}{2} - x$, $[1, 2]$

8. $f(x) = \frac{5}{12} - \frac{x}{6}$, $[-1, 2]$

9. $f(x) = \frac{x^2}{9}$, $[0, 3]$

10. $f(x) = \frac{x^2}{3}$, $[-1, 2]$

In Problems 11 to 16 compute the variance of the

random variables X with probability density function $f(x)$ defined on the given interval.

11. $f(x) = .2$, $[1, 6]$ 12. $f(x) = .5$, $[5, 7]$

13. $f(x) = .0002x$, $[0, 100]$

14. $f(x) = .5x$, $[0, 2]$

15. $f(x) = \dfrac{3x^2}{8}$, $[0, 2]$

16. $f(x) = \dfrac{x^2}{3}$, $[-1, 2]$

In Problems 17 to 40 compute the variance and the standard deviation of the random variables X with probability density function $f(x)$ defined on the given interval.

17. $f(x) = .2$, $[0, 5]$ 18. $f(x) = .5$, $[2, 4]$

19. $f(x) = .0002x$, $[0, 100]$

20. $f(x) = .5x$, $[0, 2]$

21. $f(x) = \dfrac{x^2}{9}$, $[0, 3]$

22. $f(x) = \dfrac{x^2}{3}$, $[-1, 2]$

23. $f(x) = 3x^2 - 6x + 3$, $[0, 1]$

24. $f(x) = 4x^3 - 12x + 6$, $[0, 1]$

•• 25. $f(x) = 2e^{-x}$, $[0, \ln 2]$

26. $f(x) = 3e^{-x}$, $[0, \ln 3]$

27. $f(x) = 2x^{-2}$, $[1, 2]$

28. $f(x) = \dfrac{3}{2}x^{-2}$, $[1, 3]$

29. $f(x) = 2 - 2x^{-2}$, $[1, 2]$

30. $f(x) = 1 - x^{-1/2}$, $[1, 4]$

31. $f(x) = \dfrac{3x(4 - x)}{32}$, $[2, 4]$

32. $f(x) = \dfrac{3x(4 - x)}{32}$, $[0, 4]$

33. $f(x) = 1 - x^2$, $[-1, 1]$

34. $f(x) = x - x^2$, $[0, 1]$

35. $f(x) = \dfrac{3}{4}x(2 - x)$, $[0, 2]$

36. $f(x) = x^{1/2}$, $[0, 2]$

37. A computer chip manufacturing firm determines a random variable X is the time x that a defective chip is produced, where x is measured in hours after a shift has begun production. The probability density function is $f(x) = \dfrac{3(10x - x^2)}{448}$, defined on $[0, 8]$.

38. An accounting firm uses the random variable X that measures the time x it takes for a late bill to be paid, where x is measured in months. The probability density function is
$$f(x) = \left(\dfrac{1}{12}\right)(x^2 - 6x + 10),$$
for x in $[0, 3]$.

39. A biologist determines that the random variable X measuring the time that an animal feeds within a 4-hour period has probability density function
$$f(x) = \left(\dfrac{3}{32}\right)(4x - x^2) \text{ in } [0, 4].$$

40. A manufacturing firm determines that the random variable X measuring the time x that an electronic component will fail during a four-year period has probability density function $f(x) =$
$$\left(\dfrac{4}{3}\right)(1 + x)^{-2} \text{ in } [0, 4].$$

41. One method psychologists use to study learning is to administer a test with many simple tasks that must be mastered. The test is administered many times and the number of items mastered, or learned, is recorded. If the number of subjects taking the test is large, the number of items learned is a random variable. In a study of this type of learning model, Wickens considers a test in which each item on the test has the same probability x of being mastered each time the test is administered.* Then the probability den-

*Thomas D. Wickens, *Models for Behavior: Stochastic Processes in Psychology,* San Francisco, W. H. Freeman, 1982, pp. 60–75.

sity function measuring the percentage of items mastered is given by

$$f(x) = 4(1 - x)^3$$

where x is defined on the interval [0, 1]. Find the expected value and the standard deviation.

42. The body mass, or fat reserves, of a wintering bird can be viewed as a trade-off between the probability of starvation and the probability of predation. This is because fatter birds can survive longer than thinner birds, but a leaner bird can escape a predator better than a fatter bird can. In an article comparing these attributes of wintering birds, Lima computed the optimal body mass of several types of birds.[†] In one part of the study the probability density function $f(x) = 4.13(1 - e^{-.02x})$ defined on the interval [0, 5] measured the probability of a bird being attacked while feeding for x hours. Find the expected value and the standard deviation.

[†]Steven L. Lima, "Predation Risk and Unpredictable Feeding Conditions: Determinants of Body Mass in Birds," *Ecology,* Vol. 67, 1986, pp. 377–385.

12–4 EXPONENTIAL AND NORMAL DISTRIBUTIONS

There are many types of probability distributions. In the last section the uniform distribution was mentioned. It has a constant function for its probability density function. The data are spread uniformly across the sample space. In this section we study two of the most common continuous probability distributions, the exponential distribution and the normal distribution. The exponential distribution is often used to model growth and decay. This includes reliability measures such as the lifetime of computer components and light bulbs, as well as the length of time between two occurrences of an event, such as two successive telephone calls. The normal distribution is used to model data such as height, weight, IQ (intelligence quotient) measures, and scores on standardized tests.

The Exponential Probability Distribution

The **exponential probability distribution** is defined by the following probability density function.

_____ **Definition** _____

If a is a positive constant, then the function

$$f(x) = ae^{-ax} \quad \text{for } x \text{ in } [0, \infty)$$

is called an **exponential density function.** A continuous random variable with an exponential probability density function is called an **exponential random variable.**

The graph of $f(x)$ is given in Figure 1. To see that $f(x)$ satisfies the defining properties of a probability density function, first note that $f(x) > 0$ in $[0, \infty)$ because a is positive and e^{-ax} is also positive for all x. Secondly, we must have the integral of $f(x)$ over $[0, \infty)$ equal to 1. Compute

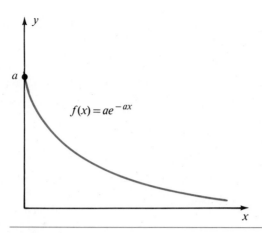

$f(x) = ae^{-ax}$

FIGURE 1

$$\int_0^b f(x) \, dx = \int_0^b ae^{-ax} \, dx = -e^{-ax} \Big|_0^b$$

$$= -e^{-ab} + 1 = 1 - \frac{1}{e^{ab}}$$

This expression approaches 1 as b approaches ∞. This shows that $f(x)$ is a probability density function.

EXAMPLE 1

Consider the exponential random variable X with the exponential density function

$$f(x) = 3e^{-3x}$$

Problem

Find $Pr(0 \le X \le 2)$.

SOLUTION From the definition of probability of a probability density function, we have

$$Pr(0 \le X \le 2) = \int_0^2 f(x) \, dx = \int_0^2 3e^{-3x} \, dx$$

$$= -e^{-3x} \Big|_0^2 = -e^{-6} + 1$$

$$\approx -.00248 + 1 = .9975$$

In examples involving an exponential density function, the values of the function can be obtained either by calculator or from the table in the Appendix.

The Expected Value and Standard Deviation of an Exponential Random Variable

The expected value and the standard deviation of an exponential random variable can be computed by using integration by parts. We give the results here and leave the computation to the exercises.

Let X be an exponential random variable with probability density function

$$f(x) = ae^{-ax} \quad \text{for } x \text{ in } [0, \infty)$$

Then the expected value of X is

$$E(X) = \frac{1}{a}$$

The standard deviation of X is

$$\sigma = \frac{1}{a}$$

EXAMPLE 2   The number of continuous days that a light bulb will burn until failure is an exponential random variable X. It is found that for a particular type of light bulb, the expected value of X is 50 days.

Problem

Find the probability that a light bulb of this type will (a) fail within the first 40 days and (b) fail between 40 and 60 days.

SOLUTION The probability density function is found by using the property that $E(X) = 1/a = 50$, so $a = 1/50 = .02$. The probability density function is

$$f(x) = .02e^{-.02x}$$

(a) The problem is to find the probability of the event that X is between 0 and 40.

$$Pr(0 \le X \le 40) = \int_0^{40} f(x)\, dx = \int_0^{40} .02e^{-.02x}\, dx$$

$$= -e^{-.02x} \Big|_0^{40} = -e^{-.8} + 1$$

$$\simeq -.4493 + 1 = .5507$$

(b) The problem is to find the probability of the event that X is between 40 and 60.

$$Pr(40 \le X \le 40) = \int_{40}^{60} f(x) \, dx = \int_{40}^{60} .02e^{-.02x} \, dx$$

$$= -e^{-.02x} \Big|_{40}^{60} = -e^{-1.2} + e^{-.8}$$

$$\approx -.3012 + .4493 = .1481$$

The Normal Probability Distribution

The **normal probability distribution** is defined by the following probability density function.

_____ **Definition** _____

Let σ be a positive constant and let $c = -\dfrac{(x - \mu)^2}{2\sigma^2}$. Here μ can be any constant. Then the function

$$f(x) = \frac{1}{\sigma\sqrt{2\pi}} e^c \qquad \text{for } x \text{ in } [-\infty, \infty)$$

is called a **normal density function.** A continuous random variable with a normal probability density function is called a **normal random variable.**

The graph of $f(x)$ is given in Figure 2. It is the well-known bell-shaped curve, called a **normal curve.** By changing the values of σ and μ, different normal curves can be derived. Figure 3 shows three examples of normal curves. They all have certain properties, which are stated without proof because their verification requires advanced techniques.

FIGURE 2

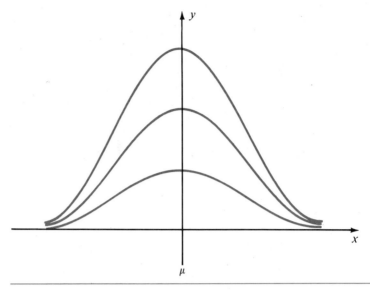

FIGURE 3   Several examples of different normal curves.

1. The expected value and standard deviation of a normal distribution are

    $$E(X) = \mu \qquad \text{standard deviation} = \sigma$$

2. A normal curve is symmetric about the line $x = \mu$, that is, about its mean.
3. Points of inflection of a normal curve occur at $x = \mu + \sigma$ and $x = \mu - \sigma$, that is, at the mean ± 1 standard deviation.
4. A normal curve never touches the x-axis; it extends indefinitely in either direction, approaching the x-axis asymptotically.
5. The area under a normal curve is 1. This is a defining property of a probability density function.

FIGURE 4

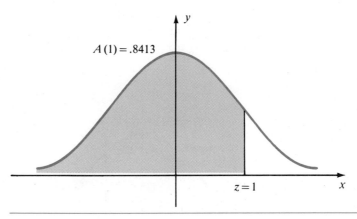

$A(1) = .8413$

$z = 1$

FIGURE 5

6. A smaller value of σ produces a ''taller'' normal curve, meaning that more values are near the mean.

Of all the different normal curves, there is one that is used the most. It corresponds to the density function with $\mu = 0$ and $\sigma = 1$. Its distribution is called the **standard normal distribution.** Its graph, given in Figure 4, is called the **standard normal curve.**

Probabilities for the standard normal distribution are defined from the integral

$$\int_a^b \left(\frac{1}{\sqrt{2\pi}}\right) e^{-x^2/2} \, dx$$

This integrand does not have an antiderivative that can be expressed as an elementary function, so numerical methods are used to evaluate the definite integral. A table in the Appendix gives values of the area under the standard normal curve. For each value of z on the x-axis the value $A(z)$ in the table is the area under the standard normal curve to the left of the number z. Figure 5 shows the value $z = 1$ and the corresponding area $A(1)$ to the left of 1 is shaded. Its value from the table is $A(1) = .8413$. This means that the shaded region in Figure 5 represents 84.13% of the area under the standard normal curve.

EXAMPLE 3 | **Problem**

Use the table in the Appendix to find the following areas under the standard normal curve: (a) to the left of $z = .50$, (b) to the right of $z = 1.50$, and (c) between $z = .50$ and $z = 1.50$.

SOLUTION (a) From the table $A(.50) = .6915$. Thus the area under the standard normal curve to the left of $z = .50$ is $.6915$. (See Figure 6.)
(b) From the table $A(1.50) = .9332$. Thus the area under the standard normal curve to the left of $z = 1.50$ is $.9332$. The area to the right of $z = 1.50$ is $1 - A(1.50) = 1 - .9332 = .0668$. (See Figure 7.)

(c) From the table $A(.50) = .6915$ and $A(1.50) = .9332$. The area under the standard normal curve between $z = .50$ and $z = 1.50$ is the difference between these two areas. It is $.9332 - .6915 = .2417$. (See Figure 8.)

FIGURE 6

FIGURE 7

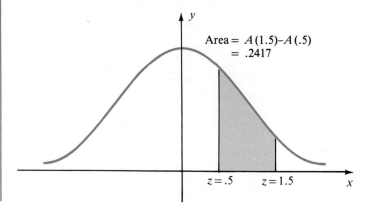

FIGURE 8

z-Scores

The standard normal distribution can be used to calculate probabilities of other normal distributions. If the normal distribution has mean μ and standard deviation σ, the values of x on the horizontal axis of its probability density curve can be converted into values on the horizontal axis of the standard normal density curve. Then the values in the table in the Appendix for the standard normal curve are used to compute the areas under the given normal curve. This conversion is done in the following way.

> If a normal distribution has mean μ and standard deviation σ, the area under this normal curve to the left of a value x is the equal to the area under the standard normal curve to the left of
>
> $$z = \frac{x - \mu}{\sigma}$$

This means that the value x is in the sample space of the given normal distribution. It is converted into a z-score, which is a value in the sample space of the standard normal curve.

EXAMPLE 4

A normal distribution has $\mu = 20$ and $\sigma = 10$.

Problem

Find the area under its normal curve (a) to the left of 25, (b) to the right of 35, and (c) between 25 and 35.

SOLUTION (a) Find the z-score corresponding to 25 by substituting $x = 25$, $\mu = 20$, and $\sigma = 10$ into equation 1 in the display.

$$z = \frac{25 - 20}{10} = \frac{5}{10} = .50$$

From the table $A(.50) = .6915$, meaning that the area under the standard normal curve to the left of $z = .50$ is .6915. Therefore the area under the given normal curve to the left of 25 is .6915.

(b) Find the z-score corresponding to 35 by substituting $x = 35$, $\mu = 20$, and $\sigma = 10$ into equation 1.

$$z = \frac{35 - 20}{10} = \frac{15}{10} = 1.50$$

From the table $A(1.50) = .9332$. Thus the area under the standard normal curve to the right of $z = 1.50$ is $1 - A(1.50) = 1 - .9332 = .0668$. Therefore the area under the given normal curve to the right of 35 is .0668.

(c) The z-scores for $x = 25$ and $x = 35$ are $z = .50$ and $z = 1.50$, respectively. From the table $A(.50) = .6915$ and $A(1.50) = .9332$. The area under the standard normal curve between $z = .50$ and $z = 1.50$ is $.9332 - .6915 = .2417$. Therefore the area under the given normal curve between 25 and 35 is .2417.

Notice the connection between Examples 3 and 4.

EXAMPLE 5   Many IQ tests are designed so that the mean score is 100 and the standard deviation is 20.

Problem

If a person is selected at random, find the probability that the IQ score is (a) less than 95 and (b) greater than 130.

SOLUTION (a) The z-score corresponding to $x = 95$ is

$$z = \frac{95 - 100}{20} = -\frac{5}{20} = -.25$$

From the table $A(-.25) = .4013$. The amount of data less than $x = 95$ corresponds to the area under the standard normal curve to the left of $z = -.25$. Therefore the probability that the IQ score is less than 95 is .4013. This means that about 40.13% of people have an IQ of less than 95.
(b) The z-score corresponding to $x = 130$ is

$$z = \frac{130 - 100}{20} = \frac{30}{20} = 1.50$$

From the table $A(1.50) = .9332$. This is the area to the left of $z = 1.5$. The amount of data greater than $x = 130$ corresponds to the area under the standard normal curve to the right of $z = 1.50$, which is $1 - .9332 = .0668$. Therefore the probability that the IQ score is greater than 130 is .0668. This means that about 6.68% of people have an IQ of greater than 130.

In many public schools the cutoff for a student to be classified as "gifted" is an IQ score of 130. Example 5 shows that about 7% of the population is "gifted," with the assumption that $\mu = 100$, $\sigma = 20$, and $x = 130$ is the cutoff.

EXERCISE SET 12–4

• In Problems 1 to 4 find the given probability for the exponential random variable X with the exponential density function $f(x) = 3e^{-3x}$.

1. $Pr(0 \le X \le 1)$

2. $Pr(0 \le X \le 3)$

3. $Pr(1 \le X \le 2)$

4. $Pr(1 \le X \le 3)$

In Problems 5 to 8 find the given probability for the exponential random variable X with the exponential density function $f(x) = 5e^{-5x}$.

5. $Pr(0 \le X \le 1)$ 6. $Pr(0 \le X \le 2)$

7. $Pr(1 \le X \le 2)$ 8. $Pr(2 \le X \le 3)$

In Problems 9 to 16 it is assumed that the number of continuous days that a light bulb will burn until failure is an exponential random variable X. It is found that for a particular type of light bulb, the expected value of X is n days. Find the probability that a light bulb of this type will fail within the given time period.

9. $n = 50$,   within the first 30 days

10. $n = 50$,   within the first 60 days

11. $n = 100$,   within the first 100 days

12. $n = 100$,   within the first 110 days

13. $n = 100$,   between 100 and 110 days

14. $n = 100$,   between 100 and 120 days

15. $n = 150$,   between 140 and 160 days

16. $n = 150$,   between 140 and 170 days

In Problems 17 to 22 use the table in the Appendix to find the area under the standard normal curve.

17. to the left of $z = .40$

18. to the left of $z = .60$

19. to the right of $z = 1.40$

20. to the right of $z = .90$

21. between $z = .40$ and $z = .90$

22. between $z = .60$ and $z = 1.10$

In Problems 23 to 26 a normal distribution has $\mu = 50$ and $\sigma = 10$. Find the given area under its normal curve.

23. to the left of 55   24. to the right of 40

25. between 55 and 65   26. between 45 and 60

In Problems 27 to 30 an IQ test is designed so that the mean score is 100 and the standard deviation is 20. If a person is selected at random, find the probability that the IQ score is within the given range.

27. less than 90   28. greater than 120

29. between 120 and 140   30. between 90 and 110

•• In Problems 31 to 36 use the table in the Appendix to find the area under the standard normal curve.

31. to the left of $z = .66$

32. to the right of $z = 1.73$

33. to the left of $z = -1.66$

34. to the right of $z = -.53$

35. between $z = .46$ and $z = .92$

36. between $z = .67$ and $z = 1.14$

In Problems 37 to 40 a normal distribution has $\mu = 60$ and $\sigma = 5$. Find the given area under its normal curve

37. to the left of 52   38. to the right of 49

39. between 51 and 62   40. between 54 and 61.5

41. The outer diameters of pipes produced by a certain machine are normally distributed with expected value 5 centimeters (cm) and standard deviation .1 cm. Find the probability that a randomly selected pipe will have a diameter less than 5.25 cm.

42. The time that an automobile battery first fails to hold to charge is normally distributed with an expected value of five years and a standard deviation of six months. Find the probability that a battery selected at random will first fail to hold a charge before six years.

43. An advertising agency determines that the annual household income of viewers of a certain television program is normally distributed with an expected value of $25,000 and a standard deviation of $5000. Find the probability that a randomly selected household watching the program has an income between $20,000 and $32,000.

44. Weights of a population of adult males are found to be normally distributed with the expected value of 170 pounds (lb) and a standard deviation of 30 lb. Find the probability that a randomly selected person from this population has a weight between 155 and 200 lb.

45. Heights of a population of basketball players are found to be normally distributed with an expected value of 80 inches (in.) and a standard deviation of 4 in. Find the percentage of these individuals whose heights are less than 72 in.

46. The combined SAT scores of students at a large university is normally distributed with an expected value of 1000 and a standard deviation of 200. Find the percentage of these students whose SAT scores are greater than 1260.

●●● 47. Use integration by parts to show that the expected value of an exponential random variable X with probability density function $f(x) = ae^{-ax}$ is $E(X) = \dfrac{1}{a}$.

48. Reliability refers to the probability that a system will not fail in a given time period. Reliability can be computed for many types of systems. Lemoine and Wenocur discussed the mortality rate of individuals who have had a heart attack. The reliability is the probability that the patient survives the heart attack.[*] The authors used an exponential model to define the reliability of the system. Find the probability the system will fail between time $t = 0$ and $t = 1$ for the probability density function $f(x) = .01e^{-.01x}$.

49. The major objective of workers' compensation insurance is to reduce legal and financial uncertainties that are caused by worker-related injuries. But many claimants with permanent injuries still contest the benefit award. In an article studying the probability that a disabled worker will hire a lawyer to contest the award, Borba and Appel used a sample of 1060 California workers with permanent injuries to identify the factors that influence such workers to seek legal help to get additional compensation.[†] In one part of the study they found that the probability that a certain type of worker with a permanent injury will contest the award is given by an exponential probability distribution $f(x) = ae^{-ax}$ where x is the time since the accident. Find the probability the worker will contest the award in the time period from $t = 0$ to $t = 2$ for $a = .03$.

50. A common method of obtaining information about an animal population is by marking or labeling some of its members and then recapturing them. This type of population estimation has been employed for centuries, but results have not been accurate when relatively small sample sizes are taken. In an article discussing methods of mark-recapture for small samples, Gazey and Staley applied their model to estimating the size of the Illinois mud turtle population in Illinois and Iowa.[‡] They took a sequence of several samples and estimated the population size. Then a computer simulation produced a larger sequence of samples. The assumption was that the limit of the sequence would give an accurate estimation. They found that the sequence produced a normal distribution with mean 10,000 and standard deviation 2000. Find the probability that population was between 8000 and 11,000.

[*]Austin J. Lemoine and Michael L. Wenocur, "On Shot-noise and Reliability Modeling," *Operations Research,* Vol. 34, 1986, pp. 320–323.

[†]Philip S. Borba and David Appel, "The Propensity of Permanently Disabled Workers to Hire Lawyers," *Industrial and Labor Relations Review,* Vol. 40, 1987, pp. 373–381.

[‡]W. J. Gazey and M. J. Staley, "Population Estimation from Mark-recapture Experiments Using a Sequential Bayes Algorithm," *Ecology,* Vol. 67, 1986, pp. 941–951.

CASE STUDY Reliability[*]

When the space shuttle *Challenger* burst into flames against the bright blue sky on January 28, 1986, more than the hopes and goals of a single space flight were shattered. The entire space program suffered a setback of major proportions: the next shuttle flight would not take place until the National

[*]Harold Ascher and Harry Feingold, *Repairable Systems Reliability,* New York, Marcel Dekker, 1984.

Aeronautics and Space Administration (NASA) endured over two years of agonizing turmoil. The agency went through upheaval in both its engineering development and its political makeup. All Americans were distressed at the accident, but the young especially were affected. They had never experienced a mishap in a U.S. space program.

What went wrong? The answer, or answers, are still being debated. Most answers are couched in probabilities. "Most likely this happened" or "Probably that happened." Certainly a critical component of *Challenger* failed, and the most likely candidates are the O-rings—rubber gaskets used to seal the joints of the booster rockets. According to an article in *Science* by R. Jeffrey Smith,* on the evening before the accident high-level engineers and scientists participated in a heated telephone conversation about the shuttle's probability of successful performance given the low temperatures predicted for the next morning. Florida was in the grip of an extraordinary cold snap, and temperatures at liftoff were expected to be in the low 20s—far below the coldest temperature for any previous launch.

Engineers from Morton Thiokol, Inc., the booster's manufacturer, emphatically warned NASA officials that several components had not undergone rigorous testing under low temperatures. In fact some test data indicated that the O-rings might not perform up to required specification at temperatures less than 50°. Company engineers recommended unanimously that the launch be postponed.

NASA's officials argued vigorously that Thiokol's data were not complete. In fact, since there were two gaskets at each joint, even if one failed, the other would succeed in containing the highly volatile fuel. Thiokol's engineers were appalled at such a suggestion, but because NASA was such an important client, they agreed to look at the data again. With the telephone button on "mute," the engineers checked the data once more and again unanimously agreed to recommend that the launch be delayed until the weather was better. But Thiokol's managers decided to give in and endorse the launch, although one senior Thiokol official refused to sign the official launch approval. A few hours later, 73 seconds into flight, *Challenger* exploded and was consumed in a fiery white debacle.

Reliability   The debate over the likely performance of the O-rings hinged on the *reliability* of the component. The probability of success of a component is called its reliability. Thiokol's engineers contended that the reliability of the O-ring component could not be adequately determined from the little experimentation at low temperatures. NASA officials contended that the data demonstrated that the reliability of the component was high enough, regardless of the weather.

The concept of reliability in business and in engineering has a firm mathematical foundation. The reliability R of a component is its probability of survival. It is related to the failure rate f, which is defined as the rate of failure

---

*R. Jeffrey Smith, "Shuttle Inquiry Focuses on Weather, Rubber Seals and Unheeded Advice," *Science*, Vol. 231, 1986, pp. 909–911.

of one component. Both are functions of time t, so we express them as $R(t)$ and $f(t)$, respectively. To determine $R(t)$ for a given component, we test a large number N of the components, and the reliability at any given time t is the ratio of the number of units $S(t)$ still operating, or surviving, divided by the constant N. Thus

$$R(t) = \frac{S(t)}{N}$$

We will use this relationship later in the following form:

$$S(t) = NR(t) \tag{1}$$

Let $F(t)$ be the number of components that have failed at time t. The number of components that have survived at time t is equal to the total number being tested minus the number that have failed: $S(t) = N - F(t)$. Therefore

$$R(t) = \frac{N - F(t)}{N} = \frac{N}{N} - \frac{F(t)}{N} = 1 - \frac{F(t)}{N} \tag{2}$$

Assume that a large number of units is tested so that the functions can be regarded as continuous functions. Take the derivative of equation 2.

$$R'(t) = -\frac{F'(t)}{N} \tag{3}$$

The function $F'(t)$ is the cumulative rate at which failures occur. The failure rate $f(t)$ is the rate at which a single component fails. They are related by the following reasoning. At time t the probability that another component will fail is the cumulative rate $F'(t)$ divided by the number that has survived until that time $S(t)$. Therefore

$$f(t) = \frac{F'(t)}{S(t)} \tag{4}$$

Substitute equation (1) into equation (4).

$$f(t) = \frac{F'(t)}{NR(t)} = \frac{1}{R(t)} \cdot \frac{F'(t)}{N} \tag{5}$$

Substitute equation (3) into equation (5).

$$f(t) = \frac{-R'(t)}{R(t)} \tag{6}$$

To solve for $R(t)$, integrate each side of equation (6) with respect to t.

$$\int f(t)\, dt = \int \frac{-R'(t)}{R(t)}\, dt = \ln R(t)$$

Express each side as a power of e, and use the fact that $e^{\ln R} = R$.

$$R(t) = -e^{\int f(t)\, dt} \tag{7}$$

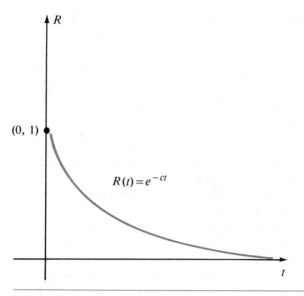

FIGURE 1

If $f(t)$ is a constant function, $f(t) = c$, then

$$R(t) = -e^{ct} \qquad (8)$$

This function is an exponential distribution function. It is characterized by $R(0) = 1$, meaning that the reliability of a component is 1 before it begins to operate. The function then decreases, indicating that the reliability of the component decreases as time increases. (See Figure 1.) The first example illustrates how this type of reliability function is used in practice.

EXAMPLE 1

In a study showing how to predict the rate of departure of individuals from a group of workers, by either promotion or cessation of employment, researchers applied their findings to a group of university faculty.* Here "failure" merely means leaving the group, so failure rate refers to the rate at which individuals leave the given professorial rank, by promotion, changing jobs, or retirement. The failure rate for the nontenured faculty at this institution was calculated to be $f(t) = .09$, where t is measured in years of employment.

Problem

The researchers used an exponential distribution. Find the reliability function for the nontenured faculty, and find the probability that an individual is still in the group of nontenured faculty after six years of employment.

*Adapted from Robert E. Stanford, "A Stochastic Model of Personnel Mobility: Application of Reliability Theory in Manpower System Analysis," *Operations Research,* Vol. 33, 1985, pp. 109–125.

SOLUTION Substitute $f(t) = .09$ into equation (8) to get

$$R(t) = e^{-.09t}$$

The probability that an individual is still in the group of nontenured faculty after six years of employment is $R(6)$.

$$R(6) = e^{-.09(6)} = e^{-.54} \simeq .58$$

System improvement A system consists of several components operating together. For example, the booster rocket of a space shuttle is a system consisting of many hundreds of components. An example of a smaller system, a subsystem of the booster rocket, is the pair of O-ring gaskets. Under normal circumstances one O-ring seal would suffice, but to improve the reliability, NASA requires two seals, the second backing up the first if it fails. The two components are said to operate *in parallel,* meaning that they do not operate together; rather, the second is called to operate only as a backup. In other words, the system still functions if one component fails. The system fails only if both components fail to function. A system with two components is said to operate *in series* if the system fails when one component fails. For example, consider a dam with two gates. If the gates operate in parallel, one backs up the other if it fails. If the first gate fails, water is still controlled by the second gate. If they operate in series, then if one gate fails, the water escapes. (See Figure 2.)

How does inserting a second component in parallel improve reliability? Suppose $R = .9$ for a given component and two such components are installed

FIGURE 2

Gates operate in parallel — 1st gate — Backup gate

Gates operate in series

in parallel. The system fails when both components fail. The probability that the first fails is .1, and the probability that the second fails is also .1. The two components operate independently, meaning that the reliability of one component does not affect the reliability of the other. To find the probability that the system fails, that is, when both components fail, we multiply the respective probabilities. This gives $(.1)(.1) = .01$. Thus the probability that both fail is .01. Therefore the system's reliability is $1 - .01 = .99$. To illustrate this idea intuitively, suppose that the component was tested 100 times. To state that $R = .9$ means that it will fail 10% of the time, or 10 times. The second component will fail 10% of the time, but it is called to operate only when the first fails. Thus it will fail only 1 out of the 10 tests in which it is called to operate. (During the other 90 tests it remains dormant.) So the system fails 1 out of the 100 tests, or 1% of the time.

EXAMPLE 2

Problem

If a component has reliability $R = .95$ and a system consists of two such components operating in parallel, find the reliability of the system.

SOLUTION The reliability of the system is

$$1 - (.05)(.05) = 1 - .0025 = .9975$$

Classes of failures Failures can be classified into three basic types: infant mortalities, random failures, and wear-out. Infant mortalities are early failures that are due to some fault in the component resulting from design or manufacturing. For many types of parts, including most electronic components, many parts are produced at once and potentially defective ones are identified by burn-in tests. Systems and subsystems are commonly "debugged" by operating them for a certain time period before delivery. When you purchase a personal computer the manufacturer recommends that you operate the unit continuously for several days. In fact, some dealers market computers by advertising that they have already subjected their machines to a 24- or 48-hour burn-in.

The random failure period of a component is the time period in which the probability of failure remains relatively constant. After the random failure period the probability of failure increases dramatically. For example, an automobile battery operates normally for about five years, depending on its quality. This means that for a batch of batteries, once the defective ones are replaced, relatively few will fail to hold a charge in the first five years, but after the fifth year the number that fail increases appreciably. The time after the random failure period is referred to as the wear-out failure period, and if a failure takes place during this time it is said to be due to wear-out. The wear-out life is defined as the point in time separating the random failure period and the wear-out period. That is, the period of time between the burn-in and the wear-out life is called the random failure period. Of course, the critical question for a manufacturer and a user of a component is to determine the wear-out life of a component. Often the component is replaced at that time rather than waiting for it to fail. This is why

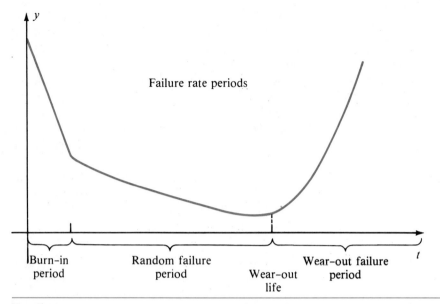

FIGURE 3

light bulbs are replaced regularly in large buildings. It is less costly to replace all bulbs that have reached their wear-out life at the same time than to replace them individually as they fail. The relationship of the three classes is shown pictorially in Figure 3. It is frequently referred to as the "bathtub curve."

Because of these three classes of failures, the reliability formula for a component is divided into three parts, each corresponding to one of the failure periods. The next example shows how different failure rates produce different reliability functions in separate failure periods.

EXAMPLE 3   The failure rates for an automobile battery with a wear-out life of five years are

1. $f(t) = .045$ during the random failure period from $t = 0$ to $t = 5$ years.
2. $f(t) = .44t - 1.055$ during the wear-out period from $t = 5$ to $t = 6$ years.

Problem

(a) Find the reliability function $R(t)$ for each period: (b) Compute $R(5)$ and $R(6)$.

SOLUTION (a) For the random failure period let $f(t) = .045$ in equation 4.

$$R(t) = e^{-\int .045 \, dt} = e^{-.045t} \tag{8}$$

For the wear-out failure period let $f(t) = .44t - 1.055$ in equation 4.

$$R(t) = e^{-\int(.44t - 1.055) \, dt} = e^{-.22t^2 + 1.055t} \tag{9}$$

(b) The value $R(5)$ should be the same for each formula. First substitute $t = 5$ into equation 8 and then into equation 9.

$$R(5) = e^{-(.045)5} = e^{-.225} \simeq .80$$
$$R(5) = e^{-(.22)25 + (1.055)5} = e^{-.225} \simeq .80$$

Since $t = 6$ is in the wear-out period, substitute it into the second formula.

$$R(6) = e^{-(.22)36 + (1.055)6} = e^{-1.59} \simeq .20$$

This means that after the burn-in period the reliability decreases from $R(0) = 1$ to $R(5) = .80$ in five years, while it decreases in the next year, the first year of the wear-out period, from $R(5) = .80$ to $R(6) = .20$. (Trade in your battery after five years.)

Trade-off analysis In any business endeavor the cost of doing business is a key factor. This is also true in reliability studies. A burden is on the manufacturer to provide reliability test data for its products, but it is often necessary for the customer to be assured that the products are of high enough quality. NASA has built reliability benchmarks into its subcontacting system for every component in the space shuttle. Ground rules for testing are explicit. For example, one rule states that the number of units to be tested should be inversely proportional to the square roots of the respective direct testing costs. This means that fewer high test–cost items are required to be tested. If the direct cost for item I is 16 times greater than that of item II, then $1/\sqrt{16} = 1/4$ the number of type I versus type II items must be tested. This shows there is a constant trade-off between reliability testing and testing cost.

The O-ring controversy NASA had been aware of problems with the O-ring gaskets for some time. During the second shuttle flight in 1981 one of the gaskets was partially eroded, and since then more than 30 others have shown wear due to compression on the inside of the booster. Test data show the O-ring has a reliability of .98. The use of a second O-ring in parallel with the first produces a system with two rubber gaskets with a reliability of $1 - (.02)(.02) = 1 - .0004 = .9996$. How could a system with such a high reliability fail?

In 1982 tests showed that rapid pressurization of the boosters widened the joints by .05 inch approximately 20 seconds after launch. This could be just enough of a gap to prevent the second O-ring from forming an effective seal. This would alter the environment of the second gasket, which could reduce its reliability appreciably. In March 1983 a memo warned that if the first gasket failed, the second might not prevent a catastrophic failure of the mission. Documents show that NASA studied the problem a year later after several seals were partially eroded and concluded that the problem was due to improper seal installation. Thus an O-ring burn-through would be an infant failure, which would be avoidable by improving installation procedures.

But the problem persisted. It was discovered that during the April 29, 1985 launch hot gases and flames from the boosters penetrated the putty around the O-ring system, blowing past the first gasket and eroding 80% of the backup. No burn-through occurred, but it prompted Thiokol executives to conduct extensive tests on gasket performance. Several solutions were proposed to improve the system. But no tests were conducted at temperatures lower than 50°. The lowest previous temperature on the launch pad was 51°F in January 1985. During that launch one O-ring was eroded and the backup charred. According to one engineer, "What little information we had in [the 45° to 50°] region indicated that the resiliency of those [gaskets] was going to hell in a handbasket, that they would essentially lie dormant at such temperatures, like a sleeping bear, with no power to think. The trend was pretty clear."

So why did the launch proceed? In all reliability practice, there is always a cost trade-off to be weighed. NASA was under tremendous pressure to demonstrate a reliable schedule of launches. One reason that Florida was chosen for a NASA launch site was the relative stability of the weather. Did the low temperature cause the mission failure? Or, was there faulty workmanship in the installation of the O-ring that was replaced at the last moment before *Challenger* was moved to the pad? Whatever the reason, and we may never be able to reach a certain conclusion, the accident has put NASA's reliability testing at the forefront of the news.

Case Study Exercises

In Problems 1 to 4 find the probability that an individual is still in the group of nontenured faculty referred to in Example 1 after the given number of years.

1. three years
2. four years
3. five years
4. seven years

In Problems 5 and 6 find the reliability of a system with two components operating in parallel, each with the given reliability.

5. $R = .92$
6. $R = .96$

In Problems 7 and 8 find the reliability of a system with three components operating in parallel, each with the given reliability.

7. $R = .90$
8. $R = .80$

In Problems 9 to 12 find $R(t)$ for the given failure rate function $f(t)$.

9. $f(t) = .05$
10. $f(t) = .08$
11. $f(t) = .05t - 1.1$
12. $f(t) = .08t - 1.5$

CHAPTER REVIEW

Key Terms

12–1   Discrete Probability Distributions

Relative Frequencies
Probability
Experiment
Sample Space
Event
Random Variable

Probability Distribution
Probability Histogram
Expected Value
Variance
Standard Deviation

Summary of Important Concepts

For discrete distributions

$$E(X) = a_1 p_1 + a_2 p_2 + \cdots + a_n p_n$$

$$\mathrm{Var}(X) = (a_1 - E)^2 p_1 + \cdots + (a_n - E)^2 p_n$$

$$\sigma = \sqrt{\mathrm{Var}(X)}$$

$$Pr(a < X < b) = \int_a^b f(x)\, dx$$

For continuous distributions

$$E(X) = \int_a^b x f(x)\, dx$$

$$V(X) = \int_a^b (x - E)^2 f(x)\, dx$$

$$\sigma = \sqrt{V(X)}$$

An exponential density function is

$$f(x) = ae^{-ax}$$

and

$$E(X) = \frac{1}{a} \quad \text{and} \quad \sigma = \frac{1}{a}$$

REVIEW PROBLEMS

In Problems 1 and 2 find the sample space of the experiment.

1. Roll a die and record whether the number on the top face is less than 2, equal to 2, or greater than 2.

2. A manufacturer chooses a sample of five items

from a shipment and records the number of defective items in the sample.

3. The experiment is to toss a die and record the number on the top face. Find $Pr(E)$ for the given event $E = \{2, 4, 6\}$.

4. For the given random variable X, find (a) the probability histogram, and (b) $E(X)$.

outcome, $X = a$	1	2	3	4
probability, p	.25	.12	.28	.35

5. Determine whether $f(x) = 3x(5 - x)$ is a probability density function on interval $[2, 4]$.

In Problems 6 and 7 find the probability that x is between a and b for the probability density function $f(x) = \dfrac{12 - 3x^2}{32}$ on the interval $[-2, 2]$.

6. $a = -1, \quad b = 1$

7. $a = -2, \quad b = 1$

8. Find a number a such that $af(x)$ is a probability density function on $[0, 1]$ for $f(x) = 4 - x^2$.

In Problems 9 and 10 find the indicated probability for the probability density function $f(x) = 1 - x^{-1/2}$ on interval $[1, 4]$.

9. $Pr(X \leq 2)$ 10. $Pr(1 \leq X \leq 2)$

11. Compute the expected value of the random variable X with density function $f(x) = 5/4 - x^2$ defined on the interval $[0, 1]$.

In Problems 12 to 14 compute the variance and the standard deviation of the random variables X with density function $f(x)$ defined on the given interval.

12. $f(x) = 4x^3 - 12x + 6, \quad [0, 1]$.

13. $f(x) = \dfrac{3}{2}e^{-x}, \quad [0, \ln 3]$

14. $f(x) = 2 - 2x^{-2}, \quad [1, 2]$

15. Find $Pr(1 \leq X \leq 2)$ for the exponential random variable X with the exponential density function $f(x) = 4e^{-4x}$.

16. It is assumed that the number of continuous days that a light bulb will burn until failure is an exponential random variable X. It is found that for a particular type of light bulb, the expected value of X is 200 days. Find the probability that a light bulb of this type will fail within the first 250 days.

17. Use the table in the Appendix to find the area under the standard normal curve to the left of $z = .46$.

18. A normal distribution has $\mu = 50$ and $\sigma = 25$. Find the area under its normal curve between 45 and 65.

In Problems 19 and 20 an IQ test is designed so that the mean score is 100 and the standard deviation is 20. If a person is selected at random, find the probability that the individual's IQ score is within the given range.

19. greater than 125   20. between 95 and 112

21. The combined SAT scores of students at a large university is normally distributed with an expected value of 1050 and a standard deviation of 200. Find the percentage of these students whose SAT scores are greater than 1260.

Appendix
Tables

TABLE 1   Areas Under the Standard Normal Curve

z	A(z)	z	A(z)	z	A(z)
−3.00	.0013	−2.48	.0066	−1.96	.0250
−2.99	.0014	−2.47	.0068	−1.95	.0256
−2.98	.0014	−2.46	.0069	−1.94	.0262
−2.97	.0015	−2.45	.0071	−1.93	.0268
−2.96	.0015	−2.44	.0073	−1.92	.0274
−2.95	.0016	−2.43	.0075	−1.91	.0281
−2.94	.0016	−2.42	.0078	−1.90	.0287
−2.93	.0017	−2.41	.0080	−1.89	.0294
−2.92	.0017	−2.40	.0082	−1.88	.0300
−2.91	.0018	−2.39	.0084	−1.87	.0307
−2.90	.0019	−2.38	.0087	−1.86	.0314
−2.89	.0019	−2.37	.0089	−1.85	.0322
−2.88	.0020	−2.36	.0091	−1.84	.0329
−2.87	.0020	−2.35	.0094	−1.83	.0336
−2.86	.0021	−2.34	.0096	−1.82	.0344
−2.85	.0022	−2.33	.0099	−1.81	.0351
−2.84	.0023	−2.32	.0102	−1.80	.0359
−2.83	.0023	−2.31	.0104	−1.79	.0367
−2.82	.0024	−2.30	.0107	−1.78	.0375
−2.81	.0025	−2.29	.0110	−1.77	.0384
−2.80	.0026	−2.28	.0113	−1.76	.0392
−2.79	.0026	−2.27	.0116	−1.75	.0401
−2.78	.0027	−2.26	.0119	−1.74	.0409
−2.77	.0028	−2.25	.0122	−1.73	.0418
−2.76	.0029	−2.24	.0125	−1.72	.0427
−2.75	.0030	−2.23	.0129	−1.71	.0436
−2.74	.0031	−2.22	.0132	−1.70	.0446
−2.73	.0032	−2.21	.0135	−1.69	.0455
−2.72	.0033	−2.20	.0139	−1.68	.0465
−2.71	.0034	−2.19	.0143	−1.67	.0475
−2.70	.0035	−2.18	.0146	−1.66	.0485
−2.69	.0036	−2.17	.0150	−1.65	.0495
−2.68	.0037	−2.16	.0154	−1.64	.0505
−2.67	.0038	−2.15	.0158	−1.63	.0515
−2.66	.0039	−2.14	.0162	−1.62	.0526
−2.65	.0040	−2.13	.0166	−1.61	.0537
−2.64	.0041	−2.12	.0170	−1.60	.0548
−2.63	.0043	−2.11	.0174	−1.59	.0559
−2.62	.0044	−2.10	.0179	−1.58	.0570
−2.61	.0045	−2.09	.0183	−1.57	.0582
−2.60	.0047	−2.08	.0188	−1.56	.0594
−2.59	.0048	−2.07	.0192	−1.55	.0606
−2.58	.0049	−2.06	.0197	−1.54	.0618
−2.57	.0051	−2.05	.0202	−1.53	.0630
−2.56	.0052	−2.04	.0207	−1.52	.0643
−2.55	.0054	−2.03	.0212	−1.51	.0655
−2.54	.0055	−2.02	.0217	−1.50	.0668
−2.53	.0057	−2.01	.0222	−1.49	.0681
−2.52	.0059	−2.00	.0227	−1.48	.0694
−2.51	.0060	−1.99	.0233	−1.47	.0708
−2.50	.0062	−1.98	.0238	−1.46	.0721
−2.49	.0064	−1.97	.0244	−1.45	.0735

TABLE 1 Areas Under the Standard Normal Curve (*continued*)

z	A(z)	z	A(z)	z	A(z)
−1.44	.0749	−0.92	.1788	−0.40	.3446
−1.43	.0764	−0.91	.1814	−0.39	.3483
−1.42	.0778	−0.90	.1841	−0.38	.3520
−1.41	.0793	−0.89	.1867	−0.37	.3557
−1.40	.0808	−0.88	.1894	−0.36	.3594
−1.39	.0823	−0.87	.1921	−0.35	.3632
−1.38	.0838	−0.86	.1949	−0.34	.3669
−1.37	.0853	−0.85	.1977	−0.33	.3707
−1.36	.0869	−0.84	.2004	−0.32	.3745
−1.35	.0885	−0.83	.2033	−0.31	.3783
−1.34	.0901	−0.82	.2061	−0.30	.3821
−1.33	.0918	−0.81	.2090	−0.29	.3859
−1.32	.0934	−0.80	.2119	−0.28	.3897
−1.31	.0951	−0.79	.2148	−0.27	.3936
−1.30	.0968	−0.78	.2177	−0.26	.3974
−1.29	.0985	−0.77	.2206	−0.25	.4013
−1.28	.1003	−0.76	.2236	−0.24	.4052
−1.27	.1020	−0.75	.2266	−0.23	.4090
−1.26	.1038	−0.74	.2296	−0.22	.4129
−1.25	.1056	−0.73	.2327	−0.21	.4168
−1.24	.1075	−0.72	.2358	−0.20	.4207
−1.23	.1093	−0.71	.2388	−0.19	.4246
−1.22	.1112	−0.70	.2420	−0.18	.4286
−1.21	.1131	−0.69	.2451	−0.17	.4325
−1.20	.1151	−0.68	.2482	−0.16	.4364
−1.19	.1170	−0.67	.2514	−0.15	.4404
−1.18	.1190	−0.66	.2546	−0.14	.4443
−1.17	.1210	−0.65	.2578	−0.13	.4483
−1.16	.1230	−0.64	.2611	−0.12	.4522
−1.15	.1251	−0.63	.2643	−0.11	.4562
−1.14	.1271	−0.62	.2676	−0.10	.4602
−1.13	.1292	−0.61	.2709	−0.09	.4641
−1.12	.1314	−0.60	.2742	−0.08	.4681
−1.11	.1335	−0.59	.2776	−0.07	.4721
−1.10	.1357	−0.58	.2810	−0.06	.4761
−1.09	.1379	−0.57	.2843	−0.05	.4801
−1.08	.1401	−0.56	.2877	−0.04	.4840
−1.07	.1423	−0.55	.2912	−0.03	.4880
−1.06	.1446	−0.54	.2946	−0.02	.4920
−1.05	.1469	−0.53	.2981	−0.01	.4960
−1.04	.1492	−0.52	.3015	0.00	.5000
−1.03	.1515	−0.51	.3050	0.01	.5040
−1.02	.1539	−0.50	.3085	0.02	.5080
−1.01	.1562	−0.49	.3121	0.03	.5120
−1.00	.1587	−0.48	.3156	0.04	.5160
−0.99	.1611	−0.47	.3192	0.05	.5199
−0.98	.1635	−0.46	.3228	0.06	.5239
−0.97	.1660	−0.45	.3264	0.07	.5279
−0.96	.1685	−0.44	.3300	0.08	.5319
−0.95	.1711	−0.43	.3336	0.09	.5359
−0.94	.1736	−0.42	.3372	0.10	.5398
−0.93	.1762	−0.41	.3409	0.11	.5438

TABLE 1   Areas Under the Standard Normal Curve (*continued*)

z	A(z)	z	A(z)	z	A(z)
0.12	.5478	0.64	.7389	1.16	.8770
0.13	.5517	0.65	.7422	1.17	.8790
0.14	.5557	0.66	.7454	1.18	.8810
0.15	.5596	0.67	.7484	1.19	.8830
0.16	.5636	0.68	.7518	1.20	.8849
0.17	.5675	0.69	.7549	1.21	.8869
0.18	.5714	0.70	.7580	1.22	.8888
0.19	.5754	0.71	.7612	1.23	.8907
0.20	.5793	0.72	.7642	1.24	.8925
0.21	.5832	0.73	.7673	1.25	.8944
0.22	.5871	0.74	.7704	1.26	.8962
0.23	.5910	0.75	.7734	1.27	.8980
0.24	.5948	0.76	.7764	1.28	.8997
0.25	.5987	0.77	.7794	1.29	.9015
0.26	.6026	0.78	.7823	1.30	.9032
0.27	.6064	0.79	.7852	1.31	.9049
0.28	6103	0.80	.7881	1.32	.9066
0.29	.6141	0.81	.7910	1.33	.9082
0.30	.6179	0.82	.7939	1.34	.9099
0.31	.6217	0.83	.7967	1.35	.9115
0.32	.6255	0.84	.7996	1.36	.9131
0.33	.6293	0.85	.8023	1.37	.9147
0.34	.6331	0.86	.8051	1.38	.9162
0.35	.6368	0.87	.8079	1.39	.9177
0.36	.6406	0.88	.8106	1.40	.9192
0.37	.6443	0.89	.8133	1.41	.9207
0.38	.6480	0.90	.8159	1.42	.9222
0.39	.6517	0.91	.8186	1.43	.9236
0.40	.6554	0.92	.8212	1.44	.9251
0.41	.6591	0.93	.8238	1.45	.9265
0.42	.6628	0.94	.8264	1.46	.9279
0.43	.6664	0.95	.8289	1.47	.9292
0.44	.6700	0.96	.8315	1.48	.9306
0.45	.6736	0.97	.8340	1.49	.9319
0.46	.6772	0.98	.8365	1.50	.9332
0.47	.6808	0.99	.8389	1.51	.9345
0.48	.6844	1.00	.8413	1.52	.9357
0.49	.6879	1.01	.8438	1.53	.9370
0.50	.6915	1.02	.8461	1.54	.9382
0.51	.6950	1.03	.8485	1.55	.9394
0.52	.6985	1.04	.8508	1.56	.9406
0.53	.7019	1.05	.8531	1.57	.9418
0.54	.7054	1.06	.8554	1.58	.9430
0.55	.7088	1.07	.8577	1.59	.9441
0.56	.7123	1.08	.8599	1.60	.9452
0.57	.7157	1.09	.8621	1.61	.9463
0.58	.7190	1.10	.8643	1.62	.9474
0.59	.7224	1.11	.8665	1.63	.9485
0.60	.7258	1.12	.8686	1.64	.9495
0.61	.7291	1.13	.8708	1.65	.9505
0.62	.7324	1.14	.8729	1.66	.9515
0.63	.7357	1.15	.8749	1.67	.9525

TABLE 1   Areas Under the Standard Normal Curve (*continued*)

z	A(z)	z	A(z)	z	A(z)
1.68	.9535	2.13	.9834	2.58	.9951
1.69	.9545	2.14	.9838	2.59	.9952
1.70	.9554	2.15	.9842	2.60	.9953
1.71	.9564	2.16	.9846	2.61	.9955
1.72	.9573	2.17	.9850	2.62	.9956
1.73	.9582	2.18	.9854	2.63	.9957
1.74	.9591	2.19	.9857	2.64	.9959
1.75	.9599	2.20	.9861	2.65	.9960
1.76	.9608	2.21	.9865	2.66	.9961
1.77	.9616	2.22	.9868	2.67	.9962
1.78	.9625	2.23	.9871	2.68	.9963
1.79	.9633	2.24	.9875	2.69	.9964
1.80	.9641	2.25	.9878	2.70	.9965
1.81	.9649	2.26	.9881	2.71	.9966
1.82	.9656	2.27	.9884	2.72	.9967
1.83	.9664	2.28	.9887	2.73	.9968
1.84	.9671	2.29	.9890	2.74	.9969
1.85	.9678	2.30	.9893	2.75	.9970
1.86	.9686	2.31	.9896	2.76	.9971
1.87	.9693	2.32	.9898	2.77	.9972
1.88	.9700	2.33	.9901	2.78	.9973
1.89	.9706	2.34	.9904	2.79	.9974
1.90	.9713	2.35	.9906	2.80	.9974
1.91	.9719	2.36	.9909	2.81	.9975
1.92	.9726	2.37	.9911	2.82	.9976
1.93	.9732	2.38	.9913	2.83	.9977
1.94	.9738	2.39	.9916	2.84	.9977
1.95	.9744	2.40	.9918	2.85	.9978
1.96	.9750	2.41	.9920	2.86	.9979
1.97	.9756	2.42	.9922	2.87	.9980
1.98	.9762	2.43	.9925	2.88	.9980
1.99	.9767	2.44	.9927	2.89	.9981
2.00	.9773	2.45	.9929	2.90	.9981
2.01	.9778	2.46	.9931	2.91	.9982
2.02	.9783	2.47	.9932	2.92	.9983
2.03	.9788	2.48	.9934	2.93	.9983
2.04	.9793	2.49	.9936	2.94	.9984
2.05	.9798	2.50	.9938	2.95	.9984
2.06	.9803	2.51	.9940	2.96	.9985
2.07	.9808	2.52	.9941	2.97	.9985
2.08	.9812	2.53	.9943	2.98	.9986
2.09	.9817	2.54	.9945	2.99	.9986
2.10	.9821	2.55	.9946	3.00	.9987
2.11	.9826	2.56	.9948		
2.12	.9830	2.57	.9949		

TABLE 2 Exponentials and Their Reciprocals

x	e^x	e^{-x}
10	22026.46580	0.00005
9	8103.08393	0.00012
8	2980.95799	0.00034
7	1096.63316	0.00091
6	403.42879	0.00248
5	148.41316	0.00674
4	54.59815	0.01832
3	20.08554	0.04979
2	7.38906	0.13534
1.9	6.68589	0.14957
1.8	6.04965	0.16530
1.7	5.47395	0.18268
1.6	4.95303	0.20190
1.5	4.48169	0.22313
1.4	4.05520	0.24660
1.3	3.66930	0.27253
1.2	3.32012	0.30119
1.1	3.00417	0.33287
1.0	2.71828	0.36788
0.9	2.45960	0.40657
0.8	2.22554	0.44933
0.7	2.01375	0.49659
0.6	1.82212	0.54881
0.5	1.64872	0.60653
0.4	1.49183	0.67032
0.3	1.34986	0.74082
0.2	1.22140	0.81873
0.1	1.10517	0.90484
0	1.00000	

TABLE 3   Natural Logarithms

x	ln x	x	ln x	x	ln x	x	ln x
0.05	− 2.99573	2	0.69315	25	3.21888	100	4.60517
0.10	− 2.30259	3	1.09861	30	3.40120	150	5.01064
0.15	− 1.89712	4	1.38629	35	3.55535	200	5.29832
0.20	− 1.60944	5	1.60944	40	3.68888	250	5.52146
0.25	− 1.38629	6	1.79176	45	3.80666	300	5.70378
0.30	− 1.20397	7	1.94591	50	3.91202	350	5.85793
0.35	− 1.04982	8	2.07944	55	4.00733	400	5.99146
0.40	− 0.91629	9	2.19722	60	4.09434	450	6.10925
0.45	− 0.79851	10	2.30259	65	4.17439	500	6.21461
0.50	− 0.69315	11	2.39790	70	4.24850	550	6.30992
0.55	− 0.59784	12	2.48491	75	4.31749	600	6.39693
0.60	− 0.51083	13	2.56495	80	4.38203	650	6.47697
0.65	0.43078	14	2.63906	85	4.44265	700	6.55108
0.70	0.35667	15	2.70805	90	4.49981	750	6.62007
0.75	0.28768	16	2.77259	95	4.55388	800	6.68461
0.80	0.22314	17	2.83321			850	6.74524
0.85	0.16252	18	2.89037			900	6.80239
0.90	0.10536	19	2.94444			950	6.85646
0.95	0.05129	20	2.99572			1000	6.90776
1.00	0.00000						

Appendix

TABLE 4   Basic Integration Formulas

1. $\displaystyle\int e^x \, dx = e^x$

2. $\displaystyle\int \frac{1}{x} \, dx = \ln |x|$

3. $\displaystyle\int a^x \, dx = \frac{a^x}{\ln a}, \ a > 0 \qquad (a \neq 1)$

4. $\displaystyle\int \frac{1}{\sqrt{a^2 + x^2}} \, dx = \ln |x + \sqrt{x^2 + a^2}|$

5. $\displaystyle\int \frac{1}{\sqrt{x^2 - a^2}} \, dx = \ln |x + \sqrt{x^2 - a^2}|$

6. $\displaystyle\int \frac{1}{a^2 - x^2} \, dx = \frac{1}{2a} \ln \left| \frac{a + x}{a - x} \right| \qquad (x^2 < a^2)$

7. $\displaystyle\int \frac{1}{x^2 - a^2} \, dx = -\frac{1}{2a} \ln \left| \frac{x + a}{x - a} \right| \qquad (a^2 < x^2)$

8. $\displaystyle\int \frac{1}{x\sqrt{a^2 - x^2}} \, dx = -\frac{1}{a} \ln \left| \frac{a + \sqrt{a^2 - x^2}}{x} \right| \qquad (0 < x < a)$

9. $\displaystyle\int x^n \ln x \, dx = x^{n+1} \left[\frac{\ln x}{n + 1} - \frac{1}{(n + 1)^2} \right]$

10. $\displaystyle\int \frac{1}{x\sqrt{a^2 + x^2}} \, dx = -\frac{1}{a} \ln \left| \frac{a + \sqrt{a^2 + x^2}}{x} \right|$

11. $\displaystyle\int \ln |x| \, dx = x(\ln |x| - 1)$

12. $\displaystyle\int \frac{x}{ax + b} \, dx = \frac{x}{a} - \frac{b}{a^2} \ln |ax + b|$

13. $\displaystyle\int \frac{x}{(ax + b)^2} \, dx = \frac{b}{a^2(ax + b)} + \frac{1}{a^2} \ln |ax + b|$

14. $\displaystyle\int \frac{1}{x(ax + b)} \, dx = \frac{1}{b} \ln \left| \frac{x}{ax + b} \right|$

15. $\displaystyle\int \frac{1}{x(ax + b)^2} \, dx = \frac{1}{b(ax + b)} + \frac{1}{b^2} \ln \left| \frac{x}{ax + b} \right|$

16. $\displaystyle\int \sqrt{x^2 \pm a^2} \, dx = \tfrac{1}{2}x \sqrt{x^2 \pm a^2} + \frac{a^2}{2} \ln |x + \sqrt{x^2 \pm a^2}|$

17. $\displaystyle\int x^n \ln x \, dx = x^{n+1} \left[\frac{\ln x}{n + 1} - \frac{1}{(n + 1)^2} \right] \qquad (n \neq -1)$

18. $\displaystyle\int x^n e^{ax} \, dx = \frac{1}{a} x^n e^{ax} - \frac{n}{a} \int x^{n-1} e^{ax} \, dx$

TABLE 4 Basic Integration Formulas (*continued*)

19. $\displaystyle \int \sin^2 u \ du = \frac{u}{2} - \frac{1}{4} \sin 2u + C$

20. $\displaystyle \int \sin^3 u \ du = -\frac{1}{3} \cos u(\sin^2 u + 2) + C$

21. $\displaystyle \int \cos^2 u \ du = \frac{1}{2}u + \frac{1}{4} \sin 2u + C$

22. $\displaystyle \int \cos^3 u \ du = \frac{1}{3} \sin u \ (\cos^2 u + 2) + C$

23. $\displaystyle \int \frac{du}{1 \pm \sin u} = \mp \tan \left(\frac{\pi}{4} \mp \frac{u}{2} \right) + C$

24. $\displaystyle \int \frac{du}{1 + \cos u} = \tan \left(\frac{u}{2} \right) + C$

25. $\displaystyle \int \frac{du}{1 - \cos u} = -\cot \left(\frac{u}{2} \right) + C$

TABLE 5 Trigonometric Functions

Deg	Rad	Sin	Cos	Tan	Deg	Rad	Sin	Cos	Tan
0	0.0000	0.0000	1.0000	0.0000	45	0.7854	0.7071	0.7071	1.0000
1	0.0175	0.0175	0.9998	0.0175	46	0.8029	0.7193	0.6947	1.0355
2	0.0349	0.0349	0.9994	0.0349	47	0.8203	0.7314	0.6820	1.0724
3	0.0524	0.0523	0.9986	0.0524	48	0.8378	0.7431	0.6691	1.1106
4	0.0698	0.0698	0.9976	0.0699	49	0.8552	0.7547	0.6561	1.1504
5	0.0873	0.0872	0.9962	0.0875	50	0.8727	0.7660	0.6428	1.1918
6	0.1047	0.1045	0.9945	0.1051	51	0.8901	0.7771	0.6293	1.2349
7	0.1222	0.1219	0.9925	0.1228	52	0.9076	0.7880	0.6157	1.2799
8	0.1396	0.1392	0.9903	0.1405	53	0.9250	0.7986	0.6018	1.3270
9	0.1571	0.1564	0.9877	0.1584	54	0.9425	0.8090	0.5878	1.3764
10	0.1745	0.1736	0.9848	0.1763	55	0.9599	0.8192	0.5736	1.4281
11	0.1920	0.1908	0.9816	0.1944	56	0.9774	0.8290	0.5592	1.4826
12	0.2094	0.2079	0.9781	0.2126	57	0.9948	0.8387	0.5446	1.5399
13	0.2269	0.2250	0.9744	0.2309	58	1.0123	0.8480	0.5299	1.6003
14	0.2443	0.2419	0.9703	0.2493	59	1.0297	0.8572	0.5150	1.6643
15	0.2618	0.2588	0.9659	0.2679	60	1.0472	0.8660	0.5000	1.7321
16	0.2793	0.2756	0.9613	0.2867	61	1.0647	0.8746	0.4848	1.8040
17	0.2967	0.2924	0.9563	0.3057	62	1.0821	0.8829	0.4695	1.8807
18	0.3142	0.3090	0.9511	0.3249	63	1.0996	0.8910	0.4540	1.9626
19	0.3316	0.3256	0.9455	0.3443	64	1.1170	0.8988	0.4384	2.0503
20	0.3491	0.3420	0.9397	0.3640	65	1.1345	0.9063	0.4226	2.1445
21	0.3665	0.3584	0.9336	0.3839	66	1.1519	0.9135	0.4067	2.2460
22	0.3840	0.3746	0.9272	0.4040	67	1.1694	0.9205	0.3907	2.3559
23	0.4014	0.3907	0.9205	0.4245	68	1.1868	0.9272	0.3746	2.4751
24	0.4189	0.4067	0.9135	0.4452	69	1.2043	0.9336	0.3584	2.6051
25	0.4363	0.4226	0.9063	0.4663	70	1.2217	0.9397	0.3420	2.7475
26	0.4538	0.4384	0.8988	0.4877	71	1.2392	0.9455	0.3256	2.9042
27	0.4712	0.4540	0.8910	0.5095	72	1.2566	0.9511	0.3090	3.0777
28	0.4887	0.4695	0.8829	0.5317	73	1.2741	0.9563	0.2924	3.2709
29	0.5061	0.4848	0.8746	0.5543	74	1.2915	0.9613	0.2756	3.4874
30	0.5236	0.5000	0.8660	0.5774	75	1.3090	0.9659	0.2588	3.7321
31	0.5411	0.5150	0.8572	0.6009	76	1.3265	0.9703	0.2419	4.0108
32	0.5585	0.5299	0.8480	0.6249	77	1.3439	0.9744	0.2250	4.3315
33	0.5760	0.5446	0.8387	0.6494	78	1.3614	0.9781	0.2079	4.7046
34	0.5934	0.5592	0.8290	0.6745	79	1.3788	0.9816	0.1908	5.1446
35	0.6109	0.5736	0.8192	0.7002	80	1.3963	0.9848	0.1736	5.6713
36	0.6283	0.5878	0.8090	0.7265	81	1.4137	0.9877	0.1564	6.3138
37	0.6458	0.6018	0.7986	0.7536	82	1.4312	0.9903	0.1392	7.1154
38	0.6632	0.6157	0.7880	0.7813	83	1.4486	0.9925	0.1219	8.1442
39	0.6807	0.6293	0.7771	0.8098	84	1.4661	0.9945	0.1045	9.5144
40	0.6981	0.6428	0.7660	0.8391	85	1.4835	0.9962	0.0872	11.4301
41	0.7156	0.6561	0.7547	0.8693	86	1.5010	0.9976	0.0698	14.3007
42	0.7330	0.6691	0.7431	0.9004	87	1.5184	0.9986	0.0523	19.0811
43	0.7505	0.6820	0.7314	0.9325	88	1.5359	0.9994	0.0349	28.6363
44	0.7679	0.6947	0.7193	0.9657	89	1.5533	0.9998	0.0175	57.2900
					90	1.5708	1.0000	0.0000	—

Answers to Odd-Numbered Exercises

CHAPTER 1

Exercise Set 1–1

1.

x	-2	-1	0	1	2	3
y	-7	-5	-3	-1	1	3
point	$(-2, -7)$	$(-1, -5)$	$(0, -3)$	$(1, -1)$	$(2, 1)$	$(3, 3)$

3.

x	-2	-1	0	1	2	3
y	6	5	4	3	2	1
point	$(-2, 6)$	$(-1, 5)$	$(0, 4)$	$(1, 3)$	$(2, 2)$	$(3, 1)$

5.

7.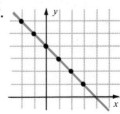

9. $2x + y = 6$ $y = 6 - 2x$

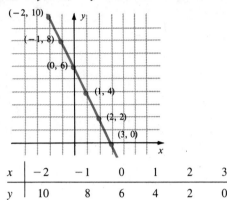

x	-2	-1	0	1	2	3
y	10	8	6	4	2	0

11. $x - y = 3$

x	y
0	-3
3	0

13. $3x - y = -1$

x	0	$-\frac{1}{3}$
y	1	0

15.

17.

19. (a) $58,100   **(b)** 10%

21.

23.

25.

27.

29.

31.

33.

35.

37.

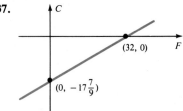

39. 37° Celsius
41. $50,000
43. **(a)**

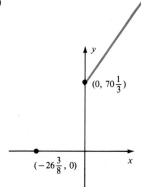

(b) 169 cm   **(c)** 43 cm
45. This is the break-even point.
47. **(a)** 72.4%   **(b)** 1978

Exercise Set 1–2

1.

3.

5. $m = 2$
7. $m = -\frac{3}{2}$
9. $m = 2$

11.

13.

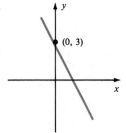

15. not parallel
17. **(a)** undefined   **(b)** $m = 0$
19. $y = -7x + 15$

21.

23.

25.

27.

29.

31.

33.

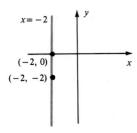

35. (a) positive **(b)** zero **(c)** negative
37. (a) $y = -\frac{3}{2}x + 6$ **(b)** $m = -\frac{3}{2}$
  (c) y-intercept $= 6$
39. (a) $y = -\frac{5}{2}x + 5$ **(b)** $y = -\frac{5}{2}x - 5$
41. $y = -4x + 1$
43. (a)

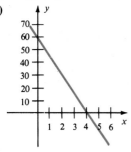

  (b) $15,000
  (c) The y-intercept represents the worth of the work
  station when it was purchased.
45. no, for example, $xy = 1$

47. (a) $y = .5x + 7.5$
  (b)

  (c) 43.5

Exercise Set 1–3
1. $1, -3, -4$ **3.** $1, 3, \frac{9}{4}$
5. (a) $2a - 3$ **(b)** $2x + 3$ **(c)** $2x + 2h - 3$
  (d) 2
7. (a) $-a^2 + a - 3$ **(b)** $-x^2 - 5x - 3$
  (c) $-x^2 - 2xh - h^2 + x + h + 3$
  (d) $-2x + 1 - h$
9. $69{,}888{,}000$ **11.** $x \geq -1$
13. $x \leq \frac{1}{2}$

15.

x	-2	-1	0	1	2
$f(x)$	-5	-6	-5	-2	3

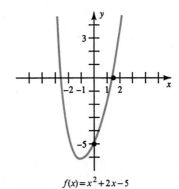

$f(x) = x^2 + 2x - 5$

17.

x	−2	−1	0	1	2
f(x)	18	4	−4	−6	−2

$f(x) = 3x^2 - 5x - 4$

19. (a) not a function   **(b)** function   **(c)** function
21. $2h$ **23.** $-2xh - h^2 + h$
25. -1 **27.** $4x + 2h - 4$
29. $x \neq 2$ **31.** $-1 \leq x \leq 1$
33. $x \geq 1$ or $x \leq -1$
35. $x \neq 0$ and $x \neq 2$
37. $x \geq 3$ but $x \neq -4$
39. (a) $\{-2, 2\}$ **(b)** none   **(c)** $\{-0.5, 0.5\}$
41. function

$f(x) = \sqrt{x}$

43. (a) yes, $d = 6t^{2/3}$ or $d = 6\sqrt[3]{t^2}$ **(b)** 49.92
miles   **(c)** 0.19 hours   or   11 minutes, 33 seconds
45. yes
47. (a) $f(x) = 0.3x - 564$ **(b)** 33   **(c)** 1977

Exercise Set 1–4

1.

x	0	1	2	3
y	0	5	4	9

$f(x) = 2x^3 - 9x^2 + 12x$

3.

x	0	1	2	−1
y	1	2	1	10

$f(x) = x^4 - 4x^3 + 4x^2 + 1$

5. 3   **7.** the line $x = 5$
9. the lines $x = 3$ and $x = -3$
11.

$f(x) = \dfrac{x}{x-5}$

13

$$f(x) = \begin{cases} 3x+2 & x \le 0 \\ x^2+2 & x > 0 \end{cases}$$

15.

$$f(x) = \begin{cases} x^2-1 & x < 0 \\ 2x-1 & x \ge 0 \end{cases}$$

17. $f(0) = 2$ **19.** 30%

21. (a) $x > 0$
  (b) Four items should be produced to minimize cost.

(4, 8)

23.

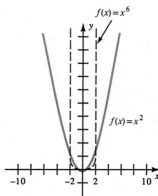

$f(x) = x^6$

$f(x) = x^2$

25. 2
27. the lines $x = 3$ and $x = -3$

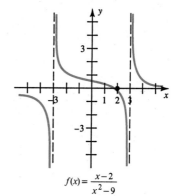

$$f(x) = \frac{x-2}{x^2-9}$$

29. the line $x = 0$

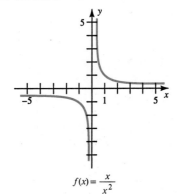

$$f(x) = \frac{x}{x^2}$$

31. the lines $x = 0$, $x = 1$, and $x = -1$

$$f(x) = \frac{1}{x(x^2 - 1)}$$

33. the lines $x = -\frac{3}{2}$ and $x = 1$

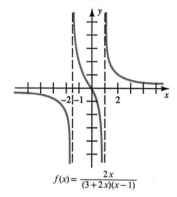

$$f(x) = \frac{2x}{(3 + 2x)(x - 1)}$$

35.

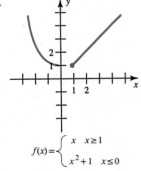

$$f(x) = \begin{cases} x & x \geq 1 \\ x^2 + 1 & x \leq 0 \end{cases}$$

37.

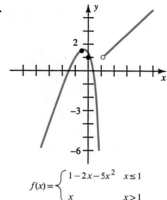

$$f(x) = \begin{cases} 1 - 2x - 5x^2 & x \leq 1 \\ x & x > 1 \end{cases}$$

39.

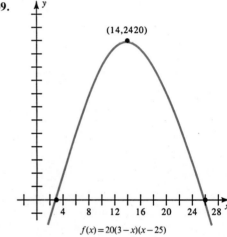

(14, 2420)

$$f(x) = 20(3 - x)(x - 25)$$

Maximum profit occurs when 14 items are produced.

41.

$$f(x) = \text{INT}(x) \text{ for } x \geq 0$$

43. 40%

45. (a)

$$f(x) = \begin{cases} .15x & \text{if } 0 \leq x \leq 29{,}300 \\ .27(x - 29{,}300) + 4350 & \text{if } x > 29{,}300 \end{cases}$$

(b)

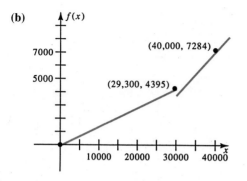

(40,000, 7284)

7000

5000

(29,300, 4395)

10000 20000 30000 40000

x

(c) 4395 **(d)** 4350.27

(e) Tax on income of $29,301 is $44.73 less than tax on income of $29,300.

47. (a) 16 minutes and 5 seconds

  (b) 16 hours, 16 minutes and 1 second

Exercise Set 1–5

1. (a) $(f + g)(x) = \dfrac{x + 2}{x - 1} + \dfrac{x}{x + 2}$

$= \dfrac{2x^2 + 3x + 4}{(x - 1)(x + 2)}$

  (b) $(f \cdot g)(x) = \dfrac{(x + 2)}{(x - 1)} \cdot \dfrac{x}{(x + 2)} = \dfrac{x}{x - 1}$

  (c) $(f/g)(x) = \dfrac{(x + 2)^2}{x(x - 1)}$

3. (a) $(f + g)(x) = \dfrac{x}{x + 1} + \dfrac{1}{x} = \dfrac{x^2 + x + 1}{x(x + 1)}$

  (b) $(f \cdot g)(x) = \dfrac{1}{x + 1}$

  (c) $(f/g)(x) = \dfrac{x^2}{x + 1}$

5. (a) all real numbers except $x = 1$ and $x = -2$

  (b) all real numbers except $x = -1$ and $x = 0$

7. (a) all real numbers except $x = 1$, $x = -2$, and $x = 0$

  (b) all real numbers except $x = -1$ and $x = 0$

9. (a) $\frac{9}{2}$ **(b)** 2 **(c)** 8

11. (a) $\frac{3}{2}$ **(b)** -1 **(c)** -4

13. (a) $(f \circ g)(x) = \sqrt{\dfrac{x}{1 - x}}$

  (b) $(g \circ f)(x) = \dfrac{\sqrt{x}}{1 - \sqrt{x}}$

15. (a) $(f \circ g)(x) = \dfrac{2(x - 2)}{(x + 2)}$

  (b) $(g \circ f)(x) = \dfrac{x - 1}{x + 1}$

17. $g(x) = \sqrt{x}$, $k(x) = x^3 + 2x - 5$, $f(x) = (g \circ k)(x)$

19. $g(x) = \dfrac{1}{x^2} + x^3$, $k(x) = x - 2$, $f(x) = (g \circ k)(x)$

21. $Q(t) = -0.005t^4 + 0.16t^3 - 3.28t^2 + 32t + 20$

23. (a) $(f + g + h)(x) = \dfrac{2x^3 - x - 10}{(x - 1)(x + 2)(x - 2)}$

  (b) $(f \cdot g \cdot h)(x) = \dfrac{x}{(x - 1)(x - 2)}$

25. (a) $(f + g + h)(x) = \dfrac{2x^2 - x + 3}{(x + 3)(x - 3)}$

  (b) $(f \cdot g \cdot h)(x) = \dfrac{x(x - 2)}{(x + 3)^2(x - 3)}$

27. 0

29. $\sqrt{x^2} = |x|$

31. $\dfrac{x + 1}{2x + 1}$

33. no

$(f \circ g)(1) = \frac{2}{3}$

$(g \circ f)(1) = 3$

35. $g(x) = \sqrt[3]{x^2}$, $k(x) = x - 1$, $f(x) = (g \circ k)(x)$

37. $\dfrac{2a^2 - 3a + 3}{a^2 - 1}$

39. $\dfrac{1}{a + 1} + \sqrt{a}$

41. no

$(f \cdot g)(x) = \dfrac{x}{x + 1}$

$(f \circ g)(x) = \dfrac{3x + 4}{-2}$

43. (a) all $x \geq 0$ except $x = 200$ **(b)** $0 \leq x \leq 100$

  (c) 10 hours **(d)** 40 hours **(e)** 66.7%

45. (a)

(b) $7 \le x \le 25$

47. (a) $\dfrac{1}{\sqrt{x + 1}}$ **(b)** $\dfrac{1}{\sqrt{x + 1}}$ **(c)** $\dfrac{1}{\sqrt{x + 1}}$

49. PEX

Exercise Set 1–6

1. $x = -4, \quad x = 2$

3. $x = -3$

5. no solution

7. domain: real numbers except $x = \frac{3}{2}, -1$
   x-intercept: $x = 2$
9. domain: real numbers except $x = \frac{3}{4}, -2$
   x-intercept: $x = -2$
11. $-2, \frac{3}{2}, 2$
13. $x = 2, \qquad x = -2$
15. $x = 4, \qquad x = -3$
17. $(1, 9)$ and $(8, 16)$
19. $(2, 12)$ and $(7, 7)$
21. $x = 2$

23. $x = -1$

25. domain: all real numbers except $x = 0$ and $x = 2$
   x-intercept: none
27. domain: all real numbers except $x = 0$ and $x = 2$
   x-intercept: none
29. $(-1, 1)$ and $(2, 4)$
31. $(0, 0)$ and $(1, 1)$
33. $(3, 13)$ and $(-3, 13)$
35. 2 seconds and 5 seconds after the object is thrown
37. 3.5 seconds after the object is thrown
39. 8 units and 24 units
41. $x = 1$
43. (a) Americans have been eating more poultry than pork since 1982.
   (b) Americans have been eating more poultry than beef since 1987.

Case Study Exercises
1. 0.58252679   **3.** 0.5826
5. no   **7.** 25.16 pounds
9.

11. (a) 1052.05   (b) 1035.30
13. Sheridan Shuttles

Review Exercises
1. $-9, -10, -9, -6, 6$
3. $2xh + h^2 - 4h$
5. all real numbers except $x = 1$
7. yes
9. the lines $x = 2$, $x = 3$, and $x = -3$.
11. (a) $\dfrac{2x^2 + 3x + 4}{(x - 1)(x + 2)}$ (b) $\dfrac{x}{x - 1}$ (c) $\dfrac{(x + 2)^2}{x(x - 1)}$
13. (a) $\frac{9}{2}$ (b) 2   (c) 8
15. $g(x) = x - \sqrt{x}$, $k(x) = x + 2$,
   $f(x) = (g \circ k)(x)$
17. no x-intercepts
19. $x = 1, 2, -2$
21. domain: all real numbers except -2 and 1
   x-intercept: $x = -1$

CHAPTER 2

Exercise Set 2–1
1. (a) 2   (b) 2
3. (a) -4 (b) -4
5. (a) 1   (b) 0.01
7. (a) 1   (b) 0.001
9. 2   **11.** 8
13. $y = 2x - 6$ **15.** $y = 8x + 18$
17. $-4x$ **19.** $2x$
21. (a) 4.1   (b) 4.01   (c) 4.001
23. (a) 8.2   (b) 8.02   (c) 8.002
25. (a) 0.24845   (b) 0.24984   (c) 0.24998
27. 4   **29.** $y = 4x - 1$
31. $4x - 4$ **33.** $3x^2 + 1$
35. $-32t + 192$ **37.** $-32t + 128$
39. (a) 26.025   (b) 36.85
41. (a) 2.05%   (b) -0.05%
43. 20%

45. (a) positive   (b) Reagan
47. $2ax + b$

Exercise Set 2–2
1. 3   **3.** 6
5. no limit   **7.** no limit
9. no limit   **11.** 3
13. 0   **15.** 0
17. 2   **19.** no limit
21. no limit   **23.** $\frac{1}{2}$
25. $\frac{1}{6}$ **27.** 0
29. -4 **31.** $\frac{1}{3}$
33. 3   **35.** 0
37. no limit   **39.** 0
41. 4   **43.** $\frac{1}{4}$
45. 18   **47.** 8
49. 0   **51.** 2

53. 2

55. 3

57. 0

59. no limit

61. 90°

63. 200 miles

65.

x	.05	.06	.07	.08
$f(x)$	.73	.86	.93	.96

$$\lim_{x \to .08} f(x) = .96$$

67. (a) 1 (b) 1

Exercise Set 2–3

1. $2x$

3. $\dfrac{1}{\sqrt{2x}}$

5. $-\dfrac{1}{2\sqrt{x}}$

7. $-\dfrac{1}{x^2}$

9. $-\dfrac{1}{x^2}$

11. $y = -2x + 6$

13. $y = -x - 6$

15. (a) 6 seconds (b) 576 feet

17. (a) $\dfrac{35}{16} = 2.1875$ seconds (b) $\dfrac{1225}{16} = 76.5625$ feet

19. 7

21. 6

23. $x = 0$

25. $x = 0$ and $x = 3$

27. $-\dfrac{1}{2x\sqrt{x}}$

29. $-\dfrac{2}{x^3}$

31. $y = -\frac{1}{4}x + 1$

33. $y = \frac{1}{4}x + \frac{3}{4}$

35. $y = 3x - 5$

37. (a) 4 seconds (b) 320 feet

39. 12 seconds

41. yes

43. no

45. Continuous everywhere
Differentiable everywhere except $x = 0$

47. Continuous everywhere except $x = 1$
Differentiable everywhere except $x = 1$

49. $2ax + b$

51. (a) 15% (b) 28% (c) 33% (d) 28%

53. (a) 6,500 rpm (b) 450 bhp

55. (a) the derivative (b) March and June (c) November

Review Exercises

1. (a) 3 (b) 2.1

3. $2x - 1$

5. 2

7. $\frac{4}{3}$

9. 0

11. $2x - 3$

13. $y = 5x - 23$

CHAPTER 3

Exercise Set 3–1

1. $3x^2$

3. $8x^3$

5. $-15x^{-4} + 5$

7. $\frac{2}{3}x^{-2/3} + \frac{3}{2}x^{-1/2} - 4$

9. $-\frac{2}{3}x^{-4/3} - \frac{7}{4}x^{-5/4}$

11. $0.3x^{-0.7} + 1.2x^{-0.6}$

13. $-\dfrac{3}{x^4} + 12x^3$

15. $-\dfrac{1.5}{x^{2.5}} - \dfrac{1}{x^{5/4}} + 9.6x^{1.4}$

17. $7x^{2.5} - \dfrac{5.6}{x^{2.4}} - \dfrac{5.2}{x^{2.3}}$

19. $\dfrac{0.125}{x^{0.5}} - \dfrac{0.192}{x^{1.4}}$

21. $\dfrac{1}{\sqrt{x}} - 2$

23. $\dfrac{5}{2\sqrt{x}} + \dfrac{1}{\sqrt[3]{x}}$

25. $-\dfrac{9}{\sqrt{3}}x^{-3/2} - \dfrac{2}{3\sqrt[3]{4}}x^{-4/3}$

27. $y = 4x - 1$

29. $y = 2x + 1$

31. (a) $v = -32t + 32$ (b) $t = 1$ second
(c) 32 feet

33. 12

35. 10

41. $-0.048x + 0.575$

Exercise Set 3–2

1. $4x^3(x + 5) + x^4 = 5x^4 + 20x^3$

3. $6x^2(3x - 7) + (3)(2x^3) = 24x^3 - 42x^2$

5. $(-3x^{-4})(3x^2 - 17) + (6x)(x^{-3} + 5)$
$$= -3x^{-2} + 51x^{-4} + 30x$$

7. $(\frac{1}{3}x^{-2/3} + 1)(3x^{1/2} - 4x) + (\frac{3}{2}x^{-1/2} - 4)(x^{1/3} + x)$

9. $(-\frac{1}{3}x^{-4/3} - \frac{7}{4}x^{-5/4})(4x^2 - 5) + (8x)(x^{-1/3} + 7x^{-1/4})$

11. $4(2x + 1)$

13. $6x^2(x^3 + 1)$

15. $\dfrac{1}{(2x + 1)^2}$

17. $\dfrac{x^3 - 10}{3x^3}$

19. $y = -3x + 12$

21. $y = 2$

23. $L = \dfrac{0.5x^2 + 3}{0.5x^2 + 5}$; $L' = \dfrac{2x}{(0.5x^2 + 5)^2}$

25. $2x(x^3 + 15)(x^2 - 17) + 3x^4(x^2 - 17)$
$$+ 2x^3(x^3 + 15) = 7x^6 - 85x^4 + 60x^3 - 510x$$

27. $-x^{-2}(x^{1/3} + 2)(x^2 - 3)$
$$+ \frac{1}{3}x^{-5/3}(x^2 - 3) + 2(x^{1/3} + 2)$$

29. $\dfrac{x^4 - 21x^2 - 70x}{(x^2 - 7)^2}$

31. $\dfrac{-x^4 - 10x^2 + 3}{x^2(x^2 - 1)^2}$

33. $C'(x) = 23 - 138x^2$

35. $-\dfrac{100}{x^2} - 0.4$

39. 80

Exercise Set 3–3

1. $15(5x + 2)^2$
3. $9x^2(x^3 + 4)^2$
5. $3(3x^2 + 6x)(x^3 + 3x^2)^2 = 9x(x + 2)(x^3 + 3x^2)^2$
7. $\frac{1}{2}x^{1/2}(x^{3/2} + x)^{-2/3}$
9. $-6(\frac{1}{2}x^{-1/2} - 2x^{-2})(x^{1/2} + 2x^{-1})^{-7}$
11. $(5x^2 + 2)^2 + 20x^2(5x^2 + 2)$
$$= (5x^2 + 2)(25x^2 + 2)$$
13. $2x(x^3 + 2x)^2 + 2x^2(x^3 + 2x)(3x^2 + 2)$
$$= 8x^3(x^2 + 1)(x^2 + 2)$$
15. $\dfrac{2x(x^3 - x)^3 - 3x^2(x^3 - x)^2(3x^2 - 1)}{(x^3 - x)^6}$
$$= \dfrac{x^2(-7x^2 + 1)}{(x^2 - x)^4}$$

17. $y = 8$
19. $25y + 7x = 12$
21. $0.16[5 - 0.2(t - 6)^2](t - 6)$
23. $L = \dfrac{x^2 + 5}{x^2 + 10}$; $L' = \dfrac{10x}{(x^2 + 10)^2}$
25. $2x^3(x^3 + 2x^2)^3(7x + 10)$
27. $x^{-1/2}(x^{3/2} + 4)^{-3} - 9x(x^{3/2} + 4)^{-4}$
29. $x(x^2 + 1)(x^3 + 1)^2(13x^3 + 9x + 4)$
31. $\dfrac{x(2x^3 + x + 3)}{(x^2 + 1)^{1/2}(x^3 + 1)^{2/3}}$

33. $\dfrac{x^{-5/2}(x^4 + 2 + 24x^{-5/2})}{(x^2 + 2x^{-2})^4}$

35. $\dfrac{-x^2 + 1}{(x^2 + 1)^2}$

37. $\dfrac{-x^2 + 4x + 5}{(x^2 + 5)^2}$

39. $-1000(200 - x)$

43. $p(x) = 0.007(16 - 0.41x^{-1/2})^2$
$$- 0.33(16 - 0.41x^{-1/2}) + 100$$
$$p'(x) = -0.02173x^{-3/2} - 0.00118x^{-2}$$

Exercise Set 3–4

1. 2
3. $24x^2$
5. $\dfrac{8}{x^3} = 8x^{-3}$
7. $\dfrac{12}{x^5} = 12x^{-5}$
9. $12x^2 - 2$
11. $\frac{8}{9}x^{-7/3} + \frac{35}{16}x^{-9/4}$
13. $0.21x^{-1.7} - 0.72x^{-1.6}$
15. 26
17. $\frac{3}{8}$
19. $2 + 12t$
21. $-\frac{1}{4}g^{-3/2}$
23. (a) $t = 1$ (b) $a(t)$ is never 0
25. (a) $t = \frac{1}{3}$ and $t = 1$ (b) $t = \frac{2}{3}$
27. -64
29. $1.8 - 0.018x$
31. $0.96x - 195.2$
33. $12x^2 + 4$
35. $\dfrac{24x^2 - 4}{(2x^2 + 1)^3}$
37. $\dfrac{1}{(x^2 - 1)^{3/2}}$
39. 0, 2
41. $1, -1$
43. $f^{(3)}(x) = 24x$; $f^{(4)}(x) = 24$
45. $f^{(4)}(x) = 48x^{-5} + 120x^{-6}$;
$$f^{(3)}(x) = -12x^{-4} - 24x^{-5}$$
47. $f^{(n)}(x) = (-1)^n(n + 1)! \, x^{-(n+2)}$
49. 0.0625

Case Study Exercises

1. 81.11
3. 84.17

Review Exercises

1. $8x + 3x^2$
3. $-20x^{-6} - \frac{1}{3}x^{-2/3}$
5. $5x + 2y = 5$
7. $\dfrac{x^4 + 6x^2}{(x^2 + 2)^2}$
9. $\dfrac{\frac{1}{3}(x^{-1/3})(x^{4/5} + 2) - (\frac{4}{5}x^{-1/5})(x^{1/3})}{(x^{4/5} + 2)^2}$
11. $4(2x^3 + x)^3(6x^2 + 1)$

13. $6x + 9(3x^2 + 4)(x^3 + 4x)^2$

15. $x(x^2 + 1)(x^3 + 1)^2(13x^3 + 9x + 4)$

17. $y = 24x - 8$

19. $\dfrac{6x(7x^3 - 8)}{(x^3 + 4)^4}$

21. $\dfrac{(-1)^n(n + 2)!\, x^{-(n+3)}}{2}$

CHAPTER 4

Exercise Set 4–1

1. Increasing on $0 < x < \infty$
Decreasing on $-\infty < x < 0$
Horizontal tangent at $x = 0$

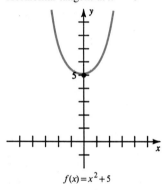

$$f(x) = x^2 + 5$$

3. Increasing on $-\infty < x < 0$
Decreasing on $0 < x < \infty$
Horizontal tangent at $x = 0$

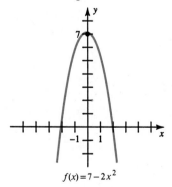

$$f(x) = 7 - 2x^2$$

5. Increasing on $0 < x < \infty$
Decreasing on $-\infty < x < 0$
Horizontal tangent at $x = 0$

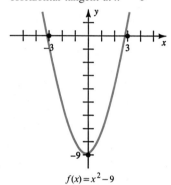

$$f(x) = x^2 - 9$$

7. Increasing on $2 < x < \infty$
Decreasing on $-\infty < x < 2$
Horizontal tangent at $x = 2$

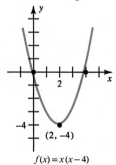

$$f(x) = x(x - 4)$$

9. Increasing on $-2 < x < \infty$
Decreasing on $-\infty < x < -2$
Horizontal tangent at $x = -2$

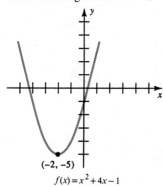

$(-2, -5)$
$f(x) = x^2 + 4x - 1$

11. Increasing on $-\infty < -3$ and $1 < x < \infty$
Decreasing on $-3 < x < 1$
Horizontal tangents at $x = -3$ and $x = 1$

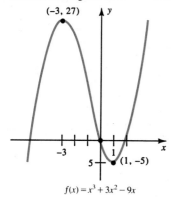

$(-3, 27)$
1
5
$(1, -5)$
-3
$f(x) = x^3 + 3x^2 - 9x$

13. Increasing on $-\infty < x < -1$ and $3 < x < \infty$
Decreasing on $-1 < x < 3$
Horizontal tangents at $x = -1$ and $x = 3$

$(-1, 5)$
$(3, -27)$
$f(x) = x^3 - 3x^2 - 9x$

15. Increasing on $-1 < x < 3$
Decreasing on $-\infty < x < -1$ and $3 < x < \infty$
Horizontal tangents at $x = -1$ and $x = 3$

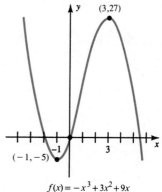

$(3, 27)$
-1
3
$(-1, -5)$
$f(x) = -x^3 + 3x^2 + 9x$

17. Increasing on $-\infty < x < -3$ and $3 < x < \infty$
Decreasing on $-3 < x < 0$ and $0 < x < 3$
Horizontal tangents at $x = -3$ and $x = 3$

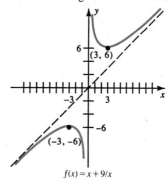

6
$(3, 6)$
-3
3
-6
$(-3, -6)$
$f(x) = x + 9/x$

19. Increasing on $-\infty < x < -2\sqrt{3}$
and $2\sqrt{3} < x < \infty$
Decreasing on $-2\sqrt{3} < x < 0$ and $0 < x < 2\sqrt{3}$
Horizontal tangents at $x = -2\sqrt{3}$
and $x = 2\sqrt{3}$

6
$(2\sqrt{3}, 4\sqrt{3})$
-3
3
$(-2\sqrt{3}, -4\sqrt{3})$
-6
$f(x) = \dfrac{x^2 + 12}{x}$

21. Increasing on $-\infty < x < -\dfrac{2\sqrt{3}}{3}$

and $\dfrac{2\sqrt{3}}{3} < x < \infty$

Decreasing on $-\dfrac{2\sqrt{3}}{3} < x < \dfrac{2\sqrt{3}}{3}$

Horizontal tangents at $x = -\dfrac{2\sqrt{3}}{3}$ and $x = \dfrac{2\sqrt{3}}{3}$

$\left(-\dfrac{2\sqrt{3}}{3}, \dfrac{16\sqrt{3}}{9}\right)$ $\left(\dfrac{2\sqrt{3}}{3}, -\dfrac{16\sqrt{3}}{9}\right)$

$f(x) = x(x^2 - 4)$

23. Increasing on $-\infty < x < 0$ and $\frac{4}{3} < x < \infty$
Decreasing on $0 < x < \frac{4}{3}$
Horizontal tangents at $x = 0$ and $x = \frac{4}{3}$

$\left(\frac{4}{3}, -\frac{32}{27}\right)$

$f(x) = x^2(x - 2)$

25. Increasing on $-\infty < x < -1$ and $-1 < x < \infty$
Never decreasing
No horizontal tangents

$f(x) = \dfrac{x}{x+1}$

27. Increasing on $-\infty < x < -1$ and $-1 < x < \infty$
Never decreasing
No horizontal tangents

$f(x) = \dfrac{x-1}{x+1}$

29. Increasing on $0 < x < 6$ and $8 < x < 10$
Decreasing on $6 < x < 8$

35. Increasing on $-\infty < x < 0$ and $4 < x < \infty$
Decreasing on $0 < x < 2$ and $2 < 4$
Horizontal tangents at $x = 0$ and $x = 4$

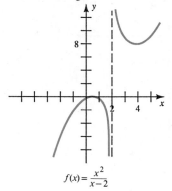

$f(x) = \dfrac{x^2}{x-2}$

37. Increasing on $-\sqrt{2} < x < 0$ and $\sqrt{2} < x < \infty$
Decreasing on $-\infty < x < -\sqrt{2}$ and $0 < x < \sqrt{2}$
Horizontal tangents at $x = -\sqrt{2}$, $x = 0$, and $x = \sqrt{2}$

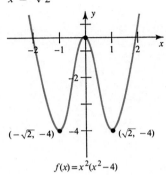

$$f(x) = x^2(x^2 - 4)$$

39. Increasing on $-\dfrac{\sqrt{10}}{2} < x < 0$

and $\dfrac{\sqrt{10}}{2} < x < \infty$

Decreasing on $-\infty < x < -\dfrac{\sqrt{10}}{2}$

and $0 < x < \dfrac{\sqrt{10}}{2}$

Horizontal tangents at $x = -\dfrac{\sqrt{10}}{2}$, $x = 0$,

and $x = \dfrac{\sqrt{10}}{2}$

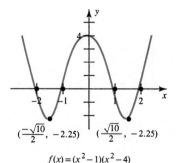

$$f(x) = (x^2 - 1)(x^2 - 4)$$

41. Increasing on $-1 < x < 2$ and $5 < x < \infty$
Decreasing on $-\infty < x < -1$ and $2 < x < 5$
Horizontal tangents at $x = -1$, $x = 2$, and $x = 5$

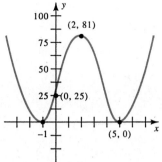

$$f(x) = (x - 5)^2(x + 1)^2$$

43. Slope of tangent line is negative for $x < 1$.
Slope of tangent line is positive for $x > 1$.

45. (a) Increasing on $(-\infty, 0)$ and $(30, \infty)$
Decreasing on $(0, 30)$
(b) It represents a percentage, and the percentage should go down.
(c) $f(20) = 25.93$
(d) When $x = 0$, $f(0) = 100$ and when $x = 30$, $f(30) = 0$. Since $f(x)$ represents a percentage, it must vary between 100% and 0%, so x must be in the interval [0, 30]; also, no shipments were made with more than 30% of the eggs below A quality.

Exercise Set 4–2

1. Relative minimum at $(0, -4)$

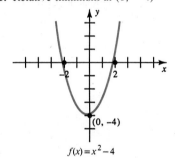

$$f(x) = x^2 - 4$$

3. Relative maximum at $(0, 2)$

$$f(x) = 2 - 3x^2$$

5. Relative minimum at $(2, -4)$

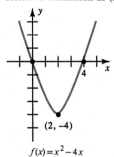

$f(x) = x^2 - 4x$

7. Relative maximum at $(-\frac{2}{3}, \frac{4}{27})$
Relative minimum at $(0, 0)$

$f(x) = x^2 (x + 1)$

9. Relative maximum at $(0, -1)$
Relative minimum at $(4, -33)$

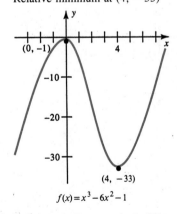

$f(x) = x^3 - 6x^2 - 1$

11. Relative maximum at $(-3, 27)$
Relative minimum at $(1, -5)$

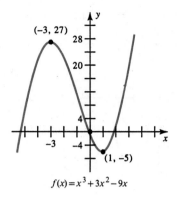

$f(x) = x^3 + 3x^2 - 9x$

13. Relative maximum at $(-1, 5)$
Relative minimum at $(3, -27)$

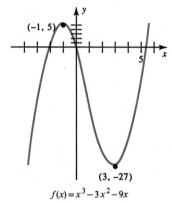

$f(x) = x^3 - 3x^2 - 9x$

15. Relative maximum at $(0, 0)$
Relative minimum at $(-1, -1)$ and $(1, -1)$

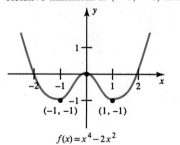

$f(x) = x^4 - 2x^2$

17. Relative maximum at $(2, 16)$ and $(-2, 16)$
Relative minimum at $(0, 0)$

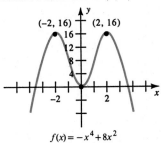

$$f(x) = -x^4 + 8x^2$$

19. Relative maximum at $(-3, -6)$
Relative minimum at $(3, 6)$

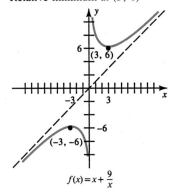

$$f(x) = x + \frac{9}{x}$$

21. Relative maximum at $(-2\sqrt{3}, -4\sqrt{3})$
Relative minimum at $(2\sqrt{3}, 4\sqrt{3})$

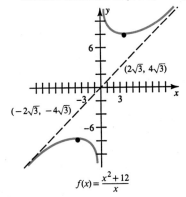

$$f(x) = \frac{x^2 + 12}{x}$$

23. No relative maximum
Relative minimum at $(3, -27)$

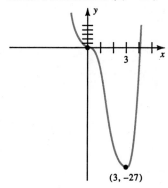

25. Relative maximum at $(0, 0)$
Relative minimum at $(-1, -1)$ and $(1, -1)$

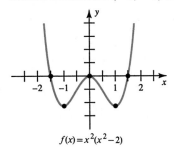

$$f(x) = x^2(x^2 - 2)$$

27. Relative minimum at $(0, 0)$
No relative maximum

$$f(x) = x^2(x^2 + 2)$$

29. No relative maximum or minimum

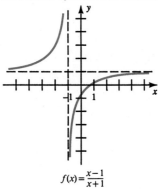

$$f(x) = \frac{x-1}{x+1}$$

31. Relative minimum at $(3, 2)$
No relative maximum

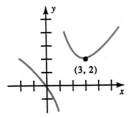

$(3, 2)$

33. Relative maximum at $(5, 4)$
Relative minimum at $(-1, 1)$

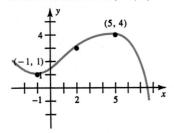

35. $x = 4$

37. Relative maximum at $(-3, 162)$
Relative minimum at $(3, -162)$

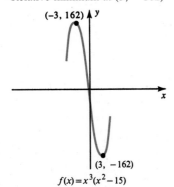

$$f(x) = x^3(x^2 - 15)$$

39. Relative maximum at $(0, 4)$
Relative minimum at $(-\frac{\sqrt{10}}{2}, -2.25)$
and $(\frac{\sqrt{10}}{2}, -2.25)$

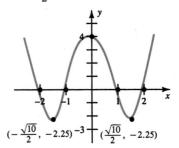

$(-\frac{\sqrt{10}}{2}, -2.25)$ $(\frac{\sqrt{10}}{2}, -2.25)$

41. Relative maximum at $(2, 81)$
Relative minimum at $(-1, 0)$ and $(5, 0)$

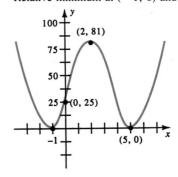

43. Relative minimum at $(5, 0)$
No relative maximum

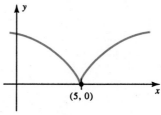

$$f(x) = (x-5)^{2/3}$$

45. 13,243 hours

Exercise Set 4–3

1. Concave down: $(-2, 0)$
Concave up: $(0, 2)$
Point of inflection: $(0, 0)$

3. Concave down: $(-1, 2)$
Concave up: $(-2, -1)$
Point of inflection: $x = -1$

5. Concave down: $(-1, 0)$
Concave up: $(-2; -1)$ and $(0, 2)$
Point of inflection: $x = -1, x = 0$

7. Concave up: $(-\infty, \infty)$
Concave down: nowhere
No point of inflection

9. Concave up: $(2, \infty)$
Concave down: $(-\infty, 2)$
Point of inflection: $x = 2$

11. Concave up: $(-1, \infty)$
Concave down: $(-\infty, -1)$
Point of inflection: $x = -1$

13. Relative maximum at $(-1, 5)$
Relative minimum at $(3, -27)$

15. Relative maximum at $(0, 0)$
Relative minimum at $(-1, -1)$ and $(1, -1)$

17. Relative maximum at $(2, 16)$ and $(-2, 16)$
Relative minimum at $(0, 0)$

19.

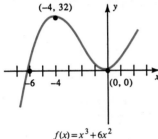

$f(x) = x^3 + 6x^2$

21.

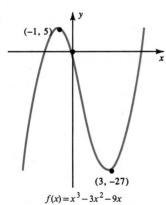

$f(x) = x^3 - 3x^2 - 9x$

23.

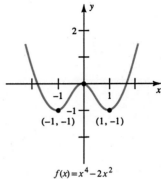

$f(x) = x^4 - 2x^2$

25.

$f(x) = 64x - 2x^4$

27.

$f(x) = x + \dfrac{9}{x}$

29.

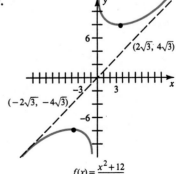

$f(x) = \dfrac{x^2 + 12}{x}$

31.

$$f(x)=x^2(x^2-2)$$

33.

$$f(x)=x^2(x^2+2)$$

35.

$$f(x)=\frac{x-1}{x+1}$$

37.

39.

41.

$$f(x)=x^3(x^2-15)$$

43.

$$f(x)=(x^2-1)(x^2-4)$$

45.

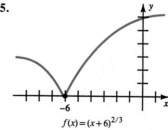

$$f(x)=(x+6)^{2/3}$$

47. 205.83

Exercise Set 4–4

1. Absolute maximum at $f(4) = 25$
Absolute minimum at $f(-1) = 0$

3. Absolute maximum at $f(-2) = 9$
Absolute minimum at $f(6) = -55$

5. Absolute maximum at $f(2) = 21$
Absolute minimum at $f(0) = 1$
7. Absolute maximum at $f(-2) = 5$
Absolute minimum at $f(0) = 1$ and $f(-3) = 1$
9. Absolute maximum at $f(-2) = 9$ and $f(2) = 9$
Absolute minimum at $f(1) = 0$ and $f(-1) = 0$
11. Absolute maximum at $f(4) = 12$
Absolute minimum at $f(-1) = -18$

13. Absolute maximum at $f(-1) = 64$
Absolute minimum at $f(1) = 0$ and $f(3) = 0$
15. 256 feet; 4 seconds
17. 10 ft $\times$ 20 ft
19. 200 and 100
21. 2000 ft $\times$ 1000 ft
23. 6.23 miles from the point he wishes to reach
25. base: 4 inches $\times$ 4 inches
side: 2 inches high
27. base: 3.175 inches $\times$ 3.175 inches
side: 3.175 inches
29. length: 18 inches
width: 14 inches
31. $\dfrac{6}{1 + \dfrac{\pi}{4}}$ feet $= 3.36$ feet

33. 6.32 $\times$ 9.49
37. base: 1.88 inches $\times$ 1.88 inches
height: 2.82 inches
39. (a) 100   (b) $P = 0.29x - 0.002x^2$ $x = 72.5$
(c)

15. $D_t y = \dfrac{2x - 3}{2y}$; $D_t x = -1$
17. 4.5 mph
19. $\dfrac{dh}{dt} = \dfrac{20}{243\pi} \dfrac{cm}{min} = 0.026 \dfrac{cm}{min}$
21. 4 ft/sec
23. $\frac{9}{7}$ ft/sec
25. 0.245 cm/sec
27. $\frac{5}{6}$ ft/sec
29. 12 ft/sec
31. 235.75 mph
33. -1.8 m^3/sec
35. 6.58 inches $\times$ 3.42 inches

Case Study Exercises
1. 29.487   **3.** 27.215

Review Exercises
1. Increasing on $0 < x < \infty$
Decreasing on $-\infty < x < 0$
Horizontal tangent at $x = 0$

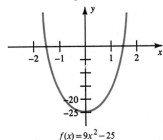

$f(x) = 9x^2 - 25$

3. Increasing on $-\infty < x < 3$ and $3 < x < \infty$
No horizontal tangent

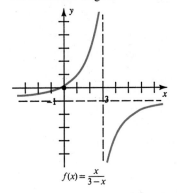

$f(x) = \dfrac{x}{3-x}$

Exercise Set 4–5

1. $y' = -\dfrac{x}{3y}$ **3.** $y' = \dfrac{-y - 2xy}{x + x^2}$

5. $y' = \dfrac{3x^2 + y^3}{-3xy^2 + 4y^3}$ **7.** $y' = \dfrac{-2x - y^3}{3xy^2 + y^{-2}}$

9. $-\dfrac{2}{3}$ **11.** 0

13. $D_t y = -\dfrac{2x}{3y^2} D_t x = -\dfrac{4}{3}$

5. Relative minimum at $(-3, -19)$

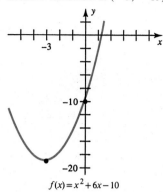

$f(x) = x^2 + 6x - 10$

7. Relative maximum at $(\sqrt{2}, 4)$ and $(-\sqrt{2}, 4)$
Relative minimum at $(0, 0)$

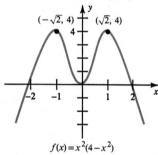

$f(x) = x^2(4 - x^2)$

9. Relative minimum at $(1, -1)$
11. Concave up: $0 < x < 1$
Concave down: $-\infty < x < 0$ and $1 < x < \infty$

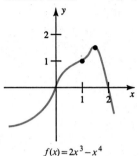

$f(x) = 2x^3 - x^4$

13. Absolute maximum of 4 at $x = 0$
Absolute minimum of -16 at $x = -2$
15. Absolute maximum of 6 at $x = 8$
Absolute minimum of -0.3849 at $x = 0.19245$

17. $y' = \dfrac{-2x - y}{x + 2y}$ **19.** $\frac{1}{24}$ m/min

---

CHAPTER 5

Exercise Set 5–1

1.

$f(x) = 5^x$

3.

$f(x) = 1.5^x$

5.

$f(x) = (\frac{1}{3})^x$

7.

$f(x) = .2^x$

9. 5   **11.** 4   **13.** 2

15.

$$f(x) = e^{x+2}$$

17.

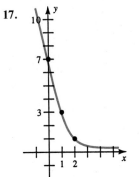

19. (a) 59.6   (b) 60
21. (a) 27.8   (b) 28
23.

$$f(x) = 1.1^x$$

25.

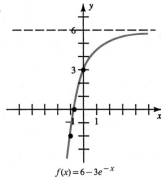

$$f(x) = 6 - 3e^{-x}$$

27.

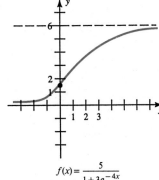

$$f(x) = \frac{5}{1 + 3e^{-4x}}$$

29.

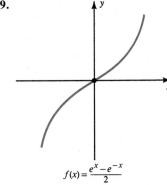

$$f(x) = \frac{e^x - e^{-x}}{2}$$

31. 3
33. 0
35.

37. 81
39. (a) 16°F   (b) 10°F
41. (a)

(b) increasing   **(c)** level off and approach 2.5 million

43. no

45. (a) $e^3 \approx 20.1$ **(b)** $e^4 \approx 54.6$

47. Figure 7 is the logistic curve, and Figure 8 is the Gompertz curve.

Case Study Exercises

1. 8 pounds

3. (a) 1925 calories   **(b)** 4 pounds

5. 7.9 pounds

7. Let w be your present weight.

    (a) $w(t) = (w - 10) + 10e^{-0.005t}$

    (b) $(w - 4)$ pounds

Exercise Set 5–2

1.

$$f(x) = \log_{10} 2x$$

3.

$$f(x) = \log_3 x$$

5.

$$f(x) = \ln (3x)$$

7.

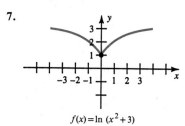

$$f(x) = \ln (x^2 + 3)$$

9. $\ln 3 + 4 \ln x$

11. $\ln 3 + \ln x + 2 \ln y - \ln(x - y)$

13. (a) $3x$ **(b)** $5 - x$

15. (a) 5   **(b)** 0.5

17. 11.9 years   **19.** 3.8 years

21.

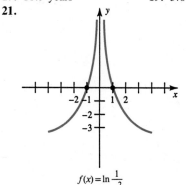

$$f(x) = \ln \frac{1}{x^2}$$

23.

$$f(x) = \ln(2x + 3)$$

25.

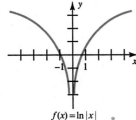

$f(x) = \ln|x|$

27. $x \neq 0$

29. $x \neq 0$

31. $\frac{1}{2}[\ln x + \ln y]$

33. $-\frac{1}{2}[3 \ln x + \ln y]$

35. $\ln(e^x) = x$

37. $\dfrac{\ln 3}{2} \approx 0.5493$ **39.** $-\ln 20$

41. $e^{0.2}$ **43.** $\dfrac{2}{\ln 3}$

45. (a) 1   (b) 4

47. (a) $\frac{1}{3}$ (b) $\frac{81}{64}$

49. (a) error   (b) error

51. about 8.99 minutes

53. the year 2022

57. (a) 0.8 hour   (b) 0.7 hour   (c) 0.48 hour
    (d) 0.23 hour

59. (a) .95 hour   (b) .92 hour   (c) .84 hour
    (d) .71 hour

61. $R(0.2) = 0.16$

Case Study Exercises

   1. $M = 3.2,$ $S \approx 2.56$

   3. \$0.25 to \$1.00

   5. \$16   **7.** \$128

Exercise Set 5–3

   1. $\dfrac{1}{x}$ **3.** $\dfrac{1}{x}$

   5. $\dfrac{1}{x + 5}$ **7.** $\dfrac{2x}{x^2 + 3}$

   9. $\dfrac{8x}{1 + 4x^2}$

   11. $\dfrac{6x^2 - \frac{1}{x^2}}{2x^3 + \frac{1}{x}} = \dfrac{6x^4 - 1}{2x^5 + x}$

   13. $1 + \ln 4x$

   15. $-\dfrac{3 \ln(x^2 - 1)}{x^4} + \dfrac{2}{x^2(x^2 - 1)}$

17. $\dfrac{\dfrac{(2x - 5)}{x} - 2 \ln 3x}{(2x - 5)^2} = \dfrac{(2x - 5) - 2x \ln 3x}{x(2x - 5)^2}$

19. $\dfrac{-5}{3 - 5x} = \dfrac{5}{5x - 3}$

21. $\dfrac{-\frac{5}{2}x^{3/2}}{4 - x^{5/2}}$ **23.** $1 + \ln|7x|$

25. Absolute minimum at $x = e^{-1}$

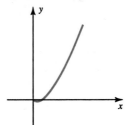

27. Absolute maximum at $x = e$

$f(x) = \dfrac{\ln x}{x}$

29. Inelastic in $[5, 8)$, elastic in $(8, 20]$

31. $2(x + \ln x)\left(1 + \dfrac{1}{x}\right)$

33. $\dfrac{2x \ln x - x}{(\ln x)^2}$

35. $\dfrac{2x + 1/x}{x^2 + \ln x} = \dfrac{2x^2 + 1}{x(x^2 + \ln x)}$

37. $\dfrac{10}{x} \cdot \dfrac{1}{\ln 10} = \dfrac{10}{x \ln 10}$

41. $f^n(x) = \dfrac{(-1)^{n+1}(n - 1)!}{x^n}$

43. $f(2) \approx 1.39$ but $g(2) \approx 0.48$;
    $f'(2) = 1$ but $g'(2) = \ln 2 \approx 0.69$

45. $q = 0.5$

Exercise Set 5–4

   1. $6e^{6x}$ **3.** $3e^{3x+5}$

   5. $28e^{7x+5}$ **7.** $2xe^{x^2+3}$

   9. $16xe^{1+4x^2}$

   11. $(x^{-2} - 6x^2)e^{2x^3 + x^{-1}}$

   13. $(4x + 1)e^{4x}$

15. $(2x^{-2} - 3x^{-4})e^{x^2 - 1}$

17. $\dfrac{(6x - 17)e^{3x}}{(2x - 5)^2}$

19. $5e^{5x} - 5e^{-5x}$

21. $\dfrac{e^{-x}}{(2 + e^{-x})^2}$

23. $\dfrac{2e^{-2x} - e^x}{(1 + e^x + e^{-2x})^2}$

25.

$f(x) = xe^x$

27.

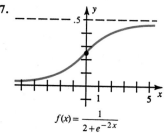

$f(x) = \dfrac{1}{2 + e^{-2x}}$

29. $\dfrac{1 + 4e^{4x}}{x + e^{4x}}$

31. $2(\ln x + e^{5x})\left(\dfrac{1}{x} + 5e^{5x}\right)$

33. $\dfrac{(2x - 1)e^{x^2} + 2x - x^2}{e^x}$

35. $\left(\dfrac{1}{x} + 3x^2 \ln x\right)e^{x^3}$

37. $f^n(x) = 2^n e^{2x}$

39.

(65, .05)

41. 1.98 months

Review Exercises

1.

$f(x) = 2^{x+3}$

3.

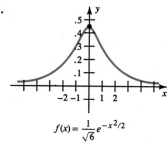

$f(x) = \dfrac{1}{\sqrt{6}} e^{-x^2/2}$

5.

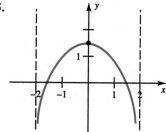

7. $-\ln(2.1) \approx -0.74$

9. $e^2 \approx 7.4$

11. $\dfrac{60[\ln(1 + 3x)^4]^4}{1 + 3x}$

13. $\dfrac{1}{2x}$

15. $x(x + 2)e^x + e^{-x}$

17. Absolute minimum at $x = 0$
No points of inflection

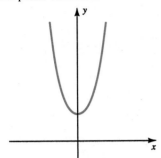

CHAPTER 6

Exercise Set 6–1

1. $\frac{3}{2}x^2 + 5x + C$

3. $\frac{2}{5}x^5 - 7x + C$

5. $-6x^{-1} + \frac{5}{2}x^2 + C$

7. $\frac{4}{3}x^{3/2} + \frac{9}{4}x^{4/3} - \frac{5}{2}x^2 + C$

9. $\frac{9}{2}x^{2/3} + \frac{28}{3}x^{3/4} - 5x + C$

11. $\dfrac{x^{1.2}}{1.2} + \dfrac{3x^{2.4}}{2.4} + C$

13. $-x^{-1} - \frac{5}{3}x^{-3} + C$

15. $-\dfrac{1}{6x^6} + \dfrac{16x^{3/4}}{3} + \dfrac{16x^{5/4}}{5} + C$

17. $\dfrac{3e^{2x}}{2} + 5\ln|x| + C$

19. $\frac{1}{4}x^4 - 2e^{-x} + \frac{2}{3}x^{3/2} - 5\ln|x| + C$

21. $\frac{10}{3}x^{3/2} + \frac{1}{2}e^{-4x} + \ln|x| + C$

23. $18\ln|x| + \dfrac{3}{\sqrt[3]{4}}x^{2/3} - \dfrac{1}{7}e^{7x} + C$

25. $y = \frac{1}{3}x^3 - 2x + \frac{2}{3}$

27. $y = 4\ln|x| - x^2 + 2 - 4\ln 2$

29. $\frac{2}{3}x^3 - 7\ln|x| + C$ **31.** $6x + \frac{5}{4}x^4 + C$

33. $\frac{4}{9}x^9 - \frac{14}{3}x^6 + \frac{49}{3}x^3 + C$

35. $S(t) = \dfrac{t^3}{3} + t^2$

37. 3258 years

41. 6,969 years   **43.** $m' = -0.068m$

Exercise Set 6–2

1. $3.125 = \frac{25}{8}$ **3.** $2.71875 = \frac{174}{64}$

5. $11.375 = \frac{91}{8}$

7.

n	2	4	8	10
area	8	7	6.5	6.4

9.

n	2	4	8	10
area	7	5.75	5.1875	5.08

11.

n	2	4	8	10
area	8	6	5.24375	4.8

13. 24   **15.** 20

17. 6   **19.** 2

21. 10.5   **23.** 6

25. $\frac{64}{3}$ **27.** $\frac{140}{3}$

29. $\frac{5}{6}$ **31.** 1.0986

37. $\dfrac{1}{5}$

Exercise Set 6–3

1. $\frac{13}{2}$ **3.** $\frac{4}{3}$

5. $-\frac{5}{6}$ **7.** $-\dfrac{476}{3}$

9. $\dfrac{171}{96}$ **11.** 0

13. 12.5   **15.** $\frac{64}{3}$

17. 1   **19.** $\frac{1}{2}$

21. 25   **23.** 20

25. 2.08   **27.** 71.5068

29. -17.2047 **31.** 5

33. $\frac{59}{12}$ **35.** 0.5754

39. 1,000   **41.** 19.84

Exercise Set 6–4

1. $\frac{14}{3}$

3. $\frac{20}{3}$

5. 10

7. $\frac{1}{6}$

9. $\frac{1}{6}$

11. $\frac{1}{6}$

13. $\frac{1}{6}$

15. $\frac{1}{6}$

17. $\frac{32}{3}$

19. $\frac{32}{3}$

21. $\frac{1}{12}$

23. 12

25. 13

27. $\frac{7}{3}$

29. 16

31. C.S. $= \frac{32}{3}$
   P.S. $= \frac{40}{3}$

33. C.S. $= \frac{76}{3}$
   P.S. $= \frac{76}{3}$

35. $1 - \ln 2 \cong 0.3069$

37. 8.107

39. 6.8795

41. C.S. $= 3166.67$

43. C.S. $= \frac{7}{32}$

Case Study Exercises

1. 4.076

3. 13.2

5. -1.3603

Review Exercises

1. $\frac{25}{4}x^{4/5} - 5x^{2/5} - \frac{3}{2}x^2 + C$

3. $y = -\frac{1}{x} - \frac{x^2}{2} + \frac{3}{2}$

5. 8

7. 344.16

9. 637.519

11. $\frac{1}{6}$

CHAPTER 7

Exercise Set 7–1

1. $\dfrac{(x^2 + 5)^4}{4} + C$

3. $\dfrac{(x^4 - 7)^3}{3} + C$

5. $\dfrac{(3x^2 + 5)^4}{24} + C$

7. $\dfrac{(3x^4 - 5)^4}{48} + C$

9. $\dfrac{(3x^6 - 5)^{5/2}}{45} + C$

11. $\dfrac{-(x^4 - 7)^{-2}}{4} + C$

13. $\dfrac{(6x^{-2} + 5)^{-1}}{12} + C$

15. $\dfrac{(x^2 + 5x)^{-2}}{-2} + C$

17. $\dfrac{(x^4 - 4x)^{3/2}}{6} + C$

19. $\frac{1}{2}e^{2x} + \frac{5}{4}e^{4x} + C$

21. $\frac{1}{4}e^{x^4} - 2e^{-x} + C$

23. $\frac{2}{3}\ln|3x + 1| + C$

25. $\frac{1}{3}\ln|x^3 - 1| + C$

27. $\frac{7}{6}$

29. 0

31. $\frac{1}{3}(10^{3/2} - 1) = 10.2076$

33. $\dfrac{19}{480}$

35. $\dfrac{(e^x + 1)^2}{2} + C$

37. $\dfrac{(\ln x)^2}{2} + C$

39. $\frac{1}{4}x^4 - \frac{2}{3}x^3 + \frac{1}{2}x^2 + C$

41. $\dfrac{-2}{75}(1 - 5x)^{3/2}(15x + 2)$

43. $\frac{3}{16}(x^2 - 1)^{8/3} + \frac{3}{10}(x^2 - 1)^{5/3} + C$

45. $\frac{1}{4}[\ln(1 + x^2)]^2$

47. $0.5(x + 1) + 0.3\ln(x + 1) + C$

Exercise Set 7–2

1. $2xe^x - 2e^x + C$

3. $\frac{1}{3}xe^{3x} - \frac{1}{9}e^{3x} + C$

5. $\dfrac{2}{15}x(3x + 5)^{5/2} - \dfrac{4}{315}(3x + 5)^{7/2} + C$

7. $-\frac{3}{10}x(5x - 2)^{-2} - \frac{3}{50}(5x - 2)^{-1} + C$

9. $\frac{1}{2}x^2 \ln x - \frac{1}{4}x^2 + C$

11. $\frac{1}{3}x^3 \ln x - \frac{1}{9}x^3 + C$

13. $x^2e^x - 2xe^x + 2e^x + C$

15. $-\dfrac{1}{2}(2x + 5)(x + 5)^{-2} - (x + 5)^{-1} + C$

17. $\dfrac{2}{3}x^2(x - 4)^{3/2} - \dfrac{8}{15}(x)(x - 4)^{5/2}$
   $+ \dfrac{16}{105}(x - 4)^{7/2} + C$

19. $-\frac{1}{3}x^2(1 - 3x)^{-1} + \frac{2}{27}(1 - 3x) + \frac{2}{27}\ln|1 - 3x| + C$

21. $\frac{1}{4}e^2 + \frac{1}{4}$

23. -1.93

25. $\frac{2}{9}$

27. $\dfrac{2}{3}x^2(x + 1)^{3/2} - \dfrac{8}{15}x(x + 1)^{5/2}$
   $+ \dfrac{16}{105}(x + 1)^{7/2} + C$

29. $\dfrac{2}{3}x^3(x + 1)^{3/2} - \dfrac{4}{5}x^2(x + 1)^{5/2}$
   $+ \dfrac{16}{35}x(x + 1)^{7/2} - \dfrac{32}{315}(x + 1)^{9/2} + C$

31. $\frac{1}{3}x^2(x^2 + 1)^{3/2} - \frac{2}{15}(x^2 + 1)^{5/2} + C$

33. $\frac{1}{2}x^2 e^{x^2} - \frac{1}{2}e^{x^2} + C$

35. -24.028

37. 0.3093

Exercise Set 7–3

1. $\frac{1}{2(3x + 2)} + \frac{1}{4}\ln\left|\frac{x}{3x + 2}\right| + C$

3. $-\frac{1}{4(3x - 4)} + \frac{1}{16}\ln\left|\frac{x}{3x - 4}\right| + C$

5. $\frac{1}{6}\ln\left|\frac{1 + 3x}{1 - 3x}\right| + C$

7. $\frac{1}{24}\ln\left|\frac{4 + 3x}{4 - 3x}\right| + C$

9. $\frac{1}{24}\ln\left|\frac{3 + x}{3 - x}\right| + C$

11. $\frac{1}{2}[\frac{1}{2}x^2\sqrt{x^4 + 9} + \frac{9}{2}\ln|x^2 + \sqrt{x^4 + 9}|] + C$

13. $\frac{1}{4}[x^2\sqrt{4x^4 + 9} + \frac{9}{2}\ln|2x^2 + \sqrt{4x^4 + 9}| + C$

15. $\frac{1}{2}[x^2\sqrt{4x^4 - 25} + \frac{25}{2}\ln|2x^2 + \sqrt{4x^4 - 25}| + C$

17. $\frac{1}{5}(\ln\frac{3}{4} - \ln 2) = -0.196$

19. $\frac{1}{4}(\ln\frac{1}{4} - \ln\frac{1}{5}) - \frac{3}{80} + C$

21. $\frac{\sqrt{2}}{4} + \frac{1}{4}\ln|1 + \sqrt{2}| + C$

23. $\frac{1}{3}x^3 e^{3x} - \frac{1}{3}x^2 e^{3x} - \frac{2}{9}xe^{3x} + \frac{2}{27}e^{3x} + C$

25. $\frac{2}{25(5x + 2)} + \frac{1}{25}\ln|5x + 2| + C$

27. $-\frac{1}{4}\ln\left|\frac{x}{x - 4}\right| + C$

29. $-\frac{1}{3}\ln\left|\frac{1 + \sqrt{1 - 9x^2}}{3x}\right| + C$

31. $\frac{1}{6}\ln|x^2 + 16 + 9x^2| + C$

33. $-\frac{1}{3}\ln\left|\frac{3 + \sqrt{9 + 4x^2}}{2x}\right| + C$

35. $\frac{1}{6}\ln\left|\frac{3 + 2x^3}{3 - 2x^3}\right| + C$

37. $x^6\left[\frac{\ln x^2}{3} - \frac{1}{9}\right] + C$

39. $\frac{-t^2}{2} - 10t - 100\ln|t - 10| + C$

Exercise Set 7–4

1. 9.5	**3.** 73
5. 22.5	**7.** 329
9. 22.67	**11.** 1.774
13. 1.683	**15.** 1.644
17. 0.9436	**19.** 0.6766
21. 2.72	**23.** 62.453
25. 37.9245	**27.** 11.5345
29. 2.748	**31.** 58.489

33. Simpson's Rule: 1.622
Trapezoidal Rule: 1.683
Fundamental Theorem: 1.609

35. 863.5   **37.** 554.6434

Case Study Exercises

1. 1.7%   **3.** 2.48%

5. 3.6% (from Example 4)

Review Exercises

1. $\frac{(x^2 + x)^{-2}}{-2} + C$ **3.** $-\frac{15}{8}$

5. $-\frac{3}{4}(1 - x)^{4/3} + \frac{6}{7}(1 - x)^{7/3} - \frac{3}{10}(1 - x)^{10/3} + C$

7. $-\frac{1}{7(5x - 7)} + \frac{1}{49}\ln\left|\frac{x}{5x - 7}\right| + C$

9. $\frac{1}{2\sqrt{14}}\ln\left|\frac{\sqrt{14} + 3x}{\sqrt{14} - 3x}\right| + C$

11. 0.8873

CHAPTER 8

Exercise Set 8–1

1. 1	**3.** -9
5. 2	**7.** 63

9. All ordered pairs for which $x \neq 2$

11. All ordered pairs for which $x > -1$

13.

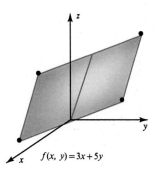

$f(x, y) = 3x + 5y$

15.

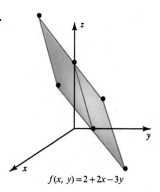

$f(x, y) = 2 + 2x - 3y$

17.

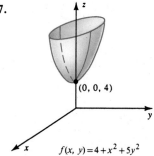

$(0, 0, 4)$

$f(x, y) = 4 + x^2 + 5y^2$

19.

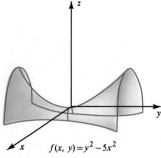

$f(x, y) = y^2 - 5x^2$

21.

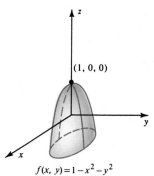

$(1, 0, 0)$

$f(x, y) = 1 - x^2 - y^2$

23.

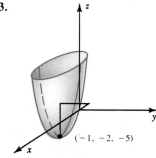

$(-1, -2, -5)$

$f(x, y) = x^2 + 2x + y^2 + 4y$

25.

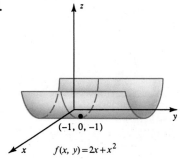

$(-1, 0, -1)$

$f(x, y) = 2x + x^2$

27. no

29. A vertical line will cut the surface in no more than one point.

31. $P = 40x + 50y + 60z$

33. 1.74

Exercise Set 8–2

1. $f_x(x, y) = 6x$
  $f_y(x, y) = \frac{5}{2}y^{-1/2}$

3. $f_x(x, y) = y^2 - 10x$
  $f_y(x, y) = 2xy$

5. $f_x(x, y) = y^2 - 6xy$
  $f_y(x, y) = 2xy - 3x^2$

7. $f_x(x, y) = -2x^{-3} + y^{-2} - 2xy^3$
$f_y(x, y) = -2xy^{-3} - 3x^2y^2 + 4$

9. $f_x(x, y) = e^{2y} - \dfrac{2}{2x + 3y}$

$f_y(x, y) = 2xe^{2y} - \dfrac{3}{2x + 3y}$

11. $f_x(x, y) = 8(2x + 3y)^3$
$f_y(x, y) = 12(2x + 3y)^3$

13. $f_x(x, y) = -2(2 + 4y)(2x + 4xy)^{-3}$
$f_y(x, y) = -8x(2x + 4xy)^{-3}$

15. $f_x(1, 0) = 0$
$f_y(2, 1) = 20.5$

17. $f_x(1, 0) = 1$
$f_y(2, 1) = -18 + e$

19. $f_{xx}(x, y) = 6$
$f_{xy}(x, y) = 0$

$f_{yy}(x, y) = \dfrac{30}{y^4}$

21. $f_{xx}(x, y) = 6y$
$f_{xy}(x, y) = 6x + 10e^{2y}$
$f_{yy}(x, y) = 20xe^{2y}$

23. $f_{xx} = 2y^2e^x + xy^2e^x + 4e^{2x}$
$f_{yy} = 2xe^x$
$f_{xy} = 2ye^x + 2xye^x$

25. $x = 1$ and $y = 1$

27. $f_{xxx} = 24x - 6$
$f_{xxy} = 0$
$f_{xyy} = 0$

29. $f_{xxx} = 0$
$f_{xxy} = 2$
$f_{xyy} = -4e^{2y}$

31. $f_x(x, y, z) = \lim\limits_{h \to 0} \dfrac{f(x + h, y, z) - f(x, y, z)}{h}$

33. $f_x = -6x^{-3} + \frac{1}{3}x^{-2/3}z - 2xz^3 + yz$
$f_y = xz$

37. $\dfrac{\partial C}{\partial Q} = 1.87Q^{0.1}F^{1.5}L^{0.36}P^{0.5}$

$\dfrac{\partial C}{\partial F} = 2.55Q^{1.1}F^{0.5}L^{0.36}P^{0.5}$

$\dfrac{\partial C}{\partial L} = 0.612Q^{1.1}F^{1.5}L^{-0.64}P^{0.5}$

$\dfrac{\partial C}{\partial P} = 0.85Q^{1.1}F^{1.5}L^{0.36}P^{-0.5}$

39. (a) $0.07D^{-0.44}A^{0.06}$ (b) $0.0075D^{0.56}A^{-0.94}$

Exercise Set 8–3
1. Relative minimum at (0, 0)
3. no

5. no   **7.** $(-3, -4)$
9. $(3, -3)$
11. $(0, +1)$ and $(0, -1)$
13. Saddle point at $(-2, 0)$
No relative maximum or minimum
15. Relative minimum at (4, 0)
Saddle point at $(4, -2)$
17. Relative minimum at $(0, -1)$
Saddle point at (0, 1)
19. Relative minimum at (4, 3)
21. Saddle point at $(-3, 1)$
23. Relative minimum at (2, 2)
Saddle point at (0, 0)
25. Saddle point at (0, 1)
Saddle point at $(0, -1)$
27. Saddle point at (4, 2)
Relative minimum at $(4, -2)$
29. Saddle point at $(4, -2.225)$
Relative minimum at (4, 0.225)
31. (12, 2) is a relative minimum.
33. $4\sqrt[3]{2} \times 2\sqrt[3]{2} \times 4\sqrt[3]{2}$

35. $x = \dfrac{a}{2b}$
y is any value.

Exercise Set 8–4
1. $(\frac{1}{2}, \frac{1}{2})$ **3.** $(-1, \frac{1}{2})$
5. $(-1, 0)$ and $(-\frac{5}{2}, \frac{9}{4})$
7. $(1.63, 0.19)$ and $(0.21, 0.87)$
9. $(19.27, 4.63)$ and $(7.44, -1.3)$
11. $(\frac{1}{4}, \frac{1}{2})$
13. $(-\frac{1}{2}, \frac{1}{6})$ **15.** $(\frac{1}{3}, \frac{2}{3}, \frac{2}{3})$
17. (0, 0, 0)
19. The numbers are 10, 10, 10.
21. (250, 25,000)
23. No maximum possible
25. $(\frac{5}{2}, -\frac{5}{2})$
27. $(\frac{1}{2}, 0)$, $(\frac{1}{6}, -\frac{2}{3})$
29. $(\frac{1}{4}, \frac{1}{6}, 1)$ **31.** (2, 0, 0)
33. (1, 2, 0)   **35.** $36 \times 36 \times 36$ inches

37. $\left(\dfrac{1.844}{C}, \dfrac{1.639}{C} \right)$

Exercise Set 8–5
1. $\dfrac{7x^2y^6}{2} + \dfrac{3x^4y^{-2}}{2} - 5x + C$

3. $xy^8 - 4x^{-4}y^5 - 3xy - \frac{1}{2}y^2 + C$
5. $31y + 2y^3 \ln 2 - 5$

7. $10x^4 + \dfrac{8}{x} - 10$

9. -2

11. 1715.09

13. $\frac{32}{3}$

15. 3

17. 12

19. $\frac{1}{15}$

21. $\frac{1}{9}$

23. 20.19

25. $\frac{2}{3}$

27. 0.186

29. $(\ln 2)(\ln 3) = 0.7615$

31. 7.189

33. 2.385

35. 0.708

37. 95.9k

5. $f_x(x, y) = y^3 - 8(2x + 3y)^3$
     $f_y(x, y) = 3xy^2 - 12(2x + 3y)^3$

7. $f_{xx} = -\frac{18}{25}x^{-8/5} + 2y^{-2}$
     $f_{yy} = 6x^2y^{-4}$
     $f_{xy} = f_{yx} = -4xy^{-3}$

9. Critical point $(\frac{10}{11}, \frac{5}{11})$

11. Relative maximum at $(4, -2)$
     Relative minimum at $(0, 2)$
     Saddle points at $(0, -2)$ and $(4, 2)$

13. $(\frac{5}{6}, \frac{5}{2})$

15. The numbers are 15, 15, and 15.

17. $7x^2y^6 + \dfrac{3x^4y^{-2}}{2} - xy + \dfrac{x^2}{2} - x$

19. -2

21. $\frac{2}{3}$

Case Study Exercises

1. $x = 111.11$
     $y = 37$

3. $x = 42.13$
     $y = 0.3555x$

5. $x = 122.22$
     $y = 43.2056$

Review Exercises

1. -1

3.

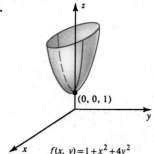

$(0, 0, 1)$

$f(x, y) = 1 + x^2 + 4y^2$

CHAPTER 9

Exercise Set 9–1

7. $f(x) = x^4 + x^2 - x + C$
     $f(x) = x^4 + x^2 - x - 1$

9. $f(x) = -\dfrac{1}{2}x^{-2} + e^x + C$

     $f(x) = -\dfrac{1}{2}x^{-2} + e^x - 1$

11. $f(x) = \dfrac{1}{4}e^{2x} + \dfrac{4}{15}x^{5/2} + Cx + K$

     $f(x) = \dfrac{1}{4}e^{2x} + \dfrac{4}{15}x^{5/2} - x + 7$

13. $f(x) = 2x^4 - x^{-2} - x$

15. $f(x) = \dfrac{1}{4}e^{2x} + \dfrac{1}{6}x^3 - \dfrac{1}{2}x - \dfrac{13}{4}$

17. 1st

19. 2nd

21. 3rd

23. $y = \dfrac{1}{2}e^{x^2} + \dfrac{1}{2}$

25. $f(x) = -xe^{-x} - e^{-x} + 1$

27. $f(x) = xe^x - 2e^x + 2x + 2$

29. $f(x) = \dfrac{x^3}{6}\ln x - \dfrac{5x^3}{36} - \dfrac{x}{4} + \dfrac{25}{18}$

31. $f(x) = \dfrac{1}{k}e^{kx}$

33. 1st order

35. $L'(x) = L(x)$, 1st order

Exercise Set 9–2

1. Yes

3. No

5. Yes

7. $f(x) = \dfrac{-2}{x^2 + C}$

9. $f(x) = 1 - Ke^{-x^2/2}$

11. $f(x) = \pm\sqrt{x^2 + \frac{8}{3}x^3 + C}$

13. $f(x) = 3 + \dfrac{1}{\sqrt{2x - 2x \ln x + C}}$

15. $f(x) = -\ln(C - e^x)$ **17.** $f(x) = \dfrac{3}{3 - x^3}$

19. $f(x) = 2e^{3x^2/2} - 1$ **21.** $f(x) = \dfrac{2}{x}$

23. $f(x) = \pm\sqrt{4 + Ke^{-2/x}}$

25. $f(x) = Ke^{(1/2)(\ln x)^2}$

27. $-(y + 1)e^{-y} = \dfrac{x^3}{3} + C$

31. $\dfrac{P}{1 - P} = Ke^{0.2t}$ or $P = \dfrac{1}{1 + Ce^{-0.2t}}$

17. $y = 2 - e^{1-x}$

19. $y = \dfrac{x^3}{2} - x \ln x + \dfrac{x}{2}$

21. $y = \dfrac{x^2 e^{-x^2}}{2} + e^{-x^2}$ **23.** $y = x + 1 - Ce^x$

25. $y = x \ln x + Cx$

27. $y = -x \ln x - xe^x + Cx$

29. $\dfrac{dy}{dt} = 5 - \dfrac{y}{16}$ **31.** $P = \dfrac{C}{a} + Ke^{-at}$

Review Exercises

3. $y = \dfrac{1}{9}e^{-3x} - \dfrac{9}{5}x^{5/3} + Cx + K;$

$y = -\dfrac{1}{9}e^{-3x} - \dfrac{9}{5}x^{5/3} + x + 1$

5. 3rd order

7. $f(x) = -xe^x + 2e^x - 2$

9. No

11. $f(x) = \frac{2}{3} + Ce^{-3/(2x^2)}$

13. $f(x) = C \ln x$ **15.** Yes

17. $f(x) = 5x + 5 - Ce^x$ **19.** $f(x) = 2e^{-1/2x^2} - 1$

Exercise Set 9–3

1. Yes   **3.** Yes

5. No   **7.** $y = -\frac{1}{2} + Ce^{-2x}$

9. $y = -\dfrac{1}{2}e^{-x} + Ce^x$ **11.** $y = 3 + 3Ce^{-x^3/3}$

13. $y = \dfrac{x^4}{3} - 5x^2 + Cx$ **15.** $y = -2 + Ce^{x^3/3}$

CHAPTER 10

Exercise Set 10–1

1. $\dfrac{\pi}{2}$ **3.** $\dfrac{7\pi}{6}$

5. $\dfrac{3\pi}{2}$ **7.** $30°$

9. $450°$ **11.** $315°$

13. $495°$, $\frac{11}{8}$ revolutions

15. $630°$, $\frac{7}{4}$ revolutions

17. $840°$, $\frac{7}{3}$ revolutions

19. $-150°$, $\frac{5}{12}$ revolutions in the counterclockwise direction

21. $-990°$, $\frac{11}{4}$ revolutions in the counterclockwise direction

23. $-840°$, $\frac{7}{3}$ revolutions in the counterclockwise direction

25. $\dfrac{13\pi}{6}$ **27.** 6π

29. 4π **31.** $\dfrac{14\pi}{3}$

33. -4π **35.** $-\dfrac{13\pi}{2}$

37. $\dfrac{13\pi}{3}$ **39.** 2π rad/sec

41. $\dfrac{\pi}{3}$ radians, $\dfrac{\pi}{6}$ radians

43. (a) $\dfrac{\pi}{90}$ radians (b) $\dfrac{\pi}{60}$ radians (c) $\dfrac{7\pi}{360}$ radians

Exercise Set 10–2

1. $\sqrt{3}$ **3.** $\frac{3}{2}$

5. $\frac{12}{13}$ **7.** $\frac{12}{13}$

9. $\frac{1}{2}$ **11.** $-\dfrac{\sqrt{3}}{2}$

13. $\dfrac{\sqrt{3}}{2}$ **15.** $-\dfrac{\sqrt{2}}{2}$

17. $-\dfrac{\sqrt{2}}{2}$ **19.** $-\frac{1}{2}$

21. 0.4794 **23.** 0.0998

25. 0 **27.** $\dfrac{\pi}{2}$

29.

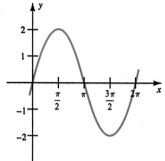

$f(x) = 2 \sin x$

31.

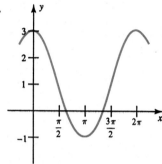

$f(x) = 1 + 2 \cos x$

33.

$f(x) = \sin 2x$

35.

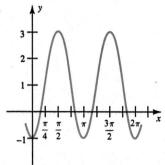

$f(x) = 1 - 2 \cos 2x$

37. Values of $f(x)$ repeat at specified intervals.
39. Amplitude $= 0.002$
41. $0.4510, \quad 0.6435$

Exercise Set 10–3

1. $f'(x) = 6 \cos 6x$
3. $f'(x) = -4x^{-5} \cos x^{-4}$
5. $f'(x) = (8x^3 - 10x)\cos(2x^4 - 5x^2)$
7. $f'(x) = 2x \cos 6x - 6x^2 \sin 6x$
9. $f'(x) = -2(x + \cos 5x)^{-3}(1 - 5 \sin 5x)$

11. $3 \sin 3x + C$ **13.** $-\dfrac{1}{3}\cos x^3 + C$

15. $-\cos^5 x + C$ **17.** $\dfrac{7}{64}$

19. $2\sqrt{2} - 1 \approx 1.828$
21. $f'(x) = e^x \cos e^x + (\cos x)(e^{\sin x})$
23. $f'(x) = (\cos 3x + e^{3x})(\sin 3x + e^{3x})^{-2/3}$
25. $f'(x) = -6x^2 \sin x^3 \cos x^3$
27. $\sin (e^x) + C$ **29.** $e^{\sin x} + C$
31. $f'(x) = 3e^{3x} \cos(e^{3x}) + 3e^{\sin 3x} (\cos 3x)$

33. $-\cos x + \dfrac{\cos^3 x}{3} + C$

35. $0, \ f(x) = 1$

Exercise Set 10–4

1. $\dfrac{5}{12}$ **3.** $\dfrac{13}{12}$

5. $f'(x) = 5 \sec^2(5x + 1)$
7. $f'(x) = (3x^2 - 5)\sec^2(x^3 - 5x)$
9. $f'(x) = 20 \tan^4 4x \sec^2 4x$

11. $f'(x) = \dfrac{1}{2}(3 + 2x \sec^2 x^2)(3x + \tan x^2)^{-1/2}$

17. $f'(x) = (2x + 3)\sec(x^2 + 3x)\tan(x^2 + 3x)$
19. $f'(x) = -(3x^{-4} + e^x)\csc^2(x^{-3} - e^x)$

21. $f'(x) = -\dfrac{3}{2}\csc^{1/2}(3x)\cot(3x)$

23. $f'(x) = \tan x$
25. $f'(x) = 6 \sec^2 (6x)e^{\tan 6x}$
27. $f'(x) = -3e^{3x} \csc^2(e^{3x})$
29. $f'(x) = 2 \sec^2 2x \cot 3x - 3 \csc^2 3x \tan 2x$

31. $\displaystyle\int \sec^2 x \, dx = \tan x + C$

$\displaystyle\int \csc^2 x \, dx = -\cot x + C$

$\displaystyle\int \sec x \tan x \, dx = \sec x + C$

$\displaystyle\int \csc x \cot x \, dx = -\csc x + C$

33. $\dfrac{\sec 4x}{4} + C$ **35.** $-\dfrac{\cot x^3}{3} + C$

37. $-\dfrac{\cot^3 x}{3} + C$

39. $a \sec^2 x + \dfrac{b \sec^2 x^{1/2}}{2x^{1/2}}$

Case Study Exercises

1. 57.2

3. 64.375

5. 36.43

7. 79.6

Chapter Review

1. (a) $\dfrac{3\pi}{4}$ (b) $\dfrac{7\pi}{6}$

3. (a) 660°, $\dfrac{11}{6}$ revolutions

(b) 750°, $\dfrac{25}{12}$ revolutions

5. $\dfrac{13\pi}{3}$

7. $-\dfrac{\sqrt{2}}{2} \approx -0.707$

9. 0.9950

11. $f'(x) = 4\left(-2x^{-3} - \dfrac{1}{3}x^{-2/3}\right) \cos(x^{-2} - x^{1/3})$

13. $f'(x) = \dfrac{-\cos x^4}{4} + C$

15. $-\tan x + \dfrac{\cos(\ln x)}{x}$

17. $-2(\sec x \tan x + 2 \sec^2 2x)(\sec x + \tan 2x)^{-3}$

19. $-\csc x \cot x (e^{\csc x})$

CHAPTER 11

Exercise Set 11–1

1. 1.2838

3. 2.10156

5. $2x$, $\quad 2x - 2x^2$, $\quad 2x - 2x^2 + \dfrac{8}{3}x^3$,

$2x - 2x^2 + \dfrac{8}{3}x^3 - 4x^4$

7. $1 + 2x$, $\quad 1 + 2x + 2x^2$,

$1 + 2x + 2x^2 + \dfrac{4}{3}x^3$,

$1 + 2x + 2x^2 + \dfrac{4}{3}x^3 + \dfrac{2}{3}x^4$

9. 0.00995

11. -0.10535

13. $1 + x + \dfrac{x^2}{2} + \cdots + \dfrac{x^n}{n!}$

15. $1 - x + x^2 - \cdots + (-1)^n x^n$

17. $1 + (x - 1) + \dfrac{(x - 1)^2}{2} + \cdots + \dfrac{(x - 1)^n}{n!}$

19. $1 + (x - 3) + (x - 3)^2 + \cdots + (x - 3)^n$

21.

x	P_3	P_5
0.001	0.000999	0.0009999
0.01	0.009999	0.009999
0.1	0.099833	0.099833
0.2	0.198666	0.198669

23. $1 - 2x^2 + \dfrac{2}{3}x^4$

25. $x + x^2 + \dfrac{x^3}{2} + \dfrac{x^4}{6}$

27. $x^2 - \dfrac{x^4}{6}$

29. $1 - \dfrac{x}{2} - \dfrac{x^2}{8} - \dfrac{x^3}{16} - \dfrac{5x^4}{128}$

31. $1 - \dfrac{x^2}{2} + \dfrac{x^4}{24} - \cdots + \dfrac{(-1)^n x^{2n}}{(2n)!}$

33. 3.0167

35. 0.9090

39. 37.5

41. 0.99

43. (a) 2 (b) 4 (c) 7 (d) 5

45. 504.999

Exercise Set 11–2

1. convergent, $\frac{1}{3}$

3. divergent

5. convergent, 2

7. convergent, $\frac{8}{7}$

9. $\dfrac{35}{99}$

11. $\dfrac{71}{99}$

13. $\dfrac{3}{10}$

15. $1,800,000

17. $600,000

19. $\dfrac{31}{32}$

21. $\dfrac{1023}{1024}$

23. convergent, $\frac{1}{2}$, 64

25. divergent, 2

27. convergent, 0.9, 100

29. convergent, 0.75, 256

31. divergent

33. divergent

35. $1, \dfrac{3}{2}, \dfrac{11}{6}, \dfrac{25}{12}$

37. $\dfrac{1}{2}, \dfrac{11}{14}, \dfrac{69}{70}, \dfrac{1037}{910}$

39. (a) $2176 **(b)** $2500

41. 128

43. It is a multiple of a divergent series (harmonic).

45. $1,250,000

Exercise Set 11–3

1. convergent   **3.** convergent
5. divergent   **7.** divergent
9. convergent   **11.** convergent
13. convergent   **15.** divergent
17. convergent   **19.** divergent
21. convergent   **23.** convergent
25. convergent   **27.** divergent
29. divergent   **31.** divergent
33. convergent   **35.** convergent
37. convergent
45. (a) convergent, **(b)** convergent, **(c)** divergent

Exercise Set 11–4

1. $1 + 2x + 2x^2 + \cdots + \dfrac{2^n}{n!} x^n + \cdots$

3. $1 - x + \dfrac{x^2}{2!} - \dfrac{x^3}{3!} + \cdots + \dfrac{(-1)^n x^n}{n!} + \cdots$

5. $1 - x + x^2 - x^3 + x^4 + \cdots + (-1)^n x^n + \cdots$

7. $1 + 2x + 4x^2 + 8x^3 + \cdots + (2x)^n + \cdots$

9. $1 + \dfrac{x}{2} - \dfrac{x^2}{8} + \dfrac{x^3}{16} - \dfrac{5x^4}{128} + \cdots$
$\qquad\qquad + \dfrac{(-1)^{n+1}\, 1\cdot 3\cdot 5\cdots}{2^n n!} x^n + \cdots$

11. $1 - 2x + \dfrac{2\cdot 3}{2!} x^2 - \dfrac{2\cdot 3\cdot 4}{3!} x^3 + \dfrac{2\cdot 3\cdot 4\cdot 5}{4!} x^4 + \cdots$
$\qquad\qquad + (-1)^n (n+1) x^n + \cdots$

13. $x - \dfrac{x^2}{2} + \dfrac{x^3}{3} - \dfrac{x^4}{4} + \dfrac{x^5}{5} + \dfrac{(-1)^n x^n}{n} + \cdots$

15. $-\infty < x < \infty$

17. $-1 < x < 1$ **19.** $-1 < x < \infty$

21. $-1 < x < 1$

23. $2x^4 + 2x^5 + 2x^6 + \cdots + 2x^{4+n} + \cdots$

25. $1 + 3x + 6x^2 + \cdots + \dfrac{1}{2} n(n-1) x^{n-2} + \cdots$

27. $x + x^2 + \dfrac{x^3}{2} + \dfrac{x^4}{3!} + \cdots + \dfrac{x^{n+1}}{n!} + \cdots$

29. $1 - x + \dfrac{x^2}{2!} - \dfrac{x^3}{3!} + \dfrac{x^4}{4!} + \cdots + \dfrac{(-1)^n x^n}{n!} + \cdots$

31. $\dfrac{1}{\sqrt{2\pi}} \left[1 - \dfrac{1}{6} + \dfrac{1}{40} - \dfrac{1}{336} + \cdots \right] = 0.3412$

33. $\dfrac{1}{\sqrt{2\pi}} \left[1 - \dfrac{(0.25)^3}{6} + \dfrac{(0.25)^5}{40} \right.$
$\qquad\qquad \left. - \dfrac{(0.25)^7}{336} + \cdots \right] = 0.3979$

35. $x - \dfrac{x^3}{3!} + \dfrac{x^5}{5!} - \dfrac{x^7}{7!} + \cdots + \dfrac{(-1)^n x^{2n+1}}{(2n+1)!} + \cdots$

37. $1 - \dfrac{(x-\pi/2)^2}{2} + \dfrac{(x-\pi/2)^4}{24} - \cdots$
$\qquad\qquad + \dfrac{(-1)^n (x-\pi/2)^{2n}}{(2n)!} + \cdots$

39. $1 - (x-1) + (x-1)^2 - (x-1)^3 + \cdots$
$\qquad\qquad + (-1)^n (x-1)^n + \cdots$

41. $x^2 - \dfrac{1}{3!} x^4 + \dfrac{1}{5!} x^6 - \dfrac{1}{7!} x^8 + \dfrac{1}{9!} x^{10} - \cdots$

43. $x^2 - \dfrac{x^6}{3!} + \dfrac{x^{10}}{5!} - \dfrac{x^{14}}{7!} + \dfrac{x^{18}}{9!} - \cdots$

45. $\dfrac{\pi^2}{4} \sin\dfrac{\pi}{4} + \left(\pi \sin\dfrac{\pi}{4} + \dfrac{\pi^3}{8} \cos\dfrac{\pi}{4} \right) \times$
$\left(x - \dfrac{\pi}{2} \right) + \cdots$. There is no simple expression for the nth term.

47. $\ln 2 + \dfrac{1}{2}(x-1) - \dfrac{1}{4\cdot 2!}(x-1)^2 + \cdots$
$\qquad\qquad + \dfrac{(-1)^{n+1}(x-1)^n}{2^n n} + \cdots$

Case Study Exercises

1. $1,333,333.33   **3.** $11,111,111.11
5. a total increase of 1.136%
7. a total increase of 1.25%
9. It represents a fraction of the additional profit. An employer would only invest a maximum of 100% of the additional profit, not more than that.

Review Exercises

1. 2.2053   **3.** 0.4010

5. $1 + (x-1) + \cdots + \dfrac{(x-1)^n}{n!}$

7. converges $(r = \tfrac{4}{9})$, $S = \tfrac{9}{5}$

9. $\tfrac{37}{99}$ **11.** $\tfrac{63}{64}$

13. convergent   **15.** convergent

17. $1 + \dfrac{x}{2} - \dfrac{x^2}{8} + \dfrac{x^3}{16} - \dfrac{5x^4}{128} + \cdots$

19. 0.4458

CHAPTER 12

Exercise Set 12–1

1. {even, odd}
3. {p, r, o, b, a, i, l, t, y}
5. {red, black}
7. {Dem., Rep., Ind.}
9. {gained, lost, same}
11. {increase, decrease, unchanged}
13. {(c, p, a), (c, a, p), (p, c, a),
(p, a, c), (a, p, c), (a, c, p)}
15. E = {1, 3, 5, 7, 9, 11},
F = {6, 7, 8, 9, 10, 11, 12}
17. $\frac{1}{3}$ **19.** $\frac{1}{2}$

21.
$E(x) = \frac{34}{9}$

23.
$E(x) = 2.74$

25. (a) 1.8 (b) 2.06 (c) 1.435
27. {0, 1, 2}   **29.** {0, 1, 2, 3}

31. (a)   (b) 12.7

33. (a)   (b) 35

35. $E(X) = \frac{5}{3}$, Var$(X) = 1.39$, $\sigma = 1.18$
37. $E(X) = 13,800$, Var$(X) = 14,160,000$, $\sigma = 3763$
39. $E(X) = 1.87$, Var$(X) = 1.6$, $\sigma = 1.26$
41. $E(X) = 14,720$
43. $E(X) = 3.09$

Exercise Set 12–2

1. No   **3.** Yes
5. No   **7.** No
9. .04   **11.** .03
13. .512   **15.** .875
17. .82   **19.** $\frac{3}{64}$
21. $\frac{3}{4}$ **23.** $\frac{3}{4}$

25. 2
29. $3 - 2\sqrt{2}$
33. $2\sqrt{3} - 3$
37. .118

27. $\dfrac{1}{(\ln 2 - \frac{1}{2})}$
31. $3 - 2\sqrt{2}$
35. $\frac{11}{28}$

Exercise Set 12–3

1. 2.5   **3.** $\dfrac{40}{3}$
5. $\dfrac{19}{12}$ **7.** $\dfrac{17}{12}$
9. $\dfrac{9}{4}$ **11.** 2.08
13. 555.56   **15.** .15
17. 2.08, 1.44   **19.** 555.56, 23.57
21. .338, .58   **23.** .0375, .194
25. .04, .2   **27.** .078, .28
29. .063, .25   **31.** 2.01, 1.42
33. .267, .517   **35.** .2, .45
37. 3.8, 1.9   **39.** 2, 2.19
41. .2, .16

Exercise Set 12–4

1. .95   **3.** .05
5. .9933   **7.** .0068
9. .45   **11.** .63
13. .035   **15.** .049
17. .6554   **19.** .0808
21. .1605   **23.** .6915
25. .2417   **27.** .3085
29. .1360   **31.** .7454
33. .0485   **35.** .1440
37. .0548   **39.** .6195
41. .9938   **43.** .7605
45. .0227   **49.** .058

Case Study Exercises

1. .76   **3.** .64
5. .9936   **7.** .99
9. $e^{-.05t}$ **11.** $e^{-.025t^2 + 1.1t}$

Review Exercises

1. {less, 2, greater}   **3.** $\frac{1}{2}$
5. no   **7.** $\frac{27}{32}$
9. $3 - 2\sqrt{2}$ **11.** $\frac{3}{8}$
13. .095, .013   **15.** .018
17. .6772   **19.** .1056
21. 14.69%

Index

CASE STUDY PREVIEWS

Models of Growth: World Population

How many humans can the earth sustain? One of the first comprehensive scientific studies of world population was done by Raymond Pearl in 1925. How do Pearl's techniques compare with the way scientists study world population today?

Melbourne's Underground Railway: Shaft Design

In the early 1980s the Melbourne, Australia, government started an extensive, controversial plan to build a complex underground system of tunnels that required a new rock-boring machine using rotating cutter heads. Trigonometric functions were used to solve several design problems.

Reliability

When the space shuttle *Challenger* burst into flames on January 28, 1986, the entire space program suffered a setback of major proportions. Reliability theory is used to test parts and components to predict their probability of failure. In particular, reliability testing of several *Challenger* parts was suspect before that critical launch.

The Multiplier Principle in Economics

A single change in demand can have far-reaching effects in various segments of an economy. This ripple effect has been known to economists for almost two centuries, but it wasn't until 1931, during the great depression, that a mathematical formulation was created to allow economists to measure these effects.